PROJECT

MANAGEMENT
BEST PRACTICES

PROJECT

MANAGEMENT
BEST PRACTICES

Achieving Global Excellence

FIFTH EDITION

HAROLD KERZNER, Ph.D.

INTERNATIONAL
Institute for Learning, Inc.

Library of Congress Cataloging-in-Publication Data applied for

Hardback ISBN: 9781394179206; E-pdf: 9781394179213; E-pub: 9781394179220

Cover Design: Wiley
Cover Image: background © ardasavasciogullari/Getty Images

SKY10054677_090523

To
my wife, Jo Ellyn
who showed me that excellence
can be achieved in
marriage, family, and life
as well as at work

Contents

Preface

For almost 50 years, project management was viewed as a process that might be nice to have but not one that was necessary for the survival of the firm. Project management practices were restricted in many companies to traditional or operational projects with well-defined scopes rather than strategic or innovation activities that may be based upon just an idea or strategic business objective. Companies reluctantly invested in some training courses simply to provide their personnel with basic knowledge of planning and scheduling. Project management was viewed as a threat to established lines of authority, and in many companies, only partial project management was used. This halfhearted implementation occurred simply to placate lower- and middle-level personnel as well as select customers.

During this 50-year period, many companies did everything possible to prevent excellence in project management from occurring. Companies provided only lip service to empowerment, teamwork, and trust. They hoarded information because the control of information was viewed as power. They placed personal and functional interests ahead of the best interest of the company in the hierarchy of priorities, including innovation and other strategic necessities, and maintained the faulty belief that time was a luxury rather than a constraint.

By the mid-1990s, this mentality began to subside, largely due to two recessions. Companies were under severe competitive pressure to create new high-quality products in a shorter period of time. Innovation was now considered a necessity for growth and survival. The importance of developing a long-term trusting relationship with the customers had come to the forefront as customers wanted more innovations. Businesses were being forced by the stakeholders to change for the better and become innovative. The survival of the firm was now at stake.

Today, businesses have changed for the better, and innovative project management was a large part of the change. Trust between the customer and contractor is at an all-time high as well as trust between management and the project teams. New products resulting from better project management practices are being developed at a faster rate than ever before. Innovation project management practices have become a competitive weapon during competitive bidding. Some companies are receiving sole-source contracts for innovative products and services because of the faith that the customer has in the contractor's ability to deliver a continuous stream of successful projects using a project management methodology that today appears more like a framework or flexible methodology than a rigid approach. All of these factors have allowed a multitude of companies to achieve some degree of excellence in project management. Business decisions are now being emphasized ahead of personal decisions.

Words that were commonplace 20 years ago have taken on new meanings today. Change resulting from better project execution is no longer being viewed as being entirely bad. Today, change implies continuous improvement. Conflicts are no longer seen as detrimental. Conflicts managed well can be beneficial. Project management is no longer viewed as a system entirely internal to the organization. Strategic partnerships may be required. Project management is now a competitive weapon that brings higher levels of quality and increased value-added opportunities for the customer. In many companies, project management is treated as a strategic competency that is one of the four or five career paths in the company that are critical for the company's future.

Companies that were considered excellent in management in the past may no longer be regarded as excellent today, especially with regard to project management. Consider the book entitled *In Search of Excellence*, written by Tom Peters and Robert Waterman in 1982 (published in New York by Harper & Row). How many of the companies identified in their book are still considered excellent today? How many of those companies have won the prestigious Malcolm Baldrige Award? How many of those companies that have won the award are excellent in project management today? Excellence in project management is a never-ending journey. Companies that are reluctant to invest in continuous improvements in project management soon find themselves with low customer satisfaction ratings.

This book covers the advanced project management topics necessary for implementation of and excellence in project management. The book contains numerous quotes from people in the field who have benchmarked best practices in project management and are currently implementing these processes within their own firms. Quotes in this book were provided by several senior corporate officers as well as others. The quotes are invaluable because they show the thought process of these leaders and the direction in which their firms are heading. These companies have obtained some degree of excellence in project management, and what is truly remarkable is the fact that this happened in less than five or six years. Best practices in implementation will be the future of project management well into the twenty-first century. Companies have created best practices libraries for project management. Many of the libraries are used during competitive bidding for differentiation from other competitors. Best practices in project management are now viewed as intellectual property.

Excellence in project management is not achieved simply by developing a project management methodology. Instead, it is how the methodology is used again and again that creates excellence and a stream of successfully managed projects. We are now trusting project managers with flexible methodologies where they can use just those components of the standard methodology that are needed for a particular project.

Project management practices and methodologies are built around the culture of companies and by determining what it takes to get people to work together, solve problems, and make decisions. Because each company most likely has its own unique culture, it is understandable that each company can have a different number of life-cycle phases, different decision points, and different success criteria. No single approach fits all companies, which is why this book discusses a variety of companies, in different industries, of different sizes, and on different continents. Hopefully, after reading this book, you will come up with ideas as to how your project management activities can improve.

Seminars and webinar courses on project management principles and best practices in project management are available using this text and my text *Project Management: A Systems Approach to Planning, Scheduling, and Controlling*, 13th edition (Hoboken, NJ: Wiley, 2022). Accompanying this text is a companion website, www.wiley.com/go/pmbestpractices4, where Instructors can access PowerPoint lecture slides, and an instructor's manual. Seminars on advanced project management are also available using this text. Information on these courses, e-learning courses, and in-house and public seminars can be obtained by contacting:

Lori Milhaven, Executive Vice President, IIL
Phone: 800-325-1533 or 212-515-5121
Fax: 212-755-0777
E-mail: lori.milhaven@iil.com

Harold Kerzner
International Institute for Learning, Inc.
2023

About the Companion Website_____

This book is accompanied by a companion website:
www.wiley.com/go/kerzner/project_management5e

The website includes the instructor's manual which contains 291 multiple choice questions, 131 true/false questions, and 170 essay/discussion questions. Answers are provided to the multiple choice and true/false questions.

1 Understanding Best Practices

1.0 INTRODUCTION

Project management has evolved from a set of processes that were once considered "nice to have" to a structured methodology that is considered mandatory for the survival of the firm. Companies are now realizing that their entire business, including most of the routine activities, can be regarded as a series of projects. Simply stated, we are managing our business through projects.

Project management is now regarded as both a project management process and a business process. Therefore, project managers are expected to make business decisions as well as project decisions. The necessity of achieving project management excellence is now readily apparent to almost all businesses.

As the relative importance of project management permeates each facet of the business, knowledge of best practices in project management is captured. Some companies view this knowledge as intellectual property to be closely guarded in the vaults of the company. Others share this knowledge in hope of discovering other best practices. Companies are now performing strategic planning for project management because of the benefits and its contribution to sustainable business value.

One of the benefits of performing strategic planning for project management is that it usually identifies the need for capturing and retaining best practices. Unfortunately, this is easier said than done. One of the reasons for this difficulty, as is seen later in the chapter, is that companies today are not in agreement on the definition of a best practice, nor do they understand that best practices lead to continuous improvement, which in turn leads to the capturing of more best practices. Many companies also do not recognize the value and benefits that can come from best practices.

Today, project managers are capturing best practices in both project management activities and business activities. The reason is simple: The best practices are intellectual property that encourages companies to perform at higher levels. Best practices lead to added business value, greater benefits realization, and better benefits management activities. Project management and business thinking are no longer separate activities.

Project management is now regarded as the vehicle that provides the deliverables that create business benefits and business values. In the last few years, there has been a tremendous growth in the need for capturing best practices related to benefits realization management and value creation.

1.1 WÄRTSILÄ[1]

Benefits Management in Operational Development Projects in Wärtsilä

Wärtsilä has a strong tradition of project-based businesses and project management practices. Because of this, a corporate-wide project management office (PMO) was established in 2007 to further strengthen the focus on project management competence within the group and to develop a project management culture, processes, competencies, and tools.

Today, project management structures and ways of working have become a fundamental part of Wärtsilä's business thinking. The business process model has gradually shifted from being a somewhat disordered process to a harmonized model, enabling the implementation of unified guidelines, targets, and terminology. The company has approached this implementation of project management practices from two different but equally important aspects. First, a project management tool providing, inter alia, more effective resource and schedule planning has been introduced and implemented. Second, the organization has been encouraged to actively participate in professional project management training and certification paths.

As the project management processes have become well-defined and have gained maturity, the emphasis has gradually shifted toward benefits management in operational development projects. The initiative to improve benefits management processes stems from the mission of the Wärtsilä PMO for Operational Development, which is to ensure synergies between Wärtsilä's business units that would help enable businesses to transform their strategic ambitions into daily operations. This would be achieved by providing management and expertise in terms of change management, business processes, and application development.

In traditional project management, projects are often measured in terms of budget, schedule, scope, or quality. Benefits management as a concept, however, focuses more on the actual value that the projects are able to deliver to the end customer. In other words, project success is not measured solely in terms of time or money. Quite the opposite, measuring the success of a project comes from the end user: Did this solution fulfill the user's needs? As the concept of value is rather vague, it is of the utmost importance that the benefits have concrete metrics and measurements. This concerns also so-called soft, intangible benefits. Although they cannot be quantified financially, they must be measured. Another important aspect of benefits planning is creating a valid

1. Material has been provided by the Wärtsilä Project Management Office. (WPMO). ©2022 by Wärtsilä Corporation. All rights reserved. Reproduced by permission.

baseline to compare the results with. Instead of comparing only to a business-as-usual situation, the results gained from the benefit realization measurements should be compared to other alternative scenarios ("Could this have been achieved some other way?").

In operational development projects, the output of the project can be, for example, an information technology (IT) tool made to improve resource planning. The most crucial part of the project, however, is making the *output* into a project *outcome*. This means that the project output (in this case, an IT tool) should become a part of the end user's way of working. In order to make this happen, the benefits planning must consider two important aspects:

1. What does the end user want and need?
2. What has to change in order to make this happen?

With proper end-user expectation management and change management, the risk of the project output becoming just another tool in the toolbox can be avoided.

In a nutshell, the benefits management system should consist of the following elements:

- *Identifying the driver for the project*. Do we really need this investment? Who else is going to benefit from it?
- *Identifying the key benefits*. What are the benefits and when will they occur? What is their proximity (how likely are they to happen)?
- *Estimating the benefits*. Defining a clear baseline for the measurements allows us to define clear metrics (which apply to the entire portfolio of projects) and provides us with consistency throughout all life-cycle phases, from project initiation to benefit realization. The critical question we must ask is: Do these metrics tolerate changes in the business environment?
- *Linking the benefits with change*. How does the organization have to change in order to enable the benefits realization? How can we enable this change? Plan the deployment and adjust it to (business) environmental changes (organizational changes, market situation changes, etc.).
- *Who is accountable for the benefit?* Define a person/organization responsible for the benefits realization.
- *Monitoring benefits*. Monitor your performance with the established metrics, improve it if needed toward the defined goal, and acknowledge risks in a proactive way.
- *Doing a postproject evaluation*. Ensure a successful deployment by communicating about the project output and honestly promoting it. Imagine yourself in the end user's position: Would you like to use this tool?
- *Learning from your mistakes*. Ensure that project success points and failures are equally handled. Focus on honest communication and learning, not blaming. Examples should come all the way from the executive level.

1.2 PROJECT MANAGEMENT BEST PRACTICES: 1945–1960 _____

During the 1940s, line managers functioned as project managers and used the concept of over-the-fence management to manage projects. Each line manager, temporarily wearing the hat of a project manager, would perform the work necessitated by his or her line organization and, when that was completed, would throw the "ball" over the fence in the hope that someone would catch it. Once the ball was thrown over the fence, the line managers would wash their hands of any responsibility for the project because the ball was no longer in their yard. If a project failed, blame was placed on whichever line manager had the ball at that time.

The problem with over-the-fence management was that the customer had no single contact point for questions. The filtering of information wasted precious time for both the customer and the contractor. Customers who wanted firsthand information had to seek out the manager in possession of the ball. For small projects, this was easy. However, as projects grew in size and complexity, this became more difficult.

During this time, very few best practices were identified. If there were best practices, then they would stay within a given functional area, never to be shared with the remainder of the company. Suboptimal project management decision-making was the norm.

Following World War II, the United States entered into the Cold War with the Soviet Union. To win the Cold War, the United States had to compete in an arms race and rapidly build weapons of mass destruction. The victor in a cold war is the side that can retaliate with such force as to obliterate the enemy. Development of weapons of mass destruction involved very large projects involving potentially thousands of contractors.

The arms race made it clear that the traditional use of over-the-fence management would not be acceptable to the Department of Defense (DoD) for projects such as the B52 bomber, the Minuteman intercontinental ballistic missile, and the Polaris submarine. The government wanted a single point of contact, namely, a project manager who had total accountability through all project phases. In addition, the government wanted the project manager to possess a command of technology rather than just an understanding of technology, which mandated that the project manager be an engineer, preferably with an advanced degree in some branch of technology. The use of project management was then mandated for some smaller weapon systems, such as jet fighters and tanks. The National Aeronautics and Space Administration (NASA) mandated the use of project management for all activities related to the space program.

Many projects in the aerospace and defense industries were having cost overruns in excess of 200–300 percent. Blame was erroneously placed on improper implementation of project management when, in fact, the real problem was the inability to forecast technology, resulting in numerous scope changes occurring. Forecasting technology is extremely difficult for projects that could last 10–20 years.

By the late 1950s and early 1960s, the aerospace and defense industries were using project management on virtually all projects, and they were pressuring their suppliers to use it as well. Project management was growing, but at a relatively slow rate, except for aerospace and defense.

Because of the vast number of contractors and subcontractors, the government needed standardization, especially in the planning process and the reporting of information. The government established a life-cycle planning and control model and a cost-monitoring system and created a group of project management auditors to make sure that the government's money was being spent as planned. These practices were to be used on all government programs above a certain dollar value. Private industry viewed these practices as an overmanagement cost and saw no practical value in project management. If any best practices were captured at that time, they were heavily focused on improvements to the standardized forms the DoD used.

Because many firms saw no practical value in project management in their early years, there were misconceptions about it. Some of the misconceptions included:

- Project management is a scheduling tool like PERT/CPM (program evaluation and review technique/critical path method) scheduling.
- Project management applies to large projects only.
- Project management is designed for government projects only.
- Project managers must be engineers, preferably with advanced degrees.
- Project managers need a command of technology to be successful.
- Project success is measured in technical terms only. (Did it work?)

1.3 PROJECT MANAGEMENT BEST PRACTICES: 1960–1985

Between 1960 and 1985, a better understanding of project management existed. Growth in the field had come about more through necessity than through desire, but at a very slow rate. Its slow growth can be attributed mainly to lack of acceptance of the new management techniques necessary for successful implementation of project management. An inherent fear of the unknown acted as a deterrent for both managers and executives.

Other than aerospace, defense, and construction, the majority of companies in the 1960s managed projects informally. In informal project management, just as the words imply, projects were handled on an informal basis, and the authority of the project manager was minimized. Most projects were handled by functional managers and stayed in one or two functional lines, and formal communications were either unnecessary or handled informally because of the good working relationships between line managers. Those individuals who were assigned as project managers soon found that they were functioning more as project leaders or project monitors than as real project managers. Many organizations today, such as low-technology manufacturing, have line managers who have been working side by side for 10 years or more. In such situations, informal project management may be effective on capital equipment or facility development projects, and project management is not regarded as a profession.

By 1970 and through the early 1980s, more companies departed from informal project management and restructured to formalize the project management process, mainly because the size and complexity of their activities had grown to a point where they were unmanageable within the current structure.

Not all industries need project management, and executives must determine whether there is an actual need before making a commitment. Several industries with simple tasks, whether in a static or a dynamic environment, do not need formalized project management. Manufacturing industries with slowly changing technology do not need project management, unless, of course, they have a requirement for several special projects, such as capital equipment activities, that could interrupt the normal flow of work in the routine manufacturing operations. The slow growth rate and acceptance of project management were related to the fact that the limitations of project management were readily apparent, yet the advantages were not completely recognizable. Project management requires organizational restructuring. The question, of course, is "How much restructuring?" Executives avoided the subject of project management for fear that "revolutionary" changes would have to be made in the organization.

Project management restructuring has permitted companies to:

● Accomplish tasks that could not be effectively handled by the traditional structure
● Accomplish one-time activities with minimum disruption of routine business

The second item implies that project management is a "temporary" management structure and, therefore, causes minimal organizational disruption. The major problems identified by those managers who endeavored to adapt to the new system all revolved around conflicts in authority and resources. Companies began to recognize the need for capturing best practices, especially those that could reduce some human behavior issues. Improvements in the methodologies were also taking place.

Another major concern was that project management required upper-level managers to relinquish some of their authority through delegation to middle managers. In several situations, middle managers soon occupied the power positions, even more so than upper-level managers.

Project management became a necessity for many companies as they expanded into multiple product lines, many of which were dissimilar, and organizational complexities grew. This growth can be attributed to four factors:

1. Technology increasing at an astounding rate
2. More money being invested in research and development (R&D)
3. More information being available
4. Shortening of project life cycles

To satisfy the requirements imposed by these four factors, management was "forced" into organizational restructuring; the traditional organizational form that had survived for decades was inadequate for integrating activities across functional "empires."

By 1970, the environment began to change rapidly. Companies in aerospace, defense, and construction pioneered the implementation of project management, and other industries soon followed, some with great reluctance. NASA and the DoD "forced" subcontractors to accept project management.

Because current organizational structures are unable to accommodate the wide variety of interrelated tasks necessary for successful project completion, the need for project management has become apparent. It is usually first identified by those lower-level and middle managers who find it impossible to control their resources effectively for the diverse activities within their line organization. Quite often, middle managers feel the impact of changing environment more than upper-level executives.

Once the need for change is identified, middle management must convince upper-level management that such a change is actually warranted. If top-level executives cannot recognize the problems with resource control, then project management will not be adopted, at least formally. Informal acceptance, however, is another story.

As project management developed, some essential factors for its successful implementation were recognized. The major factor was the role of the project manager, which became the focal point for integrative responsibility. The need for integrative responsibility was first identified in complex R&D projects.

The R&D technology has broken down the boundaries that used to exist between industries. Once-stable markets and distribution channels are now in a state of flux. The industrial environment is turbulent and increasingly hard to predict. Many complex facts about markets, production methods, costs, and scientific potentials are related to investment decisions in R&D.

All of these factors have combined to produce a king-size managerial headache. There are just too many crucial decisions to have them all processed and resolved at the top of the organization through regular line hierarchy. They must be integrated in some other way.

Providing the project manager with integrative responsibility resulted in:

1. Total project accountability being assumed by a single person
2. Project rather than functional dedication
3. A requirement for coordination across functional interfaces
4. Proper utilization of integrated planning and control

Without project management, these four elements have to be accomplished by executives, and it is questionable whether these activities should be part of an executive's job description. An executive in a Fortune 500 corporation stated that he was spending 70 hours each week working as both an executive and a project manager, and he did not feel that he was performing either job to the best of his abilities. During a presentation to the staff, the executive stated what he expected of the organization after project management implementation:

- Push decision-making down in the organization.
- Eliminate the need for committee solutions.
- Trust the decisions of your peers.

Those executives who chose to accept project management soon found the advantages of the new technique:

- Easy adaptation to an ever-changing environment
- Ability to handle a multidisciplinary activity within a specified period of time

- Horizontal as well as vertical workflow
- Better orientation toward customer problems
- Easier identification of activity responsibilities
- A multidisciplinary decision-making process
- Innovation in organizational design

As project management evolved, best practices became important. Best practices were learned from both successes and failures. In the early years of project management, private industry focused on learning best practices from successes. The government, however, focused on learning about best practices from failures. When the government finally focused on learning from successes, the knowledge of best practices came from its relationships with both prime contractors and subcontractors. Some of these best practices that came out of the government included:

- Use of life-cycle phases
- Standardization and consistency
- Use of templates (e.g., for statement of work [SOW], work breakdown structure [WBS], and risk management)
- Providing military personnel in project management positions with extended tours of duty at the same location
- Use of integrated project teams
- Control of contractor-generated scope changes
- Use of earned value measurement

1.4 PROJECT MANAGEMENT BEST PRACTICES: 1985–2016[2]

By the 1990s, companies had begun to realize that implementing project management was a necessity, not a choice. By 2016, project management had spread to virtually every industry, and best practices were being captured. In the author's opinion, the appearance of best practices by industry can be summarized as follows:

- 1960–1985: Aerospace, defense, and construction
- 1986–1993: Automotive suppliers
- 1994–1999: Telecommunications
- 2000–2003: Information technology
- 2004–2006: Health care
- 2007–2008: Marketing and sales
- 2009–Present: Government agencies, small businesses, and global acceptance of project management

---

2. Many of the comments made by executives in the remainder of this chapter and throughout the book have been taken from earlier editions of this book. The dates of the comments are not critical. But it is important to recognize how executives are now and have been viewing the growth of project management practices and the accompanying benefits.

The question now is not how to implement project management, but how fast can it be done? How quickly can we become mature in project management? Can we use the best practices to accelerate the implementation of project management?

Table 1–1 shows the typical life-cycle phases that an organization goes through to implement project management. In the first phase—the embryonic phase—the organization recognizes the apparent need for project management. This recognition normally takes place at the lower and middle levels of management, where the project activities actually take place. The executives are then informed of the need and assess the situation.

Six driving forces lead executives to recognize the need for project management:

1. Capital projects
2. Customer expectations
3. Competitiveness
4. Executive understanding
5. New project development
6. Efficiency and effectiveness

Manufacturing companies are driven to project management because of large capital projects or a multitude of simultaneous projects. Executives soon realize the impact on cash flow and that slippages in the schedule could end up idling workers.

Companies that sell products or services, including installation, to their clients must have good project management practices. These companies are usually non-project-driven but function as though they were project-driven. These companies now sell solutions to their customers rather than products. It is almost impossible to sell complete solutions to customers without having superior project management practices, because what you are actually selling is your project management expertise (i.e., your project management processes).

TABLE 1–1. FIVE PHASES OF THE PROJECT MANAGEMENT LIFE CYCLE

Embryonic	Executive Management Acceptance	Line Management Acceptance	Growth	Maturity
Recognize need	Get visible executive support	Get line management support	Recognize use of life-cycle phases	Develop a management cost/schedule control system
Recognize benefits	Achieve executive understanding of project management	Achieve line management commitment	Develop a project management methodology	Integrate cost and schedule control
Recognize applications	Establish project sponsorship at executive levels	Provide line management education	Make the commitment to planning	Develop an educational program to enhance project management skills
Recognize what must be done	Become willing to change your way of doing business	Become willing to release employees for project management training	Minimize creeping scope Select a project tracking system	

There are two situations where competitiveness becomes the driving force: internal projects and external (outside customer) projects. Internally, companies get into trouble when they realize that much of the work can be outsourced for less than it would cost to perform the work themselves. Externally, companies get into trouble when they are no longer competitive on price or quality or when they simply cannot increase their market share.

Executive understanding is the driving force in those organizations that have a rigid traditional structure that performs routine, repetitive activities. These organizations are quite resistant to change, unless it is driven by the executives. This driving force can exist in conjunction with any of the other driving forces.

New product development is the driving force for those organizations that are heavily invested in R&D activities. Given that only a small percentage of R&D projects ever make it to commercialization, where the R&D costs can be recovered, project management becomes a necessity. Project management can also be used as an early-warning system that a project should be canceled.

Efficiency and effectiveness, as driving forces, can exist in conjunction with any other driving forces. Efficiency and effectiveness take on paramount importance for small companies experiencing growing pains. Project management can be used to help such companies remain competitive during periods of growth and to assist in determining capacity constraints.

Because of the interrelatedness of these driving forces, some people contend that the only true driving force is survival. This is illustrated in Figure 1–1. When the company recognizes that survival of the firm is at stake, the implementation of project management becomes easier.

Enrique Sevilla Molina, PMP, formerly corporate PMO director, discusses the driving forces at Indra that necessitated the need for excellence in project management:

> The internal forces were based on our own history and business experience. We soon found out that the better the project managers, the better the project results. This realization came together with the need to demonstrate in national and international

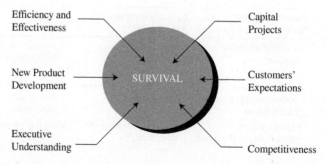

Figure 1–1. The components of survival.
Source: Reproduced from H. Kerzner, *In Search of Excellence in Project Management* (Hoboken, NJ: Wiley, 1998), p. 51/ John Wiley & Sons.

contracts, with both US and European customers, our real capabilities to handle big projects. These big projects required world-class project management, and for us managing the project was a greater challenge than just being able to technically execute the project. Summarizing, these big projects set the pace to define precise procedures on how handling stakeholders, big subcontractors and becoming a reliable main point of contact for all issues related with the project.

The speed at which companies reach some degree of maturity in project management is most often based on how important they perceive the driving forces to be. This is illustrated generically in Figure 1–2. Non-project-driven and hybrid organizations move quickly to maturity if increased internal efficiencies and effectiveness are needed. Competitiveness is the slowest path because these types of organizations do not recognize that project management affects their competitive position directly. For project-driven organizations, the path is reversed. Competitiveness is the name of the game, and the vehicle used is project management.

Once the organization perceives the need for project management, it enters the second life-cycle phase of Table 1–1, executive acceptance. Project management cannot be implemented rapidly in the near term without executive support. Furthermore, the support must be visible to all.

The third life-cycle phase is line management acceptance. It is highly unlikely that any line manager would actively support the implementation of project management without first recognizing the same support coming from above. Even minimal line management support will still cause project management to struggle.

The fourth life-cycle phase is the growth phase, where the organization becomes committed to the development of the corporate tools for project management. This includes the processes and project management methodology for planning, scheduling, and controlling, as well as selection of the appropriate supporting software. Portions of this phase can begin during earlier phases.

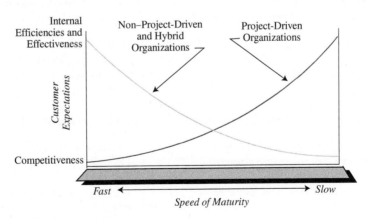

Figure 1–2. Speed of maturity.

The fifth life-cycle phase is maturity. In this phase, the organization begins using the tools developed in the previous phase. Here, the organization must be totally dedicated to project management. The organization must develop a reasonable project management curriculum to provide the appropriate training and education in support of the tools as well as the expected organizational behavior.

By the 1990s, companies finally began to recognize the benefits of project management. Table 1–2 shows the critical success factors (CSFs) and critical failure factors (CFFs) that have led to changes in our view of project management. Many of these factors were identified through the discovery and implementation of best practices.

By the 1990s, companies finally began to recognize the benefits of project management. Table 1–2 shows the critical success and CFFs that have led to changes in our view of project management. Many of these factors were identified through the discovery and implementation of best practices.

Recognizing that the organization can benefit from the implementation of project management is just the starting point. The question now becomes: How long will it take us to achieve these benefits? This can be partially answered by Figure 1–3. In the beginning of the implementation process, there will be added expenses to develop the project management methodology and establish the support systems for planning, scheduling, and control. Eventually, the cost will level off and become pegged. The question mark in Figure 1–3 is the point at which the benefits equal the cost of implementation. This point can be pushed to the left through training and education.

During the first decade of the twenty-first century, the understanding and acceptance of the benefits permeated all levels of senior management rather than just

TABLE 1–2.   CRITICAL FACTORS IN THE PROJECT MANAGEMENT LIFE CYCLE

Critical Success Factors	Critical Failure Factors
Executive Management Acceptance Phase	
Consider employee recommendations	Refuse to consider ideas from associates
Recognize that change is necessary	Unwilling to admit that change may be necessary
Understand the executive role in project management	Believe that project management control belongs at executive levels
Line Management Acceptance Phase	
Willing to place company interest before personal interest	Reluctant to share information
Willing to accept accountability	Refuse to accept accountability
Willing to see associates advance	Not willing to see associates advance
Growth Phase	
Recognize the need for a corporate-wide methodology	View a standard methodology as a threat rather than as a benefit
Support uniform status monitoring/reporting	Fail to understand the benefits of project management
Recognize the importance of effective planning	Provide only lip service to planning
Maturity Phase	
Recognize that cost and schedule are inseparable	Believe that project status can be determined from schedule alone
Track actual costs	See no need to track actual costs
Develop project management training	Believe that growth and success in project management are the same

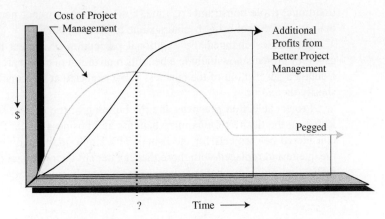

Figure 1–3. Project management costs versus benefits.

those executives that interfaced with projects on a daily basis. Three comments from senior management at American Greetings Corporation illustrate this point:

> Through project management, we've learned how to make fact-based decisions. Too often in the past we based our decisions on what we thought could happen or what we hoped would happen. Now we can look at the facts, interpret the facts honestly and make sound decisions and set realistic goals based on this information.
> **Zev Weiss, formerly chief executive officer, American Greetings**

> The program management office provides the structure and discipline to complete the work that needs to get done. From launch to completion, each project has a roadmap for meeting the objectives that were set.
> **Jeff Weiss, formerly president and chief operating officer, American Greetings**

> Through project management, we learned the value of defining specific projects and empowering teams to make them happen. We've embraced the program management philosophy and now we can use it again and again to reach our goals.
> **Jim Spira, retired president and chief operating officer, American Greetings**

When all of the executives are in agreement as to the value and benefits of project management, continuous improvements in project management occurs at a rapid pace.

1.5 PROJECT MANAGEMENT BEST PRACTICES: 2016–PRESENT

As more and more companies recognized the benefits of using project management, capturing best practices became commonplace. Perhaps the biggest change in how people viewed project management was the realization that completed projects could provide business value rather than merely deliverables. Completing projects within the traditional triple constraints of time, cost, and scope is not necessarily success if the deliverables do not bring business value to the company.

Businesses changed the traditional perception of project management. Business cases for projects now include a benefits realization plan and are often accompanied by a detailed description of the business value expected at the conclusion of the project or shortly thereafter.

Project selection practices and the building of the project portfolio are now predicated on the desire to maximize benefits and business value. Projects that were once considered pet projects for the benefit of a single individual are being removed from the queue and replaced with those that can benefit the organization as a whole. Benefits realization planning, benefits management, and business value management are now prime focuses at the executive levels of management.

1.6 BENEFITS MANAGEMENT PRACTICE AT DUBAI CUSTOMS[3]

At Dubai Customs (DC), where projects cover both core and noncore domains, effective benefits realization is critical to the achievement of the business outcomes desired from investments.

Mohammad Rashed Bin Hashim and Ajith Kumar Nair, specialists heading the IT Demand and Benefits Management section at DC, a part of the Project Delivery Department, spearheaded the work of developing a Benefits Management Framework for the Customs Development Division. Through extensive research on global benefits realization best practices, they set up a working governance process with an established methodology to capture and measure all financial and nonfinancial benefits that encapsulate overall outcomes. This process is applied in the development of business cases, benefit realization plans, and portfolio-level benefit management. It also provides decision-making support for DC Executive Development Committee in overseeing all project-related investments.

The objectives of benefits realization management at DC are to:

● Ensure benefits are identified, defined clearly at the outset, and linked to strategic outcomes (Business Needs document—Demand Outline and Business Case)
● Ensure business areas are committed to realizing their defined benefits with assigned ownership and responsibility for adding value through the realization process (Benefit Realization Plan and Activity Tracker for monitoring and measurement)
● Drive the process of realizing benefits, including benefit measurement, tracking, and recording benefits as they are realized, and manage benefits at a portfolio level to better budget and prioritize future initiatives (Benefit Realization Plan and Benefit Quadrant)

---

3. Material in this section was provided by Ajith Kumar Nair, Senior Research Head/Business Solutions Specialist. ©2022 by Dubai Customs. All rights reserved. Reproduced with permission.

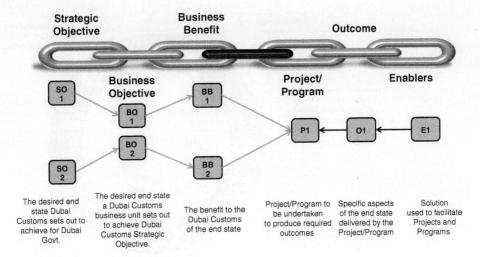

Figure 1–4. Benefits alignment map.

- Use the defined, expected benefits as a roadmap for the project/program, providing a focus for delivering change (Benefit Quadrant feeding into Portfolio Management)
- Provide alignment and clear links between the project/program (its objectives and desired benefits) as per Figure 1–4 with the strategic objectives (DC Strategic Alignment with Benefits—Benefits Alignment Map)

Benefits Realization Management Framework

The purpose of the Benefits Realization Management Framework developed at DC is:

- To provide a framework of best practice principles and concepts drawn from the latest experiences and proven best practices (Cranfield Process Model for Benefits Management and APMG International Managing Benefits: Optimizing the Return from Investments) in setting up and managing benefits for projects and programs across the project delivery department.
- To provide a standard approach for benefits realization management with the business subject matter experts (SMEs), directors, business owners, domain managers, change managers, project/program managers, business analysts, and PMO staff across Dubai Customs.
- To provide consistent terminology and benefits categorization (Revenue increase, Cost savings, Increased efficiency, Revenue Protection, and Customer Satisfaction)
- To provide an introduction and guidance for business sponsors and business benefit owners.
- Aimed at those who are involved in benefits realization, enabling them to adapt and tailor the guidance to their specific needs highlighted in Figure 1–5.

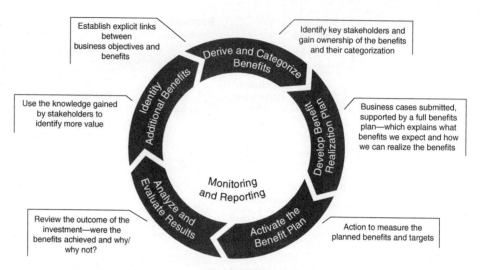

Figure 1–5. Benefits realization management framework.

- Accessible to strategy, operational business areas, and program/project teams as well as to individual practitioners and business benefit owners.
- Aimed at helping practitioners improve their decision-making and become better at implementing beneficial change.

To determine whether an initiative has succeeded and achieved its purpose, benefits management processes look at setting up an overall governance process to plan, measure, review, and evaluate results for quantifiable benefits. The processes also look at measuring qualitative benefits internally and at defining and measuring external partner benefits to DC.

The key to applying the framework is to understand the starting point.

- Have you got an approved business case, or are you still in the process of developing a business case for your project or program?
- All tasks and deliverables relevant to program or project use this standard approach to focus on developing benefits, as depicted in Figure 1–5.

Benefits Management Maturity Level

Determining the maturity level of an organization will help tailor the framework to help ensure adoption and continued use of benefits processes and templates. Low benefit management maturity will resist the introduction of a complex and comprehensive benefits framework. DC realized the need to distill the framework, processes, and templates to gain the essential mechanisms required to improve overall organization's benefits management and perform required governance reporting to plan the continued improvement of benefits management.

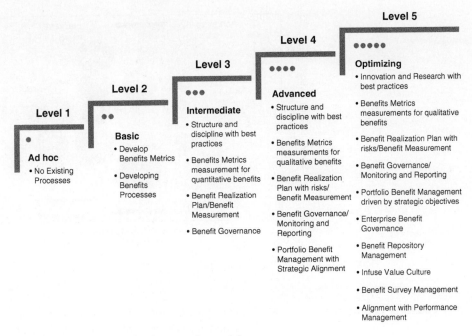

Figure 1–6. Benefits realization maturity model.

An assessment of the adequacy and effectiveness of DC benefits management practices was conducted by Stephen Jenner, a world-renowned benefits management consultancy expert in collaboration with the consultancy group International Institute for Learning, Inc. here in Dubai to gauge our maturity level. With the workshops held here at DC, we developed an internal maturity model for benefits management, depicted in Figure 1–6, to identify maturity areas achieved, to be improved, and to be further developed.

Portfolio Benefits Management (Benefits Quadrant) For many initiatives, benefits realization will only commence after project implementation is complete. At DC, benefits realization is being monitored after project completion by assigning ongoing responsibility for benefits realization to the Demand Management Section, where realizations are analyzed across the portfolio to optimize and report on a continuous basis to our higher-level governance board (Customs Development Committee). Highlighted in Figure 1–7 is a quadrant built within DC for managing benefits for the portfolio as a whole based on the ideas obtained from the managing benefits best practice of APMG. International benefits are maintained at a portfolio level and plotted with a point allocation depicting the bubble size that represents the cost of the investment around the twin dimensions of attractiveness and achievability of the benefits; this serves as a tool for management in reporting and planning future investments.

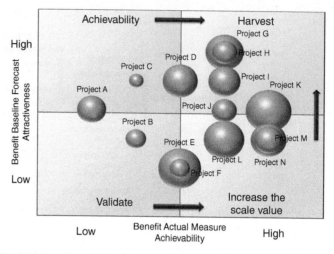

Figure 1–7. Portfolio benefits quadrant.

Key Lessons Learned

- Think about what the benefits are that you want to achieve, then come up with a program of projects that will allow you to deliver those benefits. Thinking of the projects first and then trying to align them with the corporate strategy is just wrong.
- Beware of "rogue" projects that are not strategic in nature, consume valuable resources, and distract attention from helping us deliver the organization's strategy.
- Focus on benefits with ongoing participative stakeholder management.
- Benefits forecasts and practices are driven by evidence.
- Benefits must be transparent based on open and honest forecasting and reporting, with a clear line of sight from strategic objectives to business benefits.
- Benefits must be forward-looking and evolve through learning and continuous improvement.
- Benefits must focus more on actual realization of benefits.
- Active sponsorship is essential for successful delivery of projects/programs and the expected benefits. The business owner needs to ensure that the project/program delivers the expected benefits.
- Organizations must infuse a value culture to effectively mature in the benefits realization realm.
- Benefits are often realized sometime after project completion. Even if the benefits start coming straightaway, it is essential that the long-term effectiveness of the benefits must be monitored by the business owner.
- Benefits must be managed for the full business life cycle, from identification to realization and application of lessons learned.

1.7 AN EXECUTIVE'S VIEW OF PROJECT MANAGEMENT _____

Today's executives have a much better understanding of and appreciation for project management than did their predecessors. Early on, project management was seen as simply scheduling a project and then managing the project using network-based software. Today, this parochial view has changed significantly. Project management is now a necessity for survival.

Although there are several drivers for this change, three significant reasons seem to stand out. First, as businesses downsize due to poor economic conditions or stiff competition, the remaining employees are expected to do more with less. Executives expect the employees to become more efficient and more effective when carrying out their duties. Second, business growth today requires the acceptance of significant risks, specifically in the development of new products and services for which there may not be reasonable estimating techniques or standards. Simply stated, we are undertaking more jobs that are neither routine nor predictable. Third, and perhaps most important, is that we believe we are managing our business as though it were a series of projects. Projects now make up a significant part of people's jobs. For that reason, all employees are actually project managers to some degree and are expected to make business as well as project decisions.

The new breed of executives seems to have a much broader view of the value of project management, ranging from its benefits to the selection criteria for project managers to organizational structures that can make companies more effective. This is apparent in the next comments, which were provided in an earlier edition of the book by Tom Lucas, formerly chief information officer for the Sherwin-Williams Company:

- We have all managed projects at one time or another, but few of us are capable of being project managers.
- The difference between managing projects and professional project management is like the difference between getting across the lake in a rowboat versus a racing boat. Both will get you across the lake, but the rowboat is a long and painful process. But how do people know until you give them a ride?
- Don't be misguided into thinking professional project management is about process. It is about delivering business results.
- If you don't appreciate that implementing a PMO is a cultural transition, you are destined to fail.

The next comments from other executives clearly indicate their understanding and appreciation of project management:

Over the past 15 years, ongoing transformation has become a defining characteristic of IBM—and a key factor in our success. Effective change in process and IT transformation doesn't just happen, it must be enabled by highly skilled project managers. Our project managers analyze processes, enabled by IT, in a way that allows us to innovate and eliminate unnecessary steps, simplify and automate. They help us become more efficient and effective by pulling together the right resources to get things done—on time and on budget. They are invaluable as we continue to make progress in our transformation journey.

Linda S. Sanford, formerly senior vice president,
Enterprise Transformation, IBM Corporation

Project managers are a critical element of our end-to-end development and business execution model. Our goal is to have sound project management practices in place to provide better predictability in support of our products and offerings. As a team, you help us see challenges before they become gating issues and ensure we meet our commitments to STG [System and Technology Group] and clients. . . . We continue to focus on project management as a career path for high-potential employees and we strongly encourage our project managers to become certified, not only [by] PMI, but ultimately IBM certified. . . . End-to-end project management must become ingrained in the fabric of our business.

Rod Adkins, formerly senior vice president,
IBM's System and Technology Group

Successful project management is mission critical to us from two points of view:

First, as we define and implement PLM (product life cycle management) solutions, we help customers to streamline their entire product life cycle across all functional units. This can make any large PLM project an intricate and even complex undertaking. To live up to our company mantra of "we never let a customer fail," robust and reliable project management is often the most critical component we provide aside from the PLM platform itself; the combination of the two enables our customers to achieve the business benefits they strive for by investing in PLM.

Second, Siemens itself is one of our largest customers. This is a great opportunity and, at the same time, a great challenge. Keeping a project's objectives and scope under control with our "internal" customer is at least as challenging as with external customers; yet it is critical in order to keep our development roadmaps and deployment schedules on track. Our job is to continue to successfully develop and deploy the first and only true end-to-end industry software platform. This comprehensive platform covers the entire product lifecycle from initial requirements, through product development, manufacturing planning, controlling the shop floor and even managing the maintenance, repair and overhaul of the product in question. As a result, effective project management is vital to our success.

Dr. Helmuth Ludwig, formerly Chief Information Officer, Siemens

In this age of instant communications and rapidly evolving networks, Nortel continues to maximize use of its project management discipline to ensure the successful deployment of increasingly complex projects. We foster an environment that maintains a focus on sharing best practices and leveraging lessons learned across the organization, largely driven by our project managers. We are also striving to further integrate project management capabilities with supply chain management through the introduction of SAP business management software. Project management remains an integral part of Nortel's business and strategy as it moves forward in a more services- and solutions-oriented environment.

Sue Spradley, formerly president, Global Operations, Nortel Networks

In the services industry, how we deliver (i.e., the project management methodology) is as important as what we deliver (i.e., the deliverable). Customers expect to maximize their return on IT investments from our collective knowledge and experience when we deliver best-in-class solutions. The collective knowledge and experience of HP [Hewlett-Packard] Services is easily accessible in HP Global Method. This integrated set of methodologies is a first step in enabling HPS [Hewlett-Packard Services] to optimize our efficiency in delivering value to our customers. The next step is to know what is available and learn how and when to apply it when delivering to your customers. HP Global Method is the first step toward a set of best-in-class methodologies to increase the credibility as a trusted partner, reflecting the collective knowledge and expertise of HP Services. This also improves our cost structures by customizing predefined proven approaches, using existing checklists to ensure all the bases are covered and share experiences and learning to improve Global Method.

**Mike Rigodanzo, formerly senior vice president,
HP services operations and information technology**

In 1996, we began looking at our business from the viewpoint of its core processes. . . . As you might expect, project management made the short list as one of the vital, core processes to which quality principles needed to be applied.

Martin O'Sullivan, retired vice president, Motorola

These comments clearly indicate that today's executives recognize that project management is a strategic or core competency needed for survival because it interfaces with perhaps all other business processes, including quality initiatives.

There are three business factors that have occurred in the past decade that have led to executive-level appreciation and support by senior management. First, there has been a significant increase in the growth of nontraditional projects needed to sustain the growth of the business. Believing in a business-as-usual approach is an invitation for failure. The nontraditional projects are heavily oriented around strategic business goals and objectives, as well as innovation necessities, to maintain the firm's future. Due to limited funding, executives cannot work on all the projects they desire. For the projects that are prioritized and implemented, executives need more information than in the past to assess project performance and make necessary strategic changes in direction if necessary. Effective use of project management practices accompanied by continuous improvement efforts increases the chances of success.

Second, today's business environment is heavily based on uncertainty and complexity. Decisions must be made and implemented quickly, often based on imperfect information and higher than normal levels of risk. Once again, effective project management practices improve the chances of success.

Third, a crisis such as the COVID-9 pandemic has shown many companies that they cannot expect the business to survive with a business-as-usual mentality. Hierarchical command and control from the top floor of the building will not necessarily resolve crises. When crises occur, there are usually changes in strategies. Strategies are implemented using projects, and thus the importance and need for effective project management practices appear once again.

These three topics will be discussed in detail in Sections 1.8–1.10.

1.8 THE GROWTH OF NONTRADITIONAL PROJECTS

For decades, most project managers were responsible for traditional projects with a well-defined business case and statement of work. Project managers were taught in the classroom not to begin the planning and execution phases of a project until the requirements were understood and agreed to by all parties, including stakeholders. Having well-defined requirements was believed to minimize the need for detailed risk management and tend to minimize the project's complexity. Companies created a "one-size-fits-all" methodology for managing projects. There were forms, guidelines, templates, and checklists for risk management activities, but they were all predicated upon usage on traditional projects with well-defined requirements.

Project management performance was measured by "obedience" to the singular methodology, even if the project was not a partial or complete success. Excellence in project management was defined by the continuous use of four components, namely

- The one-size-fits-all methodology
- Other organizational process assets supporting the methodology
- Project initiation using a well-defined statement of work
- Realistic assumptions, constraints, and expectations

The ability to effectively manage risk, uncertainty, and complexity was usually not one of the critical characteristics because excellence was measured only against performance of traditional projects that were evident.

Most executives were reluctant to trust project managers and were afraid that project managers would make decisions that were reserved for the senior levels of management. Therefore, using a singular methodology for all traditional projects minimized the risks somewhat and provided senior management with effective command and control from the top floor of the building. Project managers were permitted to make some decisions related to technical risk complexity, but most decisions that impacted business risk complexity were made by the project sponsors.

Projects that were strategic or complex, and with a great deal of risk, were becoming increasingly more important to executives by the turn of the century. Strategic projects, especially those involving innovation, were a necessity for sustaining the business. Unfortunately, these projects were most often assigned to functional managers because senior management trusted functional managers more than project managers. Functional managers were given the freedom to use whatever tools and techniques they wished to manage the complexity of their projects. There were no handcuffs placed on functional managers, and, as such, there were often no guidelines or standards on how to manage the complexities of nontraditional projects such as strategic or innovation projects.

As project management evolved, literature appeared discussing the best practices and lessons learned attributed to the effective use of project management. Executives began questioning whether project managers were better suited to manage nontraditional and strategic projects than functional managers.

There were several reasons for using project managers rather than functional managers. First, functional managers were receiving year-end bonuses based on the performance of their functional group. As such, they were assigning their most qualified employees to the short-term projects that impacted their bonuses, thus possibly unfavorably impacting the strategic projects that now did not have the resources with required skills. Second, functional managers were hiding some of the complexities of their projects for fear that it might impact their bonuses or, even worse, invite executives to micromanage their projects.

By the turn of the century, nontraditional and strategic projects, because of their growing importance, became the responsibility of the project managers. What became evident almost immediately was that the one-size-fits-all approach could not be used on many of the nontraditional projects. Some of the critical differences between these two types of projects are shown in Table 1–3.

Articles began appearing stating that multiple methodologies or frameworks, such as an Agile or Scrum approach, should be used that could be adapted to fit the needs of each project rather than forcing the project to use a one-size-fits-all approach that may be inappropriate. Companies understood the changes that were taking place and soon realized that all projects could have different levels of complexity and that effective risk management practices could become the most important knowledge area for future project managers.

Project managers now had a much deeper and clearer understanding of the situation at hand could make more effective decisions. More trust was being placed in the hands of the project managers, and they could not rely solely upon the project sponsors or the project's governance committees to analyze the complexity of each situation and make the appropriate decisions. Understanding risk, uncertainty, and complexity had become critical intellectual assets for project managers, along with the use of flexible methodologies.

TABLE 1–3.   DIFFERENCES BETWEEN TRADITIONAL AND NONTRADITIONAL PROJECTS

Factor	Traditional Project	Nontraditional or Strategic Project
Business case	Usually well-defined	May not exist in the early stages of the project and may be subject to change throughout the project
Statement of work	Usually well-defined	May not exist; may have only strategic goals to work with. The SOW can change throughout the project
WBS	May be developed for the entire project	May be known only for work needed for the next few weeks and subject to change
EVMS	Somewhat helpful	May have limited use based upon the number of unknowns and introduction of competing constraints
Specifications	Usually known	May not exist
Schedule	Known	Unknown and possibly subject to continuous changes
Cost	Known	Unknown at the start and may be subject to continuous changes
Risks	Usually known and can be quantified	May be unknown and subject to new risks appearing without warning
Complexity	Known with some degree of confidence	May not be known and is subject to change

1.9 THE GROWTH OF THE VUCA ENVIRONMENT

By the turn of the century, companies were realizing that project complexity was increasing on many of their projects and needed a way to better understand, define, and manage complexity. The acronym VUCA first appeared in 1987 in the writings on leadership theories by Warren Bennis and Burt Nanus to describe the environment in which leaders must make decisions. VUCA stands for volatility, uncertainty, complexity, and ambiguity. In 2002, the acronym was accepted in military education programs and began being discussed as part of risk management practices on government and military projects.

At the beginning of a project, PMs were traditionally trained to look at the enterprise environmental factors (EEFs) that could impact project decision-making. Most PMs assumed that the EEFs, once identified, would possibly remain constant for the duration of the project.

VUCA environments, which are part of the EEFs, can change continuously, and each of the four components can and will impact the other three components. Because of this interconnectivity between components, it is difficult to separate them since they can change continuously. This mandates that PMs constantly observe the VUCA environment and determine the impact it can have on project decision-making.

There are several interpretations and applications for VUCA and its components. The VUCA environment can be looked at on a global basis or just in the environment of how one company competes and makes decisions. The consensus belief is that all companies in the future must plan on how to compete and survive in a VUCA environment, regardless of the industry.

We all seem to agree that the world in which we live is dictated by VUCA, but there still exists confusion on the meaning of the components of VUCA. Some companies prefer to look at just one or two components, such as volatility and complexity, and downplay or avoid considering uncertainty and ambiguity. The application and definition of the components of VUCA are therefore industry- or even company-specific.

PMI published a study in 2014 entitled "Navigating Complexity – A Practice Guide." In the document, PMI proposed a set of 48 questions that made the project management community of practice aware that the future landscape for projects may be dictated by VUCA. Typical questions included:[4]

- Can the project requirements be clearly defined at this stage?
- Can the project scope and objectives be clearly defined?
- Are the project constraints and assumptions likely to remain stable?
- Are the stakeholder requirements likely to remain unchanged?

---

4. For an excellent discussion around some of these questions and VUCA, see Pells, D.L. (2019). Six Fresh Eggs: A half dozen new ideas for managing projects in a rapidly changing VUCA world; presented at the 13th Annual UT Dallas Project Management Symposium, Richardson, Texas, USA in May 2019; *PM World Journal*, Vol. VIII, Issue VIII, September.

The meaning of the VUCA components can change from industry to industry, company to company, and possibly project to project. In a project management environment, VUCA can be described as:

- *Volatility*. An understanding of changes that can occur, usually unfavorable changes, and the forces or events that might cause the changes. The changes could be the need for additional time or funding, poor quality, or inability to meet specifications.
- *Uncertainty*. An understanding of the issues and events that might occur but being unable to accurately predict if they will happen and when. Uncertainty could be knowing that you need more money or more time but not knowing how much.
- *Complexity*. An understanding of the interconnectivity of the events discussed under volatility and uncertainty, and any relationships between them. The greater the number of possible events, the greater the complexity. The team can become overwhelmed with information such that they are unable to decide upon a course of action. Complexity might be knowing that time is money, but not knowing the exact relationship between them.
- *Ambiguity*. Not being able to fully describe the events or misreading the risk events that can affect the outcome of the project. This may be the result of a lack of precedence, the existence of haziness, or mixed meanings concerning the events. This occurs when we are dealing with "unknown unknowns." As an example, ambiguity occurs when we do not fully understand the individual events that may cause budget and schedule issues.

The definition of the terms in VUCA analysis can change from project to project as well as in the environment in which the project takes place. The enterprise environment factors in a project can have a serious impact on VUCA analysis. Also, in the example above, only time and cost were considered. There could be 15–20, or more, different events that could create complexity and impact project leadership and decision-making.

A fifth element should be included, namely crises. In a project environment, volatility, uncertainty, and ambiguity are categorized by events, and the project managers study these events to determine their interconnectivity or complexity and the potential impact on the project. The events could be activities defined in the risk register. But what if a crisis occurs, such as product tampering or misuse, that could lead to serious health issues or even deaths? Now, the events that led up to the crisis must also be considered as part of its complexity, and you may have much less time than you thought for analyses and decision-making.

VUCA is forcing leaders and decision-makers to become managers of change. Today, there is nowhere to go where you can avoid the VUCA environment. Making long-term decisions in an environment that has an abundance of complexity and can change continuously is difficult. This will require managers at all levels, including project managers, to develop new skills necessitated by the VUCA environment and to learn how to think nonlinearly to manage complexity. Some of the new skills that managers as well as business leaders must learn are:

- Brainstorming
- Design thinking
- Creative problem-solving
- Strategic/innovation leadership
- Managing diversity
- Conflict resolution
- Understanding emotional intelligence
- Supply chain management

Reducing Complexity and Uncertainty

Reducing complexity and uncertainty requires looking at ways to minimize the unfavorable impact of all the VUCA components because of their interconnectivity. Some of the actions the project manager may perform include:

- *Volatility.* Closely monitor the EEFs using a worst-case scenario approach and build in slack if possible.
- *Uncertainty.* Continuously seek information that can support or negate the assumptions and constraints that might change the direction of the project.
- *Complexity.* Understand the connectivity and dependencies between elements of work and staff the project with highly qualified resources that can perform in a complex environment.
- *Ambiguity.* Have a willingness to conduct meaningful experimentation and understand cause-and-effect relationships.

Another activity that project managers can perform is a validation of the assumptions. For decades, we took it for granted that the assumptions identified in the business case or SOW were correct and did not need to be validated or even tracked over the duration of the project. In a VUCA environment, especially if there does not exist a business case or SOW, the more unproven assumptions you have, the greater the complexity. Also, the assumptions will most likely change over the duration of the project.

1.10 THE IMPACT OF THE COVID-19 PANDEMIC ON PROJECT MANAGEMENT

In 2020, hundreds of thousands of people across the globe were affected by the COVID-19 pandemic. Public sector, private sector, and government leaders across all businesses had to respond quickly to the effects of the crisis, often with very little time to prepare. All of this was taking place in a rapidly changing VUCA environment characterized by a high degree of risk and uncertainty and accompanied by a lack of complete information needed for decision-making.

Dealing with the crisis required that many projects be implemented quickly without having the luxury of sufficient time to use many of the traditional project management tools, techniques, and processes. The impact of the crisis affected project management leadership practices as well, especially the way that leadership should be applied and how decisions should be made. Most project teams had never been trained in how to deal with crises like this when executing both traditional and nontraditional projects. The information learned from managing projects during the pandemic has generated many best practices that are now being applied to all types of projects.

Leadership

In the early years of project management, companies believed that "business-as-usual" meant the utilization of a one-size-fits-all methodology accompanied by bureaucratic leadership. Project management was driven by rigid policies and procedures so that senior management would maintain command and control over all projects. The goal of leadership was the profitability of the project, often with little concern for the workers. It was not uncommon for workers to be treated as an expense to the project, and the goal of the PMs was to stop workers from using project charge numbers as quickly as possible. This sometimes limited the ability of the project manager to motivate workers and to be creative. Workers often did as little as possible on projects to receive their paychecks. This leadership approach was often dictated by senior management or project sponsors.

The project management community of practice wanted a project leadership template that listed all the characteristics of effective leadership needed in a project environment. Instead, the literature abounded with leadership styles such as authoritarian, participative, laissez-faire, and situational leadership styles. Each leadership style came with advantages and disadvantages. The PM made the decision as to what leadership style to use most often based upon what was in the best interest of the PM for a particular type of project and with little regard for the project team's needs or desires.

By the turn of the century, changes began to take place that would alter our view of the characteristics of effective project leadership. First and foremost, more and more projects that PMs were asked to manage were nontraditional projects to fulfill the growth in strategic and innovation needs. Second, the introduction of flexible methodologies such as Agile and Scrum made it clear that "business-as-usual" leadership may be ineffective on many types of nontraditional projects. The new methodologies required a different perception of project leadership. Third, the durations of the projects were getting longer, which meant that project managers would need to spend significantly more time collaborating with workers and stakeholders. Fourth, the importance of most of the nontraditional projects required that many workers be assigned full-time to these activities and to interface with the PMs more frequently than in the past. Fifth, project team members were allowed to evaluate the effectiveness of the PMs and project governance personnel at the end of the project.

The common bond in the five changes just described is the word "collaboration." The project management community and academia began looking at the characteristics of effective project management leadership for the future that could be described by a listing of characteristics supporting the need for better collaboration among all players.

The result is a growth in the concept of "social project management leadership," which contains the following characteristics:[5]

- Acting and interacting with the team in a socially responsible manner
- Inspiring team spirit and keeping the team energized
- Managing complexity in a VUCA environment
- Managing cultural transformation
- Participating in change management efforts
- Encouraging design thinking practices
- Understanding nonverbal cues and emotional intelligence applications
- Employ active listening practices
- Demonstrate respect for everyone
- Support quality of life factors important to team members

The pandemic has accelerated the recognition and need for significant changes in human behavior in project leadership. These characteristics are now seen as best practices and are becoming part of project management educational programs.

Trust

One of the biggest challenges in project management leadership is getting the team to trust that the decisions made by the project managers are based upon honesty, fairness, and ethical practices that consider the best interests of both the projects and the workers. Trust will encourage workers to become more engaged in the project and make it easier for the PMs to obtain their commitment. Trust can be destroyed if the PMs act in an unethical manner, abuse the power of their position, or demonstrate toxic emotions toward the team.

The pandemic placed workers under significantly higher levels of stress. Many workers quickly recognized the added complexities of their job with the reality of having to work from home. Working on virtual teams was a new experience for many workers. There was now a much heavier reliance on the project managers to provide guidance to the team members for their day-to-day activities as well as keep them updated on how the company was planning to recover the business during the crisis, especially if there was a lockdown. The need for effective collaboration to establish trust is now seen as a best practice.

To build up and maintain trust, project management leadership had to provide emotional and interpersonal support for the team on a continuous basis. This was a new experience for many project managers, who were just beginning to realize the importance of obtaining worker trust and its impact during a crisis.

---

5. For additional information on social project management leadership, see Harold Kerzner, Al Zeitoun and Ricardo Vargas, *Project Management Next Generation: The Pillars for Organizational Excellence*, 2022, Hoboken, John Wiley & Sons, Chapter 6.

Communications

Leadership during a crisis must focus on persuasion rather than formal authority and coercion. This is best achieved through effective communication. Active listening concepts should be used to understand workers' concerns.

When team members work from home for an extended amount of time, they rely on media interaction and engagement to understand what is happening during the crisis. The greater the uncertainty during the media engagement, the greater the anxiety in the listeners.

Project managers must be willing to give up the idea that information is power and support the transparency of making information available. The information should provide a clear and realistic view of what is currently happening and possibly an optimistic view of future expectations. This is now becoming a best practice on all projects.

Decision-Making

On traditional projects, PMs often adopt a wait-and-see attitude before making decisions, with the belief that all possible scenarios must be considered, including the worst-case scenario. Unfortunately, all of this takes time. During a crisis, delaying or not making a decision in a timely manner is generally a bad idea and can lead to a potentially worse outcome than expected by missing project opportunities. Decision delays can also create an environment where workers no longer have any credibility in the PM's leadership ability.

The pandemic has taught us that we must be willing to make project decisions rapidly based upon whatever information we have, even if the information is imperfect or incomplete. PMs must do the best they can with the information available to anticipate what can go wrong.

During a crisis, strategic decisions are usually made for both the long-term and short-term benefits, whereas project decisions may require a different focus. Some decisions may be designed around the organization's core values, which may be in contrast with the project's objectives. Other types of decisions may need to focus on the beliefs and values of the people affected by the decision. But regardless of the decisions made, the PM must spend sufficient time with all players, even if one-on-one in virtual meetings, to explain the meaning of the decision and what actions are expected from them.

Project Control Center

In traditional project management, with well-defined scope and expectations, critical decisions are most often made by the team in open meetings and then reviewed as needed with just the PM and either the sponsor or governance committee. However, during a crisis, executives often believe that they must take control, resulting in the removal of leadership responsibility from the project manager and possibly the project sponsor. Meetings are held behind closed doors, with only a select few in attendance. These secretive meetings often do not include the PMs or project team members, even though the resulting decisions may have a serious impact on the direction of the projects.

Using the hierarchical approach for making critical decisions during a crisis runs the risk of making bad decisions by overlooking critical information. The result is most often mistrust by the key players and a lack of commitment to the decisions.

Effective decision-making during a crisis should be done through brainstorming sessions. By selecting the right people for the session, the greater the likelihood that critical information will not be overlooked, and more perspectives will be considered. Many of the participants may not have been part of the original project team if the crisis concerned just your project. The brainstorming sessions should include suppliers, distributors, and other strategic partners that may be experiencing the same risks and uncertainties as you and must understand the rationale for whatever decisions are made and how they impact them. Their support may be essential.

There may be frequent brainstorming sessions, and workers will want to know what decisions are being made and if they will be impacted. It may be beneficial to establish a project control center or project nerve center for the control of information to the workers.

Change Management

The result of crisis decisions frequently leads to change management activities as strategies based on new or promising ideas are designed for implementation. During a crisis, companies try to reduce costs, and they often begin by abandoning teaching and educational activities. This can lead to detrimental results if the training and education are essential to fulfill and implement a strategy, especially if there is a resulting organizational change where people are removed from their comfort zone.

Conclusions

The magnitude and speed of crises are often unpredictable, as we have seen during the COVID-19 pandemic. Many of the actions that companies have taken during a crisis are now becoming standard practices when managing many types of projects and are treated as best practices.

1.11 GENERAL MOTORS AND VENTILATORS[6]

When one thinks of project risk management, the thought of a pandemic usually isn't at the top of the list, if it is on the list at all. In 2020, things really turned upside-down when the COVID-19 virus hit the world. In March 2020, General Motors idled most of its world-wide workforce in order to slow the spread of the virus and protect its workers. So here was one of the largest companies in the world, with one of the most versatile workforces, all sidelined, waiting for the next bit of news.

---

6. Material in this section was graciously provided by Jim Trela, Corp Staffs Lead Planner and Chief of Staff, and Jeffrey J. Hall, PMP, GM SPSC, Sr. Development Lead. ©2022 by General Motors Corporation. All rights reserved. Reproduced with permission.

What happened next was a brilliant pivot requiring the use of project management talent and application of manufacturing resources to help respond to the very pandemic that sidelined most of its workforce. GM's mobilization of the same project and program management resources that manufacture vehicles produced incredibly fast results for a corporate social responsibility (CSR) initiative.

Companies like General Motors today maintain a self-regulated strategy called corporate social responsibility (CSR) that is integrated into the firm's business model and identifies the ethically oriented activities the firm will undertake for the benefit of consumers, society, ecology, and government regulations. Traditional CSR activities in many companies focus upon the consumption of certain natural and renewable resources, such as water, energy, and other materials. However, as General Motors (GM) has admirably demonstrated, there are other ways to demonstrate CSR.

Included in the description of a firm's CSR is usually the term "social," which may be defined as improving or showing concern for human life without impacting the capacity of the supporting ecosystems. This leads us to the term "social innovation," which is the creation of new or innovative products and services that support CSR. The outcomes of social innovation often focus more on a concern for society instead of profitability.

GM's CSR effort regarding ventilators began on March 17 after they were approached by stopthespread.org, a coalition of CEOs, trying to organize companies to help stop the spread of the COVID-19 pandemic. The coalition contacted Mary Barra, GM's CEO, and suggested that GM team up with Ventec, a company located in Bothell, Washington, and a manufacturer of small portable ventilators. Within 24 hours, executives from GM and Ventec had their first conference call, and the following day, a team from GM flew to Seattle to meet with Ventec and begin the joint venture.

> *Senior stakeholders were quickly on board with the overall objective and remained in contact with all project workstreams on a daily basis. Empowered teams met daily and were loosely coupled to coordinate dependencies, understand the timeline, and resolve issues quickly. Issues were documented on shared folders, triaged, and resolved.*
>
> *Roadblocks did not exist for long. Decisions were made and the team moved ahead. Deliverables and approvals that would normally take weeks, only took minutes, due to the clear communication paths, team focus and executive priority. Capital was approved and contracts were put in place with minimal executive oversight/approval being required.*
>
> *All of this was possible due to the clarity of the GM vision established up front and unwavering commitment of GM people at all levels of the organization to deliver ventilators. There was plenty of Inspiration to get this done, as if someone in our family depended on it. People were excited to be involved. Each shipment was a celebration, down to the last one shipped just before midnight on the contract deadline.*

By March 20, GM had engaged its global supply base, and within 72 hours, suppliers had plans to get all the necessary parts. The UAW's national and local leadership signed off on the project, and by March 25, crews began preparing GM's 272-acre complex in Kokomo, Indiana, for production.[7] About 800 full-time and part-time GM

---

7. https://www.freep.com/story/money/cars/general-motors/2020/08/14/no-shortfall-supply-ventilators-gm-ventec-stockpile/3366678001/

employees were assigned to the project to produce 30,000 ventilators. According to a GM spokesperson:

> The Detroit automaker will produce and deliver the [30,000] ventilators to the government by the end of August, with the first 6,132 ventilators being delivered by June 1, [2020].[8]

Senior management at GM was committed to the CSR ventilator effort. "Our commitment to build Ventec's high-quality critical care ventilator, VOCSN, has never wavered," GM said. "The partnership between Ventec and GM combines global expertise in manufacturing quality and a joint commitment to safety to give medical professionals and patients access to life-saving technology as rapidly as possible. The entire GM team is proud to support this initiative."[9]

> *GM Information Technology (IT) group was a critical piece of the Ventilator production. While significantly scaling up the Ventec process to meet GM's volume expectations, each ventilator part had to be tracked through the manufacturing steps. GM reached out to Ventec's IT partners immediately and quickly worked through the contracting and architecture required to meet the business needs for quality, tracking, and throughput. This was done in parallel with setting up the initial manufacturing space to start building 'non-saleable' units within 3 weeks of GM's initial visit to Ventec.*

GM and its partner, Ventec Life Systems, delivered the last of the 30,000 ventilators owed to the U.S. government as planned. GM said the full federal order was completed in 154 days, with one ventilator made about every seven minutes.[10] According to GM CEO Mary Barra:

> . . . "the automaker's motivation to produce the critical care ventilators was fueled by thousands of people at GM, Ventec and our suppliers who all wanted to do their part to help save lives during the pandemic. It was inspiring to see so many people achieve so much so quickly."[11]
>
> *GM transitioned out of the Ventilator business after 5 months and enabled Ventec to take over the production line immediately upon completing the government contract. There was minimal down time in the transition due to careful planning. Ventec continued to build ventilators on that same production line until the ventilator demand diminished later in 2020.*

GM's commitment to CSR did not end with the delivery of the last ventilator. GM was still making face masks. The company said it will donate 2 million face masks

---

8. https://www.cnbc.com/2020/04/08/gm-to-build-30000-ventilators-for-us-for-489point4-million.html?__source=sharebar|email&par=

9. https://www.cnbc.com/2020/03/27/trump-orders-general-motors-to-make-ventilators-under-defense-production-act.html?__source=sharebar|email&par=sharebar

10. https://www.freep.com/story/money/cars/general-motors/2020/09/01/gm-and-ford-deliver-last-ventilators-amid-coronavirus/3449490001/

11. Ibid.

to Michigan public schools as part of the State of Michigan's MI Mask Aid partnership. GM's contribution includes 750,000 child-size masks for elementary students. Those will be ready for delivery by September 14, GM said. Also, 1.25 million adult-size masks for high school students, faculty, and staff will be ready for delivery by September 28.[12]

Nobody knows what will happen in the future. There could be another pandemic, acts of God, or wars that require companies to be able to pivot and redirect their resources from their primary purpose to one focused on the benefit of humanity, as did GM with the ventilators and masks. This ability will bring more than just financial benefits to the company. People will turn to GM when innovation and ingenuity are needed to solve CSR-related challenges, even on a global scale.

1.12 BEST PRACTICES PROCESS

Why capture best practices? The reasons or objectives for capturing best practices might include:

- Continuous improvements (efficiencies, accuracy of estimates, waste reduction, etc.)
- Enhanced reputation
- Winning new business
- Survival of the firm

Survival of the firm has become the most important reason today for capturing best practices. In the last few years, customers have put pressure on contractors in requests for proposals by requesting:

- A listing of the number of PMP® credential holders[13] in the company and how many will be assigned to this project
- A demonstration that the contractor has an enterprise project management methodology, whether rigid or flexible, that is acceptable to the customer, or else the contractor must use some other methodology approved by the customer
- Supporting documentation identifying the contractor's maturity level in project management, possibly using a project management maturity model for assessments
- A willingness to share lessons learned and best practices discovered on this project and perhaps previous projects from other customers

Recognizing the need for capturing best practices is a lot easier than actually doing it. Companies are developing processes for identifying, evaluating, storing, and

---

12. Ibid.

13. PMP is a registered mark of the Project Management Institute.

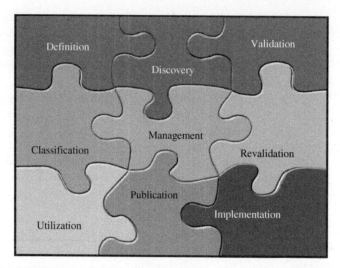

Figure 1–8. Best practices processes.

disseminating information on best practices. There are nine best practices activities, as shown in Figure 1–8, and most companies that recognize the value of capturing best practices accomplish all of these steps.

The processes answer nine questions:

1. What is the definition of a best practice?
2. Who is responsible for identifying the best practices, and where do we look?
3. How do we validate that something is a best practice?
4. Are there levels or categories of best practices?
5. Who is responsible for the administration of the best practice once approved?
6. How often do we reevaluate whether something is still a best practice?
7. How do companies use best practices once they are validated?
8. How do large companies make sure that everyone knows about the existence of the best practices?
9. How do we make sure that the employees are using the best practices and using them properly?

Each of these questions is addressed in the sections that follow.

1.13 STEP 1: DEFINITION OF A BEST PRACTICE

For more than two decades, companies have been fascinated by the expression "best practices." But now, after two decades or more of use, we are beginning to scrutinize the term, and perhaps better expressions exist.

A best practice begins with an idea that there is a technique, process, method, or activity that can be more effective at delivering an outcome than any other approach and provides us with the desired outcome with fewer problems and unforeseen complications. As a result, we supposedly end up with the most efficient and effective way of accomplishing a task based on a repeatable process that has been proven over time for a large number of people and/or projects.

But once this idea has been proven to be effective, we normally integrate the best practice into our processes so that it becomes a standard way of doing business. Therefore, after acceptance and proven use of the idea, the better expression possibly should be a "proven practice," "smart practice," "good practice," or "promising practice" rather than "best practice." Saying that something is a best practice may lead people to believe that this is the best approach and cannot be improved upon. This is just one argument why a "best practice" may be just a buzzword and should be replaced by "proven practice."

Another argument is that the identification of a best practice may lead some to believe that we were performing some activities incorrectly in the past, and that may not have been the case. The new practice may simply be a more efficient and effective way of achieving a deliverable. Another issue is that some people believe that best practices imply that there is one and only one way to accomplish a task. This may also be a faulty interpretation.

Perhaps in the future, the expression "best practices" will be replaced by "proven practices." Although for the remainder of this text, we use the expression "best practices," the reader must understand that other terms may be more appropriate. This interpretation is necessary here because most of the companies that have contributed to this book still use the expression "best practices."

As project management evolved, so did the definitions of a best practice. Some definitions of a best practice are highly complex, while others are relatively simplistic. Yet they both achieve the same purpose of promoting excellence in project management throughout the company. Companies must decide on the amount of depth to go into the best practice—should it be generic and at a high level or detailed and at a low level? High-level best practices may not achieve the efficiencies desired, whereas highly detailed best practices may have limited applicability.

Every company can have its own definition of a best practice, and there might even be industry standards on the definition. Typical definitions of a best practice might be:

- Something that works
- Something that works well
- Something that works well on a repetitive basis
- Something that leads to a competitive advantage
- Something that can be identified in a proposal to generate business
- Something that differentiates us from our competitors
- Something that keeps the company out of trouble and, if trouble occurs, the best practice will assist in getting the company out of trouble

Every company has its own definition of a best practice. There appear to be four primary reasons for capturing best practices:

1. To improve efficiency
2. To improve effectiveness
3. To achieve standardization
4. For consistency

In each of the following definitions, you should be able to identify which of the four, or combination thereof, the company targets:

> At Orange Switzerland, a best practice is defined as an experience based, proven, and published way of proceeding to achieve an objective.
>
> **Spokesperson at Orange Switzerland**

> We do have best practices that are detailed in our policies/procedures and workflows. These are guidelines and templates as well as processes that we all [members of the EPMO—enterprise project management office] have agreed to abide by as well as that they are effective and efficient methods for all parties involved. In addition, when we wrap up (conclude) a project, we conduct a formal lessons learned session (involving the project manager, sponsors, core team, and other parties impacted by the project), which is stored in a collective database and reviewed with the entire team. These lessons learned are in effect what create our best practices. We share these with other health care organizations for those vendors for which we are reference sites. All of our templates, policies/procedures, and workflows are accessible by request and, when necessary, we set meetings to review as well as explain them in detail.
>
> **Nani Sadowski, formerly manager of the Enterprise Project Management Office at Halifax Community Health Systems**

> Any tool, template, or activity used by a project manager that has had a positive impact on the *PMBOK® Guide*[14] knowledge and/or process areas and/or the triple constraint. An example of a best practice would be: Performing customer satisfaction assessments during each phase of a project allows adjustments during the project life cycle, which improves deliverables to the client, and improves overall project management. [This would be accompanied by a template for a customer satisfaction survey.]
>
> **Spokesperson for AT&T**

Generally, we view a best practice as any activity or process that improves a given situation, eliminates the need for other more cumbersome methods, or significantly enhances an existing process. Each best practice is a living entity and subject to review, amendments, or removal.

For Churchill Downs Incorporated, a best practice is any method or process that has been proven to produce the desired results through practical application. We do not

---

14. PMBOK is a registered mark of the Project Management Institute.

accept "industry" or "professional standards" as best practices until we have validated that the method or process works in our corporate environment.

Examples of some of our best practices include:

- *Charter signatures*: One of our best practices is requiring stakeholder signatures on project and program charters. This seems basic, but my experience is that a formal review and approval of a project's business objectives and goals is rarely documented. By documenting business objectives and their associated metrics, we have been able to proactively manage expectations and ensure alignment between various stakeholders.
- *Process definition*: In addition to defining the organization's project, program, and portfolio management processes, the PMO has also taken an active role in mapping all of the financial processes for Churchill Downs Incorporated, from check requests and employee reimbursement requests to procedures for requesting capital expenses and purchase orders. This practice has increased corporate-wide awareness of how standardizing processes can enhance efficiency.
- *Access to information*: The PMO developed process maps, procedures, and policies for the end-to-end budgeting processes and associated workflows and templates. These have been made available company-wide via CCN, the company's intranet site.

Comments by Chuck Millhollan, formerly director of program management, Churchill Downs Incorporated

At Indra, we consider a "best practice" in project management as a management action or activity that usually generates a positive outcome. As such, it is accepted by the management community and eventually becomes a recommended or required way of performing the task. We also consider as a "best practice," the use of predefined indicators, thresholds or metrics to make or facilitate decisions with regard to project management processes.

Comments by Enrique Sevilla Molina, PMP, formerly corporate PMO director, Indra

1.14 STEP 2: SEEKING OUT BEST PRACTICES

Best practices can be captured either within your organization or externally to it. Benchmarking is one way to capture external best practices, possibly by using the PMO as the lead for external benchmarking activities. However, there are other external sources other than benchmarking for identifying best practices:

- Project Management Institute (PMI) publications
- Forms, guidelines, templates, and checklists that can affect the execution of the project
- Forms, guidelines, templates, and checklists that can affect the definition of success in a project

- Each of the *PMBOK® Guide* areas of knowledge or domain areas
- Within company-wide or isolated business units
- Seminars and symposiums on general project management concepts
- Seminars and symposiums specializing in project management best practices
- Relationships with other professional societies
- Graduate-level theses

With more universities offering master's and doctorate-level work in project management, graduate-level theses can provide up-to-date research on best practices.

The problem with external benchmarking is that best practices discovered in one company may not be transferable to another company. In the author's opinion, most of the best practices are discovered internally and are specifically related to the company's use of its project management methodology and processes. Good project management methodologies allow for the identification and extraction of best practices. However, good ideas can come from benchmarking as well.

Sometimes it is easier to identify the drivers or metrics that affect each best practice than the best practice itself. Metrics and drivers can be treated as early indicators that a best practice may have been found. It is possible to have several drivers for each best practice. It is also possible to establish a universal set of drivers for each best practice, such as:

- Reduction in risk by a certain percentage, cost, or time
- Improve estimating accuracy by a certain percentage or dollar value
- Cost savings of a certain percentage or dollar value
- Efficiency increases by a certain percentage
- Reduction in waste, paperwork, or time by a certain percentage

There are three advantages of this approach for searching for drivers.

1. The drivers can change over time, and new drivers can emerge rapidly.
2. The best practices process is more of a science than an art.
3. We can establish levels of best practices such as those shown in Figure 1–9. In this figure, a level 4 best practice, which is the best, would satisfy 60 percent or more of the list of drivers or characteristics of the ideal best practice.

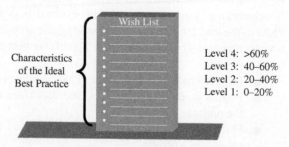

Figure 1–9. Best practices levels. Each level contains a percentage of the ideal characteristics.

Best practices may not be transferable from company to company, nor will they always be transferable from division to division within the same company. As an example, consider the following best practice discovered by a telecommunications company:

- A company institutionalized a set of values that professed that quality was everything. The result was that employees were focusing so much on quality that there was a degradation in customer satisfaction. The company then reprioritized its values, with customer satisfaction being the most important, and quality actually improved.

In this company, an emphasis on customer satisfaction led to improved quality. However, in another company, emphasis on quality could just as easily have led to an improvement in customer satisfaction. Care must be taken during benchmarking activities to make sure that whatever best practices are discovered are in fact directly applicable to your company.

Best practices need not be overly complex. As an example, the next list of best practices is taken from companies discussed in this textbook. As you can see, some of the best practices were learned from failures rather than successes:

- Changing project managers in midstream is bad, even if the project is in trouble. Changing project managers inevitably elongates the project and can make it worse.
- Standardization yields excellent results. The more standardization placed in a project management methodology, usually, the better the results are. However, this is dependent on the type of project. Some projects are better implemented with highly flexible methodologies.
- Maximization of benefits occurs with a methodology, whether rigid or flexible, based on templates, forms, guidelines, and checklists rather than policies and procedures.
- Methodologies must be updated to include the results of discovering best practices. The more frequently the methodology is updated, the quicker the benefits are realized.

As noted, best practices need not be complex. Even though some best practices seem simplistic and based on common sense, their constant reminder and use lead to excellence and customer satisfaction.

Another way to identify sources of best practices is through the definition of project success, CSFs, and key performance indicators (KPIs). Extracting best practices through the definition of success on a project may be difficult and misleading, especially if we have a poor definition of success.

Over the years, many of the changes that have taken place in project management have been the result of the way we define project success. As an example, consider the next chronological events that took place over the past several decades:

- *Success is measured by the triple constraints or competing constraints.* The triple constraints are time, cost, and performance (which include quality, scope, and technical performance). This was the basis for defining success during the birth of project management. Competing constraints can include safety, aesthetic value, benefits, safety, level of acceptable risk, and others.
- *Customer satisfaction must be considered as well.* Managing a project within the triple constraints is always a good idea, but the customer must be satisfied with the end result. A contractor can complete a project within the triple constraints and still find that the customer is unhappy with the end result.
- *Other (or secondary) factors must be considered as well.* These additional competing constraints include using the customer's name as a reference, corporate reputation and image, compliance with government regulations, strategic alignment, technical superiority, ethical conduct, business realization, value management, and other such factors. The secondary factors may end up being more important than the primary factors of the triple constraints.
- *Success must include a business component.* Project managers are managing part of a business rather than merely a project and are expected to make sound business decisions as well as project decisions. There must be a business purpose for each project. Each project is considered as a contribution of business value to the company when completed.
- *Prioritization of constraints must occur.* Not all project constraints are equal. The prioritization of constraints is done on a project-by-project basis. Sponsorship involvement in this decision is essential.
- *The definition of success must be agreed on between the customer and the contractor.* Each project can have a different definition of success. There must be up-front agreement between the customer and the contractor at project initiation or at the first meeting between them on what constitutes success.
- *The definition of success must include a "value" component.* Why work on a project that does not provide the correct expected business value at completion?

The problem with defining success as on time, within cost, and at the desired quality or performance level is that this is an internal definition of success only. Bad things can happen on projects when the contractor, customer, and various stakeholders all focus on different definitions of project success. There must be an up-front agreement on what constitutes project success. The ultimate customer or stakeholder should have some say in the definition of success, and ultimately, numerous best practices may be discovered that relate to customer/stakeholder interfacing.

Today, we recognize that the customer rather than the contractor defines quality. The same holds true for project success. Customer and stakeholder acceptance must be included in any definition of project success. You can complete a project internally within your company within time, within cost, and within quality or specification limits and yet find the project is not fully accepted by the customer or stakeholders.

Although some definitions of project success seem quite simple, many companies have elaborated on the primary definition of project success. At Churchill Downs Incorporated (CDI), success is defined more rigorously than in most companies. According to Chuck Millhollan, formerly director of program management:

> Project success is defined in our PMO charter as follows.
>
> Based on input from CDI's executive management, the PMO considers a project to be a success when the following are true:
>
> a. Predefined business objectives and project goals were achieved or exceeded.
> b. A high-quality product is fully implemented and utilized.
> c. Project delivery met or beat schedule and budget targets.
> d. There are multiple winners:
>   i. Project participants have pride of ownership and feel good about their work.
>   ii. The customer's (internal and/or external) expectations are met.
>   iii. Management has met its goals.
> e. Project results helped build a good reputation.
> f. Methods are in place for continual monitoring and evaluation (benefit realization).
>
> We do not use project management "process" indicators to define project success. While schedule and budget targets are part of the criteria, sponsor acceptance, project completion, and ultimately project success, is based on meeting defined business objectives.

Enrique Sevilla Molina, PMP, formerly corporate PMO director at Indra, provides us with his company's definition of project success and program success:

> Project success is based on achieving the proposed project targets in budget, scope, performance and schedule. Many times, the economic criteria appears as the main driving factor to measure project success, but there are other factors just as important such as building a durable relationship with the customer and building strong alliances with selected partners. Another significant criteria for project success measurement is the reliability of the project data forecast. It may be the case that, when the economic results of the project are not as good as they should be, if the fact is pointed out and reported soon enough, the success of the project is equally achieved.
>
> Program success is based on achieving the program's overall strategic targets defined during program definition and, at this level, the success is measured not only by achieving the expected economic outcomes but, most of all, reaching the expected position in the market with regard to a product or a line of products, and establishing a more advantageous position with regard to our competitors. Leadership in a product line constitutes the ultimate measure of success in a program. It is worthwhile to mention that, quite often, the success of a program is based on the partnership concept developed with our major subcontractors at the project level.
>
> Project success is defined at a business unit level by the responsible director, in accordance with the strategic goals assigned to the project.
>
> Program success is defined at the company level by the chief operations management in accordance with the program's defined mission.

AT&T defines project and program success in a similar manner. According to a spokesperson for AT&T:

> Project success is defined as a Client Satisfaction rating of "Very Satisfied" and On-Time Performance of Project Delivery of 98% or greater. The Project Management Organizational Leadership Team sets the objectives, which are tracked to determine project success. Program success is defined and tracked the same way as project success.
>
> Excellence [in project management] is defined as a consistent Project Management Methodology applied to all projects across the organization, continued recognition by our customers, and high customer satisfaction. Also, our project management excellence is a key selling factor for our sales teams. This results in repeat business from our customers. In addition, there is internal acknowledgement that project management is value-added and a must have.

Project success can be measured intermittently throughout the phase or gate review meetings that are part of the project management methodology. This allows a company to establish interim metrics for measuring success. An example of this appears in Chapter 4, "Project Management Methodologies."

Another element that is becoming important in the definition of success is the word *value*.

The ultimate definition of success might very well be when the customer is so pleased with the project that the customer allows you to use his or her name as a reference. This occurred at one company that bid on a project at 40 percent below its cost of doing the work. When asked why the bid was so low, company representatives responded that they knew they were losing money, but what was really important was getting the customer's name on the corporate resume of clients. Therefore, secondary factors may be more important than primary factors.

The definition of success can also change based on whether a company is project- or non-project-driven. In a project-driven firm, the entire business of the company is projects. But in a non-project-driven firm, projects exist to support the ongoing business of production or services. In a non-project-driven firm, the definition of success also includes completion of the project *without* disturbing the firm's ongoing business. It is possible to complete a project within time, within cost, and within quality and at the same time cause irrevocable damage to the organization. This occurs when the project manager does not realize that the project is *secondary* in importance to the ongoing business.

Some companies define success in terms of CSFs and KPIs. CSFs identify those factors necessary to meet the desired deliverables of the customer. CSFs and KPIs do not need to be elaborate or sophisticated metrics. Simple metrics, possibly based on the triple constraints, can be quite effective. According to a spokesperson from AT&T:

> The critical success factors include Time, Scope, Budget, and Customer Satisfaction. Key performance indicators include on-time performance for key deliverables. These include customer installation, customer satisfaction and cycle-time for common milestones.

Typical CSFs for most companies include:

- Adherence to schedules
- Adherence to budgets
- Adherence to quality
- Appropriateness and timing of signoffs
- Adherence to the change control process
- Add-ons to the contract

CSFs measure the end result, usually as seen through the eyes of the customer. KPIs measure the quality of the process to achieve the end results. KPIs are internal measures and can be reviewed on a periodic basis throughout the life cycle of a project. Typical KPIs include:

- Use of the project management methodology
- Establish control processes
- Use of interim metrics
- Quality of resources assigned versus planned for
- Client involvement

KPIs answer such questions as: Did we use the methodology correctly? Did we keep management informed, and how frequently? Were the proper resources assigned, and were they used effectively? Were there lessons learned that could necessitate updating the methodology or its use? Companies that are excellent in project management measure success both internally and externally using KPIs and CSFs. As an example, consider the following remarks provided by a spokesperson from Nortel Networks:

> Nortel defines project success based on schedule, cost, and quality measurements, as mutually agreed upon by the customer, the project team, and key stakeholders. Examples of key performance indicators may include completion of key project milestones, product installation/integration results, change management results, completion within budget, and so on. Project status and results are closely monitored and jointly reviewed with the customer and project team on a regular basis throughout a project to ensure consistent expectations and overall success. Project success is ultimately measured by customer satisfaction.

Here are additional definitions of CSFs and KPIs:

CSFs

> Success factors are defined at the initial stages of the project or program, even before they become actual contracts, and are a direct consequence of the strategic goals allocated to the project or program. Many times, these factors are associated with expanding the market share in a product line or developing new markets, both technically and geographically.
>
> **Enrique Sevilla Molina, PMP, formerly corporate PMO director, Indra**

Obviously, CSFs vary with projects and intent. Here are some that apply to a large variety of projects:

- Early customer involvement
- High-quality standards
- Defined processes and formalized gate reviews
- Cross-functional team organizational structure
- Control of requirements, prevention of scope creep
- Commitment to schedules—disciplined planning for appropriate level of detail and objective and frequent tracking
- Commitment of resources—right skill level at necessary time
- Communication among internal teams and with customers
- Early risk identification, management, and mitigation—no surprises
- Unequaled technical execution based on rigorous engineering.

Comments provided by a spokesperson at Motorola

KPIs

Our most common KPIs are associated to the financial projects results, for instance, project margin compliance with the allocated strategic target, new contracts figure for the business development area goals, etc. Success factors are translated into performance indicators so they are periodically checked.

By default, a first indication of projects health is provided by the schedule and cost performance indices (SPI and CPI) embedded into the PM tools. They are provided monthly by the project management information system and they are also available for historical analysis and review. These indicators are also calculated for each department, so they constitute an indicator of the overall cost and schedule performance of the department or business unit.

Enrique Sevilla Molina, PMP, formerly corporate PMO director, Indra

Postship acceptance indicators:

- Profit and loss
- Warranty returns
- Customer reported unique defects
- Satisfaction metrics

In-process indicators:

- Defect trends against plan
- Stability for each build (part count changes) against plan

Feature completion against plan:

- Schedule plan versus actual performance
- Effort plan versus actual performance
- Manufacturing costs and quality metrics
- Conformance to quality processes and results of quality audits
- System test completion rate and pass/fail against plan
- Defect/issue resolution closure rate

- Accelerated life-testing failure rates against plan
- Prototype defects per hundred units (DPHU) during development against plan
Provided by a spokesperson at Motorola

The SOW provides a checklist of basic indicators for the success of the project, but client satisfaction is also important. The SOW will indicate what the deliverables are and will provide information on costs and timelines that can be easily tracked.

Most people seem to understand that CSFs and KPIs can be different from project to project. However, there is a common misbelief that CSFs and KPIs, once established, must not change throughout the project. As projects go through various life-cycle phases, these indicators can change.

In the author's experience, more than 90 percent of the best practices that companies identify come from analysis of the KPIs during the debriefing sessions at the completion of a project or at selected gate review meetings. Because of the importance of extracting these best practices, some companies are now training professional facilitators capable of debriefing project teams and capturing the best practices.

Before leaving this section, it is necessary to understand who discovers the best practices. Best practices are discovered by the people performing the work, namely the project manager, project team, and possibly the line manager. According to a spokesperson from Motorola:

> The decision as to what is termed a best practice is made within the community that performs the practice. Process capabilities are generally known and baselined. To claim best practice status, the practice or process must quantitatively demonstrate significant improvements in quality, efficiency, cost, and/or cycle time. The management of the organization affected as well as process management must approve the new practice prior to institutionalization.

Generally, the process of identification begins with the appropriate team member. If the team member believes that he or she has discovered a best practice, they then approach their respective line manager and possibly project manager for confirmation. Once confirmation is agreed upon, the material is sent to the PMO for validation. After validation, the person who identified the best practice is given the title of "Best Practice Owner" and has the responsibility of nurturing and cultivating the best practice.

Some companies use professional facilitators to debrief project teams in order to extract best practices. These facilitators may be assigned to the PMO and are professionally trained in how to extract lessons learned and best practices from both successes and failures. Checklists and templates may be used as part of the facilitation process.

1.15 DASHBOARDS AND SCORECARDS

In our attempt to go to paperless project management, emphasis is being given to visual displays such as dashboards and scorecards utilizing and displaying CSFs and KPIs. Executives and customers desire a visual display of the most critical project performance information in the least amount of space. Simple dashboard techniques, such as traffic light reporting, can convey critical performance information. As an example:

- *Red traffic light.* A problem exists that may affect time, cost, quality, or scope. Sponsorship involvement is necessary.
- *Yellow or amber light.* This is a caution. A potential problem may exist, perhaps in the future, if not monitored. The sponsor is informed, but no action by the sponsor is necessary at this time.
- *Green light.* Work is progressing as planned. No involvement by the sponsor is necessary.

While a traffic light dashboard with just three colors is most common, some companies use many more colors. The IT group of a retailer had an eight-color dashboard for IT projects. An amber color meant that the targeted end date had passed and the project was still not complete. A purple color meant that this work package was undergoing a scope change that could have an impact on the triple constraint.

Some people confuse dashboards with scorecards. There is a difference between dashboards and scorecards. According to Eckerson:[15]

- Dashboards are visual display mechanisms used in an *operationally* oriented performance measurement system that measure performance against targets and thresholds using real-time data.
- Scorecards are visual displays used in a *strategically* oriented performance measurement system that chart progress toward achieving strategic goals and objectives by comparing performance against targets and thresholds.[16]

Both dashboards and scorecards are visual display mechanisms within a performance measurement system that convey critical information. The primary difference between dashboards and scorecards is that dashboards monitor operational processes such as those used in project management, whereas scorecards chart the progress of tactical goals. Table 1–4 and the description following it show how Eckerson compares the features of dashboards and scorecards.

TABLE 1–4. COMPARING FEATURES

Feature	Dashboard	Scorecard
Purpose	Measures performance	Charts progress
Users	Supervisors, specialists	Executives, managers, and staff
Updates	Right-time feeds	Periodic snapshots
Data	Events	Summaries
Display	Visual graphs, raw data	Visual graphs, comments

Source: W. Eckerson, *Performance Dashboards: Measuring, Monitoring and Managing Your Business* (Hoboken, NJ: Wiley, 2006), p. 13/John Wiley & Sons.

---

15. W. Eckerson, *Performance Dashboards: Measuring, Monitoring and Managing Your Business* (Hoboken, NJ: Wiley, 2006, p. 293). Chapter 12 provides an excellent approach to designing dashboard screens.
16. Ibid., p. 295.

Dashboards: Dashboards are more like automobile dashboards. They let operational specialists and their supervisors monitor events generated by key business processes. But unlike automobiles, most business dashboards do not display events in "real time," as they occur; they display them in "right time," as users need to view them. This could be every second, minute, hour, day, week, or month depending on the business process, its volatility, and how critical it is to the business. However, most elements on a dashboard are updated on an intraday basis, with latency measured in either minutes or hours.

Dashboards often display performance visually, using charts or simple graphs, such as gauges and meters. However, dashboard graphs are often updated in place, causing the graph to "flicker" or change dynamically. Ironically, people who monitor operational processes often find the visual glitz distracting and prefer to view the data in the original form, as numbers or text, perhaps accompanied by visual graphs.

Scorecards: Scorecards, on the other hand, look more like performance charts used to track progress toward achieving goals. Scorecards usually display monthly snapshots of summarized data for business executives who track strategic and long-term objectives, or daily and weekly snapshots of data for managers who need to chart the progress of their group of projects toward achieving goals. In both cases, the data are fairly summarized so users can view their performance status at a glance.

Like dashboards, scorecards also make use of charts and visual graphs to indicate performance state, trends, and variance against goals. The higher up the users are in the organization, the more they prefer to see performance encoded visually. However, most scorecards also contain (or should contain) a great deal of textual commentary that interprets performance results, describes action taken, and forecasts future results.

Summary: In the end, it does not really matter whether you use the term "dashboard" or "scorecard" as long as the tool helps to focus users and organizations on what really matters. Both dashboards and scorecards need to display critical performance information on a single screen so users can monitor results at a glance.[17]

Although the terms are used interchangeably, most project managers prefer to use dashboards and/or dashboard reporting. Eckerson defines three types of dashboards, as shown in Table 1–5 and the description that follows.

- *Operational dashboards* monitor core operational processes and are used primarily by front-line workers and their supervisors who deal directly with customers or manage the creation or delivery of organizational products and

TABLE 1–5. THREE TYPES OF PERFORMANCE DASHBOARDS

	Operational	**Tactical**	**Strategic**
Purpose	Monitor operations	Measure progress	Execute strategy
Users	Supervisors, specialists	Managers, analysts	Executives, managers, staff
Scope	Operational	Departmental	Enterprise
Information	Detailed	Detailed/summary	Detailed/summary
Updates	Intraday	Daily/weekly	Monthly/quarterly
Emphasis	Monitoring	Analysis	Management

Source: W. Eckerson, *Performance Dashboards: Measuring, Monitoring and Managing Your Business* (Hoboken, NJ: Wiley, 2006), p. 18/John Wiley & Sons.

17. Ibid., p. 13.

services. Operational dashboards primarily deliver detailed information that is only lightly summarized. For example, an online Web merchant may track transactions at the product level rather than the customer level. In addition, most metrics in an operational dashboard are updated on an intraday basis, ranging from minutes to hours depending on the application. As a result, operational dashboards emphasize monitoring more than analysis and management.

- *Tactical dashboards* track departmental processes and projects that are of interest to a segment of the organization or a limited group of people. Managers and business analysts use tactical dashboards to compare performance of their areas or projects, to budget plans, forecasts, or last period's results. For example, a project to reduce the number of errors in a customer database might use a tactical dashboard to display, monitor, and analyze progress during the previous 12 months toward achieving 99.9 percent defect-free customer data by 2007.

- *Strategic dashboards* monitor the execution of strategic objectives and are frequently implemented using a balanced scorecard approach, although total quality management, Six Sigma, and other methodologies are used as well. The goal of a strategic dashboard is to align the organization around strategic objectives and get every group marching in the same direction. To do this, organizations roll out customized scorecards to every group in the organization and sometimes to every individual as well. These "cascading" scorecards, which are usually updated weekly or monthly, give executives a powerful tool to communicate strategy, gain visibility into operations, and identify the key drivers of performance and business value. Strategic dashboards emphasize management more than monitoring and analysis.[18]

Three critical steps must be considered when using dashboards: (1) the target audience for the dashboard, (2) the type of dashboard to be used, and (3) the frequency with which the data will be updated. Some project dashboards focus on the KPIs that are part of earned-value measurement. These dashboards may need to be updated daily or weekly. Dashboards related to the financial health of the company may be updated weekly or quarterly. Figures 1–10 and 1–11 show the type of information that would be tracked weekly or quarterly to view corporate financial health.

1.16 KEY PERFORMANCE INDICATORS

Most often, the items that appear in the dashboards are elements that both customers and project managers track. These items are referred to as KPIs and were discussed previously. According to Eckerson: "A KPI is a metric measuring how well the organization or individual performs an operational, tactical or strategic activity that is critical for the current and future success of the organization."[19]

---

18. Ibid., pp. 17–19.
19. Ibid., p. 294.

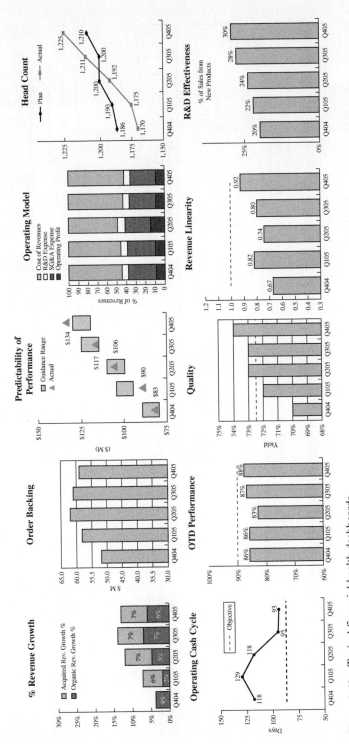

Figure 1–10. Typical financial health dashboards.
Source: Reproduced from J. Alexander, *Performance Dashboards and Analysis for Value Creation* (Hoboken, NJ: Wiley, 2007), pp. 87–88/with permission of John Wiley & Sons.

XYZ Company
Q4' 05 Week #7 of 13/54% of Q4
($ in Millions)

Bookings

Unit	Week	QTD	Forecast	% Achieved	$ Required
BU 1	0.7	15.0	30.0	50%	15.0
BU 2	—	0.9	1.0	89	0.1
BU 3	0.5	4.0	6.0	67	2.0
BU 4	0.4	1.7	4.7	37	2.9
Other	0.0	0.1	—		(0.1)
Totals	1.6	21.7	41.7	52%	$20.0

Bookings QTD

Revenue

Unit	Week	QTD	Forecast	% Achieved	Backlog	Req'd Fill
BU 1	2.0	13.0	28.0	46%	5.0	10.00
BU 2	0.4	3.0	5.0	60	1.0	1.00
BU 3	0.0	3.0	6.0	50	2.0	1.00
BU 4	2.6	3.0	7.0	43	1.0	3.00
Other	—	—	—			
Totals	5.0	22.0	46.0	48%	9.0	15.0

Revenue QTD

Receivable Collections (Cumulative)

Week	1	2	3	4	5
Actual	1.0	5.0	19.0	28.0	
Target	4.0	9.0	17.0		35.0

Cumulative AR Collections

Process Yield

Day	1	2	3	4	5
Process Yield	77%	80%	81%	68%	82%

Process Yield

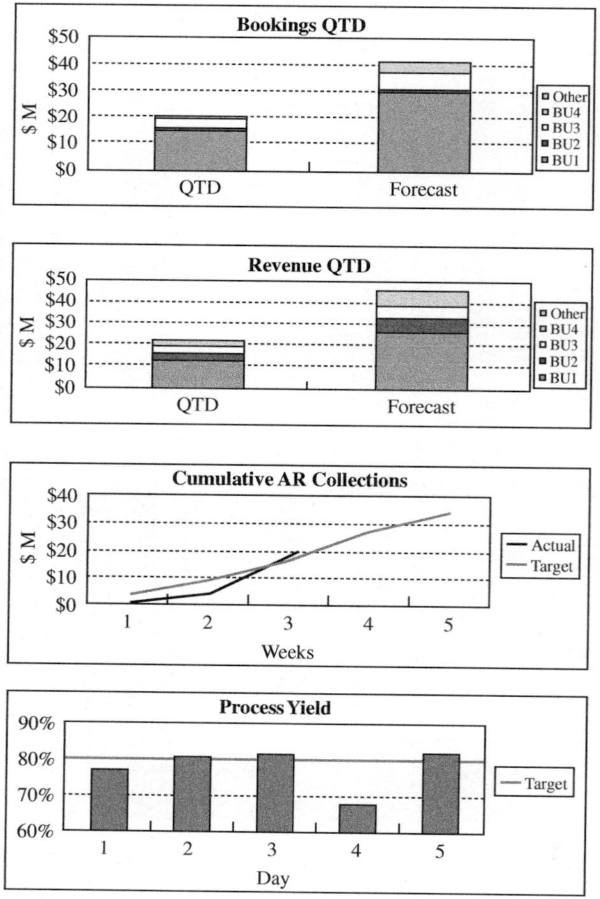

Figure 1–11. Typical financial health dashboards.

Source: Reproduced from J. Alexander, *Performance Dashboards and Analysis for Value Creation* (Hoboken, NJ: Wiley, 2007), pp. 87–88/with permission of John Wiley & Sons.

Some people confuse KPIs with leading indicators. A leading indicator is actually a KPI that measures how the work one is doing now will affect the future.

KPIs are critical components of all earned-value measurement systems. Cost variance, schedule variance, schedule performance index, cost performance index, and time/cost at completion are actually KPIs but are not referred to as such. The need for these KPIs is simple: What gets measured gets done! If the goal of a performance measurement system is to improve efficiency and effectiveness, then the KPI must reflect controllable factors. There is no point in measuring an activity if the users cannot change the outcome.

Eckerson identifies 12 characteristics of effective KPIs:

1. *Aligned.* KPIs are always aligned with corporate (or project) strategy and objectives.
2. *Owned.* Every KPI is "owned" by an individual or group on the business (or project) side that is accountable for its outcome.
3. *Predictive.* KPIs measure drivers of business (or project) value. Thus, they are "leading" indicators of performance desired by the organization.
4. *Actionable.* KPIs are populated with timely, actionable data so users can intervene to improve performance before it is too late.
5. *Few in number.* KPIs should focus users on a few high-value tasks, not scatter their attention and energy on too many things.
6. *Easy to understand.* KPIs should be straightforward and easy to understand, not based upon complex indices that users do not know how to influence directly.
7. *Balanced and linked.* KPIs should balance and reinforce each other, not undermine each other and suboptimize processes.
8. *Trigger changes.* The act of measuring a KPI should trigger a chain reaction of positive changes in the organization (or project), especially when it is monitored by the CEO (or customers or sponsors).
9. *Standardized.* KPIs are based on standard definitions, rules, and calculations so they can be integrated across dashboards throughout the organization.
10. *Context-driven.* KPIs put performance in context by applying targets and thresholds to it so users can gauge their progress over time.
11. *Reinforced with incentives.* Organizations can magnify the impact of KPIs by attaching compensation or incentives to them. However, they should do this cautiously, applying incentives only to well-understood and stable KPIs.
12. *Relevant.* KPIs gradually lose their impact over time, so they must be periodically reviewed and refreshed.[20]

There are several reasons why the use of KPIs often fails on projects, including:

- People believe that the tracking of a KPI ends at the first-line manager level.
- The actions needed to regulate unfavorable indications are beyond the control of the employees doing the monitoring or tracking.

---

20. Ibid., p. 201.

- The KPIs are not related to the actions or work of the employees doing the monitoring.
- The rate of change of the KPIs is too slow, thus making them unsuitable for managing the daily work of the employees.
- Actions needed to correct unfavorable KPIs take too long.
- Measurement of the KPIs does not provide enough meaning or data to make them useful.
- The company identifies too many KPIs, to the point where confusion reigns among the people doing the measurements.

Years ago, the only metrics that some companies used were those identified as part of the earned-value measurement system. The metrics generally focused only on time and cost and neglected metrics related to business success as opposed to project success. Therefore, the measurement metrics were the same for each project and for each life-cycle phase. Today, metrics can change from phase to phase and from project to project. The hard part is, obviously, deciding which metrics to use. Care must be taken that whatever metrics are established do not end up comparing apples and oranges. Fortunately, several good books in the marketplace can assist in identifying proper or meaningful metrics.[21]

Selecting the right KPIs is critical. Since a KPI is a form of measurement, some people believe that KPIs should be assigned only to those elements that are tangible. Therefore, many intangible elements that should be tracked by KPIs never get looked at because someone believes that measurement is impossible. Anything can be measured, regardless of what some people think. According to Hubbard:

- Measurement is a set of observations that reduces uncertainty where the results are expressed as a quantity.
- A mere reduction, not necessarily elimination, of uncertainty will suffice for a measurement.[22]

Therefore, KPIs can be established even for intangibles like those discussed later in this book in the chapter on value-driven project management (Chapter 16).

Hubbard believes that five questions should be asked before we establish KPIs for measurement:

1. What is the decision this (KPI) is supposed to support?
2. What really is the thing being measured (by the KPI)?
3. Why does this thing (and the KPI) matter to the decision being asked?

---

21. Three books that provide examples of metric identification are P. F. Rad and G. Levin, *Metrics for Project Management: Formalized Approaches* (Vienna, VA: Management Concepts, 2006); M. Schnapper and S. Rollins, *Value-Based Metrics for Improving Results: An Enterprise Project Management Toolkit* (Fort Lauderdale, FL: J. Ross Publishing, 2006); and D. W. Hubbard, *How to Measure Anything* (3rd ed.) (Hoboken, NJ: Wiley, 2014).

22. Hubbard, *How to Measure Anything*, p. 21.

4. What do you know about it now?
5. What is the value of measuring it further?[23]

Hubbard also identifies four useful measurement assumptions that should be considered when selecting KPIs:

1. Your problem (in selecting a KPI) is not as unique as you think.
2. You have more data than you think.
3. You need less data than you think.
4. There is a useful measurement that is much simpler than you think.[24]

Selecting the right KPIs is essential. On most projects, only a few KPIs are needed. Sometimes we seem to select too many KPIs and end up with some KPIs that provide us with little or no information value, and the KPI ends up being unnecessary or useless in assisting us in making project decisions.

Sometimes, companies believe that the measures that they have selected are KPIs when, in fact, they are forms of performance measures but not necessarily KPIs. David Parmenter discusses four types of performance measures:

These four measures are in two groups: result indicators and performance indicators.

I use the term result indicators to reflect the fact that many measures are a summation of more than one team's input. These measures are useful in looking at the combined teamwork but, unfortunately, do not help management fix a problem as it is difficult to pinpoint which teams were responsible for the performance or nonperformance.

Performance indicators, on the other hand, are measures that can be tied to a team or a cluster of teams working closely together for a common purpose. Good or bad performance is now the responsibility of one team. These measures thus give clarity and ownership. With both these measures some are more important, so we use the extra word "key." Thus, we now have two measures for each measure type:

1. Key result indicators (KRIs) give the board an overall summary of how the organization is performing.
2. Result indicators (RIs) tell management how teams are combining to produce results.
3. Performance indicators (PIs) tell management what teams are delivering.
4. Key performance indicators (KPIs) tell management how the organization is performing in their critical success factors and, by monitoring them, management is able to increase performance dramatically.[25]

---

23. Ibid., p. 43.
24. Ibid., p. 31.
25. David Parmenter, *Key Performance Indicators*, (3rd ed.) (Hoboken, NJ: Wiley, 2014), pp. 3–4.

Parmenter believes that:

There are seven foundation stones that need to be laid before we can successfully develop and utilize key performance indicators (KPIs) in the workplace. Success or failure of the KPI project is determined by the presence or absence of these seven foundation stones.

1. Partnership with the staff, unions, key suppliers, and key customers
2. Transfer of power to the front line
3. Integration of measurement, reporting, and improvement of performance
4. Linkage of performance measures to strategy
5. Abandon processes that do not deliver
6. Appointment of a homegrown chief measurement officer
7. Organization-wide understanding of the winning KPIs definition[26]

In a project environment, the performance measures can change from project to project and phase to phase. The identification of these measures is performed by the project team, including the project sponsor. Project stakeholders may have an input as well. Corporate performance measures are heavily financially oriented and may undergo very little change over time. The measurements indicate the financial health of the corporation.

Establishing corporate performance measures related to strategic initiatives or other such activities must be treated as a project in itself, and supported by the senior management team (SMT).

The SMT attitude is critical—any lack of understanding, commitment, and prioritizing of this important process will prevent success. It is common for the project team and the SMT to fit a KPI project around other competing, less important firefighting activities.

The SMT must be committed to the KPI project, to driving it down through the organization. Properly implemented, the KPI project will create a dynamic environment. Before it can do this, the SMT must be sold on the concept. This will lead to the KPI project's being treated as the top priority, which may mean the SMTs allow some of those distracting fires to burn themselves out.[27]

1.17 MANUFACTURING BEST PRACTICES IN ACTION[28]

Not all best practices have to be difficult or costly to find and implement. Many simple actions can be taken that will greatly improve efficiency, productivity, and quality. Take, for example, reducing Set-up Time. We are all aware that frequent set-up changes

---

26. Ibid., p. 26.

27. Ibid., p. 260. Chapter 5 of this book has excellent templates for reporting KPIs.

28. This section, "Manufacturing Best Practices in Action", has been provided by Rick Titone. ©2022 by Rick Titone. All rights reserved. Reproduced with permission.

to machinery or processes are necessary in most manufacturing operations. In such instances, the machine tool or die, the process, or the material being used must be shut down to accomplish the change. During the time that this shutdown and changeover occurs, there is no production taking place, or no product is being produced. Therefore, such time is totally wasted time. By reducing the amount of this downtime, both efficiency and productivity can be improved. However, accomplishing such a reduction involves changes to the current equipment and/or methods being used in the process.

Such changes may sometimes meet with resistance from several different sources, such as the machine or process operator, product engineers or financial personnel. We have all encountered the initial response to change, which is, "We've always done it this way." That may have been satisfactory in the past. But there may be a better way now and in the future. Therefore, we must find innovative ways to overcome such resistance and get the change implemented.

Now, some given facts are: People are naturally and inherently resistant to change. It upsets their routines or causes disruptions in their normal operations. However, there are several soundly proven methods that can be applied to overcome resistance to the innovative implementation of change that work in almost any area: People will accept change if it is their idea. There will be immediate acceptance, with no resistance. So, when change is required, talk to the people involved. Explain the change and the reasons why. Try to get their "By-In" ahead of time. Then there will be no resistance to the change itself.

People will accept change if there is something in it for them. For example, a financial or job position improvement. If the change results in such circumstances. Be sure to explain such benefits when explaining the change.

People will accept change if there is benefit to them in their current position. Will the change make their job easier, cleaner, less physical, or provide similar improvements? Include these benefits when explaining the change.

Lastly, people will accept change if it is mandated or required by a higher authority, either from corporate offices or outside forces, such as governmental authorities. Then the change will have to be implemented. Here is where the most resistance to change will occur. In some cases, operators or set-up operators will sometimes even take actions to make sure that the change, once implemented, doesn't work.

Now, let's talk about machine set-up reduction. Set-up time begins from the last good piece to the first good piece of a part on any machine. It is pure downtime, wasted nonproductive time, and all efforts must be exercised to eliminate as much of it as possible in all operations.

Back in 1988, Dr. Shigeo Shingo, working for Toyota, authored a book outlining a process for reducing machine and tool changeovers entitled; "Single Minute Exchange of Die", which won a Nobel Prize. "Single Minute of Exchange of Die" meant that all machine tool or die changes in any manufacturing operation should be able to be changed over from one part being produced to the new part to be made in less than one minute. Now, that may seem like a very short period. But just look at your wristwatch and see how long it takes for the second hand to go all the way around the dial. You probably couldn't hold your breath for that long.

Not only did Shingo propose such a feat, but he also explained how it could be accomplished. Some of his methods involve: Having tools or dies re-located from tool cribs, sometimes located far away from the machines they served, to a rack directly adjacent to the machines that used them, greatly reduced changeover times. Modifying tools or dies makes them easier to remove and reinstall by using large clamps instead of bolts, studs, and nuts. If such fasteners are absolutely required, employ breakaway bolts that will fold over after loosening just one turn of the nut. Modifying tools or dies to facilitate quick accurate insertion by employing dowel pins or adding insert tracks to eliminate any adjustments to the tool or die once it is inserted. Having duplicate tools or dies available that can be preset-up prior to the need to change over the current set-up. Teaching machine operators to make their own set-ups. Most of these simple, innovative changes can be accomplished for very little money and the modifications can usually be made in-house.

Shingo's books and methods are still actively being applied in manufacturing operations worldwide, now combined with the Just-In-Time manufacturing methods also established by Toyota.

A Case Study

At an electronic device manufacturing factory in Singapore, there was a bank of 20 winding machines winding gold wire thinner than a human hair around a ceramic core. There were many different codes being produced, each having a different number of turns or layers of wire that would designate the code number of the device and the function it served in the end product. At the end of each run of a particular code, the center section of the winding machine had to be removed by a set-up operator, taken to his bench, and the center section adjusted to accommodate the winding requirements of the next code to be manufactured. Then returned and reinstalled into the winding machine. Set-up times ranged from 12 to 15 minutes during which time the winding machine stood idle, and the operator just sat and waited for the center section of the machine to be returned and reinstalled.

After a review of the operation, the question was asked; "How much does a new center section for a winding machine cost?" The first reply was, "A lot." Which was typical of the oriental mindset. Additional inquiries resulted in a price of approximately US$1,000 each, obtained from the US machine manufacturer. A request was made to purchase five new winding machine center sections for the additional cost of approximately US$6,000 delivered. Once received, a master schedule system for tracking which codes were scheduled to be produced on each machine allowed the set-up operators to pre-set up a duplicate center section for that winding machine before the run of that code was completed. Then the exchange of the center section of that particular machine could be made in less than one minute, and production of the new code could begin. In addition, dowel pins were installed on the back of each winding machine center section to provide for immediate lock-in of the section upon reinstallation thereby eliminating any adjustments needed after the initial installation. Both set-up times and production efficiency increased tremendously for a very minimal investment.

1.18 STEP 3: VALIDATING THE BEST PRACTICE

Previously, we stated that seeking out a best practice is done by the project manager, project team, functional manager, and/or possibly a professional facilitator trained in how to debrief a project team and extract best practices. Any or all of these people must believe that what they have discovered is, in fact, a best practice. When project managers are quite active in a project, emphasis is placed on the project manager to make the final decision on what constitutes a best practice. According to a spokesperson for AT&T, the responsibility for determining what is a best practice rests with "the individual project manager that shows how it had a positive impact on their project."

Although this is quite common, there are other validation methods that may involve a significant number of people. Sometimes project managers may be removed from where the work is taking place and may not be familiar with activities that could lead to the identification of a best practice. Companies that have a PMO place a heavy reliance on the PMO for support because the approved best practices are later incorporated into the methodology, and the PMO is usually the custodian of the methodology based upon the type of PMO.

Once the management of the organization affected initially approves the new best practice, it is forwarded to the PMO, or process management for validation and then institutionalization. The PMO may have a separate set of checklists to validate the proposed best practice. The PMO must also determine whether or not the best practice is company proprietary, because that will determine where the best practice is stored and whether the best practice will be shared with customers.

The best practice may be placed in the company's best practice library or, if appropriate, incorporated directly into the company's stage gate checklist. Based on the complexity of the company's stage gate checklist process and enterprise project management methodology, the incorporation process may occur immediately or on a quarterly basis.

According to Chuck Millhollan, formerly director of program management at Churchill Downs Incorporated: "We do not label our processes or methods as 'best practices'. We simply learn from our lessons and ensure that learning is incorporated into our methodology, processes, templates, etc."

Some organizations have committees not affiliated with the PMO that have as their primary function the evaluation of potential best practices. Anyone in the company can provide potential best practices data to the committee, and the committee in turn does the analysis. Project managers may be members of the committee. Other organizations use the PMO to perform this work. These committees and the PMO most often report to the senior levels of management.

The fourth, fifth, and sixth editions of the *PMBOK® Guide* emphasize the importance of stakeholder involvement in projects. This involvement may also include the final decision on whether or not a discovery is a best practice. According to Chuck Millhollan:

> Ultimately, the final decision resides with our stakeholders, both internal and external. Another way of putting this is that the PMO does not make the decision if a method or process works. We actively seek feedback from our project stakeholders and use

their inputs to determine if our processes are "best practices" for Churchill Downs Incorporated. The specific best practices identified previously, among others, have even been accepted outside of the PMO as generally accepted practices.

Another example of stakeholder involvement is provided by Enrique Sevilla Molina, PMP, formerly corporate PMO director, Indra:

> The decision is taken by the corporate PMO responsible, the business unit manager, the local PMO authority, or even the cognizant authority, if it is the case. It depends on the subject and the scope of the task. Some of the management best practices have been established at corporate level, and they have been incorporated into the PM methodology. Many of them have also been incorporated into the Project Management Information Systems and the corporate PM tooling.

Evaluating whether something is a best practice is not time-consuming, but it is complex. Simply because someone believes that what he or she is doing is a best practice does not mean that it is in fact a best practice. Some PMOs are currently developing templates and criteria for determining whether an activity may qualify as a best practice. Some items that are included in the template might be:

- Is transferable to many projects
- Enables efficient and effective performance that can be measured (i.e., can serve as a metric)
- Enables measurement of possible profitability using the best practice
- Allows an activity to be completed in less time and at a lower cost
- Adds value to both the company and the clients
- Can differentiate us from everyone else

One company had two unique characteristics in its best practices template:

1. Helps to avoid failure
2. If a crisis exists, it helps us to get out of a critical situation

Executives must realize that these best practices are, in fact, intellectual property that benefits the entire organization. If the best practice can be quantified, then it is usually easier to convince senior management of its value.

1.19 STEP 4: LEVELS OF BEST PRACTICES

As stated previously, best practices come from knowledge transfer and can be discovered anywhere within or outside of your organization. This is shown in Figure 1–12.

Companies that maintain best practices libraries that contain a large number of best practices may create levels of best practices. Figure 1–13 shows various levels of

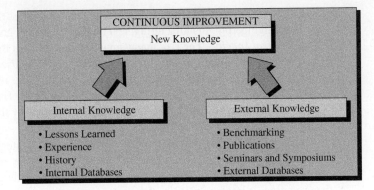

Figure 1–12. Knowledge transfer.

Figure 1–13. Levels of best practices.

best practices. Each level can have categories within the level. The bottom level is the professional standards level, which would include professional standards as defined by PMI. The professional standards level contains the greatest number of best practices, but they are more of a general nature than specific and have a low level of complexity.

The industry standards level would identify best practices related to performance within the industry. The automotive industry has established standards and best practices specific to the auto industry.

As we progress to the individual best practices in Figure 1–13, the complexity of the best practices goes from general to very specific applications, and, as expected, the number of best practices decreases. An example of a best practice at each level might be (from general to specific):

- *Professional standards.* Preparation and use of a risk management plan, including templates, guidelines, forms, and checklists for risk management.
- *Industry specific.* The risk management plan includes industry best practices, such as the best way to transition from engineering to manufacturing.

- *Company specific.* The risk management plan identifies the roles and interactions of engineering, manufacturing, and quality assurance groups during transition.
- *Project specific.* The risk management plan identifies the roles and interactions of affected groups as they relate to a specific product/service for a customer.
- *Individual.* The risk management plan identifies the roles and interactions of affected groups based on their personal tolerance for risk, possibly through the use of a responsibility assignment matrix prepared by the project manager.

Best practices can be extremely useful during strategic planning activities. As shown in Figure 1–14, the bottom two levels may be more useful for project management strategy formulation, whereas the top three levels are more appropriate for the execution or implementation of a strategy.

Not all companies maintain a formal best practices library. In some companies, when a best practice is identified and validated, it is immediately placed into the stage gate process or the project management methodology. In such a case, the methodology itself becomes the best practice. Enrique Sevilla Molina, PMP, states:

In fact, our Project Management methodology constitutes our established library of best practices applicable to every project in the company. There are additional best practices libraries in different business units. There are, for instance, detailed instructions for proposal preparation or for cost and schedule estimation purposes, which are appropriate for the specific business or operations area.

When asked how many best practices they maintain at Indra, he commented:

It is hard to say because of the subject itself and the multiplicity of business areas in the company. If we consider our PM methodology as a set of "best practices," it would be difficult to count every best practice included.

Besides our internally published *Indra Project Management Methodology Manual*, we have for instance specific guides at corporate level for WBS elaboration, project risk management, and the project's performance measurement based on earned value techniques. We have also specific instructions published for proposal preparation, costs estimation, and even detailed WBS preparation rules and formats for different business unit levels.

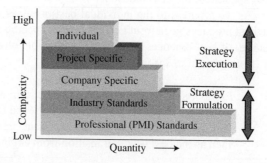

Figure 1–14. Usefulness of best practices.

1.20 STEP 5: MANAGEMENT OF BEST PRACTICES

There are three players involved in the management of the best practices:

1. The best practice's owner
2. The PMO
3. The best practices' library administrator, who may reside in the PMO

The owner of the best practice, who usually resides in the functional area, has the responsibility of maintaining the integrity of the best practice. The title "best practice owner" is usually an uncompensated and unofficial one but it is prestigious. Therefore, the owner of the best practice tries to enhance it and keep the best practice alive as long as possible.

The PMO usually has the final authority over best practices and makes the final decision on where to place them, who should be allowed to see them, how often they should be reviewed or revalidated, and when they should be removed from service.

The library administrator is merely the caretaker of the best practice and may keep track of how often people review the best practice, assuming it is readily accessible in the best practice library. The library administrator may not have a good understanding of each best practice and may not have any voting rights on when to terminate a best practice.

1.21 STEP 6: REVALIDATING BEST PRACTICES

Best practices do not remain best practices forever. Because best practices are directly related to the company's definition of project success, the definition of a best practice can change and age as the definition of success changes. Therefore, best practices must be periodically reviewed. The critical question is: How often should they be reviewed? The answer to this question is based on how many best practices are in the library. Some companies maintain just a few best practices, whereas large, multinational companies may have thousands of clients and maintain hundreds of best practices in their libraries. If the company sells products as well as services, then there can be both product-related and process-related best practices in the library.

The following two examples illustrate the need for reviewing best practices.

According to a spokesperson from EDS, "Once a practice has been nominated and approved to be a best practice, it is only sanctioned until the next yearly review cycle. Over time, best practices have the tendency to lose value and become ineffective if they are allowed to age."

A spokesperson from Computer Associates said this:

Best practices are reviewed every four months. Input into the review process includes:

- Lessons learned documents from project completed within the past four months
- Feedback from project managers, architects, and consultants

- Knowledge that subject matter experts (i.e., best practices owners) bring to the table; this includes information gathered externally as well as internally
- Best practices library reporting and activity data

There are usually three types of decisions that can be made during the review process:

1. Keep the best practice as is until the next review process.
2. Update the best practice and continue using it until the next review process.
3. Retire the best practice from service.

1.22 STEP 7: WHAT TO DO WITH A BEST PRACTICE

Given the definition that a best practice is an activity that leads to a sustained competitive advantage, it is no wonder that some companies have been reluctant to make their best practices known to the general public. Therefore, what should a company do with its best practices if not publicize them? The most common options available include:

- *Sharing knowledge internally only.* This is accomplished by using the company's intranet to share information with employees. There may be a separate group within the company responsible for controlling the information, perhaps even the PMO. Not all best practices are available to every employee. Some best practices may be password protected, as discussed below.
- *Hidden from all but a select few.* Some companies spend vast amounts of money on the preparation of forms, guidelines, templates, and checklists for project management. These documents are viewed as both company-proprietary information and best practices and are provided to only a select few on a need-to-know basis. An example of a "restricted" best practice might be specialized forms and templates for project approval where information contained may be company-sensitive financial data or the company's position on profitability and market share.
- *Advertise to the company's customers.* In this approach, companies may develop a best practices brochure to market their achievements and may also maintain an extensive best practices library that is shared with their customers after contract award. In this case, best practices are viewed as competitive weapons.

Most companies today utilize some form of best practices library. According to a spokesperson from AT&T:

> The best practices library is Sharepoint based and very easy to use both from a submission and a search perspective. Any project manager can submit a best practice at any time and can search for best practices submitted by others.

Even though companies collect best practices, not all best practices are shared outside of the company, even during benchmarking studies where all parties are expected to share information. Students often ask why textbooks do not include more information on detailed best practices, such as forms and templates. One company spokesperson commented to the author:

> We must have spent at least $1 million over the last several years developing an extensive template on how to evaluate the risks associated with transitioning a project from engineering to manufacturing. Our company would not be happy giving this template to everyone who wants to purchase a book for $85. Some best practices templates are common knowledge, and we would certainly share this information. But we view the transitioning risk template as proprietary knowledge not to be shared.

1.23 STEP 8: COMMUNICATING BEST PRACTICES ACROSS THE COMPANY

Knowledge transfer is one of the greatest challenges facing corporations. The larger the corporation, the greater the challenge of knowledge transfer. The situation is further complicated when corporate locations are dispersed over several continents. Without a structured approach for knowledge transfer, corporations can repeat mistakes as well as miss valuable opportunities. Corporate collaboration methods must be developed.

There is no point in capturing best practices unless the workers know about them. The problem, as identified earlier, is how to communicate this information to the workers, especially in large, multinational companies. Some of the techniques include:

- Websites
- Best practices libraries
- Community of practice
- Newsletters
- E-mailings
- Internal seminars
- Transferring people
- Case studies
- Other techniques

Nortel Networks strives to ensure timely and consistent communications to all project managers worldwide to help drive continued success in the application of the global project management process. Examples of the various communication methods used by Nortel include:

- The *PM Newsflash* is published on a monthly basis to facilitate communications across the project management organization and related functions.
- Project management communications sessions are held regularly, with a strong focus on providing training, metrics reviews, process and template updates, and so on.

- Broadcast bulletins are utilized to communicate time-sensitive information.
- A centralized repository has been established for project managers to facilitate easy access to and sharing of project management-related information.

The comments by Nortel make it clear that best practices in project management now permeate all business units of a company, especially multinational companies.

One of the reasons for this is that we now view all activities in a company as a series of projects. Therefore, we are managing our business by projects. Given this fact, best practices in project management are now appearing throughout the company.

Publishing best practices in some form seems to be the preferred method of communication. At Indra, Enrique Sevilla Molina, PMP, stated:

> They are published at corporate level and at the corresponding level inside the affected business unit. Regular courses and training is also provided for newly appointed project managers, and their use is periodically reviewed and verified by the internal audit teams. Moreover, the PM corporate tools automate the applications of best practices in projects, as PM best practices become requirements to the PM information systems.

According to a spokesperson from AT&T:

> We have defined a best practice as any tool, template, or activity that has had a positive impact on the triple constraint and/or any of the *PMBOK® Guide* Process or Knowledge areas. We allow the individual project manager to determine if it is a best practice based on these criteria. We communicate this through a monthly project management newsletter and highlight a best practice of the month for our project management community.

Another strategic importance of best practices in project management can be seen from the comments by Suzanne Zale, formerly global program manager at EDS.

> Driven by the world economy, there is a tendency toward an increasing number of large-scale global or international projects. Project managers who do not have global experience tend to treat these global projects as large national projects. However, they are completely different. A more robust project management framework will become more important for such projects. Planning up front with a global perspective becomes extremely important. As an example, establishing a team that has knowledge about geographic regions relevant to the project will be critical to the success of the projects. Project managers must also know how to operate in those geographic areas. It is also essential that all project team members are trained and understand the same overall project management methodology.
>
> Globalization and technology will make sound project management practice even more important.

Zale's comments illustrate the importance of extracting best practices from global projects. This could very well be the future of best practices.

1.24 STEP 9: ENSURING USAGE OF THE BEST PRACTICES

Why go through the complex process of capturing best practices if people are not going to use them? When companies advertise to their clients that they have best practices, it is understood that tracking of the best practices and how they are used must be done. This is normally part of the responsibility of the PMO. The PMO may have the authority to regularly audit projects to ensure the usage of a best practice but may not have the authority to enforce the usage. The PMO may need to seek out assistance from the head of the PMO, the project sponsor, or various stakeholders for enforcement.

Maintaining a best practices library is an excellent way of determining workers' interest in the best practices. A company inserted counters into the best practices to see how often workers viewed each of the best practices. For those best practices that appeared most viewed, the PMO worked with the best practice owners to encourage them to update the best practices as often as possible or to help identify other best practices in the same categories.

When best practices are used as competitive weapons and advertised to potential customers as part of competitive bidding, the marketing and sales force must understand the best practices and explain this usage to the customers. Unlike 10–15 years ago, the marketing and sales force today has a good understanding of project management and the accompanying best practices.

1.25 COMMON BELIEFS

There are several common beliefs concerning best practices that companies have found to be valid. A partial list follows.

- Because best practices can be interrelated, the identification of one can lead to the discovery of another, especially in the same category or level of best practices. Best practices may be self-perpetuating.
- Because of the dependencies that can exist between best practices, it is often easier to identify categories of best practices rather than individual best practices.
- Best practices may not be transferable. What works well for one company may not work for another company.
- Even though some best practices seem simplistic and based on common sense in most companies, the constant reminder and use of these best practices lead to excellence and customer satisfaction.
- Best practices are not limited exclusively to companies in good financial health. Companies that are cash-rich can make a $10 million mistake and write it off. But companies that are cash poor must be very careful in how they approve projects, monitor performance, and evaluate whether or not to cancel the project.

Care must be taken that the implementation of a best practice does not lead to detrimental results. One company decided that the organization had to recognize project management as a profession in order to maximize performance and retain qualified people. A project management career path was created and integrated into the corporate reward system.

Unfortunately, the company made a severe mistake. Project managers were given significantly larger salary increases than line managers and workers. People became jealous of the project managers and applied for transfer into project management, thinking that the "grass was greener." The company's technical prowess diminished, and some people resigned when not given the opportunity to become project managers.

Sometimes the implementation of a best practice is done with the best of intentions, but the final result either does not meet management's expectations or may even produce an undesirable effect. The undesirable effect may not be apparent for some time. As an example, consider the first best practice in Table 1–6. Several companies are now using traffic light reporting for their projects. One company streamlined its intranet project management methodology to include traffic light status reporting. Beside every work package in the WBS was a traffic light capable of turning red, yellow, or green. Status reporting was simplified and easy for management to follow. The time spent by executives in status review meetings was significantly reduced, and significant cost savings were realized.

Initially, this best practice appeared to be beneficial for the company. However, after a few months, it became apparent that the status of a work package, as seen by a traffic light, was not as accurate as the more expensive written reports. There was also some concern as to who would make the decision on the color of the traffic light. Eventually, the traffic light system was enlarged to include eight colors, and guidelines were established for the decision on the color of the lights. In this case, the company was fortunate enough to identify the disadvantage of the best practice and correct it. Not all disadvantages are easily identified, and those that are may not always be correctable.

There are other reasons why best practices can fail or provide unsatisfactory results. These include:

- Lack of stability, clarity, or understanding of the best practice
- Failure to use best practices correctly
- Identifying a best practice that lacks rigor
- Identifying a best practice based on erroneous judgment
- Failing to provide value

TABLE 1–6.   IMPROPER APPLICATION OF BEST PRACTICES

Type of Best Practice	Expected Advantage	Potential Disadvantage
Use of traffic light reporting	Speed and simplicity	Poor accuracy of information
Use of a risk management template/form	Forward-looking and accurate	Inability to see all possible risks
Highly detailed WBS	Control, accuracy, and completeness	More control and cost of reporting
Using enterprise project management on	Standardization and consistency	Too expensive on certain projects all projects
Using specialized software	Better decision making	Too much reliance on tools

1.26 THE DARK SIDE OF PROJECT MANAGEMENT BEST PRACTICES

The project management community of practice worldwide applauds the use of discovering and implementing best practices as they evolve from new and better solutions to project management practices. But what is usually not discussed in detail are the challenges that must be overcome before discovery and implementation can be successful. Some of the challenges include:

Fear of change. Some people thrive on old habits and refuse to accept a best practice for fear of being removed from their comfort zone. They may also fear the impact that the implementation of the best practice may have on their performance review or advancement opportunities, especially if the new best practice reduces their power base.

Restricted usage. Some best practices may seem appropriate to use, but there are situations where industry best practices must be used instead.

Rushing into acceptance. Sometimes people rush into the acceptance of a best practice when spending a little bit more time into its investigation may lead to a better approach.

Politics and authority. Some best practices change the political landscape in the company and may change who has the authority to make certain decisions. While the intent of a best practice is to improve decision-making, some best practices require authorization from above for certain decisions to be made, whereas previously lower-level employees may have been allowed to make the decision.

Sharing best practices. Not all best practices are appropriate or applicable to multiple projects. Some consulting companies promote the use of their ready-made templates in practices where they may be inappropriate.

Training requirements. Some organizations implement best practices too quickly and raise the bar without considering transition time. Best practice implementation may require training in their use. Rushing into the use of a best practice may lead to unexpected results and demoralize the users.

Funding requirements. Some best practices may require funding for equipment or software before effective implementation.

1.27 BEST PRACTICES LIBRARY

With the premise that project management knowledge and best practices are intellectual properties (IPs), how does a company retain this information? The solution is usually the creation of a best practices library. Figure 1–15 shows the three levels of best practices that seem most appropriate for storage in a best practices library.

Figure 1–16 shows the process of creating a best practices library. The bottom level is the discovery and understanding of what is or is not a "potential" best practice. The sources of potential best practices can originate anywhere within the organization.

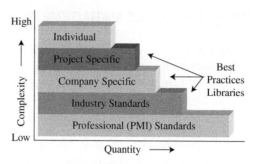

Figure 1–15. Library levels of best practices.

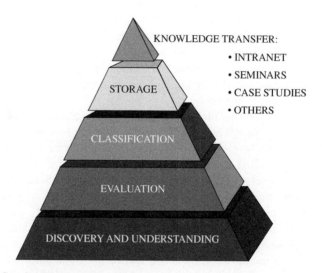

Figure 1–16. Creating a best practices library.

The next level is the evaluation level, which confirms that it is a best practice. The evaluation process can be done by the PMO or a committee, but should have involvement from the senior levels of management. The evaluation process is very difficult because a one-time positive occurrence may not reflect a best practice that will be repetitive. There must exist established criteria for the evaluation of a best practice.

Once a best practice is established, most companies provide a more detailed explanation of the best practice as well as a means for answering questions concerning its use. However, each company may have a different approach to disseminating this critical intellectual property. Most companies prefer to make maximum utilization out of their intranet websites. However, some companies simply consider their current forms and templates as part of their ongoing best practices library.

Figure 1–15 shows the levels of best practices, but the classification system for storage purposes can be significantly different. Figure 1–17 shows a typical classification system for a best practices library.

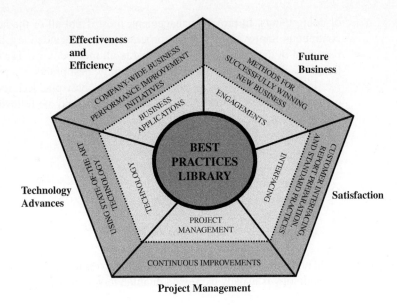

Figure 1–17. Best practices library.

The reason to create a best practices library is to transfer knowledge to the employees. The knowledge can be transferred through the company intranet, seminars on best practices, and case studies. Some companies require that the project team prepare case studies on lessons learned and best practices before the team is disbanded. These companies then use the case studies in company-sponsored seminars. Best practices and lessons learned must be communicated to the entire organization. The problem is determining how to do it effectively.

Another critical problem is best practices overload. One company started up a best practices library and, after a few years, had amassed what it considered to be hundreds of best practices. No one bothered to reevaluate whether all of these were still best practices. After reevaluation had taken place, it was determined that less than one-third were still regarded as best practices. Some were no longer best practices, others needed to be updated, and others had to be replaced with newer best practices.

1.28 DETERMINING THE VALUE OF A BEST PRACTICE

Determining the financial benefit of using a best practice can be difficult. Every industry, every company, and even functional departments within a company can have a wide range of best practices. As such, the value of a best practice is most often expressed qualitatively rather than quantitatively.

The traditional definition of a best practice is a method or technique that is superior to accomplishing the work by other means and therefore becomes the standard way of doing things. In project management, most if not all of the best practices focus on continuous improvements to the processes, forms, guidelines, templates, and checklists used in the creation of a one-size-fits-all methodology. The methodology is often treated as the most important, or even the only, best practice in project management.

In the early years of project management, best practices focused on the components needed for the creation of the methodology, and the following were identified as best practices:

- Creation of a statement of work
- Development of a work breakdown structure
- Creation of a project plan
- Types and frequency of meetings
- Stakeholder communication practices
- Standardized reporting practices

The value obtained from continuous improvements to the processes included in the methodology is usually expressed qualitatively and includes:

- Greater efficiency and effectiveness
- Less time and cost to complete work packages
- Reduction in waste and streamlining of processes
- Improvements in quality

Project management best practices and their accompanying value were oriented around a rigid framework with the belief that process improvements would increase successes and reduce failures. Unfortunately, this was predicated up traditional project with well-defined requirements at project initiation.

As discussed previously, today we have a significant growth in project management applications for nontraditional activities, including strategic and innovation projects. The result is the identification of new types of best practices that focus on strategic deliverables and outcomes created rather than processes. Examples of the new best practices include:

- Growth in market share
- Penetration into new markets
- Ability to respond to innovation opportunities quicker
- Generation of more creative and innovative ideas
- Creation of innovations that enhance the organization's image and reputation
- Creation of products and services that clearly identify the organization's CSR efforts

Companies today maintain a self-regulated strategy called corporate social responsibility (CSR) that is integrated into the firm's business model and identifies the ethically oriented activities the firm will undertake for the benefit of consumers, society, ecology, and government regulations. Environmentally friendly outcomes include the consumption of certain natural and renewable resources, such as water, energy, and other materials.

When companies can effectively integrate strategic planning, innovation, and project management practices together, the ideas generated and implemented create outstanding products, services, and a greater likelihood of supporting a CSR effort for the betterment of everyone. The results can be outstanding, as seen with Aramco in Section 1.28.

1.29 ARAMCO BOLSTERS INNOVATION THROUGH CUTTING-EDGE IDEAS[29]

"Where energy is opportunity" is the strapline of the largest energy company in the world—Saudi Aramco. It has always been assumed by many that innovation is uncommon among energy companies because the process undertaken within is just straightforward. In reality, it is the opposite.

As a global leader in providing energy, Saudi Aramco is required to lead by example and innovate. Whether it is to achieve its carbon-neutrality goal by 2050, as announced by the company's President and CEO, Amin H. Al-Nasser, or to excel in the production of oil and gas, the company believes that the most important investment is in its people and the community. Saudi Aramco aims to innovate for a better life standard, a cleaner future, and a better world.

Saudi Aramco officially inaugurated innovation in 2001, where it started to focus on IPs and patents. As a firm believer in research, technology, and innovation, the company has built 12 global research centers spreading all the way from Texas, USA to Daejeon, South Korea. Each of which focuses on a different area of interest; for example, the Houston research center focuses on upstream oil and gas technologies. Other research centers like the ones in Scotland, Paris, and Russia focus on achieving the necessary breakthroughs in technology and innovation to efficiently produce reliable and sustainable energy resources. The patents recorded by Saudi Aramco globally have significantly increased in numbers over the past 10 years, mainly due to the significant impact of innovation on the company.

As innovation is key to the Company, Saudi Aramco has established many programs that serve innovation. The largest of which is its Corporate Innovation Portal, which is accessible to all employees and where SMEs lead the efforts, interact with employees, and provide technical insights. The portal has supervisory board to assess and measure innovation. A dedicated team of specialists was brought in to manage and protect IPs.

---

29. Material in this section, "Aramco Bolsters Innovation Through Cutting-Edge Ideas", has been provided by Mohammed A. Al-Sadiq, Abdulaziz M. Anbari and Abdullah T. Aljamaan. ©2022 Saudi Arabian Oil Company. All rights Reserved. Reproduced with permission.

In terms of social responsibility, the Company launched Wa'ed Entrepreneurship Center in 2011, which focuses primarily on social welfare, digitalization, and sustainability in the kingdom. It is part of the company's effort to help innovative start-ups flourish in the Company and in the kingdom. The center was created to encourage young innovators in the country by providing them with the necessary guidance and support needed to succeed. It acts as a talent incubator for young talent and as a venture capital investment fund for promising Saudi start-ups, especially during the inception of portfolio.

Other than Wa'ed, Saudi Aramco recently invested heavily in creating LAB7. Considered as a modern research center, this will soon be the place for young, passionate innovators who will be supported and surrounded by SMEs. Adjacent to LAB7 is a green innovation park, which aims to reduce carbon emissions and produce renewable energy sources with the use of AI and machine learning. Commenting on LAB7's creation, Ahmad A. Al-Sa'adi, Senior Vice President of Saudi Aramco's Technical Services, said, "As we enter a new era of innovation, Aramco is seeking to promote prosperity, nurture potential, and encourage the adoption of the latest technologies both inside the company and across the Saudi landscape. LAB7 will help advance technology development, which will in turn provide benefits through increased operational efficiencies, improved productivity and sustainability enhancements."

Indeed, innovation is present at all company levels. Corporate-wide innovations are typically robust resulting in business change. As Saudi Aramco is the most reliable energy supplier in the world, it does not only focus on producing energy from oil and gas. Targeting its 2050 carbon-neutrality goal, the company has expanded its efforts in exploring solutions that can drive clean and sustainable energy, such as blue ammonia and hydrogen technologies. Ahmad O. Al-Khowaiter, Saudi Aramco's Chief Technology Officer, said, "The use of hydrogen is expected to grow in the global energy system, and this world's first demonstration represents an exciting opportunity for Aramco to showcase the potential of hydrocarbons as a reliable source of low carbon hydrogen and ammonia."

For Saudi Aramco, projects are major steppingstones to deploying innovative ideas. Through the execution of mega-scale projects with international partners, many innovative designs, tools, equipment, and ways of doing things are explored and adapted every day. Badr M. Burshaid, Program Manager of Saudi Aramco's Marjan and Zuluf Increment programs and President of the Project Management Institute— Kingdom of Saudi Arabia Chapter (PMI-KSA), opined, "Saudi Aramco is spreading innovative culture both internally and externally through its projects. In today's project management world, the old ways of executing projects are no longer applicable. New project management techniques involve digital solutions, robotics, virtual construction, and many other innovative approaches that preserve the environment, and enhance the safety of people and facilities while increasing efficiency and productivity. We are able to execute projects faster, cheaper, more environmentally friendly, and with zero harm to human life by supporting innovative approaches."

Not every innovative idea turns out to be a huge success. Some ideas fail, and others prove to be ineffective as planned. Whether the innovation turns out to be a huge success or a failure, it should not really matter, as both provide valuable lessons for learning and for future innovation ideas. For Saudi Aramco, to encourage innovation and promote its culture, the company has put in place incentive programs for high performers in the innovation domain. As excellence and integrity are at the core of its corporate values, the company has provided the necessary attractive environments for its employees to grow, innovate, and excel.

From Best Practice
to Migraine Headache

2.0 INTRODUCTION

For almost 40 years, project management resided in relatively few industries, such as aerospace, defense, and heavy construction. These industries were project-driven and implemented project management mainly to placate customer requests. Project management was considered something nice to have but not a necessity. As a result, best practices in project management were never really considered important.

Within the last two decades, project management has evolved into a management process that is mandatory for the long-term survival of the firm. Project management is now a necessity rather than a luxury. Project management permeates all aspects of a business. Companies are now managing their business by projects. Project management has become a competitive weapon. The knowledge learned from project management is treated as intellectual property, and project management offices (PMOs) have been established as the guardians of the project management intellectual property, reporting to the senior levels of management and being given the task of capturing best practices in project management.

As with any new project management activity, benefits are accompanied by disadvantages and potential problems. Some of the problems are small and easy to correct, while others are colossal migraine headaches and keep executives awake at night. The majority of the migraine headaches emanate from either a poor understanding of the benefits of project management or having expectations that are set too high. Other potential problems occur when an activity really is not a best practice and detrimental results occur.

2.1 GOOD INTENTIONS BECOMING MIGRAINES

Sometimes the best intentions can turn into migraine headaches. As an example, one company quickly realized the importance of project management and made it a career path position. This was certainly the right thing to do. Internally, people believed that the company considered project management to be a strategic competency, and professionalism in project management evolved. Externally, the company's customers were quite pleased seeing project management as a career path discipline, and the business improved.

These good intentions soon turned into problems. To show their support for excellence in project management, project managers were provided with 14 percent salary increases, whereas project team members and line managers received 3 to 4 percent. Within two years after implementing a project management career path, everyone was trying to become a project manager and climb the project management career path ladder of success, including critical line managers with specialized expertise. Everyone thought that the grass was greener in the project manager's yard than in his or her own. Line managers with critical skills were threatening to resign from the company if they were not given the chance to become project managers. The company eventually corrected the problem by telling everyone that every career path ladder in the company had the same career path opportunities for advancement. The large differential in salary increases disappeared and was replaced by a more equitable plan. However, the damage was done. Team members and line managers felt that the project managers exploited them, and the working relationship suffered. Executives were now faced with the headache of trying to repair the damage.

Figure 2–1 illustrates why many other headaches occur. As project management grows and evolves into a competitive weapon, pressure is placed upon the organization to implement best practices, many of which necessitate the implementation of costly internal control systems for the management of resources, costs, schedules, and quality. The project management systems must be able to handle several projects running concurrently. Likewise, obtaining customer satisfaction is also regarded as a best practice and can come at a price. As the importance of both increases, so do the risks and the headaches. Maintaining parity between customer satisfaction and internal controls is

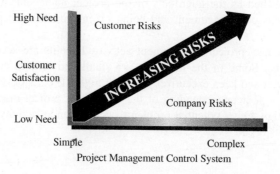

Figure 2–1. Risk growth.

not easy. Spending too much time and money on customer satisfaction could lead to financial disaster on a given project. Spending too much time on internal controls could lead to noncompetitiveness.

2.2 ENTERPRISE PROJECT MANAGEMENT METHODOLOGY MIGRAINE

As the importance of project management became apparent, companies recognized the need to develop project management methodologies. Good methodologies are best practices and can lead to sole-source contracting based on the ability of the methodology to continuously deliver quality results and the faith that the customer has in the methodology. Unfortunately, marketing, manufacturing, information systems, research and development, and engineering may have their own methodologies for project management. In one company, this suboptimization was acceptable to management as long as these individual functional areas did not have to work together continuously. Each methodology had its own terminology, life-cycle phases, and gate review processes.

When customers began demanding complete solutions to their business needs rather than products from various functional units, the need to minimize the number of methodologies became apparent. Complete solutions required that several functional units work together. Senior management regarded this as a necessity, and senior management believed that it would eventually turn into a best practice as well as lead to the discovery of other best practices.

One company had three strategic business units (SBUs), which, because of changing customer demands, now were required to work together because of specific customer solution requirements. Senior management instructed one of the SBUs to take the lead role in condensing all of the functional processes of all three units into one enterprise project management (EPM) methodology. After some degree of success became apparent, senior management tried unsuccessfully to get the other two SBUs to implement this EPM methodology that was believed to be a best practice. The arguments provided were: "We don't need it," "It doesn't apply to us," and "It wasn't invented here." Reluctantly, the president of the company made it clear to his staff that there was now no choice. Everyone would use the same methodology. The president is now facing the same challenge posed by globalization and acceptance of new methodology. Now cultural issues become important.

2.3 TRADE-OFF MIGRAINE

As project management evolved, the focus has changed to competing constraints rather than just the traditional view of looking at only time, cost, and scope. As more constraints appear on projects, new challenges and migraines appear with trade-offs. With only three constraints, it may not be difficult to determine the order of the trade-offs. But when as many as 8, 10, or 12 constraints appear on the project, the order of the trade-offs can be challenging, especially because of the dependencies that may exist between constraints.

At the onset of a project, the PM must work closely with the customer and the stakeholders to come to an acceptable definition of project success. Each project can have a different definition of success. The next step is to identify the constraints and accompanying metrics that will be used to identify performance toward meeting the definition of success.

The constraints and metrics will need to be prioritized. Based upon the definition of success, some constraints will be inflexible and others flexible. The inflexible constraints will most likely be treated as parameters not to undergo trade-offs if possible. The flexible constraints may change over the life of the project and undergo trade-offs.

If the project is being executed in a VUCA environment, all the constraints and metrics may change over the life cycle of the project. This may become a headache that organizations must live with to remain competitive.

2.4 CUSTOMER SATISFACTION MIGRAINE

Companies have traditionally viewed each customer as a one-time opportunity, and after this customer's needs were met, emphasis was placed on finding other customers. This approach is acceptable as long as there exists a large potential customer base. Today, project-driven organizations, namely those that survive on the income from a continuous stream of customer-funded projects, are implementing the engagement project management approach. With engagement project management, each potential new customer is approached in a way that is similar to an engagement in marriage where the contractor is soliciting a long-term relationship with the customer rather than a one-shot opportunity. With this approach, contractors are selling not only deliverables and complete solutions but also a willingness to make their EPM methodology, whether rigid or flexible, compatible with the customer's methodology. To maintain customer satisfaction and hopefully a long-term relationship, customers are requested to provide input on how the contractor's EPM methodology can be extended into their organization. The last life-cycle phase in the EPM methodology used by the Swedish-Swiss corporation Asea Brown and Boveri is called "customer satisfaction management" and is specifically designed to solicit feedback from the customer for long-term customer satisfaction.

This best practice of implementing engagement project management is a powerful best practice because it allows the company to capitalize on its view of project management, namely that project management has evolved into a strategic competency for the firm leading to a sustained competitive advantage. Although this approach has merit, it opened a Pandora's box. Customers were now expecting to have a say in the design of the contractor's EPM methodology. One automotive supplier decided to solicit input from one of the Big Three in Detroit when developing its EPM approach.

Although this created goodwill and customer satisfaction with one client, it created a severe problem with other clients that had different requirements and different views of project management. How much freedom should a client be given in making recommendations for changes to a contractor's EPM system? Is it a good idea to run the risk for the sake of customer satisfaction? How much say should a customer have in how a contractor manages projects? What happens if customers begin telling contractors how to do their job?

2.5 MIGRAINE RESULTING FROM RESPONDING TO CHANGING CUSTOMER REQUIREMENTS

When project management becomes a competitive weapon and eventually leads to a strategic competitive advantage, changes resulting from customer requests must be done quickly. The EPM system must have a process for configuration management for the control of changes. The change control process portion of the EPM system must maintain flexibility. But what happens when customer requirements change to such a degree that corresponding changes to the EPM system must be made and these changes could lead to detrimental results rather than best practices?

One automotive tier 1 supplier spent years developing an EPM system for the development of new products or components that were highly regarded by customers. The EPM system was viewed by both the customers and the company as a best practice. But this was about to change. Customers started trying to save money by working with fewer suppliers. Certain suppliers would be selected to become "solution providers" responsible for major sections or chunks of the car rather than individual components. Several tier 1 suppliers acquired other companies through mergers and acquisitions in order to become component suppliers. The entire EPM system had to be changed, and, in many cases, cultural shock occurred. Some of the acquired companies had strong project management cultures and their own best practices, even stronger than the acquirer, while others were clueless about project management. To make matters even worse, all of these companies were multinational, and globalization issues took center stage. Now best practices were competing.

After years of struggling, success was now at hand for many component suppliers. The mergers and acquisitions were successful, and new common sets of best practices were implemented. But, once again, customer requirements were about to change. Customers were now contemplating returning to component procurement rather than "solution provider" procurement, believing that costs would be lowered. Should this occur across the industry, colossal migraines would appear due to massive restructuring, divestitures, cultural changes, and major changes to the EPM systems. How do contractors convince customers that their actions maybe detrimental to the entire industry? Furthermore, some companies that were previously financially successful as chunk or section manufacturers might no longer have the same degree of success as component manufacturers.

2.6 REPORTING LEVEL OF THE PMO MIGRAINE

Companies may have multiple types of PMOs, but the traditional PMO is most often treated as the guardian of project management intellectual property. Included in the responsibilities of a PMO are strategic planning for project management; development and enhancement of the EPM; maintenance of project management templates, forms, and guidelines; portfolio management of projects; mentorship of inexperienced project managers; a hotline for project problem-solving; and maintaining a project management best practices library. The PMO becomes the guardian of all of the project management best practices.

While the creation of a PMO is regarded as a best practice for most companies, it places a great deal of intellectual property in the hands of a few, and information is power. With all of this intellectual property in the hands of three or four people in the PMO, the person to whom the PMO reports could possibly become more powerful than his or her counterparts. What is unfortunate is that the PMO must report to the executive levels of management, and there appears to be severe infighting at the executive levels for control of the PMO.

To allay the fears of one executive becoming more powerful than another, companies have created multiple PMOs, which are supposedly networked together and sharing information freely. Hewlett-Packard has multiple PMOs all networked together. Comau has PMOs in North America, South America, Europe, and Asia, all networked together. Star Alliance has a membership of 27 airlines, each with a PMO and all networked together with a PMO in Germany as the lead. These PMOs are successful because information and project management intellectual property are shared freely.

Allowing multiple PMOs to exist may seem like the right thing to do to appease each executive, but in some cases, it has created the headaches of project management intellectual property that is no longer centralized. Worse, what happens if every executive, including multinational executives, demands their own PMO? This might eventually lead to PMOs being viewed as an overmanagement expense, and unless the company can see a return on investment on each one, the concept of the PMO might disappear, thus destroying an important best practice because of internal politics.

2.7 CASH FLOW DILEMMA MIGRAINE

For many companies that survive on competitive bidding, the cost of preparing a bid can range from a few thousand dollars to hundreds of thousands. In most cases, project management may not appear until after the contract is awarded. The results can be catastrophic if benefit realization at the end of the project does not match the vision or profit margin expected during proposal preparation or at project initiation. When companies develop an EPM system and the system works well, most companies believe that they can now take on more work. They begin bidding on every possible contract believing that with the EPM system, they can accomplish more work in less time and with fewer resources without any sacrifice of quality.

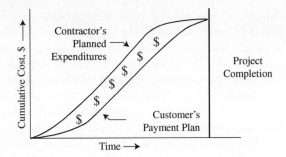

Figure 2–2. Spending curve.

In the summer of 2002, a large, multinational company set up a project management training program in Europe for 50 multinational project managers. The executive vice president spoke for the first 10 minutes of the class and said, "The company is now going to begin turning away work." The project managers were upset over hearing this and needed an explanation. The executive vice president put Figure 2–2 on the screen and made it clear that the company would no longer accept projects where profit margins would eventually be less than 4 to 6 percent because they were financing the projects for their customers. The company was functioning as a banker for its clients. Benefit realization was not being achieved. To reduce the costs of competitive bidding, the company was responding to proposal requests using estimating databases rather than time-phased labor. The cash flow issue was not identified until after go-ahead.

While project financing has become an acceptable practice, it does squeeze profits in already highly competitive markets. To maintain profit margins, companies are often forced to disregard what was told to the customer in the proposal and to assign project resources according to the customer's payment plan rather than the original project schedule provided in the proposal. While this may lead to short-term profitability, it often results in elongated schedules, potential lawsuits, and customer dissatisfaction. The balance among customer satisfaction, long-term client relationships, and profitability is creating a huge headache. The best practice of creating a world-class EPM system can lead to detrimental results if profitability cannot be maintained.

2.8 SCOPE CHANGE DILEMMA MIGRAINE

For companies that require successful competitive bidding for survival, the pot of gold is often the amount of scope changes that occur after go-ahead. The original contract may be underbid in the hope that lucrative customer or contractor-generated scope changes will occur. For profit maximization, a best practices scope change control process must be part of the EPM system.

Over the years, project managers have been encouraged by their superiors to seek out any and all value-added scope changes to be funded by the customers. But these scope changes are now playing havoc with capacity-planning activities and the assigning of critical resources needed for the scope changes and other projects. As companies mature in project management, the EPM systems become web based. All individual project schedules are rolled up into a master schedule such that senior management can get a realistic picture of resources committed for the next 90 or 180 days. This allows a company to determine how much additional work it can undertake without overtaxing the existing labor base. Furthermore, if a resource bottleneck is detected, it should be relatively clear how many additional resources should be hired and in which functional groups.

As capacity planning changes from an art to a science, the problems with obtaining qualified resources for unplanned scope changes grow. Maximization of profits on a particular project may not be in the best interest of the company, especially if the resources can be used more effectively elsewhere in the organization. Organizations today are understaffed, believing that it is better to have more work than people than more people than work. Executives must find a way to balance the need for added project resources, scope changes, portfolio selection of projects, and the strain on the working relationship between project and line managers. How do executives now convince project managers that scope changes are unnecessary and forget profit maximization?

2.9 OUTSOURCE OR NOT MIGRAINE

One of the responsibilities of a PMO is debriefing the project team at the completion of the project. This includes capturing lessons learned, identifying opportunities for improving the EPM system, and updating the estimating database. As the estimating database improves, companies realize that they can outsource some project work at a significantly lower cost than performing the same work internally.

While outsourcing can become an important best practice and can save the company some money, there may be detrimental results. A bank received significant negative publicity in local newspapers when it was discovered that the information systems division would be downsized in conjunction with cost-effective outsourcing. Another organization also outsourced its information systems work to such an extent that it had to begin providing its suppliers and contractors with company-proprietary data. Headaches occur when executives must balance short-term profitability with the long-term health of the corporation and community stakeholder needs and expectations.

Best practices are designed to benefit both the company and the workers. When the implementation of best practices leads to loss of employment, the relative importance of best practices can diminish in the eyes of the employees.

2.10 DETERMINING WHEN TO CANCEL A PROJECT MIGRAINE

Virtually every EPM system is based on life-cycle phases. Each life-cycle phase terminates with an end-of-phase gate review meeting designed to function as a go/no-go decision point for proceeding to the next phase. Very few projects seem to be terminated at the early gate review meetings. One reason for this is that project managers do not necessarily provide all of the critical information necessary to make a viable decision. Project managers provide information in forecast reports on the estimated cost of completion and time of completion. What is missing is the expected benefits at completion, and this value may be more important than time and cost. While it is understandable that this value may be difficult to obtain during early life-cycle phases, every attempt should be made to present reasonable benefits-at-completion estimates.

If a project comes in late or is over budget, the expected benefits may still be achievable. Likewise, if a project is under budget or ahead of schedule, there may be no reason to believe that the vision at the project's initiation will be met at completion. One company has initiated a concept called "map days," when periodically the team maps its performance to date. The maps are reviewed with senior management to make sure that the project should continue. This concept can be expanded to include possible benefits at completion.

While good project management methodologies are best practices and provide valuable information for the management of projects, the system must also be capable of providing the necessary information to senior management for critical decision-making. All too often, EPM systems are developed for the benefit of project managers alone rather than for the best interest of the entire company.

2.11 PROVIDING PROJECT AWARDS MIGRAINE

Perhaps the biggest headache facing senior management is the establishment of an equitable project award/recognition system that is part of the wage and salary administration program. Companies have recognized that project management is a team effort and that rewarding project teams may be more beneficial than rewarding individuals. The headache is how to do it effectively.

Several questions need to be addressed:

- Who determines the magnitude of each person's contribution to the project's success?
- Should the amount of time spent on the project impact the size of the award?
- Who determines the size of the award?
- Will the award system impact future estimating, especially if the awards are based on underruns in cost?
- Will the size of the awards impact future personnel selection for projects?
- Will employees migrate to project managers who have a previous history of success where large awards are provided?
- Will people migrate away from high-risk projects where rewards may not be forthcoming?
- Will employees avoid assignments to long-term projects?
- Can union employees participate in the project award system?

Providing monetary and nonmonetary recognition is a best practice as long as it is accomplished in an equitable manner. Failure to do so can destroy even the best EPM systems as well as a corporate culture that has taken years to develop.

2.12 MIGRAINE FROM HAVING THE WRONG CULTURE IN PLACE

Creating the right corporate culture for project management is not easy. However, when a strong corporate culture is in place and it actively supports project management such that other best practices also develop, the culture is very difficult to duplicate in other companies. Some corporate cultures lack cooperation among the players and support well-protected silos. Other cultures are based on mistrust while others foster an atmosphere where it is acceptable to persistently withhold information from management.

A telecommunications company funded more than 20 new product development projects, which all had to be completed within a specific quarter to appease Wall Street and provide cash flow to support the dividend. Management persistently overreacted to bad news, and information flow to senior management became filtered. The project management methodology was used sparingly for fear that management would recognize early on the seriousness of problems with some of the projects.

Not hearing any bad news, senior management became convinced that the projects were progressing as planned. When it was discovered that more than one project was in serious trouble, management conducted intensive project reviews on all projects. In one day, eight project managers were either relieved of their responsibilities or fired. But the damage was done, and the problem was really the culture that had been created. Beheading the bearer of bad news can destroy potentially good project management systems and lower morale.

In another telecommunications company, senior management encouraged creativity and provided the workforce with the freedom to be creative. The workforce was heavily loaded with technical employees with advanced degrees. Employees were expected to spend up to 20 percent of their time coming up with ideas for new products. Unfortunately, this time was being charged back to whatever projects the employees were working on at the time, thus making the cost and schedule portion of the EPM system ineffective.

While management appeared to have good intentions, the results were not what management expected. New products were being developed, but the payback period was getting longer and longer while operating costs were increasing. Budgets established during the portfolio selection of the project's process were meaningless. To make matters worse, the technical community defined project success as exceeding specifications rather than meeting them. Management, in contrast, defined success as commercialization of a product. Given the fact that as many as 50–60 new ideas and projects must be undertaken to have one commercially acceptable success, the cost of new product development was bleeding the company of cash, and project management was initially blamed as the culprit. Even the best EPM systems are unable to detect when the work has been completed other than by looking at money consumed and time spent.

It may take years to build up a good culture for project management, but it can be destroyed rapidly through the personal whims of management. A company undertook two high-risk R&D projects concurrently. A time frame of 12 months was established for each in hope of making a technology breakthrough; even if it could happen, both products would have a shelf life of about one year before obsolescence would occur.

Each project had a project sponsor assigned from the executive level. At the first gate review meeting, both project managers recommended that their projects be terminated. The executive sponsors, in order to save face, ordered the projects to continue to the next gate review rather than terminate them while the losses were small. The executives forced the projects to continue on to fruition. The technical breakthroughs occurred six months late, and virtually no sales occurred with either product. There was only one way the executive sponsors could save face—promote both project managers for successfully developing two new products and then blame marketing and sales for their inability to find customers.

Pulling the plug on projects is never easy. People often view bad news as a personal failure, a sign of weakness, and a blemish on their career path. There must be an understanding that exposing a failure is not a sign of weakness. The lesson is clear: Any executive who always makes the right decision is certainly not making enough decisions, and any company where all of the projects are being completed successfully is not working on enough projects and not accepting reasonable risk.

2.13 MIGRAINES DUE TO POLITICS

The completion of a project requires people. But simply because people are assigned to the project does not necessarily mean that they will always make decisions for what is in the best interest of the project. When people are first assigned to a new project, they ask themselves, "What's in it for me? How will my career benefit from this assignment?"

This type of thinking creates severe migraines and can permeate all levels of management on a project, including those responsible for project governance. People tend to play politics to get what they want, and this gamesmanship creates barriers that the project manager must overcome. People are motivated by the rewards they can receive from the formal structure of the company and also from the informal political power structure that exists. Barriers are created when an individual's rewards from either structure are threatened. The barriers lead to conflicts and can involve how the project will be planned; who will be assigned to specific activities, especially those activities that may receive high-level visibility; which approach to take to solve a problem; and other such items that are often hidden agenda items. Some people may even want to see the project fail if it benefits them.

Political savvy is an essential skill for today's project manager. Project managers can no longer rely solely on technical or managerial competence when managing a project. They must understand the political nature of the people and organizations they must deal with. They must understand that politics and conflicts are inevitable and are a way of life in project management. Project managers of the future must become politically astute. Unfortunately, even though there are some books published on politics in project management, there has been limited research conducted on project management politics compared to other areas of the *PMBOK® Guide*.[1]

Political Risks

On large and complex projects, politics are often treated as a risk, especially when the project is being conducted in the host country and is subject to government interference or political violence. The factors often considered as part of political risks include:

- Political change, such as a new party elected into power.
- Changes in the host country's fiscal policy, procurement policy, and labor policy.
- Nationalization or unlawful seizure of project assets and/or intellectual property.
- Civil unrest resulting from a coup, acts of terrorism, kidnapping, ransom, assassinations, civil war, and insurrection.
- Significant inflation rate changes resulting in unfavorable monetary conversion policies.
- Contract failure, such as license cancellation and payment failure.

We tend to include many of these risks within the scope of enterprise environmental factors that are the responsibility of the project sponsor or the governance committee. But when the project is being conducted within the host country, it is usually the project manager who has to deal with the political risks.

The larger and more complex the project, the larger the cost overrun, and the larger the cost overrun, the greater the likelihood of political intervention. In some countries, such as the United States, escalating problems upward usually imply that the problem ends up in the hands of the project's sponsor. But in other countries, especially in emerging market nations, problems may rise beyond the governance committee and involve high-level government officials. This is particularly true for megaprojects that are susceptible to large cost overruns.

Reasons for Playing Politics

There are numerous reasons why people play political games. Some of the common reasons include:

- Wanting to maintain control over scarce resources.
- Seeking rewards, power, or recognition.
- Maintaining one's image and personal values.
- Having hidden agendas.
- Fear of the unknown.
- Control over who gets to travel to exotic locations.
- Control over important information since information is a source of power.

---

1. PMBOK is a registered mark of the Project Management Institute, Inc.

- Getting others to do one's work.
- Seeing only what one wants to see.
- Refusing to accept or admit defeat or failure.
- Viewing bad news as a personal failure.
- Fear of exposing mistakes to others.
- Viewing failure as a sign of weakness.
- Viewing failure as damage to one's reputation.
- Viewing failure as damage to one's career.

All of these are reasons that may benefit you as project manager personally. There are also negative politics where political games are played with the intent of hurting others, which in turn may end up benefiting you personally. Some examples are:

- Wanting to see the project fail.
- Fear of change if the project succeeds.
- Wanting to damage someone else's image or reputation especially if they stand in the way of your career advancement.
- Berating the ideas of others to strengthen your position.

Situations Where Political Games Will Occur

While politics can exist on any project and during any life-cycle phase, history has shown us that politics are most likely to occur as a result of certain actions and/or under certain circumstances:

- Trying to achieve project management maturity within a conservative culture.
- During mergers and acquisitions, when the "landlord" and the "tenant" are at different levels of project management maturity.
- Trying to get an entire organization to accept a project management methodology that was created by one functional area rather than a committee composed of members from all functional areas (i.e. the "not invented here" syndrome).
- Not believing that the project can be completed successfully and wanting to protect oneself.
- Having to change work habits and do things differently if the project is a success.
- Not knowing where problems that occur will end up for resolution.
- Assuming that virtual teams are insulated from project politics.
- Failing to understand stakeholder effective relations management practices.
- The larger and more complex the project, the greater the chances of political interference.
- The larger the size of the governance committee, the greater the chance for disagreements and political issues to appear.
- The more powerful the people are on the project, the greater the chance that they will be involved in project politics.
- Employees who are recognized as prima donnas are more prone to play political games than the average worker.

The Governance Committee   Project politics usually ends up pushing the project in a direction different from the original statement of work (SOW). The push can originate within your own senior management, among some of your project team members, with the customer, and even with some stakeholders. Each will want a slightly different outcome, and your job is to try to find a way to appease everyone.

On the surface, the simplest solution appears to be the creation of a governance committee composed of senior managers from the company, representation from the customer's company, and representatives from various stakeholder groups. It seems that you can let governance committee members resolve all of the conflicts among themselves and present a unified direction for the project. Gaining support from a higher power certainly seems like the right thing to do. Unfortunately, there is still the possibility that the committee cannot come to an agreement, and even if the members appear to be in agreement, certain members may still try to play politics behind the scenes. The existence of the governance committee does not eliminate the existence of project politics. People who serve on a governance committee often play the political game in order to enhance their power base.

Most companies have limited funds available for projects. The result is an executive-level competition for project funding that may serve the best interest of one functional area but not necessarily the best interest of the entire company. Executives may play political games to get their projects approved ahead of everyone else, viewing this as an increase to their power base. But the governance committee may include executives from those functional areas that lost out in the battle for project funding, and these executives may try to exert negative political influence on the project, even going so far as to hope that the project will fail. In these cases, the result is a project manager who is assigned to such projects and brought on board after the project is approved never fully understanding the politics that were played during project approval and initiation.

Friends and Foes   It is often difficult to identify quickly which people are friends and which are foes. Not all people who have political agendas are enemies. Some people may be playing the political game for your best interest. It is therefore beneficial to identify, if possible, whether people are friends or foes based on their personal agendas. Doing this means that you must communicate with them, perhaps more informally than formally, to understand their agendas. Reading body language is often a good way to make a first guess if someone is a friend or foe.

One possible way to classify people might be:

- *True supporters.* These are people who openly demonstrate their willingness to support you and your position on the project.
- *Fence-sitters.* These are people whom you believe will support you down the road as long as you prove to them that you are deserving of their trust and support. You may need to spend extra time with them to show them your position and to gain their support.

- *True unknowns.* Unlike fence-sitters, who may be won over to your way of thinking, these are people who may have hidden agendas that are not in your best interest, but they are relatively quiet and may have not yet expressed their concerns. These people could pose a serious threat if they are adamantly opposed to the direction in which the project is proceeding.
- *True enemies.* These are people who have made it quite clear that they are unlikely to support your views. You understand their position and probably are quite sure how they will respond to you and the direction the project is taking.

Attack or Retreat   When people play political games on projects, we seem to take for granted two facts. First, these people are most likely experienced in playing such games, and second, they expect to win. Based on whom the conflict is with, you must decide whether to aggressively attack them or retreat. Simply taking no action is a form of withdrawal, and you are sure to lose the battle.

The first rule in battle is to gather as much intelligence as you can about your enemy. As an example, as part of stakeholder relations management, we can map project stakeholders according to Figure 2–3. Stakeholder mapping is most frequently displayed on a grid comparing each stakeholder's power and level of interest in the project.

- *Manage closely.* These are high-power, interested people who can make or break your project. You must put forth the greatest effort to satisfy them. Be aware that there are factors that can cause them to change quadrants rapidly.
- *Keep satisfied.* These are high-power, less interested people who can also make or break your project. You must put forth some effort to satisfy them but not with excessive detail that can lead to boredom and total disinterest. They may not get involved until the end of the project approaches.
- *Keep informed.* These are people with limited power but keen interest in the project. They can function as an early warning system of approaching problems and may be technically astute to assist with some technical issues. These are the stakeholders who often provide hidden opportunities.

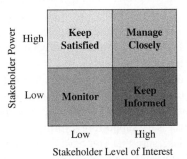

Figure 2–3. Stakeholder mapping.

- *Monitor only*. These are people with limited power and who may not be interested in the project unless a disaster occurs. Provide them with some information but not with so much detail that they will become disinterested or bored.

When you go on offense and attack the people playing politics, you must have not only ammunition but also backup support if necessary. You must be prepared to show how the political decision might affect the constraints on the project as well as the accompanying baselines. Based on the power and influence level of your opponent, according to Figure 2–3, you may need other stakeholders to help you plead your case. It is highly beneficial to have supporters at the same level of position power or higher than the people playing the political game.

Not all political battles need to be won. People who play politics and possess a great deal of power may also have the authority to cancel the project. In such cases, retreating may be the only viable option. If you truly alienate the people playing power games, the situation can deteriorate even further. There is always the chance that you may have to work with the same people in the future. In any case, the best approach is to try to understand the people playing politics, the reason why they are playing politics, and how much power and influence they have over the final decision.

The Need for Effective Communication

While it is not always possible to tell when someone is playing or intends to play the political game on your project, there are some telltale signs that this may be happening. Some of the signs include:

- People do not care about your feelings.
- People avoid discussing critical issues.
- People never ask you about your feelings on the matter.
- People procrastinate about making decisions.
- People have excuses for not completing action items.
- People discuss only those items that may benefit them personally.

Although project managers may not have any control over these telltale signs, ineffective communication can make the situation worse. To minimize the political impact on a project, the project manager should consider using the following practices:

- Listen carefully before speaking and do not jump to conclusions.
- Make sure you understand what others are saying and try to see the issue from their point of view.
- Follow up all informal communications with a memo outlining what was discussed and to make sure that there were no misunderstandings.
- Before stating your point of view, make sure that you have gathered all of the necessary supporting information.
- Make sure that you have a clear understanding of how culture affects the way that people are communicating with you.

- If you must provide criticism, make sure that it is constructive rather than personal criticism.
- When resolving political issues, there will be winners and losers. It is not a matter of just picking a winner. You must also explain to everyone why you selected this approach and why the other approaches were not considered. This must be done tactfully.
- If the situation cannot be managed effectively, do not be embarrassed to ask senior management for advice and assistance.
- Ineffective communication encourages lying, which, in turn, generates additional political games to be played accompanied by a great deal of mistrust.

Project managers must be careful when discussing politics with team members, clients, and stakeholders. The information could be misunderstood or filtered, especially if people hear what they want to hear. The result could be additional unexpected politics, and friends could easily turn into foes.

Power and Influence
Effective communication skills alone cannot resolve all political situations. To understand why, we must look at how project management generally works. If all projects stayed within the traditional hierarchy, someone would have the ultimate authority to resolve political issues. But since most projects are managed outside the traditional hierarchy, the burden for the resolution of conflicts and political issues usually falls upon the shoulders of the project manager even if a governance committee is in place. The governance committee may very well be the cause of the conflict.

On the surface, it seems like the simplest solution would be to give the project manager sufficient authority to resolve political issues. But projects are usually executed outside of the traditional hierarchy, thus limiting the authority that the project manager will possess. This lack of formal authority makes the project manager's job difficult. While project charters do give project managers some degree of authority for a given project, most project managers still have limitations because they:

- Must negotiate with functional managers for qualified resources.
- May not be able to remove employees from a project without the functional manager's concurrence.
- Generally have no direct responsibility for wage and salary administration.
- May possess virtually no reward or punishment power.
- May not be able to force the employees to work on their projects in a timely manner, if employees are assigned to multiple projects.

With a lack of position power that comes from the traditional hierarchy, and without the ability to reward or punish, project managers must rely on other forms of power and the ability to influence people. Behavioral skills such as effective communication, motivation techniques, conflict management, bargaining, and negotiating are essential to resolve political disputes. Unfortunately, most project managers lack political savvy and have poor conflict resolution skills.

Managing Project Politics   While project politics are inevitable, there are actions project managers can take to minimize or control political issues. Some of these actions include:

- Gather as much information as possible about the political situation.
- Make sure that everyone fully understands the impact of the political situation on the project's baselines.
- Try to see the picture through the eyes of the person playing politics.
- Try to form a coalition with the people playing politics.
- See if the sponsor or the governance committee can insulate you from the political games.
- Have a structured decision-making process as part of your project management methodology to reduce some of the political games.
- Try to determine people's political positions by reading their body language.
- If the political situation cannot be resolved quickly, demonstrate a willingness to compromise as long as the integrity of the project is not sacrificed.

Power breeds politics, and politics in turn breeds power. Expecting to manage a project without any political interference is wishful thinking rather than reality. We cannot predict customer and stakeholder behavior. Sometimes the political situation occurs without any early warning signs.

No one can agree on a definition of organizational or project politics. Politics can appear in many shapes, forms, and sizes. Therefore, project managers must develop superior behavioral skills to deal with political situations. The danger in not being able to manage political situations correctly is project redirection or misdirection.

2.14 MIGRAINES CAUSED BY THE SEVEN DEADLY SINS[2]

For more than 40 years, the project management landscape has seen textbooks, journal articles, and papers discussing the causes of project failures. Unfortunately, many of the failure analyses seem to look at failure superficially rather than in depth. When trying to discover the root cause of a failure, we usually look first to the contractor's company for someone to blame rather than to our own company. If that doesn't work, then we begin climbing the organizational hierarchy in our own company by focusing on the

---

2. Portions of the section "The Seven Deadly Sins" are adapted from Wikipedia contributors, "Seven Deadly Sins," *Wikipedia, The Free Encyclopedia,* https://en.wikipedia.org/w/index.php?title=Seven_deadly_sins& oldid=802218381 (accessed August 2017).

project team, followed by the project manager. Once we find someone to blame, the search ends, and we feel comfortable that we have discovered the cause of the failure.

It is human nature to begin finger-pointing at the bottom of the organizational hierarchy, rather than at the top. Yet, more often than not, the real cause of failure is the result of actions (or inactions) and decisions made at the top of the organizational chart rather than at the bottom. It is also human nature to make decisions based on how we are affected by the Seven Deadly Sins, namely: envy, anger, pride, greed, sloth, lust, and gluttony. Decisions based on the Seven Deadly Sins, whether they are made at the top or bottom of the organization, can have dire consequences for projects. Sometimes the sins are hidden and not easily recognized by us or others. We simply do not see or feel that we are committing a sin.

The Seven Deadly Sins affect all of us sooner or later, even though we refuse to admit it. Some of us may be affected by just one or two of the sins, whereas others may succumb to all seven. What is unfortunate is that the greatest damage can occur on projects when the sins influence the way those at senior levels of management interface with projects, whether as a project sponsor or as a member of a governance group. Bad decisions at the top, especially if based on emotions rather than practicality, can place the project on a destructive path even before the day the project is kicked off.

The Seven Deadly Sins

The term "Seven Deadly Sins," also referred to as the "capital vices" or "cardinal sins," is a classification of objectionable vices. They were originally part of Christian ethics and have been used since early Christian times to educate and instruct Christians concerning fallen humanity's tendency to sin. Part or all of the sins have been discussed over the past four centuries from different perspectives in the religious writings of Christianity, Hinduism, Islam, Buddhism, and Judaism. Over the years, the sins have been modified and discussed by the clergy, philosophers, psychologists, authors, poets, and educators.

A brief description of each of the Seven Deadly Sins appears in Table 2–1. Each of the sins can be related to an animal, a specific color, and even a punishment in hell for committing the sin.

In a project environment, any or all of these sins can cause rational people to make irrational decisions, and this can occur at any level within the organizational hierarchy. At some levels, the existence of the sins may have a greater impact on project

TABLE 2–1. THE SEVEN DEADLY SINS

Sin	Traits	Animal	Color	Punishment in Hell
Envy	The desire to possess what others have	Snake	Green	Placed in freezing water
Anger/Wrath	A strong feeling of displeasure	Lion	Red	Dismembered alive
Pride	The need for inward emotional satisfaction	Peacock	Violet	Broken on the wheel
Greed	The desire for material wealth or gain	Toad	Yellow	Put in cauldrons of boiling oil
Sloth	The avoidance of work	Snail	Light blue	Thrown into a snake pit
Lust	A craving, but not necessarily sexual	Goat	Blue	Smothered in fire and brimstone
Gluttony	The desire to consume more than needed	Pig	Orange	Forced to eat rats, toads, and snakes

performance than at other levels. If a sin is apparent in the beginning of a project, then poor decisions in the initiation phase can have detrimental consequences in all of the downstream phases.

Envy

Envy is the art of counting the other fellow's blessing instead of your own.

—Harold Coffin

Envy is ignorance. Imitation is suicide.

—Ralph Waldo Emerson

When men are full of envy, they disparage everything, whether it be good or bad.

—Publius Cornelius Tacitus

Envy is the desire to have what others have. Resentful emotions occur when one lacks another's superior qualities, such as status, wealth, good fortune, possessions, traits, abilities, or position. Envy may encourage someone to inflict misfortune on another person and try to undo someone else's advantage or deprive the other of obtaining the advantage. Envy can also affect the relationship between people, as when people ignore a person of whom they are envious. Envy is often synonymous with jealousy, bitterness, greed, spite, and resentment.

Envy can be malicious or benign. Malicious envy has all of the characteristics just mentioned. Benign envy can be a positive motivational force if it encourages people to act in a more favorable manner such that the desires are attainable. Benign envy usually exists at the bottom of the organizational hierarchy, whereas malicious envy appears most frequently at the top.

Four situations illustrate how envy can lead to project disasters:

Situation 1: Reorganizational Failure

A company had four divisions, each headed by a senior vice president. In the past, most projects had stayed entirely within one division. Each division had its own project management methodology, and the number of project successes significantly outnumbered the failures. As the marketplace began to change, the company began working on projects that required that more than one division work together on the same project. Using multiple methodologies on the same project proved to be an impossible task.

The president decreed that there must be one and only one methodology and that all of the divisions must use the same methodology for managing projects. The company created a PMO, and the president assigned one of the vice presidents with control over the PMO. Employees from each of the other three divisions were then assigned on a dotted-line relationship to the PMO for the development of the singular methodology.

The people in the PMO seemed to work well together, but the four vice presidents demanded that they have final signature authority over the adoption of the singular methodology. Each vice president believed that the project management approach used in their division should be the driving force for the creation of a singular methodology.

Regardless of what design the PMO came up with, each vice president demonstrated envy and resentment, finding fault with the others– ideas, and playing out the not-invented-here syndrome. While this was happening, the number of project failures began to increase, because of the lack of structure for project execution.

It also became obvious to each of the four vice presidents that whichever one had control of the PMO would become more powerful than the other three vice presidents because of the control over all of the project management intellectual property. Information is power and envy for control of the information had taken its toll on the ability to manage and control projects effectively. Eventually, the president stepped in and allowed each vice president to have a PMO. However, the PMOs had to be networked together. This helped a little, but even after they agreed on a common methodology, each PMO tried to seduce the other PMOs into their way of thinking. As expected, continuous changes were being introduced to the methodology, with the projects still suffering from a lack of direction. Envy prevented decisions from being made that would have been in the best interest of the entire company.

Situation 2: Reward Failure

Believing that an effective reward/bonus system would motivate project teams, senior management announced that bonuses would be given to each project team based on the profitability of their projects. The company survived on competitive bidding to win contracts, and most of the projects were in the millions of dollars. Project managers quickly learned that large bonuses could be awarded if the project's cost estimates were grossly inflated during the bidding process so the contracts could be won. This way, the actual profits on some contracts could exceed the targeted profits.

Although the company lost a few contracts it expected to win because of the inflated costs, some bonuses given to project managers at the end of the contracts were similar in size to the bonuses given to some executives. Many executives were now envious of the people below them who were receiving such large bonuses. Due to envy, the executives then changed the bonus policy, whereby part of the bonus fund would be distributed among the executives, even though the executives were not functioning as project sponsors. The bonuses given to the workers and the project managers were then reduced significantly. Some workers then sabotaged some projects rather than see the executives receive bonuses that were awarded at the expense of the workers.

Situation 3: Failure Due to Inflicting Misfortune

Paul was the director of operations for a medium-size company. He was in the process of establishing a PMO that would report directly to the chief executive officer. Paul desperately wanted the new position of director of the PMO, believing it would be a stepping stone to becoming a vice president. His major competitor for the position of director of the PMO was Brenda, a 20-year company veteran who was considered the company's best project manager. Because of Brenda's decision-making skills, she was almost always empowered with complete decision-making authority on her projects.

When Brenda was assigned to her latest project, Paul requested and was granted the position of project sponsor for her project. Paul was envious of Brenda's abilities and good fortune and believed that, if he could somehow sabotage Brenda's project without hurting himself in the process, he could easily be assigned director of the PMO. Paul placed limits on Brenda's authority and demanded that, as the sponsor, he approve any and all critical decisions. Paul continuously forced Brenda to select non-optimal alternatives when some decisions had to be made. Brenda's project was nearly a disaster, and Paul was later assigned as director of the PMO.

Envy can force us to inflict pain on others to get what we desire. Paul received his promotion, but the workers and Brenda knew what he had done. Paul's working relationship with the functional subject matter experts deteriorated.

Situation 4: The Relationship Failure

Jerry and two of his friends lived near each other and joined the company at exactly the same time. Jerry worked in project management, and the other two worked in engineering. They formed a carpool and traveled to and from work together every day. They also socialized when not at work.

Two years after joining the company, Jerry received his second promotion, whereas the other two workers had not received any promotions. The other two workers were envious of Jerry's success to the point where they stopped socializing and carpooling with him. The jealousy became so strong that the two workers even refused to work on Jerry's projects. The workers never visibly displayed their jealousy of Jerry, but their actions spoke louder than words and made it clear how they really felt.

Anger (or Wrath)

For every minute you are angry, you lose sixty seconds of happiness.

—Ralph Waldo Emerson

Speak when you are angry—and you'll make the best speech you'll ever regret.

—Dr. Lawrence J. Peters

Anger is never without reason, but seldom a good one.

—Benjamin Franklin

Anger is one letter short of danger.

—Anonymous

Anger, if not restrained, is frequently more hurtful to us than the injury that provokes it.

—Seneca

Anger or wrath is a strong feeling of displeasure. Sometimes we become angry because the actions of others on the project have offended us. Other times we use anger unnecessarily, as a means to stop a behavior that threatens the project, such as continuous schedule slippages or cost overruns. Anger is often synonymous with ire, annoyance, irritation, rage, and resentment.

When we get angry, we often lose our objectivity. The anger we feel and demonstrate can appear suddenly, or it can be deliberate. There are ranges of anger. On the soft end of the spectrum, anger can be just a mild irritation, whereas, on the hard end, anger can result in fury and rage. Not all anger is readily visible. For example, passive anger can be seen as a phony smile, giving someone the cold shoulder, overreacting to something, or constantly checking things. Aggressive anger can appear as bullying, expressing mistrust, talking too fast, or destructiveness.

Here are some examples of how anger can affect projects.

Situation 5: Failure Due to Unjust Anger

While selecting the portfolio of 20 projects for the upcoming year, senior management established the budgets and schedules without any supporting data on what might or might not be realistic. To make matters worse, the executive sponsors on each project emphatically stated that they would not tolerate schedule slippages or cost overruns. The project teams developed the detailed project plan and, on eight of the 20 projects, the teams determined that the budgets and schedules provided by senior management were unrealistic. Rather than inform senior management immediately that their budget and schedule perceptions might be wrong, the teams began executing the projects and hoping for a miracle. The teams felt that this was a better approach than incurring the wrath of senior management when they were apprised of the situation.

The eight teams were unsuccessful in their quest for a miracle. After a few months, senior management performed a health check on one of the eight projects and discovered the truth: The project was in bad shape. A health check was then performed on all 20 projects, and it became apparent that eight of the projects were in trouble both financially and technically. Senior management became enraged that they had not been informed of this previously, canceled the eight troubled projects, and fired all eight project managers in one day.

Part of the blame certainly falls on the shoulders of the project teams for not informing senior management early on. However, a lot of the blame must rest with senior management, especially when they have a history of demonstrating irate behavior that may have been unjustified. When project teams believe that they will encounter anger rather than support for problems from the top of the organizational hierarchy, project management may not succeed, and projects will fail. Bad news is often filtered to prevent the occurrence of anger.

Situation 6: Failure Due to a Hidden Agenda

The chief information officer (CIO) became the project sponsor for a $25 million information technology (IT) project scheduled to last about one year. The CIO established October 1 as the "go live" date for the project. During a July review of the status of the project, the CIO was informed by the project manager that the go-live date was unrealistic. The CIO became furious and asked, "How much of the software would be operational by October 1?" The project manager responded, "Perhaps 10 percent."

The CIO stormed out of the meeting after demonstrating anger, calling the project team "incompetent fools." The CIO then authorized significant overtime and awarded the prime contractor almost $5 million in additional costs if they could get at least 50 percent of the software operational by October 1 and 70 percent or more by November 1. The CIO knew that his year-end corporate bonus was partly aligned with implementation of this project, and with 70 percent of the software operational, his bonus would be significant. When the project was finally completed in February, the executive committee viewed the project as a partial failure because of the $5 million cost overrun and the project manager was reprimanded. However, the CIO received his bonus.

Situation 7: Failure Due to Information Filtering

Senior management in a government agency established a culture that bad news would be filtered as the news proceeded up the organizational hierarchy. Allowing bad news to reach the top would be an invitation for fury and rage coming from the top back down to the projects. Therefore, by the time the information reached the top, much of the bad news had disappeared, and the risks associated with the project were buried. The result of one project was just as the technical risk experts predicted: Seven astronauts were killed when the space shuttle *Challenger* exploded during liftoff.[3] There were other factors, as well, that led to this disaster. During a congressional committee meeting reviewing the cause of the fatality, one subject matter expert was asked by the committee, "Why didn't you explain to senior management what the risks were?" The subject matter expert asserted, "I didn't report administratively to senior management. My responsibility was to report this to my boss and he, in turn, should have reported it higher up."

Situation 8: Failure Due to a Collective Belief

A collective belief is a fervent, and often blind, desire to achieve—regardless of the cost and consequences. When a collective belief exists, especially at the senior levels of management, rational organizations begin making irrational decisions, and any deviation from the collective belief is met with anger. People who question the collective belief or challenge progress are removed from the project or severely reprimanded. In order to work on these projects, people must suppress their anger and go with the flow, regardless of the outcome. These projects can be technical successes but financial failures, never totally fulfilling the corporate business strategy.

A good example of this is the Iridium Project, an 11-year project that missed the service launch date by one month.[4] The service was a network of 66 satellites circling the earth, allowing people to talk to anyone anywhere. The project management activities performed by Motorola and Iridium LLP were outstanding, especially when we consider that the project resulted in more than 1,000 patents and 25 million lines of software code. Technically, the project was a success, but financially it was a disaster, invoking anger when it became apparent that the company could not get the number of subscribers it needed to break even. Throughout the project, the threat of severe anger from above, as well as the existence of the collective belief that management is always right made it almost impossible for people to challenge the projections on the number of subscribers.

Anger need not be demonstrated to inflict pain on a project. Just the implied threat or fear of anger can limit a team's performance significantly.

---

3. For additional information on this case study, see "Case Study: The Space Shuttle *Challenger* Disaster" in Harold Kerzner, *Project Management Case Studies* (6th ed.) (Hoboken, NJ: Wiley, 2022), pp. 363–410.

4. For additional information, see "Case Study: The Rise, Fall and Resurrection of Iridium; A Project Management Perspective," in Kerzner, *Project Management Case Studies*, 6e, (Hoboken, NJ. Wiley, 2022), pp. 261–291.

Pride

> A proud man is always looking down on things and people; and, of course, as long as you are looking down, you can't see anything above you.
>
> **—C. S. Lewis**

> The blind cannot see—the proud will not.
>
> **—Russian proverb**

> Vanity and pride are different things, though the words are often used synonymously. A person may be proud without being vain. Pride relates more to our opinion of ourselves; vanity, to what we would have others think of us.
>
> **—Jane Austen**

> We are rarely proud when we are alone.
>
> **—Voltaire**

Pride is an inward emotion that leads to personal satisfaction or meeting personal goals. Pride can be a virtue or simply love of oneself or an inflated sense of one's accomplishments, which leads to exhilarating emotions. Pride can have both negative and positive connotations. In a negative sense, pride can cause us to grossly inflate what we have accomplished. In a positive sense, it can be an attachment to the actions of others or a fulfilled feeling of belonging, such as national or ethnic pride or being a member of the team on an important project.

Pride is often seen as a virtue. Overinflated pride can result in a disagreement with the truth, which sometimes comes with self-gratification. The antonyms of pride are "humility" and "guilt."

Here are some examples of how pride can affect a project.

Situation 9: Failure Due to Too Much Expertise

Peter was one of the most experienced engineers in the company. His technical expertise was second to none. Peter was asked to solve a problem on a project. Even though there were several possible options, he chose the option that was the costliest and resulted in the addition of unnecessary features, which we refer to as bells and whistles. Peter asserted that his solution was the only practical one, and the project manager reluctantly agreed. Peter saw this project as a way of increasing his reputation in the company regardless of the impact on the project. The bells and whistles increased the final cost of the deliverable significantly. They also inflated Peter's self-esteem.

Situation 10: Failure Due to the Wrong Sponsor

Nancy was the director of marketing. Her superior, the vice president for marketing, had requested the development of a rather sophisticated IT project for the Marketing Division. It was customary for the IT department to act as the project sponsor on all IT projects once the business case for the project was approved. Nancy knew that this project would get the attention of the seniormost levels of management. She had never served in the capacity of a project sponsor but believed that, if she could be the sponsor for this project, her identification with this project could result in a promotion.

Nancy's campaign to become the sponsor was a success. Unfortunately, numerous IT issues had to be resolved at the sponsor level and, because of Nancy's lack of expertise in IT, she made several wrong decisions. The project ended up being late and over budget because many of Nancy's decisions had to be changed later in the project. Nancy's quest for pride ended up having detrimental results.

Greed (Avarice)

Ambition is but avarice on stilts, and masked.

—**Walter Savage Landor**

Avarice has ruined more souls than extravagance.

—**Charles Caleb Colton**

Avarice is the vice of declining years.

—**George Bancroft**

Avarice is generally the last passion of those lives of which the first part has been squandered in pleasure, and the second devoted to ambition.

—**Samuel Johnson**

Poverty wants much, but avarice, everything.

—**Publilius Syrus**

Poverty wants some things, luxury many things, avarice all things.

—**Benjamin Franklin**

Love is always a stranger in the house of avarice.

—**Andreas Capellanus**

To hazard much to get much has more avarice than wisdom.

—**William Penn**

Greed is a strong desire for wealth, goods, and objects of value for oneself. Greed goes beyond the basic levels of comfort and survival, asking for more than we actually need or deserve. Greed can also appear as the desire for power, information, or control of resources. Synonyms for greed are "avarice" and "covetousness."

The following are several examples of how greed can affect projects.

Situation 11: The Failure of Too Many Resources

Karl was placed in charge of a two-year project that required 118 people, many of whom were needed on just a part-time basis. Karl convinced senior management that this project required a colocated team, with everyone assigned full time, and that the team should be housed in a building away from the employees' functional managers. Senior management knew this was a bad idea but reluctantly agreed to it, knowing full well that the project was now overstaffed and overmanaged.

At the end of the first year, it became obvious that none of the employees on Karl's project had received promotions or salary merit increases. The functional managers were rewarding only those employees who sat near them and made them look good on a daily basis. The employees on Karl's project now felt that assignment to this project was a nonpromotable assignment. Several employees tried to sabotage the project just to get off it. Later, Karl discovered that several of the other project managers now had a strong dislike for him because his greed for resources had affected their projects.

Situation 12: The Failure of Power

Carol was a department manager. She was proud of the fact that she finally had become a department manager. Word spread throughout the company that senior management was considering downsizing the company. Carol was afraid that her department might be eliminated and that she would lose her position as a department manager.

To protect her power position, Carol began giving conflicting instructions to the people in her department. The workers kept coming back to her for clarification of the conflicting instructions. Carol then told her superiors that the people in her department needed daily supervision or else the department's performance would suffer. While this technique seemed to prevent the downsizing of Carol's department, it did have a detrimental effect on the work the employees were doing on the projects. Carol's greed for power and resources proved detrimental to the company but beneficial to her own personal needs.

Situation 13: The Failure of Greed for the Bonus

The vice president for engineering was assigned as the project sponsor for a multimillion-dollar Department of Defense contract, and Ben was the project manager. A large portion of the vice president's bonus was based on the profitability of the projects directly under his control and of which he was the sponsor. This large project, headed by Ben, was scheduled to be completed in November; the follow-on contract, which was also quite large, was scheduled to begin in February.

The vice president and Ben agreed that a large management reserve should be established to support the project team between contracts. If the team was to be disbanded in November, there would be no guarantee that the same people would be available for the follow-on contract that would begin in February. When the project came to fruition in October, the remaining management reserve was large enough to support the resources with critical skills between October and February. These people would be working on some activities that would be needed for the follow-on contract, such as preliminary planning activities and procurement planning.

When the contract finally ended in October, the vice president told the financial people to book the management reserve as additional profit on the project. This increased the vice president's bonus significantly. However, without the management reserve, the critically skilled resources were reassigned back to their functional departments, and many were not available to work on the follow-on contract. The follow-on contract suffered from cost overruns and schedule slippages because it had different resources and a new learning curve. The damage from the vice president's greed was now apparent.

Sloth

Nothing irritates me more than chronic laziness in others. Mind you, it's only mental sloth I object to. Physical sloth can be heavenly.

—**Elizabeth Huxley**

We excuse our sloth under the pretext of difficulty.

—Marcus Fabius Quintilian

Diligence overcomes difficulties, sloth makes them.

—Benjamin Franklin

Sloth and silence are a fool's virtues.

—Benjamin Franklin

All things are easy to industry; all things are difficult to sloth.

—Benjamin Franklin

Sloth is the act of being physically, mentally, and/or emotionally inactive and often is characterized as laziness. Sloth can result in extreme waste in the effective use of people, things, skills, information, and even time. Sloth often forces us to overestimate the difficulty of the job.

The following are examples of how sloth can affect projects.

Situation 14: The Failure of Laziness

Becky was placed in charge of a one-year project that was relatively easy to accomplish and low risk. When negotiating with the functional managers for project staff, Becky overestimated the complexity and risk of the project so that she could request more experienced people. That would certainly make Becky's job easier. The functional managers were not sure if Becky's estimates of risk and complexity were valid, but they decided that it would be better to grant her request than to provide mediocre resources and find out later that she was correct.

There wasn't much for Becky to do on the project because the subject matter experts did it all. Eventually, the experienced people on Becky's project reported back to their respective functional managers that lower pay-grade resources should have been assigned. While Becky's project was considered a success and there was not much for her to do, the other projects that really could have used the more experienced resources suffered. Sloth usually benefits a single individual is at the expense of the greater good.

Situation 15: Failure Due to the Union Standard

A company had a powerful union that discouraged new employees who were eager to show what they could do by producing more units than the standard agreed to by the union. New workers were told to slow down and enjoy life.

The company soon became uncompetitive in the marketplace, and its business base began to deteriorate. Senior management then told the union that either the standards must be updated or people might lose their jobs. The union maintained its complacency and refused to budge on the standards. When management threatened to outsource much of the work and lay people off, the union workers went on strike.

Management personnel and nonunion workers began doing the work that was previously done by the union workers. They turned out 70 percent of the work using 10 percent as many nonunionized employees. Human resources personnel were running drill presses and lathes, and salespeople worked on the assembly line. Management now had a clear picture of what the sin of sloth had been doing to the company for years and had no intention of negotiating an end to the strike. Eventually, the union conceded and returned to work. However, more than 160 of the union workers were laid off after the new standards were adopted. The company was now competitive again.

Lust

> Lust is to other passions what the nervous fluid is to life; it supports them all, ambition, cruelty, avarice, revenge are all founded on lust.
>
> **—Marquis de Sade**

> Of all the worldly possessions, lust is the most intense. All other worldly passions seem to follow its train.
>
> **—Buddha**

> Society drives people crazy with lust and calls it advertising.
>
> **—John Lahr**

> Their insatiable lust for power is only equaled by their incurable impotence in exercising it.
>
> **—Winston Churchill**

> Hell has three gates: lust, anger, and greed.
>
> **—Bhagavad Gita**

> It is not power itself, but the legitimation of the lust for power, which corrupts absolutely.
> **—Richard Howard Stafford Crossman**

> The lust of avarice is so totally seized upon mankind that their wealth seems to rather possess them than they possess their wealth.
>
> **—Pliny the Elder**

Lust is the emotion or feeling of intense desire in the body. Although lust is usually described in a sexual content, it can also appear as a strong desire for power, knowledge, or control. It can lead to great eagerness or enthusiasm, which may be good, especially if it fulfills the need to gratify the senses.

Two examples of how lust can affect projects are given here.

Situation 16: Failure Due to the Lust for Power

Ralph was elated to be assigned as the project manager for a new project that was won through competitive bidding. The chance for significant follow-on work from this client was highly likely. This would be Ralph's chance to become more powerful than the other project managers and possibly be promoted and given a corner office. Corner offices with large windows were signs of power and prestige. For this to happen, Ralph had to slowly build his project into an empire of resources, regardless of the consequences.

By the end of the initial contract, Ralph had more resources assigned full time than planned for during project initiation. The project was significantly overstaffed, and this had an adverse effect on profits. But Ralph explained to his superiors that this would lead to increased profits in the future.

When the follow-on contract appeared, Ralph argued that he needed even more resources and that a projected organizational structure was needed with Ralph as its head. The company agreed. The projected structure allowed Ralph to have all remaining part-time workers assigned to his project full time. Partway through the project, the company was notified that there would be additional follow-on contracts, but these would all be awarded through competitive bidding. Ralph's power was now at an all-time high.

Unfortunately, because of the need to support his empire, all of the profits from the follow-on contract that Ralph was finishing up were going to workers' salary. Once again, Ralph argued that significant profits would be forthcoming. During competitive bidding for new follow-on work, Ralph's superiors significantly increased the price of the bid. Unfortunately, the company was now uncompetitive. Ralph and part of the empire he had built up were laid off. The lust for power resulted in that power, which had taken two years to develop, vanishing in one day.

Situation 17: Revisiting the Failure Due to the Lust for Power

This project would be Kathy's first chance to function as a project sponsor. Kathy believed that her lust for power would thrive if she micromanaged the project team and demanded to make any and all decisions. Senior management would certainly notice this. At least that's what she thought.

Kathy was correct in that senior management saw that she was making all of the decisions. Unfortunately, the subject matter experts assigned to the project, as well as the project manager, knew that Kathy had very limited knowledge regarding some of the technical decisions that needed to be made. They were also quite unhappy with being micromanaged. Many of Kathy's decisions were wrong, and team members knew it, but they went along with the bad decisions without questioning them. Management also saw the bad decisions that Kathy had made, and eventually, she was removed as the project's sponsor.

Gluttony

Glutton: one who digs his grave with his teeth.

—**French proverb**

Gluttony is the source of all our infirmities, and the fountain of all our diseases. As a lamp is choked by a superabundance of oil, a fire extinguished by excess of fuel, so is the natural health of the body destroyed by intemperate diet.

—**Robert Burton**

The miser and the glutton are two facetious buzzards; one hides his store and the other stores his hide.

—**Josh Billings**

Gluttony is an emotional escape, a sign something is eating us.

—**Peter De Vries**

Gluttony kills more than the sword.

—**George Herbert**

Gluttony is usually defined in terms of food with terms, such as "gulp down" or "swallow." We see it as an overconsumption of food. In a business environment, gluttony is the desire to consume more than what is required. It is extravagance or waste.

The example that follows shows how gluttony can lead to both success and failure.

Situation 18: The Success of Gluttony of Resources

Jerry was one of the directors of manufacturing reporting to the vice president for manufacturing. As technology began to change, manufacturing personnel recognized the need to create several new departments to take advantage of new technologies. Jerry had a thirst for resources. He convinced the vice president for manufacturing that these new departments belonged under his control. Within the next two years, all new departments were under Jerry's supervision. Jerry now controlled more than 75 percent of the resources in the Manufacturing Division.

When the vice president for manufacturing retired, Jerry was promoted to vice president. Jerry's first action was to break up the empire he created so that no one could ever become as powerful as he had been. In Jerry's eyes, he now had control over all resources, regardless of where they resided in the Manufacturing Division.

We have painted a bleak picture here of how the Seven Deadly Sins can have a negative impact on projects. From a project perspective, some of the sins are closely related and cannot be separated and discussed as easily as psychologists and philosophers would have us believe. This can be seen from some of the situations presented previously, for example, where the desire for control of vast resources could be considered as some form of lust, gluttony, or avarice.

It is true that, in some situations, the sins can produce positive results. They can force us to become more aggressive, take risks, accept new challenges, and add value to the company. Our fascination with pride and lust can help us turn around a distressed project and make it into a success so that we can get corporate-wide recognition. The greed for wanting a large bonus can likewise encourage us to make our project successful. The downside risk of the vices is that they most certainly can have a negative effect on our ability to establish our interpersonal skills and our relationships with the project teams and functional departments.

So, should we train project managers and team members on how to identify and control the sins? Perhaps not as long as beneficial results are forthcoming. Once again, we all succumb to some or all of these sins, but to varying degrees.

The Roman Catholic Church recognizes seven virtues that correspond inversely to each of the Seven Deadly Sins:

Vice	Virtue
Envy	Kindness
Wrath	Patience
Pride	Humility
Greed	Charity
Sloth	Diligence
Lust	Chastity
Gluttony	Temperance

From a project management perspective, perhaps the best solution would be to teach the virtues in project management training courses. It is even possible that in future editions of the *PMBOK® Guide*, the Human Resources Management chapter may even discuss vices and virtues. Time will tell.

2.15 SOURCES OF SMALLER HEADACHES

Not all project management headaches lead to migraines. The following list identifies some of the smaller headaches that occurred in various companies but do not necessarily lead to major migraines:

- *Maintaining original constraints.* As the project team began working on a project, work began to expand. Some people believed that within every project there was a larger project just waiting to be recognized. Having multiple project sponsors all of whom had their own agendas for the project created this problem.
- *Revisions to original mission statement.* At the gate review meetings, project redirection occurred as management rethought its original mission statement. While these types of changes were inevitable, the magnitude of redirections had a devastating effect on the EPM system, portfolio management efforts, and capacity planning.
- *Lack of metrics.* An IT organization maintained a staff of over 500 employees. At any given time, senior management was unable to establish metrics on whether the IT group was overstaffed, understaffed, or just right. Prioritization of resources was being done poorly, and resource management became reactive rather than proactive.
- *More metrics.* In another example, the IT management team, to help identify whether projects were being delivered on schedule, recently implemented an IT-balanced scorecard for projects. After the first six months of metric gathering, the conclusion was that 85 percent of all projects were delivered on time. From executive management's perspective, this appeared to be misleading, but there was no way to accurately determine whether this number was accurate. For example, one executive personally knew that none of his top five projects and all 10 of an IT manager's projects were behind schedule. Executive management believed the true challenge would be determining appropriate metrics for measuring a project's schedule, quality, and budget data.
- *Portfolio management of projects.* When a PMO was reviewing project portfolios or individual projects, all of the plans were at different levels of detail and accuracy. For example, some plans included only milestones with key dates, while other plans had too much detail. The key issue became "What is the correct balance of information that should be included in a plan, and how can all plans provide a consistent level of accuracy across all projects?" Even the term "accuracy" was not consistent across the organization.

- *Prioritization of projects and resources.* In one company, there were no mechanisms in place to prioritize projects throughout the organization, and this further complicated resource assignment issues in the organization. For example, the CIO had his top five projects, one executive had his top 10 projects, and an IT manager had his top 10 projects. Besides having to share project managers and project resources across all of these projects, there was no objective way to determine that the CIO's #3 project was more/less important than an executive's #6 project or an IT manager's #1 project. Therefore, when competing interests developed, subjective decisions were made, and it was challenging to determine whether the right decision had been made.

- *Shared accountability for success and failure.* The organization's projects traditionally were characterized as single-resource, single-process, and single-platform projects. Today almost every project was cross-team, cross-platform, and cross-process. This new model not only increased the complexity and risk for many projects but also required increased accountability by the project team for the success/failure of the project. Unfortunately, the organization's culture and people still embraced the old model. For example, if one team was successful on its part of a project and another was not, the attitude would be "I am glad I was not the one who caused the project to fail" and "Even though the project failed, I succeeded because I did my part." While there was some merit to this, overall, the culture needed to be changed to support an environment where "If the project succeeds, we all succeed" and vice versa.

- *Measuring project results.* Many of the projects that were completed were approved based on process improvements and enhanced efficiency. However, after a process improvement project was completed, no programs were in place to determine whether the improvements were achieved. In fact, because the company was experiencing double-digit growth annually, the executive team questioned whether approved process improvements were truly scalable in the long term.

- *Integrating multiple methodologies.* Application development teams adopted the software development methodology and agile methodology for software development. Both of these methodologies had excellent approaches for delivering software components that met quality, budget, and schedule objectives. The challenge the organization faced was whether components from both methodologies could be adapted to projects that were not software development related and, if so, how this could be accomplished. This debate had elevated to upper management for resolution, and upper management had been reluctant to make a decision one way or the other. This difference in views on how projects should be managed, regardless of whether the project was software development related or not, led to several different groups lobbying for others to join their efforts to support software development methodology and agile for all projects. Overall, the lobbying efforts were not adding value to the organization and were wasted efforts by key resources.

- *Organizational communications.* Although there was a lot of communication about projects throughout the organization, many shortcomings existed with the existing process. For example, one executive stated that when he had his monthly status meeting with his direct reports, he was amazed when a manager was not aware of another manager's project, especially if the project was getting ready to migrate into production. The existing process led many managers to react to projects instead of proactively planning for projects. Additionally, the existing communication process did not facilitate knowledge sharing and coordination across projects or throughout the organization. Instead, the existing communication process facilitated individual silos of communication.

- *Meaning of words.* A project was initiated from the staff level. The SOW contained numerous open-ended phrases with vague languages, such as "Develop a world-class control platform with exceptional ergonomics and visual appeal." The project manager and his team interpreted this SOW using their own creativity. There were mostly engineers on the team with no marketing members, and the solution ended up being technically strong but a sales/marketing disaster. Months were lost in time to market.

- *Problem with success.* A project was approved with a team charter that loosely defined the project's boundaries. During the course of the project, some early successes were realized, and word quickly spread throughout the organization. As the project moved forward, certain department managers began "sliding" issues into the project scope, using their own interpretation of the SOW, hoping to advance their own agendas with this talented group. The project eventually bogged down and the team became demoralized. Senior management disbanded the group. After this, management had real trouble getting people to participate in project teams.

- *Authority challenges.* A new cross-functional project team was assembled involving technical experts from numerous departments. The project manager was a consultant from an outside contractor. During this large project, resource conflicts with production schedules began to arise. Inevitably, the line managers began to draw resources away from the project. The consultant promptly reported pending delays due to this action, and the staff reiterated the consultant's concerns and the need for the organization to support the project. The struggles continued through the entire length of the project, creating stressful situations for team members as they tried to balance their workloads. The project finished late with significant cost overruns and indirectly caused a great deal of animosity among many of the participants.

- *Open-ended deliverables.* A project was launched to redesign and deploy the engineering change management system. The team received strong support throughout its duration. At a project closure meeting with the executive staff, the project manager presented the team's interpretation of the deliverables. Much to his surprise, the staff determined that the deliverables were not complete. In the end, this particular team worked on "spider webs" spawning off of their original SOW for over three years (the original closing presentation took place after nine months). The team was frustrated working on a project that never seemed to have an end, and staff members grew impatient with a team they felt was milking a job. The project management process at the company came under fire, threatening future efforts and staff support.

- *Cost overruns.* Soon after a major product renovation project was commissioned, the project manager reported that the cost of completion was grossly understated. Unfortunately, the marketing department, in anticipation of a timely completion, had already gone to the marketplace with a promotion blitz, and customer expectations were high in anticipation of the product's release. Senior staff was faced with a decision to either have a runaway cost issue to complete the project on time or endure loss of face and sales in the marketplace to delay the project's completion.

Despite all of these headaches, project management does work and works well. But is project management falling short of expectations? Some people argue yes because project management is not some magic charm that can produce deliverables under all circumstances. Others argue that project management works well and nothing is wrong except that executives' expectations are overinflated. Project management can succeed or fail, but the intent, commitment, and understanding at the executive level must be there.

2.16 TEN UGLIES OF PROJECTS[5]

Introduction

Project management methodologies, classes, and books are adequate at explaining the mechanics of running projects and the tools used to do so. Understanding these mechanics is essential, but it is the experience that distinguishes successful project managers. More specifically, it is the sum of all of the negative experiences that project managers have in their careers that teaches them what not to do. As Vernon Law explains, "Experience is a hard teacher because she gives the test first, the lesson afterwards."

In my many years of project management experience, I have come across several areas that consistently cause projects to experience difficulties. I call these the "uglies" of projects since these are the things that make projects turn ugly. These are also usually the things that, once recognized, are hard to fix easily.

This section will discuss the 10 project uglies and propose some solutions. There are definitely other uglies out there, but these 10 are the ones that seem to be the most common and have the biggest impact based on my experience.

The 10 Uglies

The following are the 10 uglies with a description of each and some symptoms that indicate that these uglies may be happening.

1. *Lack of maintained documentation.* Often when projects are in a crunch, the first thing that gets eliminated is documentation. Sometimes documentation is not created even when projects do have the time. Even when documentation is created properly, as projects continue to progress it is a rarity to see the documentation maintained.

---

5. Section 2.16 was provided by Kerry R. Wills, PMP, formerly director of portfolio management, Infrastructure Solutions Division, The Hartford. © 2022 by Kerry R. Wills. Reproduced by permission of Kerry R. Wills.

Symptoms
- Requirement documents that do not match what was produced.
- Technical documents that cannot be used to maintain the technology because they are outdated.
- No documentation on what decisions were made and why they were made.
- No audit trail of changes made.

These symptoms are a problem since documentation provides the stewardship of the project. By this, I mean that future projects and the people maintaining the project once it has been completed need the documentation to understand *what* was created, *why* it was created, and *how* it was created. Otherwise, they wind up falling into the same traps that happened before—in this case "He who ignores history in documentation is doomed to repeat it."

2. *Pile phenomenon*. "What is that under the rug?" is a question often asked toward the end of a project. The mainstream work always gets the primary focus on a project, but it is those tangential things that get forgotten or pushed off until "later," at which point there are several piles (swept under the rug) that need to be handled. I call this the "pile phenomenon" because team members think of it as a phenomenon that all this "extra" work has suddenly appeared at the end.

Symptoms
- Any work that gets identified as "we will do this later" but is not on a plan somewhere.
- Growing logs (issues, defects, etc.).
- Documentation assumed to be done at the end.

There is no "later" accounted for in most project plans, and therefore these items either get dropped or there is a mad rush at the end to finish the work.

3. *No quality at source*. Project team members do not always take on the mantra of "quality at the source." There is sometimes a mentality that "someone else will find the mistakes" rather than a mentality of ownership of quality. Project managers do not always have the ability to review all work, so they must rely on their team members. Therefore, the team members must have the onus to ensure that whatever they put their name on represents their best work.

Symptoms
- Handing off work with errors before reviewing it.
- Developing code without testing it.
- Not caring about the presentation of work.

Several studies show that quality issues not found at the source have an exponential cost when found later in the project.

4. *Wrong people on the job*. Project roles require the right match of skills and responsibilities. Sometimes a person's skill set does not fit well with the role that he or she has been given. I also find that work ethic is just as important as skills.

Symptoms
- Team members being shown the same things repeatedly.
- Consistent missing dates.
- Consistent poor quality.

As project managers, all we have are our resources. Not having the right fit for team members will result in working harder than necessary and impacts everyone else on the team who has to pick up the slack. There is also a motivational issue here: When team members are in the wrong roles, they may not feel challenged or feel that they are working to their potential. This has the impact of those persons not giving their best effort, not embodying a solid work ethic when they normally would, feeling underutilized, and so on.

5. *Not involving the right people*. The people who know how to make the project successful are the team members working on the project. Not involving the right team members at the right time can set the project up for failure before it begins.

Symptoms
- Having to make changes to work already completed.
- Constant scope changes from the customer.
- Lack of team buy-in to estimates.
- Lack of ownership of decisions.

Not involving the right people up front in a project always results in changes to work. Not involving team members in decisions and estimates causes them to feel like they have no control over their work or the outcomes of the project.

6. *Not having proper sponsorship*. Projects need internal and customer executive sponsorship to be successful. Sponsors act as tiebreakers and eliminate organizational politics/roadblocks that are holding up the project.

Symptoms
- Inadequate support from different areas of the organization and from customer stakeholders.
- Issues taking very long before being resolved.
- Decisions not being made efficiently.

Not having proper sponsorship can result in projects spinning their wheels. Also, when a change effort is involved, not having proper sponsorship can keep impacted employees from buying into a project (i.e. not cascading the messages from the top down to the masses).

7. *No rigor around process.* Almost every company uses a methodology for implementing projects. The success of these methodologies depends on the amount of rigor used in the project. Often, processes are not adhered to and projects run awry.

Symptoms
- Incomplete/nonexistent deliverables.
- Inconsistencies within the project.
- Lack of understanding of the project's big picture.
- Lack of repeatable processes (reinventing the wheel unnecessarily).

Processes are only as valuable as the rigidity placed on them. In some companies, there are too many project management methodologies used. Some are necessary due to the varying nature of work, but basic project management practices and principles (and even tools, i.e. using Project versus Excel) could easily be standardized but are not. When one manager has to transfer a project to another project manager, this creates an extra layer of complexity, because a common language is not being used between the two people. (It is like trying to interpret someone else's code when they have not followed the standards you have been using.)

8. *No community plan.* Project managers spend a significant amount of time on planning, estimating, and scheduling activities. If these results are not shared with team members, then they do not know what they are working toward and cannot manage their own schedules. This includes the communication of goals and items that are a big picture for the team.

Symptoms
- Lack of knowledge about what is due and when it is due.
- Missed dates.
- Lack of ownership of deliverables.
- Deliverables get forgotten.

Not having a community plan will result in not having an informed community. Having a shared plan and goals helps to build a cohesiveness and a greater understanding of how the work of each individual fits overall.

9. *Not planning for rework.* Estimation techniques often focus on the time that it takes to create units of work. What usually gets left out is the time spent on rework. By this, I mean work that was done incorrectly and needs to be revisited as opposed to scope management. When rework is required, it either takes the place of other work, which now comes in late, or it is pushed off until later (see ugly #2).

Symptoms
- Missed dates
- Poor quality

Never assume that anything is going to be done right the first time.

10. *Dates are just numbers.* Schedule is a major driver of project success. I am amazed at the number of people who think of dates as suggestions rather than deadlines. Because of interdependencies on projects, a missed date early on could ripple through the schedule for the remainder of the project.

Symptoms
- Consistently missed dates.
- Items left open for long periods of time.
- Incomplete/nonexistent deliverables.
- Lack of a sense of urgency on the project team.

Without structure around the management of dates, success requires a lot more effort. One other issue here is that of communication—these dates need to be communicated clearly, and people must agree that this is their target. Also, they must understand what is on the critical path and what has slack, so if they slip on a critical path item, they know there is an impact on the project or another project within the same program.

Possible Remedies

Upon analyzing the uglies, I observed that they are all interrelated. For example, not having rigor around processes (#7) can result in not having a shared plan (#8), which can result in people not caring about dates (#10), and so on. (See Figure 2–4.) I also realized that a few remedies could mitigate these uglies. The trick here is to proactively resolve them rather than react to them since by the time you realize that there is an ugly, *your project is already ugly.*

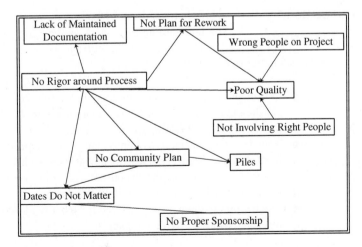

Figure 2–4. Observed interrelationships.

Proactive Management
Proactive management means spending the appropriate amount of time up front to minimize the number of fires that need to get put out later. Proactive management includes the following actions:

- Creation of a detailed plan.
- Always looking at the plan to see what is coming up and preparing for it.
- Thinking about the upcoming work and running down any issues that may be coming. I think of the team as running a marathon and it is my job to clear the road in front of them so they can keep on running.
- Setting up logistics. Something as trivial as not having a conference room booked in advance can cause a schedule delay.
- Lining up the appropriate people to be ready when the work comes their way.
- Know people's vacation schedules.
- Constant replanning as information becomes more available.
- Understanding what is going on with the project. I see so many project managers in the ivory tower mode where they find out about things about *their* project for the first time on a status report. By this time, as much as a week has gone by before the project manager is aware of issues.

There will always be unexpected issues that arise, but proactive management can help to mitigate those things that are controllable. This is an investment of time, in that you will spend far more time (and money) reacting to problems than you will focus on ensuring that the process is followed properly. This is difficult for some project managers because it requires the ability to always look ahead to the current state of the project rather than just focusing on the problem of the day. A key element of proactive management is having the ability to make decisions efficiently.

"Do It While You Do It"
Now that you are not reacting to fires, you can focus team members on maintaining their work as they go. This means staying focused on all aspects of the current work and thinking of implications. Characteristics of this include:

- Documenting as work is being done and not at the end. I am sure that this will get the knee-jerk "We don't have time" reaction, but I really believe (and have proved) that documenting as you go takes far less time than doing it at the end.
- Thinking of implications as things change on the project. For example, if a document changes, the owner of that document should think about any other deliverables that may be affected by the change and communicate it to the appropriate person.
- Check all work before passing it on to others.
- Use the process/plan as a guideline for what work has to be done. I have heard this referred to as "living the plan."

The result of this technique will be an even distribution of work across the project and minimal spikes at the end. Rather than the notorious "death march," the worst case could be considered an "uncomfortable marathon."

Empower the Team
Project managers must realize that project structures resemble an inverse pyramid where the project manager works *for* the team. It is the team members who do the work on the project, so the project manager's primary role is to support them and address obstacles that may keep them from completing their work. This includes:

- Involving team members in project planning, so they cannot say that they were just given a deadline by management.
- Asking team members how things are doing and then acting on their concerns. Asking for feedback and then doing nothing about it is worse than not asking at all because it suggests an expectation that concerns will be addressed.
- Celebrating the successes of the team with the team members.
- Being honest with the team members.

I am a big fan of W. Edwards Deming, who revolutionized the manufacturing industry. His 14 points of management revolve around empowerment of the team and apply very much to projects.[6] Excerpts are noted in Table 2–2 with my opinion of how they relate to project management.

Empowering the team will enable the project manager to share information with team members and will also enable team members to feel like they have control over their own work. The result is that each team member becomes accountable for the project.

Results of the Remedies
The results of applying these remedies to the uglies are shown in Table 2–3. I call my vision of the new way of doing things the "attractive state" since it attracts people to success.

TABLE 2–2. DEMING POINTS OF MANAGEMENT

Deming Point	Observation
8. Drive out fear, so that everyone may work effectively for the company.	This means that the iron-fist technique of project management is not such a great idea. People will be averse to giving their opinions and doing a quality job.
10. Eliminate slogans, exhortations, and targets for the workforce asking for zero defects and new levels of productivity. Such exhortations only create adversarial relationships, as the bulk of the causes of low quality and low productivity belong to the system and thus lie beyond the power of the workforce.	I take this to mean that project managers should not just throw out targets but rather involve the team members in decisions. It also means that project managers should look at the process for failure, not at the team members.
12. Remove barriers that stand between the hourly worker and his [or her] right to pride in workmanship.	This is my marathon metaphor—where project managers need to remove obstacles and let the team members do their work.
13. Institute a vigorous program of education and self-improvement.	Allow the team members to constantly build their skill sets.

6. See W. E. Deming, *Out of the Crisis: Quality, Productivity and Competitive Position* (Cambridge: MIT Press, 1982, 1986), pp. 23–24.

TABLE 2–3. ATTRACTIVE STATE CHARACTERISTICS

Ugly Number	Ugly Name	Ugly State Characteristics	Proactive Management	Do It While You Do It	Empower
1	Maintained documentation	• No record of what decisions were made • No explanation of why decisions were made • Cannot rely on accuracy of documents • Cannot use on future projects	• Updated documentation • Documentation will be planned for. • Anyone can understand decisions.	• Done during the project • No extra work at the end of the project	Team members will own documentation.
2	Piles	• Put off until later • May never get done	• Work is manageable. • If piles do exist, they will be scheduled in the plan	Will be worked on as people go, so they should never grow out of control	Are minimized because people will take ownership of work
3	Quality at the source	• No ownership of work • Poor quality • Expensive fixes	Better quality because you have spent appropriate time up front	Will be focused on as people do their work rather than assumed at a later time	Will be upheld as people take ownership of work
4	People fit	• Bad project fit	The project manager has the ability to recognize resource issues and resolve them before they seriously impact the project. Proper resource fit from the start is necessary.	Manage work so resource issues are identified early.	Other team members may take on work for failing colleagues.
5	People involvement	• Changes after work has been done • No ownership of work • No accountability for results	Involve the right people up front to avoid rework later.	Involve people during work rather than have them react to it later.	Empowered team members take ownership for work.
6	Sponsorship	• Cannot resolve problems • Caught up in organizational politics	Engaging stakeholders early will enable you to access their support when really needed.	Rapid and effective decisions as needed	May be improved due to better understanding of issues
7	Process rigor	• No rigor • Poor quality • Inconsistent work	• Proper rigor is the essence of proactive management. • Repeatable processes. • Looking ahead will ensure proper attention to process.	• Occurs as team members follow the process • Ensures that process steps are not missed	Ownership of work will enable better rigor around process.
8	Community plan	• No idea what is due and when. • Team members do not take accountability for work—the plan is for the project manager.	Project managers have the ability to share a plan and goals with the team.	Everyone is working on the same plan and knows where they are going.	• Everyone is informed—shared goals. • People can manage their own work.
9	Rework	• No plan for trade-off between doing other work or fixing issues	Anticipating areas where there may be rework or scope creep and working with key stakeholders early to address those planned for	• Will be accounted for as the team members work • By staying on top of the project, the magnitude of potential rework is known and can be planned for as needed.	Should be minimized due to motivation and ownership of work
10	Dates	• Dates do not matter • No accountability • Missing deliverables	Dates (and impacts of missing them) clearly communicated	Will matter, and items will be closed when they are due	Team members take ownership of dates.

Conclusion   Focusing on proactive management, keeping up with work, and empowering your teams are key to running a successful project.

There is nothing in this section that has not been written of or spoken of hundreds of times before. Nothing should sound new to a project manager. And yet we keep seeing the uglies over and over again. That leads me to conclude that it is the application of these concepts that is the challenge. I find that after I read a good paper or attend a management course, I have great enthusiasm to try out the new techniques, but at the first signs of trouble, I revert to my comfort zone. Therefore, I propose that there is a fourth remedy for the uglies—being conscious. This is nothing more than being aware of what is going on and how you are managing your project.

I come to work every morning a little earlier than the rest of the team so I can have my quiet time and think about what work needs to be done (not just for that day but in the upcoming days). I also give myself reminders that trigger my step-back-and-think mode. An excellent series that goes into this technique is the "emotional intelligence" books by Daniel Goleman.[7]

There will always be uglies on your projects, but if you are conscious of them, then you can identify them when they are happening, and you may be able to prevent them from throwing your projects into chaos. Best of luck.

---

7. See D. Goleman, *Working with Emotional Intelligence* (New York: Bantam Books, 1998).

3 Journey to Excellence

3.0 INTRODUCTION

Every company has its own motivations, or driving forces, as we discussed in Chapter 1, that impel the company to embark on a journey for excellence in project management. Some companies complete the journey in two or three years, while others may require a decade or more. In this chapter, we discuss the approaches taken by a variety of companies. Each company took a different path, but they all achieved some degree of excellence in project management.

Some companies embark on the journey at the request of their own workers, whereas other companies are forced into it by the actions of competitors and customers. In any event, there are driving forces that propagate the quest to excel in project management.

The driving forces for excellence, as discussed previously, include:

- Capital projects
- Customer expectations
- Competitiveness
- Executive understanding
- New product development
- Efficiency and effectiveness

Even the smallest manufacturing organization can conceivably spend millions of dollars each year on capital projects. Without good estimating, good cost control, and good schedule control, capital projects can strap the organization's cash flow, force the organization to lay off workers because the capital equipment either was not available or was not installed properly, and irritate customers with late shipment of goods. In non-project-driven organizations and manufacturing firms, capital projects are driving forces for maturity.

Customers' expectations can be another driving force. Today, customers expect contractors not only to deliver a quality product or quality services but also to manage this activity using sound project management practices. These practices include effective periodic reporting of status, timely reporting of status, and overall effective customer communications. It should be no surprise that low bidders may not be awarded contracts because of poor project management practices on previous projects undertaken for the client.

The third common driving force behind project management is competitiveness. Many of today's multinational companies view project management as a competitive weapon. Project-driven companies that survive on contracts (i.e., income) from external companies market their project management skills through virtually every proposal sent out of house. The difference between winning and losing a contract could very well be based on a firm's previous history of project management successes and failures.

The most common form of competitiveness is when two or more companies are competing for the same work. Contracts have been awarded based on previous project management performance, assuming that all other factors are equal. It is also not uncommon today for companies to do single-source procurement because of the value placed on the contractor's ability to perform. A subset of this type of competitiveness is when a firm discovers that outsourcing is cheaper than insourcing because of the maturity of their contractor's project management systems. This can easily result in layoffs at the customer's facility, disgruntled employees, and poor morale. This creates an environment of internal competition and can prevent an organization from successfully implementing and maturing in project management.

A fourth driving force toward excellence is executive buy-in. Visible and participative executive support can reduce the impact of many obstacles. Typical obstacles that can be overcome through executive support include:

- Line managers who do not support the project.
- Employees who do not support the project.
- Employees who believe that project management is just a fad.
- Employees who do not understand how the business will benefit.
- Employees who do not understand customers' expectations.
- Employees who do not understand the executives' decisions.

Another driving force behind project management is new product development. The development of a new product can take months or years and may well be the main source of the company's income for years to come. The new product development process encompasses the time it takes to develop, commercialize, and introduce new products to the market. By applying the principles of project management to new product development, a company can produce more products in a shorter period of time at lower cost than usual with a potentially high level of quality and still satisfy the needs of the customer.

In certain industries, new product development is a necessity for survival because it can generate a large income stream for years to come. Virtually all companies are involved in one way or another in new product development, but the greatest impact may very well be with aerospace and defense contractors. For them, new product development and customer satisfaction can lead to multiyear contracts, perhaps for as long as 20 or more years. With product enhancements, the duration can extend even further.

Customers will pay only reasonable prices for new products. Therefore, any methodology for new product development must be integrated with an effective cost management and control system. Aerospace and defense contractors have become experts in earned value measurement systems (EVMS). The cost overruns we hear about on new government product development projects often are attributed not necessarily to ineffective project management or improper cost control but more to scope changes and enhancements.

Improvement in the overall efficiency and effectiveness of the company is sometimes difficult, if not impossible. It often requires change in the corporate culture, and culture changes are always painful. The speed at which such changes accelerate the implementation of project management often depends on the size of the organization. The larger the organization, the slower the change.

Obviously, the most powerful force behind project management excellence is survival. It could be argued that all of the other forces are tangential to survival (see Figure 3–1). In some industries, such as aerospace and defense, poor project management can quickly lead to going out of business. Smaller companies, however, certainly are not immune.

Sometimes there are additional driving forces:

- Increase in project size mandated by the necessity to grow.
- Customers demanding faster implementation.
- Customers demanding project management expertise for some assurance of success completion.
- Globalization of the organization mandated by the need to grow.
- Consistency in execution in order to be treated as a partner rather than as a contractor.

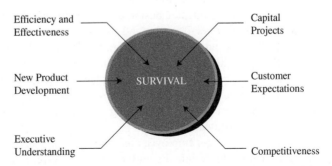

Figure 3–1. Components of survival.

3.1 STRATEGIC PLANNING FOR PROJECT MANAGEMENT

Over a period of five decades, project management has matured, from what was once considered just a fad that would soon disappear, into a strategic competency and career path necessary for the growth and survival of the firm. Project management is now being used in virtually every industry and in all parts of the business. We have matured to the point where we believe that we are managing our business as though it is a series of projects and where project managers (PMs) are expected to make both project decisions and business decisions. PMs are now considered to be businesspeople rather than just PMs.

Today, project management is recognized as a series of processes that can be used on every project, regardless of its length or complexity, the project's dollar value, or the project's exposure to risk. Yet the one part of the business where project management has been slow in being accepted, at least to now, has been in strategic planning execution projects. We can always argue that managing strategic planning execution projects is no different from managing any other type of project. While this argument may have merit, there are several important differences that must be considered. Specifically, PMs must think strategically rather than tactically or operationally, and they may have to change from traditional project management leadership to strategic leadership based on the complexity of the project.

Why Strategic Plans Fail To understand how project management can benefit strategic planning, it is important to understand why some strategic plans fail. Some of the common reasons for failure, as seen through the eyes of PMs, include:

- Neglecting to understand that strategies are achieved through completed projects.
- Neglecting to understand how the enterprise's environmental factors can influence senior management's vision of the future.
- Inadequate understanding of consumer behavior or the client's actions.
- Improper research prior to project approval.
- Poorly defined or ill-defined scope.
- Poorly documented business case resulting in the approval of the wrong project.
- Failing to get executive and stakeholder buy-in right from the start.
- Poor executive governance once the strategy begins to be implemented.
- Constantly changing the membership of the governance team.
- Overestimating resource competencies needed for project execution.
- Poor capacity planning efforts resulting in understaffed projects.
- Functional managers refusing to commit the proper resources for the duration of the strategic project.
- Failing to get employee commitment to the project.
- Failing to explain the importance of the project to the project execution team.
- Failing to explain to the execution team the incentives or financial benefits of working on this long-term project.

- Failing to understand the magnitude of the organizational change needed for the project to be a success.
- Unable to manage change effectively.
- Failing to consider the impact of changes in technology during the execution of the project.
- Poor estimating of time and cost.
- Having an execution team that is unable to work with ill-defined or constantly changing requirements.
- Poor integration of the project across the entire organization.
- Inadequate communications.

There are numerous other reasons for the failure of strategic planning execution projects. These causes could occur on any project, but on strategic planning execution projects, the potential damage to the firm could be quite severe.

Project Management: An Executive Perspective

With the ability to produce repeated successes on projects, it is no wonder that executives are now realizing the value of using project management for the execution of a strategic plan. There are several reasons why executives see value in using project management for these activities:

- Execution takes significantly more time than planning and consumes more resources. Executives do not have the time to spend possibly years coordinating and integrating work across a multitude of functional areas.
- Without a successful implementation plan, strategic planning cannot succeed.
- PMs can successfully manage the dysfunctional separation between planning and execution.
- Long-term strategic objectives must be broken down into short-term objectives to simplify execution. This can be done easily using project management tools and a work breakdown structure (WBS).
- Project management staffing techniques, possibly with the use of a project management office (PMO), can match the proper resources to the projects. This is critical when establishing a portfolio of projects.
- The organizational process assets (which include dashboard reporting systems) used in project management can keep senior management updated on project status.
- Strategic planning objectives, because of the long-time duration, are highly organic and subject to change. PMs know how to manage and control change.

Strategic Planning: A Project Management Perspective

Strategic planning is an organization's process of defining where and how it would like to be positioned in the future. The future may be measured in a three-year, five-year, or ten-year (or longer) window. The strategic plan is based on the firm's vision, mission, social consciousness, and values. Strategic planning requires an understanding of the firm and its environment. Executives, more so than PMs, have a better understanding of the enterprise

TABLE 3–1. THE *PMBOK® GUIDE* AND THE EXECUTION OF STRATEGIC PROJECTS

Area of Knowledge	Strategic Planning Project Impacts
Integration Management	The integration of the effort may very well span the entire organization both domestically and globally.
Scope Management	The scope can change as technology changes. The length of the project makes it imperative that an effective scope change control process exists. The scope baseline may appear as a moving window requiring constant updates.
Time Management	Matching the right people and their availability to the constantly changing scope will play havoc with scheduling. Losing people due to firefighting in functional areas may have a serious impact.
Cost Management	Predicting the true cost of the project is almost impossible. Reestimation must take place on a routine basis to make sure that the benefits and business value still exceeds the cost.
Quality Management	Customers' expectations of quality and competitive forces can cause major changes to the direction of the project.
Human Resource Management	The longer the project, the greater the likelihood that changes in resources will occur, possibly for the worse. Long-term motivation may be difficult to maintain.
Communication Management	Communication requirements can span the entire company. Changes in stakeholders will also have a serious impact on the communication plan.
Risk Management	The project may need to have a dedicated risk management team.
Procurement Management	The length of the project may make it difficult to accurately determine procurement costs up front.
Stakeholder Management	Because of the length of the project, the PM may end up interfacing with a different set of stakeholders at the end of the project and at the beginning.

environmental factors, namely products offered, markets served, present and future technologies, supplier base, labor markets, economic conditions, the political environment, and regulatory requirements.

Executives establish high-level objectives for *what they want done*. Often this is nothing more than a wish list that may or may not border on reality. The role of the PM is to determine *if it can be done*. This requires a clear business case for each project, a scope statement, and use of the WBS to break down the high-level objectives into sub-objectives, or lower-level objectives that are easier to understand and accomplish. If the PM and the project team believe that is can be done, then a formalized project action plan is created.

On the surface, it may appear that strategic planning execution projects can be treated like any other projects. However, if we look at the Areas of Knowledge in the *PMBOK® Guide*,[1] we can see some significant differences mostly attributed to project length. A few of these differences are shown in Table 3–1.

The Benefits of Project Management

Perhaps the primary benefit of using project management that makes it extremely attractive for strategic planning projects is to provide executives and clients with a single point of contact for

---

1. PMBOK is a registered mark of the Project Management Institute, Inc.

TABLE 3–2. BENEFITS OF USING PROJECT MANAGEMENT

Attribute	Benefit
Efficiency	Allows an organization to take on more work in less time without any increase in cost or degradation of quality.
Profitability	With all other things being equal, profitability should increase.
Scope changes	Allows for better up-front planning, which should reduce the number of scope changes downstream and prevent unwanted changes from happening.
Organizational stability	Focuses on effective teamwork, communication, cooperation, and trust rather than organizational restructuring.
Quality	Quality and project management are married, with both emphasizing effective up-front planning.
Risks	Allows for better identification and mitigation of risks.
Problem-solving	Project management processes allow for timely, informed decision-making and problem-solving.

TABLE 3–3. ADDITIONAL BENEFITS FOR STRATEGIC PLANNING EXECUTION PROJECTS

Attribute	Benefit
Alignment	Better alignment of projects to corporate strategic objectives.
Underperformance	Earlier identification of underperforming investments.
Capacity planning	Better analysis of corporate resource planning and availability of qualified resources.
Prioritization	Combining capacity planning efforts and project management allows for better prioritization of the portfolio of projects.
Risk mitigation	Allows for better mitigation of business risks by using more what-if scenarios.
Time to market	Allows for quicker time to market.
Decision making	More informed and timely decisions due to availability of essential information.
Efficiency and effectiveness	Allows the organization to work on more projects without increasing headcount.
Better information flow	Elimination of duplication of efforts by managers who are unaware of what others are doing.
Selection of projects	Better analysis of what is and what is not a good idea.

status reporting. Most of today's strategic planning projects are so complex that they cannot be managed effectively by one functional manager, who may have a conflict between functional duties and project duties. These projects require the coordinated effort of several functional areas, such as sales, marketing, engineering, and manufacturing. Without having a single point of contact for status reporting, executives would need to do the coordination and integration themselves, and it is highly unlikely that they would have the time to do this in addition to their other duties. Likewise, functional managers do not have sufficient time to manage their functional areas and perform integration work on various projects. The need for project management is quite clear.

There are many other benefits of using project management, some of which are shown in Table 3–2.

The benefits shown in Table 3–2 apply to just about all projects including strategic planning execution, complex, and traditional projects. But some additional benefits affect strategic planning execution projects more so than other types of projects. These are illustrated in Table 3–3.

Dispelling the Myths

When we look at Tables 3–2 and 3–3 and see all of the advantages, we must ask ourselves, "Why is there still resistance to the acceptance of project management, especially for strategic planning execution projects?" The answer is quite clear; there are still myths about the use of project management for activities related to strategic planning.

Myth #1: Project managers have strong technical knowledge but limited knowledge about the business. While it is true that historically PMs came out of technical disciplines and many even possessed master's degrees and doctorates in technical disciplines, today's PM has more of an understanding of technology than a command of technology but has an excellent knowledge of the business. Business knowledge is essential to bridge strategy and execution effectively. PMs who are considered "global" PMs must have a good understanding of the client's business as well as their own firm's business. This is a necessity to compete in a global marketplace. These global PMs are also being trained in stakeholder relations management, politics, culture, and religion since all of these topics can have an impact on the client's project.

We believe today that we are managing our firm's business as if it is a series of projects, where PMs are expected to make both project decisions and business decisions. Some companies are requiring their PMs to become certified in the company's business processes or to take coursework leading to certification as a business analyst.

Myth #2: Project managers should be assigned to a project after the project is approved and the business case is developed. Years ago, PMs were brought on board a project at the end of the initiation rather than at the beginning. We believed that, because PMs had limited knowledge of the business, they could not contribute anything worthwhile during the initiation process. After the projects were selected, PMs would be brought on board and told to begin execution. Today, PMs are brought on board at the beginning of the project initiation and selection process and are expected to make a valuable contribution because of their understanding of the business.

Myth #3: If we implement project management, project managers will begin making decisions that should be made at the executive levels of management. Strategic planning and the accompanying necessary decisions are made by executives, not by someone else for them. However, in some cases, strategic planning execution decisions may be made for executives rather than by them. Executives have always been fearful of having to empower PMs with authority and responsibility regarding project decision-making. This myth has been a great impediment to the successful implementation of project management.

The problem was partially resolved with the creation of the position of executive sponsor or project sponsor. PMs were allowed to make technical decisions, but project sponsors reserved the right to make any and all business-related decisions. This approach worked well for projects of reasonably short duration. But for strategic planning execution projects, which can be five to 10 years in length, the number of decisions that must be made can be overwhelming. Therefore, to overcome this myth, it is beneficial to clearly define the empowerment of the PM with regard to responsibilities and decision-making authority.

Myth #4: Project managers do not know how to use the organization process assets effectively for controlled measurement systems needed for informed decision-making to take place. For the past five decades, the two primary metrics used by PMs for status reporting purposes were time and cost. This was because of the rule of inversion, which states that we often select the easiest metrics to measure and report, even though they may not provide us with a clear picture of the health of the project. Time and cost alone cannot predict the success of a project or whether the value will be there at the completion of the project. This is particularly true for strategic planning execution projects.

There are seminars in the marketplace today on measurement techniques. There are also textbooks on measurement techniques, which argue that anything can be measured if only the information at your disposal is understood. The result has been the creation of additional metrics for project management. There is a belief that we should consider the following as core metrics for today's projects:

- Time
- Cost
- Resources
- Scope
- Quality
- Action items

These core metrics apply to all projects, but additional core metrics must be added based on the size, nature, scope, and importance of the project. Because strategic planning implementation projects can be long in duration, significant changes can take place. Therefore, we must allow metrics to change over the course of the project. Establishing a set of core metrics that can be used on every project may be difficult.

Ways That Project Management Helps Strategic Planning

There are often special situations where project management can significantly benefit an organization. In a company that manufactures household appliances, each functional area was allowed to perform its own strategic planning. The problem occurred when functional units had to work together on the same project. In this company, new products were introduced at trade shows, and there were two trade shows each year. Missing a trade show product launch could easily result in lost revenue for six months until the next trade show.

The launching of new products was the highest priority in marketing's strategic plan. Research and development (R&D), in contrast, had more than 300 projects in the queue. The new products that marketing needed for trade shows were low on the R&D list of priorities. Battles between marketing and R&D ensued.

In another company, marketing was allowed to prioritize projects as part of its strategic planning activities. For each project, marketing also prioritized the attributes of the project/product that had a direct bearing on the way the product would be advertised and marketed. But when the project/product went into manufacturing, the manufacturing people often had a different set of priorities for the attributes. Battles over priorities between marketing and manufacturing ensued.

In both examples, the issues were resolved when project management personnel requested that the organization create a single priority list for all company projects. The result was that R&D, engineering, and manufacturing would meet once every three months and come to an agreement on the priorities of the projects. However, there were too many projects in the queue to prioritize each one. The decision was then made that only 20 projects at a time would be prioritized. This greatly benefited the project staffing process because everyone was now working off the same priority list.

Another effective use of project management is gap analysis and gap closure. Gap analysis is used to strengthen a company's competitive position or to reduce the competitive position of competitors by reducing gaps. Projects are established to take advantage of best practices and lessons learned on other projects, by which gaps can be compressed. The gaps can be:

- Speed with which new products are introduced (time to market).
- Competitiveness on cost.
- Competitiveness on quality.
- Introduction of new technology or product performance.

Strategic Project Management Leadership   We have shown some of the ways that project management can benefit the execution of a strategic plan. For this to happen, PMs may need to change their leadership styles from traditional project management leadership, which focuses heavily on situational leadership oriented toward the project team, to a strategic project management leadership style, where the end result can affect organizational change across the entire company.

"Strategic leadership" is a term usually reserved for the seniormost levels of management. It entails an executive's ability to express a strategic vision for the company's future and then motivate or persuade the organization to acquire or follow that vision. Strategic leadership requires the development of action plans, and this is where project management takes on paramount importance. Visions do not help much unless plans can be developed and implemented to make the visions a reality. Managing projects that involve the implementation of a strategy are significantly different from managing traditional projects. Unlike action plans for traditional projects that are based on well-defined SOWs, the PM has to develop action plans that may be based on complexity, ambiguity, uncertainty, and volatile knowledge. Because of the large number of unknowns and their ability to change constantly, PMs must understand that managing these projects requires consequential decisions that must involve the managers who have ultimate control of the resources for executing these decisions.

If the projects require certain degrees of innovation, leadership skills must be designed around getting the team to be innovative and creative. Brainstorming and problem-solving sessions could occur each week. Facilitation skills are also a necessity. The leadership skills needed for long-term innovation projects may be significantly different from the skills needed to provide a client with a simple deliverable.

To be effective in strategic project management leadership, PMs must realize that they are now managers of organizational change and, as such, may have to build prepared minds on a large scale. Each PM, and possibly the entire team, must now function as a cheerleader and enforcer at the same time in order to get people across the entire organization to agree on a common sense of purpose. For these types of projects, the project team is usually referred to as a strategy support team (SST). For this team to function effectively, its members must be willing to coach and guide the strategy process as it unfolds. The most difficult challenge for this team will be on projects that require organizational change. Significant roadblocks must be overcome. The SST members must also be innovators and change agents, capable of seeing the big picture and thinking strategically rather than operationally or tactically. They must give up short-term thinking and focus on the distant future.

The main objective of strategic leadership is to make the organization more strategically productive and inventive as well as efficient and effective. Workers must be encouraged to follow their own ideas when feasible and provide feedback on technical or behavioral innovations that can be captured through lessons learned and best practices. Lessons learned and best practices allow companies to focus only on the right energies that will help them profit in the long run.

Traditionally, PMs were expected to capture project-related knowledge and send it to the PMO for analysis and storage. But with strategic leadership, more business-related knowledge must be captured and fed into a corporate knowledge repository.

Strategic Project Management Leadership Traits

For over four decades, we examined the skills needed to be a PM and provide effective project leadership. The analyses were made focusing on traditional projects, which may last 12 to 18 months or less. In addition, the SOW is reasonably well defined, many of the people may be full time but only for a few weeks, and the outcome of the project may affect only a small number of people. The potential long-time frames for strategic projects are now forcing us to revise some of these leadership skills.

It is almost impossible to create an all-inclusive list of the competencies required for a PM to provide strategic project management leadership. However, some of the possible changes in leadership that will be needed can be shown (see Table 3–4). There is a valid argument that all PMs need these skills, but they may be more critical on strategic projects.

The Project Manager as a Manager of Change

The role of the project is continuously evolving. As stated in Table 3–4, some strategic planning projects are designed for organizational change, and the change may affect the entire company worldwide. An example might be the implementation of a new corporate-wide security system, information system, or secured email system.

The new question becomes "Who will manage the implementation of the change once the project is ready for implementation?" Historically, PMs created the deliverables, and someone from the management ranks took the lead for the implementation of the change. Today, PMs are being asked to take the lead role or at least participate in organizational change management. There may also be a project sponsor from the seniormost levels of management with specialized knowledge in organizational change management.

For years, some of us have been managing strategic projects without realizing that we were doing so, and we may not have recognized the possible need for a different leadership style. But as the use of project management began to grow in terms of its

TABLE 3–4.   DIFFERENCES BETWEEN TRADITIONAL AND STRATEGIC PROJECT MANAGEMENT LEADERSHIP STYLES

Traits	Differences
Authority	From leadership without authority to significant authority.
Power	From legitimate power to judicious use of power.
Decision making	From some decision-making to having authority for significant decision-making.
Types of decisions	From project-only decisions to project and business decisions.
Willingness to delegate	The length and size of the project force PMs to delegate more authority and decision-making than they normally would.
Loyalty	From project loyalty to corporate vision and business loyalty.
Social skills	Strong social skills are needed since we could be working with the same people for years.
Motivation	Learning how to motivate workers without using financial rewards and power.
Communication skills	Communication across the entire organization rather than with a selected few.
Status reporting	Recognizing that the status of strategic projects cannot be made from time and cost alone.
Perspective/outlook	Having a much wider outlook, especially from a business perspective.
Vision	Must have the same long-term vision as the executives and promote the vision throughout the company.
Compassion	Must have much stronger compassion for workers, since they may be assigned for years.
Self-control	Must not overreact to bad news or disturbances.
Brainstorming and problem-solving	Must have very strong brainstorming and problem-solving skills.
Change management	Going from project to corporate-wide change management.
Change management impact	Going from project to organizational change management effects.

application to strategic planning execution projects, we may need to conduct more research on the specific leadership skills needed. We are in the infancy states of strategic project management applications, but we do expect this trend to take hold over the next decade or longer.

3.2 ROADBLOCKS TO EXCELLENCE

"Excellence" is often defined as a stream of successfully managed projects. Some people believe that excellence in project management is really a strategic goal that can never be achieved because not all projects we work on will be successful. Therefore, perhaps a better definition might be a continuous growth in successfully managed projects where the ratio of project successes to project failures increases year after year.

Regardless of how good we become in project management, there are always roadblocks that can appear, causing us to revisit how we manage projects. The most common roadblock is when something happens that removes people from their comfort zone and requires that they work differently. Based on the size of the roadblock, the definition of excellence may change, and companies may have to revise a lot of the processes they put in place.

Examples of roadblocks might include:

- The company decides to track additional metrics other than just time, cost, and scope. Workers are worried about how they will be required to identify, measure, track, and report the additional metrics.
- Your company has reasonably mature project management processes. Your customers are now demanding that you use agile and scrum approaches to your projects, and this requires that workers learn new processes.
- Your company decides to use agile and scrum on some projects but the traditional waterfall methodology on others. Workers become confused as to which approaches to use and when.
- A new technology has appeared that will change the way project estimating takes place. Some workers are apprehensive about being held accountable for new estimates.
- Your company either acquired another firm or participated in a joint venture with another firm. Both firms have different approaches to project management, and a common ground must be developed.
- The company restructures and changes some worker roles and responsibilities.

Obviously, other roadblocks could be discussed.

3.3 PAIN POINTS

Background

All too often, people of all ages enter the world of project management and are usually impressed with what they see and hear, especially in the benefits that project management can bring to a company along with possibly a sustainable competitive advantage. But what most people do not see or hear about are the pain points that companies had to endure and overcome to achieve their current level of project management maturity and excellence. Almost everyone involved in project management can most likely identify pain points that they needed to overcome.

Roadblocks are usually major obstacles that may require detours from a desired path. Pain points are frequently just small annoyances that can be overcome quickly. Pain points can become major roadblocks if not resolved in a timely manner.

Pain points began to appear shortly after the end of World War II when the Department of Defense (DoD) invested heavily in the number of projects given out to aerospace and defense contractors. DoD was the pioneer in developing many of the processes, tools, and techniques that became the foundation elements for today's project management approaches. Project management was also used in the construction industry although DoD was seen as the primary creator of project management practices.

For many of the contractors, new DoD projects brought with them new types of pain points. Even with the birth of the Project Management Institute (PMI), and the eventual publication of updated editions of the *PMBOK® Guide* and *Standards for Project Management,* many of the pain points were still there. And as new approaches to the processes, tools, and techniques for project management appeared, they were accompanied by the emergence of new types of pain points that needed to be resolved and mitigated.

Understanding Pain Points   Historically, pain points have been used most frequently by business analysts or marketers to identify recurring problems, annoyances, or other obstructions that may be inconveniencing their customers. Identifying pain points thereby provides you with the opportunity to sell products or services to your customers to relieve the pressure or distress caused by the pain points and position your company as a pain point eliminator.

Today, pain points are also being identified in the way that contractors perform the processes needed to satisfy both their company's business model and deliverables expected by their clients. Pain points can be identified in all forms of management and leadership activities, including project and program management, and can create the outcomes shown in Figure 3–2. Pain points can create brick walls that impede successful project management practices and can lead to project failure if not mitigated.

The challenge is in the identification and agreement that some repetitive occurrence is a pain point. What one person perceives as a pain point, another individual may not see it in the same light and not believe that it constitutes an issue needed to be addressed.

Pain points may appear as simple problems, but a deeper analysis may lead to the need for establishing pain point categories. In a project management environment with emphasis on the processes, tools, and techniques, typical categories might be:

- *Display Pain Points:* Determining the best mixture of metrics and KPIs necessary to provide a true meaning of the project's status.
- *Budgetary Pain Points:* Determining the best way to predict the expected cost of the project and potential scope changes that may occur.

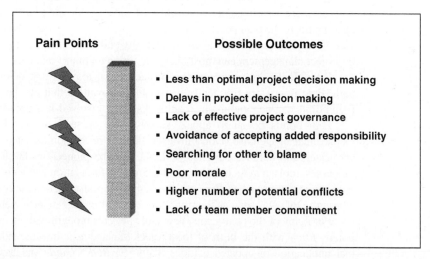

Pain Points   **Possible Outcomes**

- Less than optimal project decision making
- Delays in project decision making
- Lack of effective project governance
- Avoidance of accepting added responsibility
- Searching for other to blame
- Poor morale
- Higher number of potential conflicts
- Lack of team member commitment

Figure 3–2.   The impact of Pain Points on Project Management.

- *Scheduling Pain Points:* The inability to eliminate waste and unproductive time from the schedules.
- *Governance Pain Points:* The lack of a structured help line or decision-making process for the timely identification and resolution of critical issues.
- *Methodology Pain Points:* The belief that all of the projects can fit into a one-size-fits-all methodology.

Pain Point: Customer Communications as Seen by the Contractor

During most of the 1900s, corporate strategic planning was built around the product-market element, namely products offered, and markets served. This implied that the marketing and sales organizations were the dominant players in the formulation of a strategic plan. Most companies appeared to be sales- or marketing-driven because these people were seen as being the primary revenue generators.

Salespeople believed that they "owned" the rights to their customers and should be the primary communications link with the customers even though many companies assigned PMs to the activities to support the salespeople. PMs communicated with the sales team, and they relayed the information to the customers. Senior management allowed this to continue although senior management often participated in the communications processes. The relationship between the sales personnel and the customers was seen as a strategic necessity and almost everyone dismissed the fact that it could also become a serious pain point.

Pain Point: Customer Communications as Seen by DoD

As the number of contracts began to increase, DoD recognized that talking primarily with the sales force about project issues was time-consuming and not necessarily productive. Whenever DoD had technical questions that needed to be answered, the sales force responded that they would eventually get back to DoD with answers. This often took weeks for an effective response to the customers.

DoD wanted to talk directly with the PMs and the technical people that could provide them with immediate answers to their technical questions. The sales force did everything possible to prevent this from happening because they were afraid that they would then have to share yearend bonuses with PMs.

Management persisted in supporting the sales force as the primary communications link. DoD then made the decision to invoke the Golden Rule by holding up the government checkbook to the eyes of senior management and stating, "He who has the gold, rules!" Senior management now felt the pain of possibly having to restructure the company to satisfy the needs of their customers or lose business.

Pain Point: Project Management Becomes a Career Path Position

In the early years of project management, most companies did not treat project management as a career path position. Project management was seen as a part-time position to be filled on a temporary basis while performing one's normal functional responsibilities. PMs had their performance reviews conducted by their functional managers and their performance evaluations were often heavily based upon their overall service to their functional unit rather than based upon the success or failure of the project they were managing on a part-time basis.

Government agencies began insisting that project management become a career path position. This created a serious pain point for senior management. Traditionally, contractors were awarded lucrative contracts through competitive bidding. Technical personnel, and sometimes the PMs, would provide input to the sales force that had the responsibility for preparing the final competitive bidding package to be submitted to the client.

The relationship between the sales force and PMs was becoming tenuous. In 1970, an aerospace and defense contractor made the decision to fire everyone in marketing and sales except for one marketer. PMs were then asked to write the proposals and perform the selling to the clients. When asked what the primary skills were for selecting someone to fill a PM position, an executive commented "communications and effective writing skills."

The government's pressure of invoking the Golden Rule had forced senior management into making project management a career path position. But the pain point was still there. Because many of the contractors were in the aerospace and defense industry, most of the PMs were engineers with advanced degrees in some technical field and often possessing poor or marginal communication and writing skills. Aerospace and defense contractors created technical writing departments to assist the PMs with proposal preparation.

Making project management a career path position also brought with it the pain of having PMs that had never taken any courses in interpersonal skills training or effective leadership. On short-term projects, management endured the pain and instructed the PMs to remove people from the project as quickly as possible after the team members completed their job to keep the project costs as low as possible. It was not uncommon for the PMs to have very little contact with the team members and to rely heavily upon functional managers to provide the necessary day-to-day leadership and direction to their functional employees assigned to project teams.

On long-term projects or those that may have behavioral issues, several aerospace companies assigned organizational development (OD) specialists to assist the PMs. Several years ago, when one of these OD specialists was asked what his role was on projects, he stated that his job was to help the PM resolve conflicts and other behavioral issues. He also stated that he knew very little about the technology on any of the projects and that his role was mainly mitigation of behavioral issues.

Allowing engineers with advanced degrees to manage long-term high-technology projects brought with it additional pain points. Some highly technical PMs viewed their projects as a chance for fame and notoriety. As such, their goal was to exceed the specifications rather than simply meet them, regardless of the cost overrun. Military personnel that provided the funding for the cost overruns knew this was happening and believed that, after their two to three-year tour of duty was completed on this assignment, they would be transferred to another assignment shortly and their replacement would then have to deal with the cost overruns which were often greater than 300–400 percent.

Pain Point: Project Sponsorship

Making project management a career path position was certainly recognized as a pain point that management knew would happen.

However, there was an accompanying pain point that needed to be addressed quickly, namely the chance that the PMs would make some decisions that were reserved for the senior levels of management. How could senior management control or influence the decisions made by the PMs, including those decisions that have a serious impact on the business or may lead to unwanted cost overruns?

The answer was in project governance by providing a project sponsor to oversee and participate in project decision-making. The pain point was mitigated by creating project sponsorship positions for all the critical projects and staffing the positions with senior and middle-level managers.

As the number of projects increased, senior managed tried to decrease the number of projects they personally sponsored because it became a time-consuming effort that distracted them from their other duties. Unfortunately, this brought to the surface another pain point on customer (specifically DoD) interfacing. Many of the military officers that controlled the funding for the projects did not consider the PMs and even some of sponsors from middle-level management and lower as being equal to them in rank and status. The motto was "Rank Has Its Privileges," and government personnel persisted in wanting to communicate only to the senior level of management believing that these people were equal to them in rank. As such, senior management was forced to remain as sponsors on many critical projects.

Another related pain point to sponsorship was the impact that a failed project could have on the sponsor's career. If an executive believed that having their name attached to a project that could potentially fail would damage their career path opportunities, they would assign people beneath them in rank as sponsors. If they were still forced to remain as sponsors, they would create a plan whereby others could be blamed if a failure occurred.

As an example, two executives in a telecom company acted as sponsors on two "pet" projects to create new products that they believed would increase sales and generate larger executive bonuses. Both projects required innovation and were costly endeavors. At each of the project review meetings, the PMs recommended canceling the projects because of the significant costs of developing two products that might not generate the expected revenue streams. This occurred at all of the review meetings. Both sponsors believed that canceling the projects could impact their careers because of the funds expended. Therefore, at each project review meeting, they allowed the projects to continue to the next gate review meeting. Eventually, both projects were completed at significant cost overruns. To avoid the embarrassment of having to explain what happened when the marketplace was not interested in purchasing these products, the sponsors promoted the PMs for having developed the products but them blamed marketing and sales personnel for not finding customers for the new products. This reduced and even eliminated the sponsorship pain points.

Pain Point: Standardization of Processes

As the number of projects funded increased, DoD realized that effective control over the continuously increasing number of projects was troublesome. Each contractor had their own way of performing the work and their own reporting systems. DoD then had the painful experience of having to interpret the data in each status report.

The solution was the creation of the EVMS and the establishment of standardized status reporting. DoD developed a series of publications encouraging contractors to use the EVMS and government-related life-cycle phases and major milestones. Contractors initially saw this as a new pain point but soon realized its potential benefits. Companies created a one-size-fits-all methodology that was to be used on all projects. This provided the necessary standardization that executives wanted and made sponsorship and control easier. Performance reviews of PMs and team members were heavily based upon how well they followed and used the forms, guidelines, checklists, and templates associated with the one-size-fits-all approach rather than the success or failure of the project.

Pain Point: Finding Other Applications for Project Management

From the 1970s to the turn of the century, books and articles appeared citing the benefits of implementing project management correctly and the successes achieved. But what most people failed to realize was the type of projects that were analyzed to make this determination.

Historically, project management was used on traditional or operational projects that were initiated with a well-defined statement of work (SOW) and WBS. Projects for government agencies and most customers were initiated with well-defined requirements. PMs were taught that they should refrain from planning, scheduling, and pricing out a project unless the requirements were very well defined, and techniques were published for finding ways to improve the requirements definition processes. The use of the EVMS worked well on the traditional or operational projects that possessed clear requirements.

But what about the strategic projects, such as those involving innovation activities, that were initiated based upon an idea rather than rigid requirements, and were therefore subject to possible continuous changes? As stated previously, executives were afraid that PMs might make decisions that were reserved for the senior levels of management. With well-defined requirements and the use of project sponsors, these risks were minimized. Therefore, PMs would be allowed to manage the traditional or operational projects and functional managers, whom executives trusted more so than PMs, would manage the strategic projects.

Giving functional managers control of most of the strategic projects seemed like a good idea, but it eventually brought to the surface a serious pain point that was hidden from view. Functional managers in many companies received year-end bonuses based upon the success of the company or their functional unit over the past twelve months. As such, functional managers were retaining their best resources for the short-term projects that affected their bonuses, and the long-term strategic projects were suffering.

There were several reasons why this pain point had been hidden for years. First, functional managers were given the freedom to use whatever processes and techniques they wished to use on their projects. Executives had trust in their functional managers. This allowed them to fudge or alter the true status of some of their projects in their reporting to senior management. Second, even if the functional managers used the

EVMS, which they mostly avoided, the reporting was on time, cost, and scope and no information was provided on the quality or capabilities of the assigned resources. Forecast reports, using KPI information, were often an exaggeration rather than reality.

21st-century Pain Points Appear

The pain points discussed previously focused on pain points that appeared in the past. Simply because we have been using project management practices for decades, and there have been many successful continuous improvement (CI) efforts in project management, does not mean that, in the future, the same or new pain point will not occur.

By the turn of the century, there was a significant growth in the number and types of projects that companies needed to implement for a sustainable and successful future, as seen in Figure 3–3. The greatest change was in the growth of strategic projects as companies realized that continuing to conduct business the same old way was an invitation for disaster.

The importance of strategic projects became apparent. Executives asked two questions:

- Do we manage strategic projects the same way we manage traditional projects?
- Do we use the same people, processes, tools, and techniques?

Because strategic projects were significantly more complex than traditional projects, other approaches such as using flexible agile and scrum methodologies became obvious. Strategic projects worked better using flexible methodologies and required different skill set training for PMs.

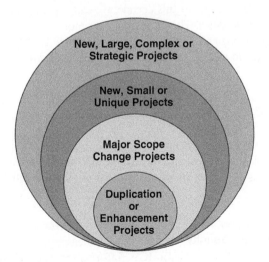

Figure 3–3. New Types of Projects.

Pain Point: The EVMS Becomes a Dinosaur

For more than 50 years, the EVMS has been used with reasonable success supporting the one-size-fits-all methodology, but primarily on traditional or operational projects. The EVMS focuses mainly on time and cost metrics that can be looked at in various ways to create several approaches for determining project status. But as project management practices are being applied to other types of projects, additional information will be required so that management can make decisions based upon facts rather than guesses and intuition. The EVMS system may not become entirely extinct as did dinosaurs, but it is inevitable that it will undergo radical changes.

Pain Point: Executive Support for the New Metrics Management Programs

Sharing information on new metrics, especially business and strategic metrics, requires executive support and discarding the old belief that "information is power." Some executives will not take ownership for a new metrics management system for fear of looking bad in the eyes of their colleagues if the metrics reporting system is not accepted by the workers or fails to provide meaningful results. Executives will not support a metrics management system that looks like pay for performance for executives that can affect their bonuses and chances for promotion. If they were to support such a system, the executives may then select only those metrics that make them look good.

Pain Point: The Growth of New Flexible Methodologies

The growth and acceptance of agile and scrum methodologies have made some people believe that these two flexible methodologies are the "light at the end of the tunnel." It is more likely that these two flexible approaches are the beginning of things to come, and a potential pain point executives must face.

By the end of this decade, we can expect to have 20–30 different types of flexible methodologies. The days of having a one-size-fits-all methodology have disappeared. At the beginning of each project, the PM will look at the type of project he/she has been asked to manage, and then determine the best flexible approach to be taken.

3.4 HITACHI LTD.

When strategic planning for project management is done correctly, the beneficial use of project management can permeate the entire company, and project management may be integrated into virtually all business areas. An excellent example of this is seen in Hitachi.

Initiatives to Strengthen Project Management Capacity at Hitachi[2]

Hitachi Group's business covers a wide variety of fields encompassing the development, manufacturing, sales, and provision solutions for ICT [information and communications technology] systems,

2. Section 3.4 has been provided by the PM Technical Committee, Hitachi Ltd. © 2017 by Hitachi Ltd. Reproduced by permission.

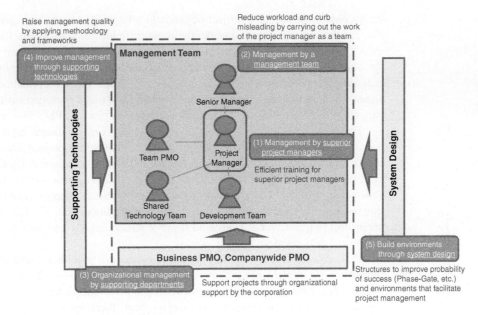

Figure 3–4. Project management support components.

power systems, social infrastructure and industrial plant systems, high-functional materials, rail systems, elevators and escalators, automotive products and components, construction machinery, digital media, and consumer products, as well as related consulting and services. For each line of business, improvements to engineering technologies to support the quality of the business and improvements to project management are essential. This is the context for the initiatives to strengthen the project management capacity for every line of business.

From the viewpoint of project management, the five perspectives indicated in Figure 3–4 are the support components for taking a project to its successful completion.

Perspective (1) refers to initiatives to provide continuous and effective training for superior PMs. Since the success or failure of a project depends to a large degree on the capabilities of the PM, it is important to train superior personnel. To do so, it is necessary not only to build the educational systems for training personnel but to train according to individual characteristics. In terms of PM skills, the Project Manager Competency Development Framework analyzes the relationship between the PM's individual characteristics and performance[3] and carries out initiatives to boost personnel training by leveraging the individual characteristics of PMs.[4]

Perspective (2) refers to support for team building and aims to reduce the PM's workload by building a team to carry out the work of the PM. As mentioned earlier,

---

3. Takeshi Yokota et al., "Strengthening of Personnel Training Process of Project Managers," *Journal of the Society of Project Management* 15, no. 2 (2013).

4. Takafumi Kawasaki et al., "Practice Action of Project Managers: The Difference between Highly Competent PM and Moderately Competent PM," *Proceedings of 13th National Conference of the Society of Project Management*, 2007, pp. 373–377. http://ci.nii.ac.jp/naid/110007602747; Hitoshi Yamadera et al., "Relations between Achievement and Characteristics of Project Managers," *Proceedings of 16th National Conference of the Society of Project Management*, 2009, pp. 209–212. http://ci.nii.ac.jp/naid/110007602894.

the success or failure of a project depends to a large degree on the capabilities of the PM, but the larger the project scale, the more difficult it is for a single PM to cover all areas. Therefore, the management skills of a project management team that includes not only the PM but also a senior manager, a PMO, a shared technology team, and a development team are required. Studies are under way of evaluation models for management structures to evaluate management skills as a management team, and not only from the viewpoint of the individual characteristics of a PM.[5]

Perspective (3) aims to support projects through organizational support at the corporation with the PMO or other organizations evaluating the project situation, offering advice, and carrying out assessments from a third-party standpoint. By developing an external understanding of the situation as the project progresses, risks that have gone undetected by those who are involved may be identified. By providing organizational support and risk assessment not only by the management team, including the PM for the project in question but from a third-party standpoint, the probability of success for the project may increase.[6]

Perspective (4) uses supporting technologies, methodologies, and frameworks to support project implementation activities. The support domain includes risk management, requirement identification, communication support, knowledge management, PMO support, and so on. In terms of risk management, there are initiatives to support risk identification and to design countermeasures in a range of business fields.[7] In terms of requirement identification support, requirement identification for clients is supported through ethnographic surveys based on human-centered design processes and initiatives such as building construction management systems based on the results.[8] For communication support, there are initiatives to identify management issues by visualizing communication about project progress using sensor systems.[9] In terms of knowledge management, there are methods for extracting empirical knowledge[10] and techniques

5. Akiyuki Onaka, "Model of Project Team Assessment to Make Projects Succeed," *ProMAC2010*, 2010.

6. Kenji Hatsuda et al., "PMO Information System as a Support of Project Management Office Activities," *Journal of the Society of Project Management* 5, no. 4 (2003): 28–31. http://ci.nii.ac.jp/naid/110003726282.

7. Toyama Minamino, "An Application of Modern Project Management 'IT' System Development Projects", *ProMAC2002*, 2002. Takeshi Yokota et al., "Development of a Contract Risk Assessment Support System (CRARIS)," *Journal of the Society of Project Management* 7, no. 3 (2005): 20–25. http://ci.nii.ac.jp/naid/110003726628; Takeshi Yokota et al., "Development of a Risk Management System for Construction Projects," *Journal of the Society of Project Management* 8, no. 5 (2006): 36–41. http://ci.nii.ac.jp/naid/110006278350; Takeshi Yokota et al., "Upgrade of Risk Management Technique for IT System Development Project," *Journal of the Society of Project Management* 14, no. 3 (2012): 25–30. http://ci.nii.ac.jp/naid/110009495477. Yoshinobu Uchida, "Development of the Risk Management System for Construction Projects," *ProMAC2011*, 2011.

8. Hisako Okada et al., "An Approach to Advance Construction Management System for Large-Scale Power Plant Projects," *Journal of the Society of Project Management* 15, no. 1 (2013): 8–13.

9. Hideyuki Maeda et al., "Visualization of Communication using the Team Activity Measuring System and Its Application to the Project Management," *Journal of the Society of Project Management* 12, no. 1 (2010): 5–10. http://ci.nii.ac.jp/naid/110007573280; Yoshinobu Uchida et al., "Development of a Project Review Technique Employing Risk Propagation Models," *Proceedings of 24th National Conference of the Society of Project Management*, 2014, pp. 105–110.

10. Yoshinobu Uchida et al., "An Approach of Knowledge Extraction via Empirical Failure Knowledge in Project Management," *Journal of the Society of Project Management* 12, no. 4 (2010): 27–32. http://ci.nii.ac.jp/naid/110007880184.

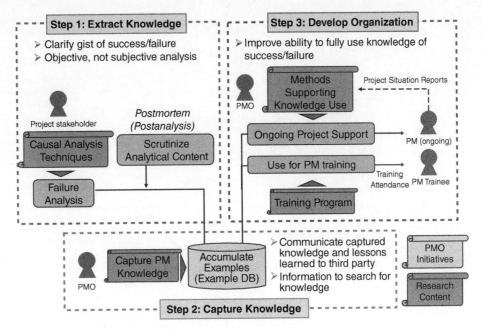

Figure 3–5. The across-the-organization circulation process for empirical knowledge in project management.

for using knowledge[11] as a system for circulating empirical knowledge gained from projects across the organization. The circulation of knowledge is implemented as illustrated in Figure 3–5 by collaboration with the system design of perspective (5). There are also information system initiatives to support not only the PM but also the PMO.[12]

Furthermore, initiatives associated with perspectives (3) and (4) are currently under way in relation to loss-cost management. Here, "loss-cost" means unnecessary or wasteful expenses. Reducing loss costs is a key corporate task in view of their direct relationship with corporate performance. At the Hitachi Group, we have long been working to reduce such loss costs, with a focus on hardware manufacturing. Such endeavors are also incorporated into IT projects, and the entire process from loss-cost tracking (visualization) to the analysis and drafting of response measures is being systematized and implemented as a means of loss-cost management.[13] Our support departments have

11. Yoshinobu Uchida et al., "Proposal for Risk Management Support Method Using Failure Knowledge," *Journal of the Society of Project Management* 7, no. 6 (2005): 3–8. http://ci.nii.ac.jp/naid/110006278374; Yoshinobu Uchida et al., "Proposal of Utilization of the Failure Experience in Project Management," *Proceedings of 15h National Conference of the Society of Project Management*, 2008, pp. 140–143. http://ci.nii.ac.jp/naid/110007602790.

12. Hatsuda et al., "PMO Information System as a Support of Project Management Office Activities."

13. Kenji Hatsuda et al., "Loss-Cost Management for IT Projects," *Proceedings of 28th National Conference of the Society of Project Management*, 2016, pp. 43–44.

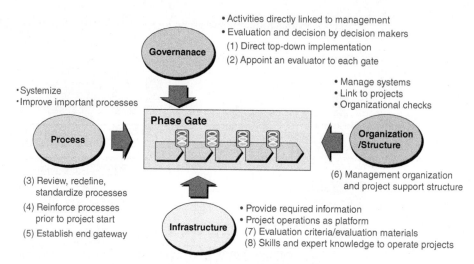

Figure 3–6. The Hitachi Phase-Gate Management Process.

been monitoring loss-costs, drafting/implementing measures for reducing loss costs incurred as an organization, and holding training and activities to raise awareness on the subject. Additionally, we have been developing technologies that help us keep track of loss costs, analyze their causes, and draft measures.[14]

Perspective (5) builds structures for increasing the probability of success as a system, certification systems for PMs, and structures for project governance. One of these structures is the use of phase-gate management (a structure that divides the product process into several phases and erects gateways to review whether or not the conditions have been met before moving to the next phase) illustrated in Figure 3–6 to make decisions about continuing or stopping projects.[15] By using phase-gate management, it is possible to optimize decision-making to lower risk, improve design quality, and maximize management gains.

Further, a business improvement project called D-WBS (Denryoku Work Breakdown Structure) is under way at Power Systems company,[16] with an eye to linking perspectives (1) to (5) in an organic fashion rather than treating them as independent from each other. The D-WBS Project has a platform (the D-WBS Platform) for promoting business synergy in relation to project management as shown in Figure 3–7 that is informed by management, knowledge, and a variety of other perspectives and constructs business processes that unite perspectives (1) to (5) on that platform.[17]

14. Ibid. Yoshinobu Uchida et al., "Proposal of Failure Prediction Method Employing Loss Cost Generation Mechanisms—Loss-Cost Management for IT Projects," *Proceedings of 28th National Conference of the Society of Project Management*, 2016, pp. 45–50.

15. Koji Okada et al., "Applying Phase-Gate Management for Diverse Business Types," *Journal of the Society of Project Management* 13, no. 6 (2011): 29–34. http://ci.nii.ac.jp/naid/110009425403

16. Tomoyuki Aoki et al., "The Case Study of Business Process Reengineering for EPC Project Management," *Journal of the Society of Project Management* 14, no. 6 (2012): 5–10.

17. Kazuhito Shibata and Natsuko Sato, "Development of Integrated Project Management Framework and Practical Platform for EPC Project in Power Plant Business," *ProMAC2015*, 2015.

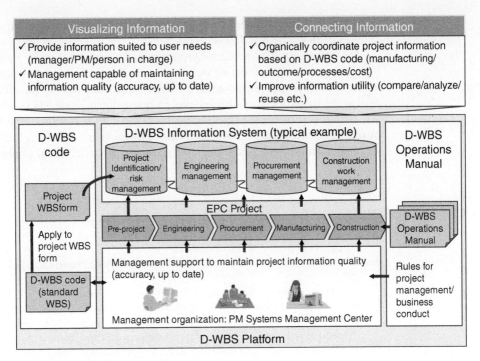

Figure 3–7. D-WBS project outline.

The abovementioned support components are required components for implementing projects without relying on business units. Even though each field of application has its own characteristics, comparison of project management techniques at Power Systems company and the Information and Telecommunication Systems Division and ICT Systems Division shows several aspects that can serve as useful references for both sides. Considering that sharing and utilizing knowledge of initiatives to strengthen project management in the wide-ranging business areas of the Hitachi Group will become a source of competitive excellence for Hitachi, the internal technical committee is creating opportunities for information exchange.

This technical committee is called the Project Management Technical Committee, and it started with working-level discussions at the Power Systems Division and ICT Systems Division around 2000. Then a technical committee open to the whole company was set up at a meeting of volunteers in 2005, and today, many business units are participating with the focus on the PMO. The Technical Committee is promoting stronger project management across the whole company. As well as exchange of information, it supports the research institutes by investigating solution strategies for shared issues.[18] Examples include the structure for sharing project management

---

tasks across business units[19] or company-wide activities based on efforts to leverage knowledge in the field of ICT systems.[20]

As well as organizing internal forums with the aim of educating staff about project management or communicating technical committee activities for professionals to the whole Hitachi Group, the technical committee also conducts regular surveys of awareness of the issues for strengthening project management at each business unit.

In this way, there are initiatives under way to deliver synergy at the Hitachi Group by rolling out expertise laterally across the business units, identifying shared issues, and studying solution strategies through the activities of the Technical Committee.

REFERENCES

Abstracts provided where available.

Akiyuki Onaka, "Model of Project Team Assessment to Make Projects Succeed." *ProMAC2010*, 2010.

In our company, the management ability of the project managers of IT system projects is evaluated and their PM rank—divided into three levels, "small," "middle," or "large," according to project size—is certified based on our own criteria, the "Project Manager Accreditation System." Certified project managers are appointed to a project whose size corresponds to their PM rank according to the Project Manager Appointment System.

Some projects, especially large-size ones, went wrong even though the project managers were properly qualified by abovementioned systems. This fact caused us to infer the importance of the behaviors of project members, particularly those expected to help project managers. The members to be considered are, for example, those who reduce manager's burdens and those who supervise and advise project managers. However, our insufficient discussion about such points of view made it difficult to assess project teams, including support members, in the same way as we assessed project managers. In order to improve such situations, we analyzed previous project data to conclude how to organize a project team. Based on the conclusion, we created an assessment model for project management teams that takes project size and development type into account. This paper describes the result of our research for the model of project management team assessment.

Hideyuki Maeda et al., "Visualization of Communication using the Team Activity Measuring System and Its Application to the Project Management." *Journal of the Society of Project Management* 12(1), 5–10, 2010-02-15 (http://ci.nii.ac.jp/naid/110007573280).

By advancing sensor and analysis technologies, a system that can automatically measure and visualize a team activity has been developed. We have applied this system to a large-scale project and have experimentally measured and analyzed the condition of a communication within it. As a result, we have successfully quantified and visualized the condition of a project communication. A result of a quantification analysis shows that there is a strong

---

19. Koji Okada et al., "Challenge for Extracting Project Management Knowledge across Business Units," *Journal of the Society of Project Management* 10, no. 3 (2008): 23–28. http://ci.nii.ac.jp/naid/110006950594

20. Koji Okada et al., "An Analysis Method for Extracting Project Lessons Learned which Are Sharable across Business Units," *Journal of the Society of Project Management* 12, no. 6 (2010): 21–26. http://ci.nii.ac.jp/naid/110008592927

relationship between the productivity and the time spent [in] face-to-face communication. By continually monitoring the face-to-face communication time, we have been able to help expose a project's problem in the early phase. It has also helped us offer useful information to [the] project manager for solving this exposed problem.

Hisako Okada et al., "An Approach to Advance Construction Management System for Large-scale Power Plant Projects." *Journal of the Society of Project Management* **15** (1), 8–13, 2013–02-15.

Power plant construction project is large-scale and complex involving many stakeholders. To ensure "quality, schedule, cost" of this huge project, Hitachi has applied IT and system to construction area. Its focal point is (1) realization of huge project consistent and coordinated control, and (2) realization of improving construction field efficiency and quality, for reducing risk and cost. From 1990s, it was applied to actual projects and achieved an effect, but further improvement was reaching a limit in management and system-centered approach. Hence, by reconsidering the basis "Construction is a Production by Human," Hitachi has conducted research construction management system based on human-centered approach by focusing user/human side and has conducted reflecting the results on projects management itself.

Hitoshi Yamadera et al. "Relations between Achievement and Characteristics of Project Managers." In *Proceedings of 16th National Conference of The Society of Project Management*, 2009, 209–212, 2009-03-10 (http://ci.nii.ac.jp/naid/110007602894).

Kazuhito Shibata and Natsuko Sato, "Development of Integrated Project Management Framework and Practical Platform for EPC Project in Power Plant Business." *ProMAC2015*, 2015.

In Hitachi Power Systems Company, an integrated and practical project management platform plays a crucial role in the success of power plant construction projects. Project Management Office (PMO) established a company project management framework covering the project life cycle with the aim of improving the project management quality at practical level. The basic elements of the framework are project management standards, project management process definition, measurement and audit system, and facilitating system of learning by experience. In this paper, we report the structure of the framework which is built by PMO in Hitachi Power Systems Company and the effective application to the project management process with the project management platform.

Kenji Hatsuda et al. "Loss-Cost Management for IT Projects." In *Proceedings of 28th National Conference of The Society of Project Management*, 2016, 43–44, 2016-09-02.

A key reason for vendors to work on IT project management lies in the reduction of loss costs. Completely eliminating loss costs while taking on challenges may not be feasible; nevertheless, loss costs must be kept within appropriate limits in order to ensure management stability. Accordingly, efforts have been made to establish loss-cost management techniques. As part of this effort, we have developed risk propagation models, ways to identify failure patterns/project failure scenarios, and other techniques. This paper provides an overview of these techniques and stresses the importance of engaging in loss-cost management.

Kenji Hatsuda et al., "PMO Information System as a Support of Project Management Office Activities," *Journal of the Society of Project Management* **5**(4), 28–31, 2003–08-15 (http://ci.nii.ac.jp/naid/110003726282).

According to the increase in recognizing the importance of project management, the project management office (PMO) has also come to play important roles as the organization for project management promotion. Project management needs to be the task to be strategically promoted, and to be systematically deployed by PMO. The roles of PMO are common base developments such as project management procedures, personnel training, and technical developments, as well as the project supports across the organizations. It is effective to build the PMO information system which supports the activities of PMO. This paper focuses on the activities based on practical PMO and considers of development, utilization, and expected effects of the PMO information system as a support of PMO.

Kichie Matsuzaki, "Hitachi, Ltd. 100th Anniversary Series: Genealogy of the Pioneers (20) Inheriting and Reforming the Heart of Monozukuri at Hitachi: Companywide Activities toward Monozukuri." *Hitachi Review* **92**(2), 136–143, 2010-02 (http://digital.hitachihyoron.com/pdf/2010/02/2010_02_pioneers.pdf).

Koji Okada et al., "An Analysis Method for Extracting Project Lessons Learned which are Sharable across Business Units." *Journal of the Society of Project Management* **12**(6), 21–26, 2010-12-15 (http://ci.nii.ac.jp/naid/110008592927).

Improvements in project management activities are desired in every business domain. Sharing project lessons learned across business units can be an effective way to improve project management activities in an enterprise composed of various business units. In order to share project lessons learned across business units, we collected 31 failed project cases analyzed in each business unit from 8 business units, reanalyzed them, and extracted 50 sharable project lessons learned. Moreover, we designed an analysis method to extract sharable project lessons learned that reflect analysis know-how gathered from actually performed reanalysis activities.

Koji Okada et al., "Applying Phase-Gate Management for Diverse Business Types." *Journal of the Society of Project Management* **13**(6), 29–34, 2011-12-15 (http://ci.nii.ac.jp/naid/110009425403).

Global competition has become harder and harder in every business domain. In order to wipe out unprofitable projects and to improve business profits under such situation, we started a corporate initiative to deploy phase-gate management, which has produced successful results in leading business units, into every business unit broadly. At first, the concept of "Hitachi Phase-Gate Management," which is applicable to diverse business types, was made clear. Then common fundamental enablers, such as (1) operation guides, (2) training materials, (3) phase-gate maturity model, (4) a KPI setting guide, and (5) sharing knowledge contents, are developed/established and provided broadly Moreover, ten model business units were selected and supported on both developing their improvement action plans of phase-gate management and performing them. According to the results, improvement of both phase-gate maturity levels and some KPIs were demonstrated in all selected model business units.

Koji Okada et al., "Challenge for Extracting Project Management Knowledge across Business Units." *In Journal of the Society of Project Management* **10**(3), 23–28, 2008-06-15 (http://ci.nii.ac.jp/naid/110006950594).

In order to prevent project trouble or to repeat project success, enterprises have developed their own QMS (Quality Management System) for project management. However, these improvement activities are performed in individual business units; obtained knowledge was not shared across business units. In this paper, we describe methodology for extracting and organizing project management knowledge, which is developed through real practice for extracting and organizing them. In particular, we devised fundamental concepts based on commonalities, differences, and specialties. Also, we developed a knowledge-extracting worksheet, a knowledge description sheet, and a knowledge map as supporting tools, as well as a procedure for extracting and organizing knowledge through three trial cycles of real practice.

Minamino, Toyama, "An Application of Modern Project Management "IT" System Development Projects." ProMAC2002, 2002.

In recent years, each IT system development project has come to be diversified and complicated, and its exploitation is required in the short term. Also, changes to the requests during the development have also increased. IT system projects have characteristics that the whole image of the system cannot be observed as a concrete shape directly, either during or after the development. The authors have attempted the application of modern project management, especially for risk management and scope management in such system development projects. The authors provide some examples of application of modern project management to IT system development projects and describe future aspects on the application of model project management.

Takafumi Kawasaki et al., "Practice Action of Project Managers: The Difference between Highly Competent PM and Moderately Competent PM." *Proceedings of 13th National Conference of The Society of Project Management*, 2007, 373–377, 2007-03-15 (http://ci.nii.ac.jp/naid/110007602747).

Conventionally, the knowledge and skills that effective project managers possess have been the explanation of successful performance. But highly complex situations require professional project managers to create adaptive and useful practices. This practice is referred to as "knowing," that is, creating new knowledge and adaptive actions. This study analyzed actions of superior project managers and less superior project managers and explained the difference in terms of promoting member's coknowing.

Takeshi Yokota et al., "Development of a Contract Risk Assessment Support System(CRARIS)," *Journal of the Society of Project Management* **7**(3), 20–25, 2005-06-15 (http://ci.nii.ac.jp/naid/110003726628).

A business process necessary to support a contract risk assessment in overseas projects was examined, and ContRAct RISk assessment support system (CRARIS) that was a knowledge management system concerning the contract risk management was developed. CRARIS is based on a contract checklist made in a legal affairs section, and it characterizes in a presentation of knowhow information that relates to each check item and automatic evaluation of risks according to content of checklist inputs. Moreover, about 2000 [pieces of] knowhow information has been extracted from hearing results to specialists in an operation division and the legal affairs section, the minutes of the evaluation of actual projects, and so on. In addition, an examination of a business process and an organizational structure effective to evaluate the contract risk was executed.

Takeshi Yokota et al., "Development of a Risk Management System for Construction Projects," *Journal of the Society of Project Management* **8**(5), 36–41, 2006-10-15 (http://ci.nii.ac.jp/naid/110006278350).

In order to support the risk management of a construction project, we have developed a system that uses progress simulation technology and supports evaluation of the problem of a project and the decision of countermeasures to [the] problem. This system is characterized by having progress evaluation simulation logic, which evaluates detailed progress of each work of a project serially per week. Moreover, it also has the [capability] to take into consideration situations, such as change in working efficiency, and increase the number of workers, in simulation logic. This system was evaluated using the data of an actual project, and the validity of the project evaluation result using the various functions of a system was checked.

This study investigated relations between achievement of project managers and personality, work attitude, and type of project activity. The results showed that extroversion, problem consciousness, and learn from others were related to achievement. Concerning project activity five types of expertise were identified as for normal-level contribution managers. Though they fully contributed to their projects, it was revealed that high-level contributors equally and consciously leveled their behavior up to promote organizational evolution.

Takeshi Yokota et al., "Strengthening of Personnel Training Process of Project Managers." *Journal of the Society of Project Management* **15**(2), 2013-04-15.

To improve a strike rate of construction projects, we have been developing personnel training process for project managers. We have developed the method, which evaluates characteristics of project managers quantitatively. This method has a basic data, which consists of replies of about 200 items of questionnaire. It evaluates project manager's characteristics in some viewpoints (project experience, behavioral trait, knowledge, etc.). This method defines the target score of project management work, and by comparing project manager's characteristics with target score, it clarifies their strength and weak points for education. Furthermore, we will support the organization of project formation by evaluating a result of comparison.

Takeshi Yokota et al., "Upgrade of Risk Management Technique for IT System Development Project," *Journal of the Society of Project Management* **14** (3), 25–30, 2012-06-15 (http://ci.nii.ac.jp/naid/110009495477).

To support the effective introduction of IT systems, we have constructed a business justification analysis support system for IT system development. It evaluates the benefit of systems, investment effects, risk factors, and the justification of development systems. By using this information, it clarifies the appropriateness and priority level of development investment and supports the risk management process of development phase. To clarify characteristics of projects more accurately, we classified risk score by considering whether project managers can manage or not. By applying this technique to real projects, we verified that this risk classification technique is effective for project management.

Tomoyuki Aoki et al., "The Case Study of Business Process Reengineering for EPC Project Management." *Journal of the Society of Project Management* **14**(6), 5–10, 2012-12-15.

In 2009, the BPR (business process reengineering) project called "D-WBS project" was begun at Hitachi Power Systems Company, and we are driving forward this project's first phase for completion by FY 2013. In the D-WBS project, we would like to achieve the improvement of project management capability by developing a standard management platform, which can be used among our business segments. This platform's target is an EPC project that constructs power plant and advanced medical system.

In this paper, we first introduce the background and overview of the D-WBS project. Then we explain our project management platform that consists of four domains, a WBS code system, an IT system, an operation standard, and an operational department. We also explain the application methodology of D-WBS code for the EPC project's planning. Finally, we share our BPR approach.

Yoshinobu Uchida, "Development of the Risk Management System for Construction Projects." *ProMAC2011*, 2011.

To support the risk management of construction projects, we have developed a risk management system that identifies project risks and supports decisions on countermeasures to these risks. The risk management system includes a project evaluation system, a risk register, and a risk management web portal. The project evaluation system provides a checklist suitable for a project and supports project evaluation. The risk register provides the worksheet and supports project risks identification and response planning development. The risk management web portal visualizes evaluation results through the project evaluation system. In this paper, we report each subsystem of the risk management system.

Yoshinobu Uchida et al., "An Approach of Knowledge Extraction via Empirical Failure Knowledge in Project Management." *Journal of the Society of Project Management* **12**(4), 27–32, 2010-08-15 (http://ci.nii.ac.jp/naid/110007880184).

One way to make a project successful is to have an understanding the essence of past experiences. To develop a scheme for learning from the past experiences, it is important to accumulate the valuable knowledge of organization. The knowledge is extracted by analyzing and interpreting it after information obtained from the experience is arranged again.

But it is difficult to derive an objective and profitable knowledge according to the following obstruction factor. (1) The analysis based on superficial or a local situational awareness is done. (2) The consideration of buck-passing work. To solve these problems, we developed causal analytical method including visualization of the decision sequence to support the knowledge extraction and defined knowledge form to understand knowledge. We evaluate the analytical method and the knowledge form and show the effectiveness of the knowledge extraction.

Yoshinobu Uchida et al. "Development of a Project Review Technique Employing Risk Propagation Models." In *Proceedings of 24th National Conference of The Society of Project Management*, 2014, 105–110, 2014-03-13.

In order to make effective use of knowledge within an organization, it is important to accumulate knowledge on a continuing basis and prevent obsolescence. In this study, we developed a risk-propagation model (RPM)-based project review technique. This entailed creating a model of the processes leading to failures from over 300 events that occurred previously based on data linking causes and effects. In our review technique, a propagation process for risks that may lead to project failure events is prepared based on checklist questionnaire entries in coordination with a checklist-based risk assessment system. By using the results of evaluation and risk propagation models during review, our technique proposes failure scenarios that may derive from questionnaire entries. This technique was applied to a real-world project and was found to be effective.

Yoshinobu Uchida et al. "Proposal for Risk Management Support Method Using Failure Knowledge." *Journal of the Society of Project Management* **7**(6), 3–8, 2005-12-15 (http://ci.nii.ac.jp/naid/110006278374).

The failure of projects has a major influence on corporate performance. Many SI enterprises need to rebuild project management. We work on the extermination of the deficit project and are researching the method of using the failure knowledge in project management to aim to prevent the same failure. In this paper, we define *process information for risk* (PIR) as information on the correspondence process on the risk, and propose the method of using the information in the project management process. Advantages of our proposal are the following. (1) PIR is extracted from the periodic report automatically; (2) A past similar case is presented as a failure case; (3) The project member deliberates [about] measures based on the failure case. It is thought that our proposal can support [preventing] the project from failing from the same cause.

Yoshinobu Uchida et al. "Proposal of Failure Prediction Method Employing Loss Cost Generation Mechanisms: Loss-Cost Management for IT Projects." In *Proceedings of 28th National Conference of The Society of Project Management*, 2016, 45–50, 2016-09-02.

Loss costs are additional costs incurred as a result of divergence from the original plan. We have been working on loss-cost management with an eye to reducing loss costs deriving from design reworks and quality improvement work. Loss-cost management requires process improvement and other organization-level work, as well as project-level work to pick up on the signs of failures that may lead to loss costs in the course of the project's implementation and take appropriate measures accordingly. This paper proposes a failure prediction method that entails modeling mechanisms of loss-cost generation based on failure analysis and presenting possible failure scenarios based on data that is predictive of loss costs as a means of supporting project-level activities.

Yoshinobu Uchida et al. "Proposal of Utilization of the Failure Experience in Project Management." In *Proceedings of 15th National Conference of the Society of Project Management*, 2008, 140–143, 2008-03-14 (http://ci.nii.ac.jp/naid/110007602790).

Understanding the essence of the failure experience and learning lessons from it are important to creating knowledge. We think that our organization can strengthen the project management by learning from failure experiences in past projects and sharing precepts in the organization. To learn from failure experiences, we should utilize the failure experience in project management.

Our approach assumes the assessment activity by the Project Management Office in an ongoing project. We surveyed on current assessment activity, and found the following issues:

1. How should assessors understand the situation in the project? Not every assessor has the ability to clarify all the aspects in the project. Often an assessor does not have enough information to verify the effectiveness of a countermeasure.
2. How should assessors find the information the project manager needs in the project?

To solve these issues, we defined a format to describe the situation in the project and search strategy for information the project manager needs.

3.5 FARM CREDIT MID-AMERICA BEST PRACTICES[21]

Farm Credit Mid-America Introduction

The Farm Credit System comprises over 70 independent financial institutions, referred to as Associations, with a unified goal of securing the future of rural communities and agriculture. Our member-owned cooperative structure ensures our business decisions remained focused on the customers we support. "Our loans and related financial services support farmers and ranchers, farmer-owned cooperatives and other agribusinesses, rural homebuyers, and companies exporting U.S. ag products around the world (Farm Credit 2021)."

Farm Credit Mid-America, one of the largest Associations in the Farm Credit System, serves the credit needs of farmers and rural residents across Indiana, Ohio, Kentucky, and Tennessee. We provide a wide range of financial services including real-estate loans for land purchases, operating loans designed to meet the feed, seed, and fuel needs for daily operations, business loans, equipment financing, and crop insurance products. We have approximately 80 retail offices spread throughout rural communities in our four-state area of operations.

In 2012, Farm Credit Mid-America was 19.7 billion earning asset Association (Farm Credit Mid-America 2021 Annual Report 2012). Led by the desire to serve their member owners, Farm Credit Mid-America stood up their Operations team in early 2012 with the objective of mapping their business processes and identifying opportunities to not only gain efficiencies but also enhance its customer experiences through an in-depth understanding of its core business processes and how they delivered value to their members. The Operations team, staffed with CI professionals with backgrounds in Six Sigma and Lean Manufacturing, conducted extensive Voice of the Business (VOB) workshops to elicit not only a deep understanding of the current state of business processes but also to identify and prioritize needs of our team members related to process stability and enhancement.

The Operations team identified hundreds of process improvement opportunities, and Farm Credit Mid-America senior leadership identified the need for structured

---

21. Material in this section provided by Dr. Chuck Millhollan, Chief Operating Officer, Farm Credit Mid-America. ©2022 by Farm Credit Mid-America. All rights reserved. Reproduced with permission.

approach for not only prioritizing, selecting, and implementing the projects, but also making decisions based on strategic alignment and value. In 2014, Farm Credit Mid-America established the Process Improvement and Execution team with the responsibilities for managing the Association's strategic project portfolio, including the allocation of resources and prioritization of projects to meet organizational strategic imperatives. As of June 2022, Farm Credit Mid-America is backed by the strength of more than 80,000 customers and $33.3 billion in earning assets, both owned and managed.

Due to realized successes founded in the application of project management best practices, the Operational Process Excellence team was established in 2017 and expanded the Process Improvement & Execution team's scope to include providing leadership for the Association's strategic planning processes in collaboration with the Executive Committee and Board of Directors, leading business owners for key business processes, provide real-time business process support, and the collection and management of business process-related knowledge management.

The following outlines how Farm Credit Mid-America applied project management best practices as we designed our structured approach to strategic project portfolio management and CI in a financial services environment.

Farm Credit Mid-America's Starting Point

The first step in applying project management best practices in any environment is to understand the current state, the organization's culture, and what processes would offer immediate value to not only enhance project delivery but also to gain buy-in to the structure introduced through project management tools and techniques. Better stated, the approach at Farm Credit Mid-America was, to begin with, a few focused best practices that would contribute to providing the desired business results in lieu of "imposing" a complex project management governance structure that, in my experience, is perceived more as administrative overhead than valued-added methodology. Regardless if the latter is true or not, business buy-in essential for the long-term sustainability of a PMO and the associated practices.

This "crawl, walk, run" strategy does not imply there was not a long-term objective of maturing our project management processes over time. On the contrary, each step on our infinite journey to enhancing the value provided to the association was based on establishing a culture of project leadership discipline that was flexible enough to adapt to each project and started from the perspective of business value delivery versus project completion.

It is worth highlighting that we had an advantage when designing the project management methodology based on best practices due to the Association's culture of collaboration and CI. The Operations team mentioned earlier, had demonstrated the value of team member engagement in the CI space, and to their credit, applied several best practices in that effort. For example, each CI event began with clearly defining the business problem or objective and identifying the appropriate business performance metrics used to measure solution efficacy. Additionally, the team socialized and gained acceptance of leveraging VOB and Voice of the Customer (VOC) throughout the CI event planning and execution process.

The Operations team and the CI event processes created both a culture of engagement and a culture of learning that set a strong foundation for establishing the Process

Improvement and Execution team in 2014. The best way to describe the Process Improvement and Execution team is a hybrid PMO and quality improvement office. This leads to the first best practice applied in maturing the association's project leadership, the hiring process.

Adopting Project Management Best Practices

Hiring Practices and Flexible Project Management Methodologies

Interpersonal skills are at the top of the list of the most important attributes for PM efficacy as it relates to project success (Millhollan 2015). A sound methodology based on valued-added best practices, coupled with a project leader with a combination of strategic thought and focus, leadership skills, and expert facilitation and communication skills increases the probability of project success. This combination of skills directly contributes to value delivery through the project management framework. Linking strategic thought to the project leadership skill set tends to create a culture of focusing on business metrics instead of project management-related metrics such as schedule, budget, and scope.

Another focus during the hiring process was to understand the depth and breadth candidates' toolkit. Given we are the Process Improvement and Execution team, it is important for project and program managers to have the ability to leverage tools and techniques from various methodologies. For example, project leaders may leverage parts of the DMAIC (define, measure, analyze, improve, and control), process early in the project discovery phase if there is not a predetermined solution to a business problem. They may then shift to an agile development and delivery model during project planning and implement phases and finally combine elements of project benefit realization processes with DMAIC improve and control processes as they transition the deliverables into ongoing operations. This requires that project leaders understand and embrace methodological diversity.

Accordingly, we structured the interview process to include scenarios that would place candidates in various situations to apply critical thinking, explain proper application of tools and techniques given the situation, and justify why they have modified the process to deliver the intended value. In addition to the project scenarios, interview panel members would also ask situational, behavioral questions to better understand how candidates apply the key interpersonal skills necessary to thrive in a project leadership role. Table 3–5 provides descriptive quotes from senior leaders during research conducted in 2014 and 2015 on factors most important from PM efficacy (Millhollan 2015) that was used to design the project scenarios and interview questions.

Value-Delivery Focus

One of the foundational best practices that we focused on early in our journey was creating a culture of focusing on realized benefits from strategic projects. Establishing a top-down ethos for benefits realization began with the Executive Committee and the Board of Directors. Using our strategic imperatives (see Table 3–6), we dedicate time during our annual strategic planning kick-off in July discussing strategic projects and their value proposition through the eyes of our Board of Directors and executive leaders (Millhollan and Mott 2022).

TABLE 3–5. IMPORTANT PROJECT MANAGER ATTRIBUTES

Project Manager Attribute	Sample Descriptive Quotes
Facilitation skills	1. What makes them good is like having the agenda ready, being prepared ahead of time, making sure the right people are in the room, that the room is set up before people get in there, being able to capture decisions that are made, ensure appropriate documentation, and having meeting minutes sent out on a timely basis, and that kind of stuff. 2. I think it < facilitation > is the ability to understand that you might have a conflicting need for resources or a timeline issue and then working that through with stakeholders and being able to come up with a solution that everyone could live with. You know, negotiating an agreement. When facilitating the discussion, you got to be able to get the real issue out on the table so the solution addresses the problem. It is not about a win-win compromise, it is about leading them to the right solution.
Individual personality traits (attitude, trustworthiness, unbiased)	1. Give me a negative-minded project manager, and I will show you a failed project before it starts. The project manager needs to be a cheerleader for both the project goals and the team. 2. A project manager has to be trustworthy and respected. They do not have direct control of the people, so their power comes through what the team members think of them. 3. A positive person makes everyone else positive, even with things are hard. Project work can be hard. Who wants a negative, mean, or disrespectful person in a leadership role.
Communication skills	1. She < project manager> takes the time to know the team members and talks on their level. I do not mean she talks down to them but uses terms and examples they are familiar with to ensure they understand. 2. She chooses the tool < medium > best suited for the message. In other words, she is not stuck in email or conference calls. 3. < Project manager name > always listens before she talks. She asks more questions than anything else. You just know she is actually listening and wants to hear you. 4. They have to know more than just how to draft a communications plan. I had a project manager that drafted and plan and sent it to everyone via email. They could not figure out why no one read the plan. 5. There are a lot of different ways you can communicate a message and you have to be very careful, especially in email that you do not come across poorly. I have seen many examples of people getting upset for really no reason just because it was a poor choice of words.
Leadership skills	1. I think that a project manager's team-building skills need to be stronger than what I expect from my IT managers. Their < project manager's > teams are constantly changing, and the team members often come from different departments and do not work together on a regular basis. Naturally, this would lead to the potential for greater conflict within the team. 2. Part of leadership is the approach you use to lead up the chain too. How do you keep the decision-makers engaged? A project manager needs to know how to lead their sponsors. I think building a relationship with them is the best approach.

TABLE 3–6. FARM CREDIT MID-AMERICA STRATEGIC IMPERATIVES

Provide an exceptional customer experience	Add value beyond customer expectations so that customers desire to repeat the experience and tell others about it.
Grow constructively	It is important that the portfolio continues to grow and effectively serve the marketplace in a manner that benefits customers now and in the future.
Maintain credit quality and administration	Doing the right things in the right ways allows us to work with customers as they experience challenges. Make quality new loans, provide solid loan administration, and responsively service distressed loans.
Maintain sustainable financial operations	Ongoing diligence will help provide financial strength and position the Association for future success despite economic challenges and changes in regulatory and government policy.
Live our purpose and values	Our purpose and values are reflected in Our Compass (see Figure 2) and everything we do. We work together, care for one another, and coach and grow talent to better serve our customers.

TABLE 3–7. VALUE PROPOSITION DEFINITIONS

Increased productivity	Doing more with the same number of resources, e.g., increasing the number of loans a credit analyst can decision. This implies a backlog of loans ready for work (available input).
Cost reduction	Removing unwarranted expenses (operating or capital) through gained efficiencies, e.g., eliminating waste in operational processes.
Decreased turn time	Reducing the amount of time it takes to complete tasks in the loan origination process, e.g., the average time it takes to decision loans is reduced through automating obvious decisions (approvals and denials).
Enhanced customer experience	Measurable increase in volume, margin, or fees, e.g., creating an experience that incentivizes members to consolidate debt held by competitors.
Enhanced employee engagement	Measure of the emotional connection to our Association, workplace relationships and collaboration, commitment to our purpose, and commitment to the success of our organization, e.g., reduced turnover or higher engagement surveys.
Long-term business sustainability	Holistic perspective with the objective of long-term growth, financial strength, and Association stability to sustain operating capabilities through regulatory and financial challenges, e.g., focus on long-term business impact vs short-term profit, price, or results.

Using our strategic imperatives to brainstorm metrics as a group, we defined a set of value propositions to lead a discussion related to each strategic project and the expected benefits and benefit realization timelines. Table 3–7 lists the identified value propositions and their definitions (Millhollan and Mott 2022).

This strategy related to creating a culture of selecting projects based on quantifiable business benefits and, more importantly, establishing a culture of measuring realized benefits post-solution implementation was designed to supplement traditional project management metrics such as schedule, budget, and scope. The latter provides valuable information throughout the project; however, focusing on strategy execution, i.e., value-delivery verses project completion, sets the foundational best practice of business metrics over metrics that measure project management methodology.

REFERENCES

Farm Credit. (2021). Retrieved from farmcredit.com/overview-and-mission on April 14, 2021.

Farm Credit Mid-America 2012 Annual Report. (2012). Retrieved from https://e-farmcredit.com/fcs/media/Assets/Media%20Downloads/2012-Annual-Report.pdf on April 14, 2021.

Millhollan, C. (2015). A phenomenological study of factors that influence project manager efficacy: The role of soft skills and hard skills in IT-centric project environments. (Unpublished dissertation). Syracuse, NY: Syracuse University.

Millhollan, C. and Mott, D. (2022). "Excellence in action: Farm Credit Mid-America, Chapter 1." In Kerzner, H., Zeitoun, A., and Vargas, R. V. (eds.). *Project Management Next Generation: The Pillars for Organizational Excellence*. John Wiley & Sons, New York, pp. 49–64.

3.6 NCS INTEGRATED DELIVERY METHODS (IDM) & PROJECT MANAGEMENT METHOD (PMM)[22]

NCS Integrated Delivery Methods (IDM)

About NCS

NCS, a subsidiary of Singtel Group, is a leading technology services firm with presence in Asia Pacific and partners with governments and enterprises to advance communities by harnessing technology. Combining the experience and expertise of its 12,000-strong team across 66 specializations, NCS provides differentiated and end-to-end technology services to clients with its NEXT capabilities in digital, data, cloud, and platforms, as well as core offerings in application, infrastructure, engineering, and cybersecurity. NCS also believes in building a strong partner ecosystem with leading technology players, research institutions, and start-ups to support open innovation and co-creation. For more information, visit ncs.co.

What is NCS Integrated Delivery Methods (IDM)?

The delivery of the end-to-end technology services is supported by NCS Integrated Delivery Methods (IDM)—the foundation for high quality and consistent service to our clients. They

- provide a well-established, disciplined approach, and a common set of vocabulary.
- contain our cumulative wisdom and sound practices acquired through many years of implementing Infocomm Technology solutions.

NCS IDM Suite of Methods

IDM is a suite of delivery methods for delivering IT-enabled solutions as seen in Table 3–8:

Combining the Methods for Delivery

The NCS IDM suite of methods spans from initial presales engagement, project implementation to post-implementation support and operations, as seen in Figure 3–8. Project Management Method (PMM) integrates the project end-to-end during Implementation and Integrated Architecture Method (IAM) integrates the solution end to end. Mandatory quality assurance (QA) and quality control (QC) activities are built into our methods for better internal controls.

Depending on the services committed to the client, the PM will select the relevant methods to combine into a tailored approach for delivering the solution. This gives us the flexibility needed as a Systems Integrator serving many clients in different industries while having a framework to ensure consistency, integration, and quality. For example, if a client requires a network system to be implemented, we will use Infrastructure Implementation Method (IIM) with PMM and IAM.

TABLE 3–8.   NCS INTEGRATED DELIVERY METHODS (IDM)

Method	What is this method?
Innovation – design thinking (NCS DRIVE)	A design thinking method to design human-centric solutions, facilitating the acceleration of enterprise digital transformation and addressing new market opportunities.
Integrated architecture method (IAM)	A set of repeatable and iterative processes that prescribe • the solution consisting of technology, delivery and contractual components, their interrelationships, and the principles, risks, and considerations identified at the initial solution design stage; and • the structure of system components, their interrelationships, and the principles and guidelines governing their design and evolution in delivery stage.
Project management method (PMM)	A method for Project Managers in NCS to ensure projects are successfully delivered on time, within budget, and meet the mutually agreed business and quality objectives.
Application development method (ADM)	A method for developing custom-built business application software through the stages of requirements analysis, design, incremental builds, test, and commission.
Agile method (agile)	A software development method based on the fast-paced, iterative, and incremental development nature of agile. The NCS Agile Method utilizes scrum as the overall base framework, with Agile Best Practices and Principles incorporated to create a pragmatic approach to deliver software.
Packaged software method (PSM)	A pragmatic approach to implement configurable application packaged solution to meet and integrate with organizational business needs. It may be used for implementing shrink-wrapped software requiring minimal configuration, highly configurable commercial software, or software as a service.
Infrastructure implementation method (IIM)	An integrated approach to requirements, design, and implementation of turn-key ICT solutions which includes hardware, system software, network, communications, system security, and database management systems.
Service management method (SMM)	A rich set of processes incorporating the ITIL best practices and ISO20000 requirements to plan, deliver and manage ICT managed services and operations.

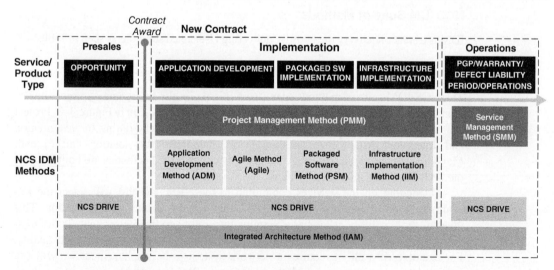

Figure 3–8.   Using a combination of Methods to support projects.

NCS Project Management Method (PMM)

Projects bring together resources, skills, technologies, and ideas to deliver business benefits and achieve specific business objectives. Good project management helps to ensure that these benefits/objectives are achieved within budget, within schedule, and meet the required quality.

What is NCS Project Management Method (PMM)?

The NCS PMM is a set of coordinated and well-tested processes designed to provide a consistent framework for PMs in NCS.

Why Use PMM?

Effectiveness of project management processes contributes significantly to the success of the project and leads to the quality of the delivered products. The NCS PMM recognizes this and provides both a framework and a systematic approach to managing a project to ensure it meets its intended objectives.

The NCS PMM ensures that projects are undertaken in manageable steps and continue to stay on track till their successful completion. Control points and quality planning activities are also built throughout the project process so that proactive and timely actions are taken where necessary.

The NCS PMM has been well tested, having been consistently and successfully applied to numerous projects of varying complexity, size, duration, and environment.

NCS PMM Framework of Project Stages and Dimensions

The PMM is a systematic and consistent framework for planning and managing a project from Initiation to Closure, to ensure that the project is successfully delivered on time, within budget, and meets agreed business objectives and quality objectives.

The PMM comprises Stages and Dimensions for PMs to apply the appropriate project management knowledge and skills throughout the project life cycle. We use the following frame in Figure 3–9 to guide PMs in managing the project through the project life cycle. The Dimension of 'Tools' allows us to align the many project teams in NCS and increase the practice consistency, and the Dimension of 'Security'

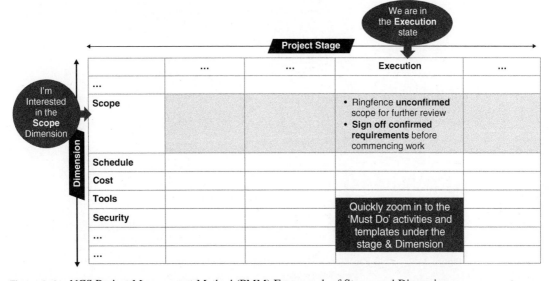

Figure 3–9. NCS Project Management Method (PMM) Framework of Stages and Dimension.

highlights the importance of having appropriate security measures in place when managing IT projects.

NCS PMM on Mobile-Enabled Web Application

Being in a company of over 12,000 people and managing numerous projects, we want to reach out to many of our people and make the NCS PMM easy to use. This is done via a mobile-enabled web application as seen in Figure 3–10.

"Checklist by Stage" allows the PM to zoom in to a particular stage, e.g., Planning stage, and scan through the activities for the different Dimensions. "Checklist by Dimension" allows PM to focus on a dimension and what needs to be done across the project life-cycle stages. Best practices and landmines are incorporated to make the method more practical.

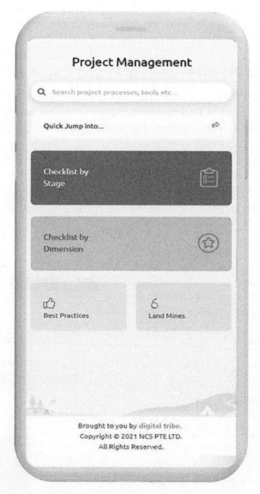

Figure 3–10. NCS PMM on mobile-enabled web application.

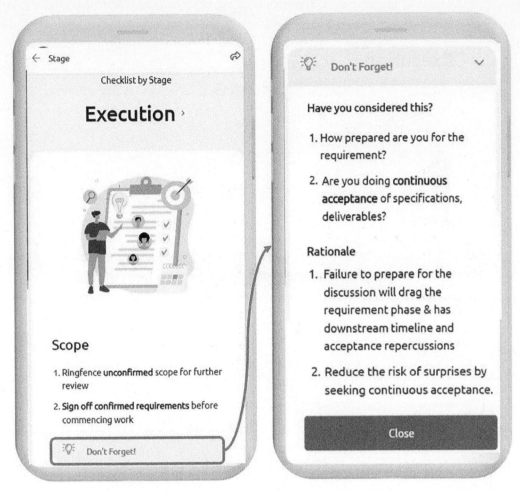

Figure 3–11. "Must Do" activities in the Stage and Dimension.

The "Must Do" activities are listed in bite-sized cards organized by the Stage and Dimension as seen in Figure 3–11. PM can scan through each card on the "Must Dos" activities. Selecting "Don't Forget" will show the important considerations that should not be missed out and "Rationale" explains why they are important.

Selecting "Useful Tools" will list the related templates, forms, and checklists that are available, as seen in Figure 3–12.

Change Control Process
An important aspect of the Integration Management dimension is Change Control Process.

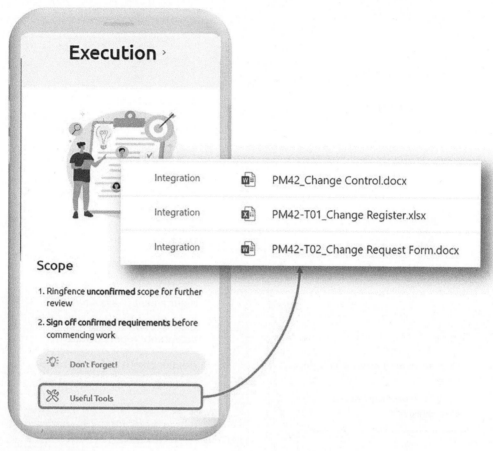

Figure 3–12.　Useful Tools.

Why is Change Control Important?

A project could be derailed due to uncontrolled changes. The Change Control Process

- provides the project team with a governance structure and process for identifying, documenting, and tracking all project changes with the client.
- encourages active management of contractual rights and obligations and helps to keep both NCS and the client on the same page for all changes.
- provides an avenue to discuss and address changes and associated impact on scope, resources, project timeline, and costs with the client, and
- helps to mitigate project scope issues and prevents possible disputes/conflicts.

Establishing Change Control Process at Start of Project

It is important to establish a Change Control Process at the start of the project at Planning Stage, which could include what constitutes a change, the notification process, the authority for approval, pricing mechanism, and time frame for processing the change. It is typically defined in the Project Management Plan and agreed with the client. This Change Control Process should be followed by all levels in the project organization and at all stages of the project to manage changes in the project. The Change Control Process in Figure 3–13 may be adapted to suit the project nature and project organization structure.

Effective Enforcement of the Change Control Process

Effective enforcement of the Change Control Process involves the steps in Figure 3–14.

Establishing a common understanding and managing expectations with clear impact analysis, assumptions, and dependencies are key to effective Change Control Process. Comprehensive impact analysis needs to be done. Some of the less obvious impact include impact on subcontractor's work areas and terms, warranty, and maintenance extensions. Agreement or approval of the change to be implemented must be sought and the change is implemented only after written agreement is obtained. All changes should be recorded and tracked till closure.

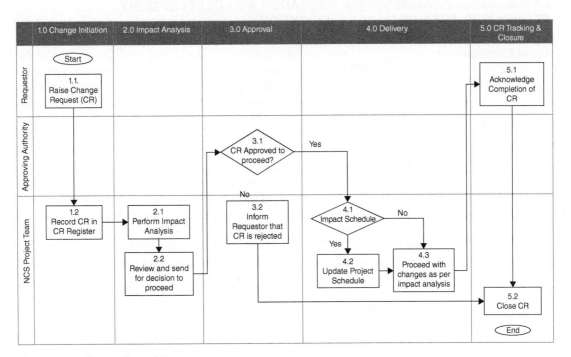

Figure 3–13. Change Control Process.

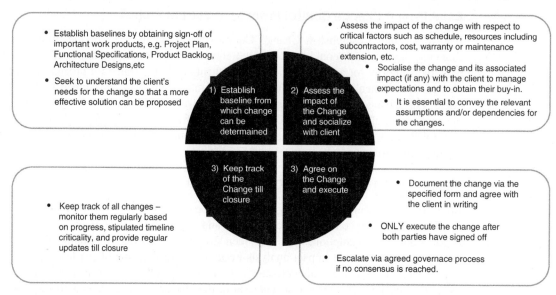

Figure 3–14. Enforcing the Change Control Process.

3.7 MANAGING CHANGE WITHIN RESEARCH AND DEVELOPMENT AT BUSINESS AREA NETWORKS, ERICSSON[23]

About Ericsson

Ericsson enables communications service providers to capture the full value of connectivity. The company's portfolio spans Networks, Digital Services, Managed Services, and Emerging Business and is designed to help our customers go digital, increase efficiency and find new revenue streams. Ericsson's investments in innovation have delivered the benefits of telephony and mobile broadband to billions of people around the world. The Ericsson stock is listed on Nasdaq Stockholm and Nasdaq New York. www.ericsson.com

Everything is moving faster, for every year.

Just look at the uptake of new products and services. It took 68 years to reach 50 million passengers for the airline industry. The same journey to reach 50 million users took 62 years for the car industry, 50 years for the telephone, 46 years for electrified homes, 22 years for the television, 14 years for the computer, 12 years for mobile phones, 7 years for the internet, 3 years for Facebook, 1 year for WeChat and 19 days for Pokémon Go.

The fourth industrial revolution, which experts say mark our times, is the trend toward automation and data exchange in manufacturing technologies and processes that include cyber–physical systems (CPS), internet of things (IoT), cloud computing, cognitive computing, and artificial intelligence. Never has managing change been more important and relevant than now!

---

23. This section was provided by Marianne Rimbark, Improvement and Performance manager at Business Area Networks, Ericsson. © 2022. All rights reserved. Reproduced with permission.

In such a rapidly changing world, with a complicated to complex environment, I argue that the only way for companies to survive is to have an agile mindset and to work agile. And we need to have a sense–analyze–respond or probe–sense–respond strategy (according to Cynefin model introduced in 1999 by Dave Snowden to aid in decision-making) when developing products and implementing changes.

During my years at Ericsson, I have thought a lot about managing change and how to manage lasting change within large-scale R&D organizations. Now, based on my company's experiments and learnings, I propose an update to Knoster's model of "Managing Complex Change." Knoster's model is built on five dimensions and for a successful transformation to occur, all five had to be in place: vision, skill, incentive, resource, and action plan.

After a few years, I came in contact with Grant Lichtman's model that he uses when transforming schools and which he calls a "stairway to successful innovation." Grant Lichtman is an internationally recognized thought leader in the drive to transform education. The "stairway to successful innovation" is built on eight dimensions vision, leadership, resource, timeline, skill, communication, inclusiveness, and commitment.

My updated version of Knoster's model incorporates some of Lichtman's dimensions. It builds on seven dimensions, or *change capabilities*: **vision, leadership, people, management, skills and capabilities, communication,** and **processes and tools**. If all seven change capabilities are in place, your transformation will progress as it should and the change you put in place will also stick, i.e, you will have success!

But if one or more change capabilities are missing, the change will not stick.

If you lack:

- Vision there will be confusion, lack of need, aim, and urgency.
- Leadership there will be failure to start, implement and have a result that sticks.
- People the initiative will not be taken seriously, there will be frustration, and failure to implement and deploy.
- Management there will be false start, drifting in time, and in the worst case standing still (vacuum) and lack of commitment.
- Skills and capabilities there will be a slow start, anxiety, and it will be ineffective.
- Communication there will be disconnected silos and people are unaware.
- Process and tools there will be lack of institutionalization, if it is not easy/simple I will rather do it the old way.

Now, let me go through each of the seven change capabilities I believe must be in place for a successful change to occur.

Vision

I believe that any change needs to start with a clear answer as to why we need this change. What is the intention, the big opportunity, the drive, and what do we want to achieve with this change? We also need to create a sense of urgency. In the best of all worlds, all this can be clearly formulated in a vision, that should be compelling, simple, and easy to grasp.

With this vision, we want to grab hearts and minds of all those that will be affected by the change. Signs that the vision is clear and easy to understand are if you get reactions such as:

- "I trust this."
- "I think this is going to be good for me and others."
- "This will be beneficial for me, my team, and my family; I understand why, so I want to contribute."

For inspiration on setting a vision, I recommend watching the online TEDx talk by Simon Sinek, *Start with why—how great managers inspire action* (2009).

The vision should be complemented with a strategy, that can be further broken down into manageable pieces to implement and deploy and that is measured/followed up by Objectives and Key Results (OKRs).

Leadership

Without good leadership, no change. And good managers also need to have the support of *their* managers. I cannot stress enough the importance of having the commitment of leaders, at all levels of your organization, if you want to succeed in implementing a change. There are official leaders and unofficial leaders, and both are equally important. My opinion is that servant leadership is the leadership style of the future.

To have a successful change, I believe that leadership is when you:

- Establish direction or purpose by means of a compelling vision, by stimulating, inspiring, and aligning people.
- Show interest by motivating and mobilizing people.
- Take action by empowering and through alignment to support autonomy.
- Pull results and set the direction moving forward.
- Promote and encourage continuous learning.
- Managers are role models for the change and sponsors for the change.
- Managers set the tone of an organization, its values, and standards.

I also agree with management consultant John Kotter who in 2012 stated that managers should be able to remove barriers to change, work with structure, skills and capabilities, and communication to bolster the change.

To be a manager in a complicated to complex domain, the traits you need are:

- Respect for people.
- Fail fast and learn.
- Looking at the whole: flow and throughput (value optimization over people optimization).
- Servant leadership, leading through intention.

For inspiration regarding leadership, a good book is *Greatness* by David Marquet (2013).

Another good book regarding working Agile and Servant Leadership is *It is all about the spirit: The core of Agile and Care & Growth* by Anna Pucar Rimhagen (2021).

And I cannot stress enough the importance of changing views, when thinking about change initiatives. From a top management point of view, it might sometimes look like there are not many change initiatives happening at any one time. Meanwhile, staff might feel like they are drowning in change initiatives. Many change initiatives might be ongoing for a long time, and we might experience change fatigue at times. It is therefore important that managers always reflect on their organization's change capacity. How many changes can we truly handle simultaneously? Remembering also that it is not only changed at work that affect us; it is changed in our private lives and in our society, country, and in the world at large.

The role of stakeholder(s) in change management: The change will not fully start until the top management is committed and clearly communicates the importance of the change. Top management needs to put the spotlight on the change. And as we learned in our transformation journey at Ericsson, if the change starts from the bottom up, it is not until management commitment is obtained that the change process will truly gain momentum.

But in addition to top management, there are normally many other stakeholders who also need to put their weight behind the proposed change. Such stakeholders can be the manager that is sponsoring the change; organizations that are affected or impacted by the change at different levels represented by a manager; or tools/automation owners that are enabling a change. You could say that while top management gives the impetus for change, it is the stakeholders that provide an arena for the change. They set the aim, approve the OKRs, fund the change out of their budgets—and they also decide when a change or OKR is done. Controlling the OKRs is a powerful stance to have in any change project.

If there is a lack of stakeholder and top management commitment, there will always be managers and people questioning and denying the change, so the change starts to happen in silos.

When the change program is up and running, the stakeholder(s) of the change normally start showing interest and check the various dashboards for information—or they "pull the results," as we say. They do this fairly frequently at first because it is hard to get the transformation going. After a while, the pull frequency can be reduced but it is important to keep up a cadence of regular pulling.

Now, at a certain point, the change is implemented, and we have a result. The stakeholder(s) need to decide about the future. Should we start the next change initiative because there is more potential to be realized? Or do we just want to maintain the results we have now reached? In that case, the processes and tools that we used for the change should be adapted for continuous usage. If we do nothing, or if we did not achieve the definition of done (DoD), then the change might even revert or deteriorate.

Another aspect to bear in mind, when it comes to stakeholders, is that their engagement in a change process may—and probably should—be more intense at the start and end of the process, but less so in the middle. Because once the change process is underway, it should be up to the people actually implementing the change program to move the process forward.

People

The people working in an organization are the organization's key asset. I do not think anyone will argue against that.

In a change process, it is the people—me, we, us—who ultimately will implement whatever change that is required. So, in my experience, we need early adopters "to get the snowball rolling." They need to work on developing best practices, build and share knowledge and skills, communicate, inspire, and collaborate.

Then, to get our people to work well together, we need to build high-performing teams, and we need an early adopter in every team to ensure engagement, understanding and to achieve fast change.

And for change to happen, we have to address, in our teams and groups, how to change behaviors and mindsets. My summary is that I will change my behavior if:

- I see others doing it; I have role models to follow.
- I understand why we are changing; this makes me committed.
- I know how to do it because I have the skills and capabilities.
- I get the information I need to change; and I can share it with and learn from others.
- My tools and routines make it easy.

When the above aspects are in place, a change in mindset is much more likely to follow.

For more inspiration, I recommend searching on YouTube for *First Follower: Leadership from a Dancing guy* (Derek Sivers, 2010).

So, in short, get out of your comfort zone, be courageous enough to follow a transformation leader. Create an awesome team! And remember to have fun!

Management

In my model, I separate leadership from management. Because there is a difference, in my opinion, between change leadership and change management. Change management is all about how you plan, structure, and execute the work (risk management, budgeting for the change, action list). Meanwhile, change leadership is about servant leadership, talking about the intent behind the change, creating a sense of urgency, being a role model, listening to the challenges, empowering people, and encouraging autonomy.

For more on this, I recommend a short video by John Kotter, that you will find on YouTube if you search for *Change Management versus Change Leadership: What is the Difference?*

Now, beyond the difference between leadership and management, it is probably worth defining more exactly what management is. And I know there are many definitions of management out there. For that purpose, I have used John Kotter (2012) who defines management as "setting the structure, planning, budgeting, organizing, staffing, problem solving, measuring, and producing dependable and reliable results."

I believe that in order to live up to everything that is in that definition and be successful in management, it is key to work according to lean and agile values and principles and to use agile ways of working.

Further, I recommend that any transformation or change manager considers the McKinsey 7S model (strategy, structure, systems, shared values, skills, staff, and leadership style).

With those definitions and models in mind, my own main tools for managing change projects have developed to be:

- Create transparency by analyzing and documenting your customer value flows (with methods such as value stream mapping or using SIPOC).
- Add facts (data) to your customer value flows.
- Make integration plans or anatomy plans to show how far you have come in the development and deployment.
- Work Agile:
  - Make it visible what you want to achieve when and by whom (for instance by using OKRs). Such plans will make it—painfully—visible when things start drifting in time to take corrective actions.
  - Make a backlog with user stories in order to find the right level. If you are at too high a level, things will never be ready. If the level is too low, people find themselves being ticket driven and lose the helicopter view. (User stories are from the Scrum method, it describes functionality that will be valuable to a user, it consists of one or more sentences in the everyday or business language of the user that capture what the user wants to achieve.)
  - Work in sprints to achieve subgoals to enable quick wins, which are motivating. Be clear about the DoD.
- Perform risk management monthly, involve your team, and highlight the most important risks to your stakeholders to get support. Find ways to avoid, reduce, mitigate, and sometimes accept the risks.
- Keep your eye on the budget: Clear reporting on costs and deviations from the budget.

Skills and Capabilities

In our lean and agile transformation, we had to start with courses to build everyone's knowledge about this new way of working. Because without the skills and capabilities needed for the change, no change will ever happen. Similarly, when we introduced the operational performance management system, we needed to build additional skills and capabilities among our teams.

Acquiring new skills does also not mean reading a book. Although a book often is a good place to start, it is rarely enough. Just because you have read a recipe for cookies does not make you a professional baker. No, to develop skills and capabilities, you need to work with the task at hand. So, try it out and accept that you will fail a few times before you have learned the new skill or acquired a new capability. As we always say: fail fast and learn. In any change initiative, leave room for this to happen too. If you are working with sprints, for example, make sure to include time in each sprint to build these new skills and capabilities.

A good model, when thinking about how far an organization has come in acquiring new capabilities, is the Capability Maturity Model (CMM). It was originally developed, many decades ago, by Carnegie Mellon University, as a tool to help organizations analyze how mature they were and to trigger reflection. If there is a gap in where the organization wants to be, you need to trigger an action. It has five levels for how far an organization has come in acquiring new capabilities: initial, repeatable, defined, capable, and efficient.

My main tools for building skills and capabilities are:

- Make a learning plan with push and pull activities.
- Provide courses and evaluate and improve the course material provided.
- Plan for working with the newly acquired knowledge in your daily work, until it is repeatable, defined and capable, and finally efficient (following the CMM levels).
- Use early adopters as change champions, and as their skills and capabilities evolve, they can train others in return.
- Continuous learning evaluations to improve learning materials and understand the level of learning being consumed in the organization.
- CI to refine the skills and capabilities based on learnings acquired when working.

For more inspiration on skill and capabilities, find the YouTube video *Sharpen your saw—7 Habits of Highly Effective People* by Stephan R. Covey—or read his book with the same name.

Communication

Communication in a transformation is vital. It has to happen continuously, in many different ways, and include the entire organization. This is one of the hardest things to manage in a transformation. In a large organization, it is already hard to know who is impacted by a change and then to figure out how to reach them, whether they understand the messages you send, if they have ways of providing feedback and comments, and then in the end, if it is clear to them what actions that are required of them.

Communication is therefore not just about passing information onward, but also about collaboration, sharing, and learning together. Communication needs to be thought of as a push-and-pull system. We push the form of the change, and we need to pull the feedback to be able to iterate and revise the form.

I personally believe that every time you communicate it should be clear to you what you want your audience to **know, feel,** and **do**. And when communicating a change, you need to be clear about:

- **Why** the change is needed.
- **What** the change is.
- **How** the change will affect you, changes in ways of working or tools, etc.
- **Who** is affected and who you can contact.
- **Where** the change will happen.
- **When** it will happen.

Processes and Tools

In my model for managing change, I have broken out processes and tools as a separate category. Because having processes, tools, and routines in place will simplify everything around your transformation.

Our main tool is value stream mapping. I recommend that any change project start with a visualization of the current situation (AS-IS), using a value stream map. Then develop a value stream map for where you want to go (your TO-BE map). Do this exercise first at a high level, but then, as the change progresses and you become more detail oriented, repeat the exercise as often as you see a need, so that the details of the process are properly described and understood by the change champions in the organization. Processes are often connected to tools, they can be Excel sheets, PowerPoints, or different applications that interact with a database. Include the tools' view in the value stream map to increase the transparency and what the expectations are on the different roles in the process.

When driving change the end goal for processes and tools is that they should be easy to use and support the work. If not, the change is not made properly.

A favorite on this topic is the short sketch *Medieval helpdesk* from the Norwegian public service broadcaster NRK. Search for it on YouTube.

Summary

When you are leading change there will never be perfect circumstances, you need to do risk management continuously and find ways to avoid, reduce, mitigate, and sometimes accept the risks. Have a tight dialog and collaboration with your stakeholders.

Change takes time, be perseverant.

Remember the 7 *change capabilities*: **vision** – setting the scene to why we are doing this, **leadership** – our role models, so that I see others do it**, people** – who are leading the change, **management** – what, how, when, and where, **skills and capabilities** – I can do it, **communication** – collaboration, change management, sharing and learning, and **processes and tools** – my tools and routines make it easy (how).

If you want to read the full story of how global telecom group Ericsson transformed into a lean and agile organization in the early 2010s you can read the book *Transforming an organization – from the Lean and Agile movement at Ericsson* by Marianne Rimbark. This is where the publication of this change model first was publicized. The book provides a comprehensive overview of the process, full of both practical and theoretical insights. A handbook on how companies can improve their operations—and as such meant as a contribution to the global lean and agile community and the Change management community.

> *"It is not the strongest of the species that survives, nor the most intelligent, but the one most responsive to change"*
>
> *Charles Darwin*

3.8 INTEL CORPORATION AND "MAP DAYS"

The introduction of project management methodologies several decades ago was structured around end of phase or gate reviews. A typical methodology would have about four or five gate reviews, and each gate review was mainly an examination of the trends in budgets and schedules from which a go or no-go decision would be made. Companies found it difficult to cancel or even redirect a troubled project and preferred to let projects go through to completion in case a miracle would occur.

There were many project management horror stories where executives would identify the assumptions and constraints for a given project. Sometimes not all assumptions and constraints were identified, and, to make matters worse, PMs would assume that these would not change over the life of the project. Then, at project completion, people would become upset that the deliverables no longer satisfied the firm's strategic business objectives.

Several years ago, Intel introduced a concept called "Map Days," which, among other characteristics, included periodic review of the project toward desired business objectives.[24] For many companies, Intel's "Map Days" was a vision of the future of project management performance review practices. It showed interested parties that tracking other items, such as changes in assumptions and constraints, are just as important as tracking time and cost.

Today, many of the characteristics of "Map Days" are being used in techniques such as agile and scrum as well as in traditional project management practices. Agile and scrum use short time blocks, called sprints. At the end of each time block, the direction of the project can be reviewed for possible course changes. Many of the concepts of Intel's "Map Days" are still being used by companies but perhaps by a different name.

3.9 APPLE COMPUTER AND CELL PHONES

For decades, written reports were the prime method for project performance reporting. The reports were time-consuming and costly to prepare. Decision-making was often predicated on the timeliness of the report. Without frequent reports, decisions were made based on best guesses rather than facts or evidence.

The introduction of the cell phone opened the door for real-time status reporting. PMs could now update the status of their project on cell phones (or, today, any mobile device) and transmit the data to just about any location in the world.

Decision-making is now based on evidence and facts and can be done in real time. Today's cell phones can display images and metrics that can be easily read. For many companies, cell phones accompanied by social media software can significantly reduce the cost of written reports as well as the number of meetings and costly travel expenses.

---

24. For more information, see Harold Kerzner, *Advanced Project Management: Best Practices on Implementation*, 2nd ed. (Hoboken: Wiley, 2004), pp. 115–116.

3.10 THE LIGHT AT THE END OF THE TUNNEL

Most people seem to believe that the light at the end of the tunnel is the creation of an enterprise project management (EPM) methodology, either flexible or inflexible, that is readily accepted across the entire organization and supports the need for survival of the firm. Actually, the goal should be to achieve excellence in project management, and the methodology is the driver for this. According to a spokesperson at AT&T, excellence can be defined as:

> A consistent project management methodology applied to all projects across the organization, continued recognition by our customers, and high customer satisfaction. Also, our project management excellence is a key selling factor for our sales teams. This results in repeat business by our customers. In addition, there is internal acknowledgment that project management is value-added and a must-have.

While there may be some merit to this belief that excellence begins with the creation of a methodology, other elements must be considered, as shown in Figure 3–15. Beginning at the top of the triangle, senior management must have a clear vision of how project management will benefit the organization. The two most common visions are for the implementation of project management to provide the company with a sustained competitive advantage and for project management to be viewed internally as a strategic competency.

Once the vision is realized, the next step is to create a mission statement, accompanied by long- and short-term objectives that clearly articulate the necessity for project management. As an example, look at Figure 3–16. In this example, a company may wish to be recognized by its clients as a solution provider rather than as a supplier of products or services. Therefore, the mission might be to develop a customer-supported EPM methodology that provides a continuous stream of successful solutions for

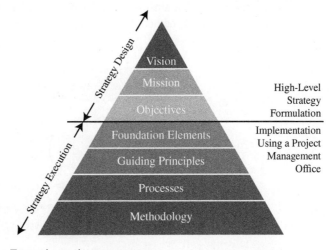

Figure 3–15. Enterprise project management.

Figure 3–16. Identifying the mission.

customers whereby customers treat the contractor as a strategic partner rather than as just another supplier. The necessity for the EPM methodology may appear in the wording of both the vision statement and the mission statement.

Mission statements can be broken down into near- and long-term objectives. For example, as seen in Figure 3–17, the objectives might begin with the establishment of metrics from which we can identify the critical success factors (CSFs) and the key performance indicators (KPIs). The CSFs focus on customer satisfaction metrics within the product, service, or solution. The KPIs are internal measurements of success in the use of the methodology. The CSFs and KPIs are the drivers for project management to become a strategic competency and a competitive advantage. Notice also in the figure that the CSFs and KPIs can be based on best practices.

The top three levels of the triangle in Figure 3–15 represent the design of the project management strategy. The bottom four levels involve the execution of the strategy beginning with the foundation elements. The foundation elements are the long- and short-term factors that must be considered perhaps even before beginning with the development of an EPM methodology (Table 3–9). While it may be argumentative as

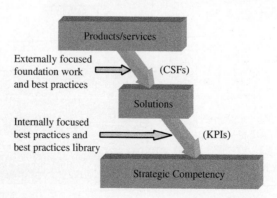

Figure 3–17. Identifying the metrics.

TABLE 3–9. FOUNDATION ELEMENTS

Long Term	Short Term
Mission	Primary and secondary processes
Results	Methodology
Logistics	Globalization rollout
Structure	Business case development
Accountability	Tools
Direction	Infrastructure
Trust	
Teamwork	
Culture	

to which factors are most important, companies seem to have accelerated to excellence in project management when cultural issues are addressed first.

To achieve excellence in project management, first the driving forces that mandate the need for excellence must be understood. Once the forces are identified, it is essential to be able to identify the potential problems and barriers that can prevent successful implementation of project management. Throughout this process, executive involvement is essential. In the following sections, these points will be discussed.

3.11 MANAGING ASSUMPTIONS

Whenever we discuss the journey to excellence, people expect to see a chronology of events as to how the company matured in project management. While this is certainly important, there are other activities that happen that can accelerate the maturity process. One such factor is an understanding of the assumptions that were made and a willingness to track the assumptions throughout the project. If the assumptions were wrong or have changed, then perhaps the direction of the project should change, or it should be canceled.

Planning begins with an understanding of the assumptions. Quite often, the assumptions are made by marketing and sales personnel and then approved by senior management as part of the project selection and approval process. The expectations for the final results are based on the assumptions made.

Why is it that, more often than not, the final results of a project do not satisfy senior management's expectations? At the beginning of a project, it is impossible to ensure that the benefits expected by senior management will be realized at project completion. While project length is a critical factor, the real culprit is changing assumptions.

Assumptions must be documented at project initiation using the project charter as a possible means. Throughout the project, the PM must revalidate and challenge the assumptions. Changing assumptions may mandate that the project be terminated or redirected toward a different set of objectives. The journey to excellence must include a way to revalidate assumptions. The longer the project, the greater the chance that the assumptions will change.

A project management plan is based on the assumptions described in the project charter. But there are additional assumptions made by the team that are inputs to the project management plan. One of the primary reasons that companies use a project charter is that PMs are brought on board well after the project selection process and approval process are complete. As a result, PMs need to know what assumptions were considered.

3.12 PROJECT GOVERNANCE

Most companies begin the journey to excellence with the development of a project management methodology. The purpose of the methodology is not only to provide a road map of how to proceed but also to provide the PM with necessary and timely information for decision-making. Decision-making requires some form of governance, and too often this need for governance is discovered late in the journey toward excellence.

A methodology is a series of processes, activities, and tools that are part of a specific discipline, such as project management, and designed to accomplish a specific objective. When the products, services, or customers have similar requirements and do not require significant customization, companies develop methodologies to provide some degree of consistency in the way that projects are managed. These types of methodologies are often based on rigid policies and procedures.

As companies become reasonably mature in project management, the policies and procedures are replaced by forms, guidelines, templates, and checklists. These provide the PM more flexibility in how to apply the methodology to satisfy a specific customer's requirements and lead to a more informal application of the project management methodology.

Today, we refer to this informal project management approach as a framework. A framework is a basic conceptual structure that is used to address an issue, such as a project. It includes a set of assumptions, concepts, values, and processes that provide the PM with a means for viewing what is needed to satisfy a customer's requirements. A framework is a skeleton support structure for building the project's deliverables.

Frameworks work well as long as the project's requirements do not impose severe pressure on the PM. Unfortunately, in today's chaotic environments, this pressure appears to be increasing because:

- Customers are demanding low-volume, high-quality products with some degree of customization.
- Project life cycles and new product development times are being compressed.
- Enterprise environmental factors are having a greater impact on project execution.
- Customers and stakeholders want to be more actively involved in the execution of projects.
- Companies are developing strategic partnerships with suppliers, and each supplier can be at a different level of project management maturity.
- Global competition has forced companies to accept projects from customers that are all at a different level of project management maturity.

These pressures tend to slow down the decision-making processes at a time when stakeholders want the processes to be accelerated. This slowdown is the result of:

- The PM being expected to make decisions in areas where he or she has limited knowledge.
- The PM hesitating to accept full accountability and ownership of projects.
- Excessive layers of management being superimposed on top of the PMO.
- Risk management being pushed up to higher levels in the organization hierarchy.
- The PM demonstrating questionable leadership ability.

These problems can be resolved using effective project governance. Project governance is actually a framework by which decisions are made. Governance relates to decisions that define expectations, accountability, responsibility, the granting of power, or verifying performance. Governance relates to consistent management, cohesive policies and processes, and decision-making rights for a given area of responsibility. Governance enables efficient and effective decision-making to take place.

Every project can have different governance even if each project uses the same EPM methodology. The governance function can operate as a separate process or as part of project management leadership. Governance is designed not to replace project decision-making but to prevent undesirable decisions from being made.

Historically, governance was provided by the project sponsor. Today, governance is most frequently by committee. Membership of the committee can change from project to project and industry to industry. Membership may also vary based on the number of stakeholders and whether the project is for an internal or external client.

3.13 SEVEN FALLACIES THAT DELAY PROJECT MANAGEMENT MATURITY _____

All too often, companies embark on a journey to implement project management only to discover that the path they thought was clear and straightforward is actually filled with obstacles and fallacies. Without sufficient understanding of the looming roadblocks and how to overcome them, an organization may never reach a high level of project management maturity. Their competitors, in contrast, may require only a few years to implement an organization-wide strategy that predictably and consistently delivers successful projects.

One key obstacle to project management maturity is that implementation activities are often spearheaded by people in positions of authority within an organization. These people often have a poor understanding of project management yet are unwilling to attend training programs, even short ones, to capture a basic understanding of what is required to successfully bring project management implementation to maturity. A second key obstacle is that these same people often make implementation decisions based on personal interests or hidden agendas. Both obstacles cause project management implementation to suffer.

The fallacies affecting the maturity of a project management implementation do not necessarily prevent project management from occurring. Instead, these mistaken beliefs elongate the implementation time frame and create significant frustration in the project management ranks. The seven most common fallacies are explained here.

Fallacy 1: Our ultimate goal is to implement project management. Wrong goal! The ultimate goal must be the progressive development of project management systems and processes that consistently and predictably result in a continuous stream of successful projects. A successful implementation occurs in the shortest amount of time and causes no disruption to the existing work flow. Anyone can purchase a software package and implement project management piecemeal, but effective project management systems and processes do not necessarily result. Furthermore, successfully completing one or two projects does not mean that only successfully managed projects will continue.

Additionally, purchasing the greatest project management software in the world cannot and will not replace the necessity of people having to work together in a project management environment. Project management software is not:

- A panacea or quick fix to project management issues.
- An alternative for the human side of project management.
- A replacement for the knowledge, skills, and experiences needed to manage projects.
- A substitute for human decision-making.
- A replacement for management attention when needed.

The right goal is essential for achieving project management maturity in the shortest time possible.

Fallacy 2: We need to establish a mandatory number of forms, templates, guidelines, and checklists by a certain point in time. Wrong criteria! Project management maturity can be evaluated only by establishing time-based levels of maturity and by using assessment instruments for measurement. While it is true that forms, guidelines, templates, and checklists are necessities, maximizing their number or putting them in place does not equal project management maturity. Many project management practitioners—me included—believe that project management maturity can be accelerated if the focus is on the development of an organization-wide project management methodology that everyone buys into and supports.

Methodologies should be designed to streamline the way the organization handles projects. For example, when a project is completed, the team should be debriefed to capture lessons learned and best practices. The debriefing session often uncovers ways to minimize or combine processes and improve efficiency and effectiveness without increasing costs.

Fallacy 3: We need to purchase project management software to accelerate the maturity process. Wrong approach! Purchasing software just for the sake of having project management software is a bad idea. Too often, decision-makers purchase project management software based on the bells and whistles that are packaged with it, believing that a larger project management software package can accelerate maturity. Perhaps a $200,000 software package is beneficial for a company building nuclear power plants, but what percentage of projects require elaborate features? PMs in my seminars readily admit that they use less than 20 percent of the capability of their project management software. They seem to view the software as a scheduling tool rather than as a tool to proactively manage projects.

Consider the following example that might represent an average year in a mid-size organization:

- Number of meetings per project: 60
- Number of people attending each meeting: 10
- Duration of each meeting: 1.5 hours
- Cost of one fully loaded man-hour: $125
- Number of projects per year: 20

Using this information, the organization spends an average of $2.25 million (U.S.) for people to attend team meetings in one year! Now, what if we could purchase a software package that reduced the number of project meetings by 10 percent? We could save the organization $225,000 each year!

The goal of software selection must be the benefits to the project and the organization, such as cost reductions through efficiency, effectiveness, standardization, and consistency. A $500 software package can, more often than not, reduce project costs just as effectively as a $200,000 package. What is unfortunate is that the people who order the software focus more on the number of packaged features than on how much money using the software will save.

Fallacy 4: We need to implement project management in small steps with a small breakthrough project that everyone can track. Wrong method! This works if time is not a constraint. The best bet is to use a large project as the breakthrough project. A successfully managed large project implies that the same processes can work on small projects, whereas the reverse is not necessarily true.

On small breakthrough projects, some people will always argue against the implementation of project management and find numerous examples of why it will not work. Using a large project generally comes with less resistance, especially if project execution proceeds smoothly.

There are risks with using a large project as the breakthrough project. If the project gets into trouble or fails because of poorly implemented project management, significant damage to the company can occur. There is a valid argument for starting with small projects, but the author's preference is for larger projects.

Fallacy 5: We need to track and broadcast the results of the breakthrough project. Wrong course of action! Expounding a project's success benefits only that project rather than the entire company. Illuminating how project management caused a project to succeed benefits the entire organization. People then understand that project management can be used on a multitude of projects.

Fallacy 6: We need executive support. Almost true! We need *visible* executive support. People can easily differentiate between genuine support and lip service. Executives must walk the walk. They must hold meetings to demonstrate their support of project management and attend various project team meetings. They must maintain an open-door policy for problems that occur during project management implementation.

Fallacy 7: We need a project management course so our workers can become PMP® credential holders.[25] Once again, almost true! What we really need is lifelong education in project management. Becoming a PMP® credential holder is just the starting point. There is life beyond the *PMBOK® Guide*. Continuous organization-wide project management education is the fastest way to accelerate maturity in project management.

Needless to say, significantly more fallacies than discussed here are out there, waiting to block your project management implementation and delay its maturity. What is critical is that your organization implements project management through a well-thought-out plan that receives organization-wide buy-in and support. Fallacies create unnecessary delays. Identifying and overcoming faulty thinking can help fast-track your organization's project management maturity.

3.14 MOTOROLA

"Motorola has been using project management for well over 30 years in 2005," according to a spokesperson at Motorola. The forces that drove the company to recognize the need to become successful in project management were increasing complexity of projects coupled with quality problems, and schedule and cost overruns, which drove senior management to seek an alternative management solution to what previously existed. A chronology of what Motorola did early on to get where it is today as well as some of the problems encountered are as follows:

- 1995: Hire a director of project management.
- 1996: First hire PMs—formal role definition and shift in responsibilities for scheduling and ship acceptance.
- 1998: Formal change control instituted—driven by PMs.
- 1998: Stage gates rolled out and deployed across all projects.
- 2000: Deployment of time-tracking tool.
- 2001: Deployment of a more formal resource tracking.
- 2002: Improved resource planning and tracking.
- 2004: Project cost accounting.

---

25. PMP is a registered mark of the Project Management Institute, Inc.

Initially, program management was viewed as an overhead activity, with engineering managers reluctant to give up program control and status communication. It was only through senior management's commitment to formal project management practices that a PMO was created and roles and responsibilities shifted. Full engineering management acceptance did not occur until after several years of project management demonstrating the value of structured program management practices, which resulted in consistent on-time product delivery. These include formal, integrated, and complete project scheduling, providing independent cross-functional project oversight, communicating unbiased program status, coordinating cross-functional issue resolution, and the identification and management of program risks. Later, project management responsibilities increased to include other key areas such as customer communications, scope control and change management, cost containment, and resource planning.

Executive support was provided through sponsorship of the development of the program management function. The reporting structure of the function has been carefully kept within an appropriate area of the organization, ensuring independence from undue influences from other functional areas so that objective and independent reporting and support would be provided.

3.15 TEXAS INSTRUMENTS[26]

A critical question facing companies is whether the methodology should be developed prior to establishing a project management culture. Companies often make the fatal mistake of believing that the development of a project management methodology is the solution to their ailments. While this may be true in some circumstances, the excellent companies realize that people execute methodologies and that the best practices in project management might be achieved quicker if the focus is on the people rather than the tools. One way to become good at project management is to develop a success pyramid as shown in Figure 3–18. Every company has its own approach as to what should be included in a success pyramid.

Texas Instruments recognized the importance of focusing on people as a way to accelerate project success. Texas Instruments developed a success pyramid for managing global projects that focused first on people.[27] The challenges on global projects are often more difficult than on national projects because of team diversification, different religions, national politics, and different laws.

Understanding and trust became the driver for establishing global project management practices. Texas Instruments realized that a culture that was based upon understanding and trust was a necessity whereby team members could feel free to state

---

26. Material in Section 3.14 is reproduced from H. Kerzner, *Advanced Project Management: Best Practices in Implementation* (Hoboken, NJ: Wiley, 2004), pp. 46–48.

27. For a detailed description of the Texas Instruments Success Pyramid, see H. Kerzner, *Advanced Project Management: Best Practices in Implementation* (Hoboken, NJ: Wiley, 2004), pp. 46–48.

Figure 3–18. Success pyramid.

their opinions without fear of reprimand and this would lead to a reduction in project conflicts and better engagement by the team. For this to work well, it must be supported by senior management, which is the second level of the success pyramid in Figure 3-15.

At Texas Instruments, the emphasis on culture became a best practice. It is unfortunate that more companies do not realize the importance of this.

3.16 NAVIAIR: ON TIME—ON BUDGET[28]

How to Make Big and Complex Programs a Success

Recognize the Setting

Air navigation service provision in Europe is one of the last market segments that has not been liberalized to any larger extent Figure 3–19). Air navigation is—with the exception of the tower area—still a monopoly for the 37 air navigation service providers and as Siim Kallas, vice president of the European Commission, expressed in his opening speech at the Single European Sky— The Time for Action conference, in Limassol, Cyprus, on October 10, 2012: "We are moving towards a regulatory environment which is more streamlined, coherent and based on a market economy."

In parallel, this industry has been heavily regulated in the same way as the railway and the medical sectors. New demands are scoped as European Union (EU) regulations, national legislation, and new or updated ICAO [International Civil Aviation Organization] standards are continuously rolled out with tight target dates to be met.

---

28. Material in this section, "Naviair: On Time – On Budget", has been provided by Mikael Ericsson, Director ATM Projects & Engineering, Steen Myhre Taschner Erichsen, Director/Manager Project Office, ATM Projects & Engineering (B.Sc.E.E.) and Michael Wibelius, Tactical Management (M. Sc. Planning and Management). ©2022 by Naviair. All rights reserved. Reproduced by permission.

Figure 3–19. How to make big and complex programs a success.

Significant investments are made in order to meet the regulatory requirements. At the same time, traffic is stagnating and even decreasing in the Danish airspace due to the fifth year of consecutive recession reported in the first quarter of 2013. In other words, limited resources are available from a service provider point of view to meet the ongoing complex and demanding nature of the aviation sector.

Based on the growing number of EU regulations provided by the European Commission, there is an expectation that the air navigation service provision in Europe shall develop more efficient ways to perform air traffic control. In this context, Naviair has formed COOPANS in cooperation with the Swedish, Irish, Austrian, and Croatian air navigation service providers and the French supplier Thales. This cooperation shares the necessary costs and resources for the development, implementation, and mainte-nance of a state-of-the-art air traffic management (ATM) system that is compliant with existing and future EU regulations. So far the COOPANS program has been very suc-cessful, and today it is operational in four countries and in six air traffic control centrals. The seventh control central located in Zagreb became operational in operations in 2014.

In this setting, there is a strong need for success. Scarce resources and external pressure make this endeavor challenging. However, when we benchmarked ourselves with other similar market segments, we were proud of how successfully we actually performed our programs and projects. There is no room for failure, and in Naviair we have a hit rate of nearly 100 percent when talking about delivering on time and on budget.

Naviair's ability to cope with the setting and at the same time delivering on time and on budget is based on six main principles (see Figure 3–20):

1. Build confidence
2. Plan for success
3. Manage performance and culture
4. Tailor processes
5. Organize and report
6. Communicate everywhere

Since the main principles are not strictly interrelated and since success does not necessarily rely on a full rollout of every principle, the user level of the principles can be tailored to the organization in question as some parameters may be more useful in some organizations than in others. Therefore, the senior management and the project/ program managers specifically (as they are the target audience of this section) are free to pick and choose from the ideas contained in the description of each of the principles. However, one should bear in mind that it is recommended to maximize the use of each of the principles as described in this section.

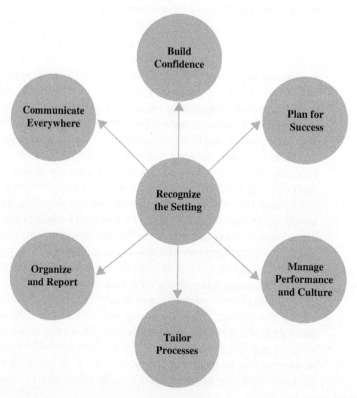

Figure 3–20. On-time–on-budget framework (Naviair).

Build Confidence

Change management is too often not prioritized or not taken into account when a big program is performed. Many companies have had a negative experience with previous projects, and therefore management simply does not expect internal organizations to be able to run a big program smoothly. In Denmark, analyses of IT projects performed by the government revealed that as many as 75 percent of the projects did not deliver on time. Furthermore, a significant amount of the projects did not deliver on budget, and 40 percent of these were heavily overspent.

When a program is initiated in Naviair, we start by ensuring that the organization understands the changes that are about to come. Questions as to why the changes are necessary are welcome, as are discussions concerning alternatives. This supports a demystification of the changes in the organization and is an important initial step toward avoiding time that has to be invested in this at a later stage when things either cannot be changed or are accompanied by great difficulties and/or expenses.

A key area of concern is identification of problems while at the same time recognizing the fact that these may be populated by people with different backgrounds and concerns. In this context, it is important to avoid reacting protectively and allowing groups with different professions to express their opinions. It is our experience that this makes the change process smoother and allows for fine-tuning of the direction in order to mitigate different risks that otherwise might turn into problems. You should make sure that you listen to all parts of the organization and create a common view, even if this will change the scope slightly. It is very easy to change the scope at this stage compared to doing it at later stages of the program/project. In order to ensure that all involved internal stakeholders have the same understanding of the changes, a high-level project frame forming the main benefits and measurable objectives should be agreed on as the first thing, before any actual project preinvestigations are performed.

The governance structure must also allow the different stakeholders to discuss and get the appropriate level of information during all phases of the program/project. Naviair performed a very large program containing more than 50 interrelated projects that went into operation at the end of 2007 and led to a completely new ATM system in Denmark. The responsibility for integrating all technical solutions from many different suppliers was put on our shoulders. Although the technological challenges were great, the change management was even greater. In fact, it is a mental challenge to pull through such a program if you expect to meet the targets spot on. Naviair managed to do so, but we had to invest a lot of time and concern in order to implement this governance structure and to ensure that all stakeholders, internally as well as externally, were involved. We also had to perform regular surveys to make sure that everybody supported the changes, and sometimes certain groups had concerns that had to be addressed immediately. The mantra in this context is that such concerns are very useful in the process of making the program successful. We never tried to defend ourselves or to make difficult comments go away, and this has become a permanent practice in Naviair today.

Plan for Success

A golden rule in Naviair is to define a date for going operational with the new system as soon as possible. If possible, we even set an exact time. In the abovementioned program, we also had a countdown clock on the Naviair intranet front page. It is much easier if you have the courage to define a very visible target for the organization. The pitfall is, however, that the date cannot be changed. A professional tennis player like Roger Federer does not think about possible failure when he enters the tennis court, and you must do the same: Be a professional each and every day with one focus—on time and on budget.

If you succeed in getting your organization behind such a date, which is an achievable goal, you can start to plan backward. If you have a gate-driven approach, which is strongly recommended, you will immediately find yourself and your teams very busy even if you have a multiyear program. You should always remember that in the beginning, the time schedule is a qualified guess. The schedule will gradually improve and be more detailed as the program moves onward. A program manager who is able to follow the time schedule has a lot to gain and all work related to revision of the schedule is avoided. When the program is ready, the time schedule will be a perfect plan. However, you should never use this argument to fool yourself into postponing planning. As long as the operational date is not changed, milestones can be adjusted if necessary, which is often the case with most programs.

The expected outcome from late activities, such as verification and validation, training, or live tests, must be addressed early. Your organization, if not mature, will, for example, most probably argue that the training cannot be planned before the system is physically in place. Such arguments should be taken seriously due to the fact that they express that the stakeholders do not know how to proceed in this early phase of the program. Once the different parts of the organization learn to address the topics on the right level, the work can be initiated early and the targets can be met. Inexperienced members of the program must be supported by a PMO or similar in order to learn how to plan the activities before they enter the solution mode.

You have to communicate the plan in your governance structure repeatedly. The key to success is to obtain buy-ins from all stakeholders, and some facts must be spelled out. At the same time, all fora should be taken into consideration as described in the principle "Build Confidence" discussed earlier. Different governance parties must be addressed at the appropriate level, and some external stakeholders might be satisfied with the going-operational date if they are not affected by your tests, and so on.

One of the most important key messages from Naviair is never to operate with a plan B including an alternative date for operation of the system. You are allowed and advised to implement mitigating actions in relation to the risk of missing the operational date or "O-date" (e.g., by having a well-tested roll-back plan and other similar action plans). However, only one plan should be available, and management and internal stakeholders must agree on this plan and communicate the following: We will make it!

Manage Performance and Culture

A team is not automatically stronger than the individuals but with a high-performance team culture, the outcome can be fantastic. A common method used by sports teams,

Special Forces, or the like is seldom used in program management. When a SWAT (special weapons and tactics) team is gathered for the first time or when the team is changed, members usually spend 11 weeks getting to know each other. At this point, the task to solve as such is not even on the agenda. What is then the purpose of such a social event?

When a task is performed by a SWAT team, the participants are totally depending on each other. In order to be able to trust each other 100 percent, it takes much more than just a number of professional individuals. You also have to know the persons behind you or supporting you, the social factors, and parts of their life histories. In a SWAT team, you are about to put your life in another person's hands and that would not work with a total stranger. The same goes for a challenging program that might affect your night's sleep, your family life, and your leisure activities. When a program is pulled off successfully, most of the participants would say the same: "It has been hard work but an experience of a lifetime!"

The process of building a high-performance team, as shown in Figure 3–21, should start with the social interaction in an environment that is protected from the daily interference from the office or the factory floor. In this environment, the first step would be asking *Why* this change? At the same time, the team members should get acquainted with each other. Many different methods could be used when socializing; one used in Naviair is to ask every participant to bring along a very important personal item and make a speech about this. You will find new sides of your colleagues that you never thought existed. You are now at the second step called: *Who*. At this stage you build up the trust among the team members.

Stay at step 1 and/or 2 as long as you can, at least during a seminar and at a follow-up meeting. Now you can go on to step 3, which is *What*. At this step, you scope the changes and the task. This step and step 4: *How*, is straightforward for a program organization where you have established governance, terms of reference, etc. In most

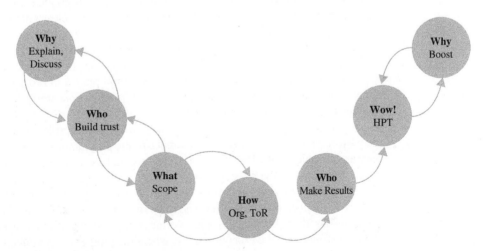

Figure 3–21. High-performance teams.

organizations, *what* and *how* are the starting points. Using *what* and *how* will work but will, however, only bring a mediocre performance. If you start at step 3 or 4, the process cannot be reversed due to the fact that it is very hard for most people (and at least a couple in your team) to leave the "solution mode" once it is initiated.

The next step is starting to work with your scope and your team. If you did start out the right way, it is very likely that you will experience the *wow* step, where the team is high performing. This must be maintained so that the last step will repeat the first step, *why*, again, which you have to go through at least once a year or immediately after you have replaced one of your team members. If you replace one of your team members, you will by definition have a new team so do not be misled into thinking that a high-performance culture will continue forever.

In multicultural teams, as most teams are today, you must as a program manager be knowledgeable concerning cultural differences. Having some knowledge of home countries, history, religion, political scenarios, and culture (e.g., a male- or female-dominant culture) will bring about successful team building and support the achievement of a high-performance team.

Tailor Processes

Being a program manager is like being "between the devil and the deep blue sea." Though you are on the top of your own governance, you have many people and circumstances to consider. You have to cope with the environment as the program moves along. Your target will be affected economically, technologically, and by market fluctuations, but often your program may also be affected politically due to cooperation or alliances that your company may be a part of. These factors may add further complexity to the program.

The Naviair project model, which contains the project processes and the templates that are used for initiating, executing, delivering, and closing projects, is based on the PRINCE2 principles. However, it is tailored to the organizational setup, nature, and setting of our projects. As such, the Naviair project model is pragmatic in its nature with a lean paper flow and a simple phase structure with a very clear go/no-go decision to be made by the steering group (please refer to the "Organize and report" principle) between two phases. The project processes are clearly linked to the surrounding company processes such as the annual budget process, maintenance procedures, and so on.

The Naviair project model focuses on the initial phases in order to ascertain that the project is justified and that the right decision is made concerning the product specifications and the scope of the program/projects before proceeding to the execution phase. The project initiation phase is based on a high-level project frame compiled by the project owner forming the main reference of the program/project with clearly stated measurable objectives. The PM is assigned to analyze possible solutions—if any— within the scope of the project frame. In this context and as a final step in the initial phases, the PM makes a relatively detailed project assessment containing estimates concerning budget, resources, time, and main risks to form the basis for a recommended solution. Based on this analysis, the steering group decides whether or not the project should continue into the execution phase where the progress is continuously monitored (please refer to the Organize and Report principle). If the project is no longer justifiable, it can be terminated at any time during the project life cycle. Once the project deliverables are complete, the project handover phase is carried out before the actual closure, and the lessons-learned phase is initiated. The latter phase provides for knowledge sharing and benefits realization, which in turn may lead to a new project being initiated. This approach has been very successful with a very good track record of being on time and budget with every investment made throughout the process.

The project realization phase is a lean one focusing on monitoring the progress and mitigating risks and problems.

The portfolio prioritization is performed in accordance with the Naviair prioritization standards. These standards were implemented in order to make sure that we only realize programs and projects that will support and strengthen our business values.

Organize and Report Your governance is very important, and a rule of thumb is to place the sponsorship fairly high in the organization. This person should be a member of the executive management and on the business value side, such as the chief operating officer or the chief executive officer. If the sponsorship is held by the chief financial officer or others at the senior levels of management, it normally brings another type of focus to the program, a very strong focus on either the financial side or the technological side.

The sponsor should be the chairman of the program steering group, and that group should be manned by management representatives from each organizational key area in order to ensure that decisions concerning prioritization and program/project phase shifts are holistic and aligned within the whole organization. The program steering group should meet regularly. The frequency of the steering group meetings very much depends on the program, but it can be an advantage to meet more frequently as the going-operational date approaches. If your organizational setup allows for forming one steering group for all programs and projects, this would be advantageous as you can prioritize the full portfolio at once and thus benefit from one holistic view.

A complex program should have its own administrative and planning support. Richard Branson, the entrepreneur and founder of the brand Virgin, stated, "I prefer a brilliant assistant." We are of the same opinion, and for Naviair's most complex program, COOPANS, we do have such an organization and support.

The subgroups in the program organization should be balanced in such a way that it reflects the different internal stakeholders in a positive way and is equipped with sufficient competencies to make decisions related to their area of expertise in order to ensure progress. External stakeholders can be part of the organization, but it is more likely that they are part of a group of interest or a user group. It is not important what kind of subgroups you define; what is important is how the subgroups interact with each other and with you as a program manager.

In Naviair we prefer a pragmatic view on reporting. Our template used for status reporting is made as a simple Excel tool and is based on traditional traffic light reporting. The report itself consists of six parameters, some of which represent business KPIs linked to Naviair's overall balanced scorecard. Depending on the complexity of the portfolio, the reporting frequency varies from once a week to once a month. The most important thing is, however, not the current traffic light status. That is all history and old information. The program risks and problems and the proactivity to mitigate for these are much more important. If you have a large portfolio with many programs and projects, you will need more complex tools and risk management. In Naviair we use pragmatic tools for risk management, experienced and certified managers, and a lot of physical meetings to interact in relation to the risks and problems. If you have the experience and have performed many projects previously, you can use your gut feeling to decide where to use your efforts. Therefore, we focus on physical one-to-one meetings, discussions, and interactions rather than on extensive reporting.

In Naviair we have learned that it is very hard for an organization to decide whether the results of a program are satisfactory. Often an organization turns shaky and too detail-oriented before finally going operational with a new system—often with delays as a consequence. In Naviair we have developed an accept-criteria matrix to decide whether a program is satisfactory to be put into operation. We have two levels: one with detailed milestones and descriptions per criteria and one steering group level that can make a quick report on a two-slide PowerPoint presentation. When all criteria are met, we are ready to go into operation. There will be no hassles or discussions as to whether we are ready or not.

Communicate Everywhere

Successful communication concerning a program is not performed without a communication plan. The communication plan should be based on a stakeholder analysis, a SWOT analysis, and/or similar in order to get a clear picture of the target audience and how this audience may react to certain statements. The communication plan based on the abovementioned analyses will provide a more targeted communication, which in the end will see that you achieve the result you were looking for.

You should consider any possible media as well as the frequency and timing of addressing the different internal and external stakeholders and at which level of information. When addressing the different stakeholders, the key message must be clear, consistent, and easy to understand and relate to. Use the producers of great beverages, cars, or service companies and how they communicate their product values as prominent examples. You should address your business values and not the technological advantages, which to most people are useless and only seem expensive. You should repeat the business values and key messages until the program has been executed.

You can target your communication toward the different key fora in many different ways, and you should not confine yourself only to use well-known traditional communication tools. A successful program manager will have to use nearly half of his or her working hours just to communicate and lobby in order to ensure the success of the program. A program manager who prioritizes communication correctly will never refuse a possibility to present the program and the related business values.

It is good practice in communication to submit articles to magazines, define a project portal to which the internal organization will have access, publish news on both the internet and the intranet, arrange kick-off meetings and open-house events, and of course, provide physical presentations whenever a possibility occurs. One of the keys to effective communication is to vary the means of communication and to find new creative ways of addressing the stakeholders, for example, arrangements for merchandisers where the key message is displayed, involving the canteen, broadcast interviews with key members of the program/project on the news portal or display banners with key messages at often-visited places, for instance, by the coffee machine. The latter also provides for a more informal way of communication, since the coffee machine and/ or similar places represent "safe zones" where communication flows freely between employees. A banner with a positive message may lead the conversation in a more positive direction or simply provide more "airtime" and visibility among the employees. Informal ways of communicating have been utilized with success in Naviair. An example, among others, is the weekly breakfast meetings chaired by the PM on Fridays where employees meet and discuss the progress of the program/project while enjoying a Danish pastry and a cup of coffee/tea. Since no high-level management is present at these meetings, discussions of worries and information that otherwise might not have been brought up may occur. The same could be the case concerning rumors, which the PM will have the opportunity to spot and react to in order to avoid the undermining of progress and key business values of the program/project. The PM may follow up on the breakfast meetings with an informal status email to keep the people who were not present at the meeting up to date on the program/project and invite people to present comments if they possess any other or contradicting information.

Lack of communication and information will make people develop their own information when meeting at the coffee machine. This will lead to rumors and worries that have to be handled seriously since rumors can take over facts. The simple cure is to attack all rumors when they are heard; they should be challenged in order to see if there are any facts behind them. Often there are no facts behind occurring rumors. However, if the rumors are based on facts, a solution has to be found.

You should not forget to celebrate any important milestone that has been achieved. Celebrate in an acceptable way that is compatible with the company culture. Some company cultures have no problems with providing free dinners or tickets to the opera whereas others cannot accept this for tax reasons or simply because it is not *comme il faut*. An inexpensive toy race car for an achieved site acceptance test could signal gratefulness from the organization. In Naviair some program managers have nice collections of toy Ferraris and Lamborghinis that they proudly display very visibly in their offices.

The most important and effective thing you can do in relation to projects is to praise and give credit and recognition to the staff who are involved in the project. The most powerful recognition you could give is the one you communicate to a person from another department.

Summary

The six main principles presented in this section are continuously fine-tuned based on lessons learned, external inputs, and as the industry for air navigation service provision in Europe develops in order to maximize Naviair's project performance and ability to deliver on time and budget. The key message here is that we have to recognize the setting and adapt to it and realize that we are living in a dynamic world—no matter how good your best practices may be, they may become yesterday's news if focus is not put on continuous development with a willingness to change and adapt. Finally, as the late Irish playwright, socialist, and a cofounder of the London School of Economics, George Bernard Shaw, put it: "Those who cannot change their minds cannot change anything."

3.17 AVALON POWER AND LIGHT

Avalon Power and Light (a disguised case) is a mountain states utility company that, for decades, had functioned as a regional monopoly. All of this changed in 1995 with the beginning of deregulation in public utilities. Emphasis was now being placed on cost cutting and competitiveness.

The Information Systems Division of Avalon was always regarded as a thorn in the side of the company. The employees acted like prima donnas and refused to accept any of the principles of project management. Cost-cutting efforts at Avalon brought to the surface the problem that the majority of the work in the Information Systems Division could be outsourced at a significantly cheaper price than performing the work internally. Management believed project management could make the division more competitive, but would employees be willing to accept the project management approach?

According to a spokesperson for Avalon Power and Light:

> Two prior attempts to implement a standard application-development methodology had failed. Although our new director of information systems aggressively supported this third effort by mandating the use of a standard methodology and standard tools, significant obstacles were still present.
>
> The learning curve for the project management methodology was high, resulting in a tendency of the leads to impose their own interpretations on methodology tasks rather than learning the documented explanations. This resulted in an inconsistent interpretation of the methodology, which in turn produced inconsistencies when we tried to use previous estimates in estimating new projects.
>
> The necessity to update project plans in a timely manner was still not universally accepted. Inconsistency in reporting actual hours and finish dates resulted in inaccurate availabilities. Resources were not actually available when indicated on the departmental plan.
>
> Many team leads had not embraced the philosophy behind project management and did not really subscribe to its benefits. They were going through the motions, producing the correct deliverables, but managing their projects intuitively in parallel to the project plan rather than using the project plan to run their projects.
>
> Information systems management did not ask questions that required use of project management in reporting project status. Standard project management metrics were ignored in project status reports in favor of subjective assessments.

The Information Systems Division realized that its existence could very well be based on how well and how fast it would be able to develop a mature project management system. By 1997, the sense of urgency for maturity in project management had permeated the entire Information Systems Division. When asked what benefits were achieved, the spokesperson remarked:

> The perception of structure and the ability to document proposals using techniques recognized outside of our organization has allowed Information Systems to successfully compete against external organizations for application development projects.
>
> Better resource management through elimination of the practice of "hoarding" preferred resources until another project needs staffing has allowed Information Systems to actually do more work with less people.
>
> We are currently defining requirements for a follow-on project to the original project management implementation project. This project will address the lessons learned from our first two years. Training in project management concepts (as opposed to tools training) will be added to the existing curriculum. Increased emphasis will be placed on why it is necessary to accurately record time and task status. An attempt will be made to extend the use of project management to non-application-development areas, such as network communications and technical support. The applicability of our existing methodology to client–server development and internet application development will be tested. We will also explore additional efficiencies such as direct input of task status by individual team members.
>
> We now offer project management services as an option in our service-level agreements with our corporate "customers." One success story involved a project to implement a new corporate identity in which various components across the corporation were brought together. The project was able to cross department boundaries and maintain an aggressive schedule. The process of defining tasks and estimating their durations resulted in a better understanding of the requirements of the project. This in turn provided accurate estimates that drove significant decisions regarding the scope of the project in light of severe budget pressures. Project decisions tended to be based on sound business alternatives rather than raw intuition.

3.18 ROADWAY EXPRESS

In the spring of 1992, Roadway Express realized that its support systems (specifically information systems) had to be upgraded in order for the company to be well-positioned for the twenty-first century. Mike Wickham, then president of Roadway Express and later chairman of the board, was a strong believer in continuous change. This was a necessity for his firm because rapid changes in technology mandated that reengineering efforts be an ongoing process. Several of the projects to be undertaken required a significantly larger number of resources than past projects had needed. Stronger interfacing between functional departments would also be required.

At the working levels of Roadway Express, knowledge of the principles and tools of project management was minimal at best in 1992. However, at the executive levels, knowledge of project management was excellent. This would prove to be highly beneficial. Roadway Express recognized the need to use project management on a two-year project that had executive visibility and support and that was deemed strategically critical to the company. Although the project required a full-time PM, the company chose to appoint a line manager who was instructed to manage his line and the project at the same time for two years. The company did not use project management continuously, and the understanding of project management was extremely weak.

After three months, the line manager resigned from his appointment as a PM, citing too much stress and an inability to manage his line effectively while performing project duties. A second line manager was appointed on a part-time basis, and, like his predecessor, he found it necessary to resign as PM.

The company then assigned a third line manager, but this time released her from all line responsibility while managing the project. The project team and selected company personnel were provided with project management training. The president of the company realized the dangers of quick implementation, especially on a project of this magnitude, but was willing to accept the risk.

After three months, the PM complained that some of her team members were very unhappy with the pressures of project management and were threatening to resign from the company if necessary simply to get away from project management. But when asked about the project's status, the PM stated that the project had met every deliverable and milestone thus far. It was quickly apparent to the president, Mike Wickham, and other officers of the company that project management was functioning as expected. The emphasis now was how to "stroke" the disgruntled employees and convince them of the importance of their work and how much the company appreciated their efforts.

To quell the fears of the employees, the president assumed the role of the project sponsor and made it quite apparent that project management was here to stay at Roadway Express. The president brought in training programs on project management and appeared at each training program.

The reinforcement by the president and his visible support permeated all levels of the company. By June of 1993, less than eight months after the first official use of project management, Roadway Express had climbed farther along the ladder to maturity in project management than most other companies accomplish in two to three years due to the visible support of senior management.

Senior management quickly realized that project management and information systems management could be effectively integrated into a single methodology. Mike Wickham correctly recognized that the quicker he could convince his line managers to support the project management methodology, the quicker they would achieve maturity. According to Wickham:

> Project management, no matter how sophisticated or how well trained, cannot function effectively unless all management is committed to a successful project outcome. Before we put our current process in place, we actively involved all those line managers who thought it was their job to figure out all of the reasons a system would never work! Now the steering committee says, "This is the project. Get behind it and see that it works." It is a much more efficient use of resources when everyone is focused on the same goal.

3.19 KOMBS ENGINEERING

While some companies are fortunate enough to be able to identify crises early and take corrective action, others are not as fortunate. Although the next two companies appear to be outdated, there are valuable lessons that can be learned about what not to do when embarking on the path to maturity. Consider the Michigan-based Kombs Engineering (name of the company is disguised at company's request).

In June 1993, Kombs Engineering had grown to a company with $25 million in sales. The business base consisted of two contracts with the Department of Energy (DoE), one for $15 million and one for $8 million. The remaining $2 million consisted of a variety of smaller jobs for $15,000 to $50,000 each.

The larger contract with the DoE was a five-year contract for $15 million per year. The contract was awarded in 1988 and was up for renewal in 1993. The DoE had made it clear that, although it was very pleased with the technical performance of Kombs, the follow-on contract had to go through competitive bidding by law. Marketing intelligence indicated that the DoE intended to spend $10 million per year for five years on the follow-on contract with a tentative award date of October 1993. On June 21, 1993, the solicitation for proposal was received at Kombs. The technical requirements of the proposal request were not considered to be a problem for Kombs. There was no question in anyone's mind that on technical merit alone Kombs would win the contract. The more serious problem was that the DoE required a separate section in the proposal on how Kombs would manage the $10 million/year project as well as a complete description of how the project management system at Kombs functioned.

When Kombs won the original bid in 1988, there had been no project management requirement. All projects at Kombs were accomplished through the traditional organizational structure. Only line managers acted as project leaders.

In July 1993, Kombs hired a consultant to train the entire organization in project management. The consultant also worked closely with the proposal team in responding to the DoE project management requirements. The proposal was submitted to the DoE during the second week of August. In September 1993, the DoE provided Kombs with a list of questions concerning its proposal. More than 95 percent of the questions involved project management. Kombs responded to all questions.

In October 1993, Kombs received notification that it would not be granted the contract. During a postaward conference, the DoE stated that it had no "faith" in the Kombs project management system. Kombs Engineering is no longer in business.

Kombs Engineering is an excellent case study to give students in project management classes. It shows what happens when a subcontractor does not recognize how smart the customer has become in project management. Had Kombs been in close contact with its customers, the company would have had five years rather than one month to develop a mature project management system.

3.20 WILLIAMS MACHINE TOOL COMPANY

The strength of a culture can not only prevent a firm from recognizing that a change is necessary but also block the implementation of the change even after need for it is finally realized. Such was the situation at Williams Machine Tool Company (another disguised case).

For 75 years, the Williams Machine Tool Company had provided quality products to its clients, becoming the third largest U.S.-based machine tool company by 1980. The company was highly profitable and had an extremely low employee turnover rate. Pay and benefits were excellent.

Between 1970 and 1980, the company's profits soared to record levels. The company's success was due to one product line of standard manufacturing machine tools. Williams spent most of its time and effort looking for ways to improve its bread-and-butter product line rather than to develop new products. The product line was so successful that other companies were willing to modify their production lines around these machine tools rather than asking Williams for major modifications to the machine tools.

By 1980, Williams was extremely complacent, expecting this phenomenal success with one product line to continue for 20 to 25 more years. The recession of 1979 to 1983 forced management to realign its thinking. Cutbacks in production had decreased the demand for the standard machine tools. More and more customers were asking either for major modifications to the standard machine tools or for a completely new product design.

The marketplace was changing, and senior management recognized that a new strategic focus was necessary. However, attempts to convince lower-level management and the workforce, especially engineering, of this need met strong resistance. The company's employees, many of them with over 20 years of employment at Williams, refused to recognize this change, believing that the glory days of yore would return at the end of the recession.

In 1986, the company was sold to Crock Engineering. Crock had an experienced machine tool division of its own and understood the machine tool business. Williams was allowed to operate as a separate entity from 1985 to 1986. By 1986, red ink had appeared on the Williams balance sheet. Crock replaced all of the Williams senior managers with its own personnel. Crock then announced to all employees that Williams would become a specialty machine tool manufacturer and the "good old days" would never return. Customer demand for specialty products had increased threefold in just the last 12 months alone. Crock made it clear that employees who would not support this new direction would be replaced.

The new senior management at Williams recognized that 85 years of traditional management had come to an end for a company now committed to specialty products. The company culture was about to change, spearheaded by project management, concurrent engineering, and total quality management.

Senior management's commitment to project management was apparent by the time and money spent on educating the employees. Unfortunately, the seasoned 20-year veterans still would not support the new culture. Recognizing the problems, management provided continuous and visible support for project management in addition to hiring a project management consultant to work with the people. The consultant worked with Williams from 1986 to 1991.

From 1986 to 1991, the Williams Division of Crock Engineering experienced losses in 24 consecutive quarters. The quarter ending March 31, 1992, was the first profitable one in over six years. Much of the credit was given to the performance and maturity of the project management system. In May 1992, the Williams Division was sold. More than 80 percent of the employees lost their jobs when the company was relocated over 1,500 miles away.

Williams Machine Tool Company did not realize until too late that the business base had changed from production-driven to project-driven. Living in the past is acceptable only if you want to be a historian. But for businesses to survive, especially in a highly competitive environment, they must look ahead and recognize that change is inevitable.

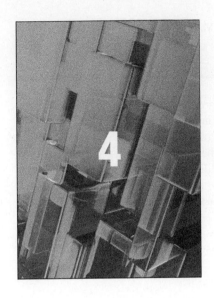

4 Project Management Methodologies

In Chapter 1, we described the life-cycle phases for achieving maturity in project management. The fourth phase was the growth phase, which included the following:

- Establish life-cycle phases.
- Develop a project management methodology.
- Base the methodology on effective planning.
- Minimize scope changes and scope creep.
- Select the appropriate software to support the methodology.

The importance of a good methodology cannot be overstated. Not only will it improve performance during project execution, but it will also allow for better customer relations and customer confidence. Good methodologies can also lead to sole-source or single-source procurement contracts.

Creating a workable methodology for project management is no easy task. One of the biggest mistakes made is developing a different methodology for each type of project. Another is failing to integrate the project management methodology and project management tools into a single process, if possible. When companies develop project management methodologies and tools in tandem, two benefits emerge: First, the work is accomplished with fewer scope changes. Second, the processes are designed to create minimal disturbance to ongoing business operations.

This chapter discusses the components of a project management methodology and some of the most widely used project management tools. Detailed examples of methodologies at work are also included.

4.1 EXCELLENCE DEFINED

Excellence in project management is often regarded as a continuous stream of successfully managed projects. Without a project management methodology, repetitive successfully completed projects may be difficult to achieve.

Today, everyone seems to agree somewhat on the necessity for a project management methodology. However, there is still disagreement on the definition of excellence in project management, the same way that companies have different definitions for project success. In this section, we discuss some of the different definitions of excellence in project management.

Some definitions of excellence can be quite simple and achieve the same purpose as complex definitions. According to a spokesperson from Motorola:

Excellence in project management can be defined as:

- Strict adherence to scheduling practices
- Regular senior management oversight
- Formal requirements change control
- Formal issue and risk tracking
- Formal resource tracking
- Formal cost tracking

A spokesperson from AT&T defined excellence in this way:

Excellence [in project management] is defined as a consistent Project Management Methodology applied to all projects across the organization, continued recognition by our customers, and high customer satisfaction. Also, our project management excellence is a key selling factor for our sales teams. This results in repeat business from our customers. In addition, there is internal acknowledgement that project management is value-added and a must have.

4.2 RECOGNIZING THE NEED FOR METHODOLOGY DEVELOPMENT

Simply having a project management methodology and following it do not lead to success and excellence in project management. The need for improvements in the system may be critical. External factors can have a strong influence on the success or failure of a company's project management methodology. Change is a given in the current business climate, and there is no sign that the future will be any different. The rapid changes in technology that have driven changes in project management over the past two decades are not likely to subside. Another trend, the increasing sophistication of consumers and clients, is likely to continue, not go away. Cost and quality control have become virtually the same issue in many industries. Other external factors include rapid mergers and acquisitions and real-time communications.

Project management methodologies are organic processes and need to change as the organization changes in response to the ever-evolving business climate. Such changes, however, require that managers on all levels be committed to them and have a vision that calls for the development of project management systems along with the rest of the organization's other business systems.

Today, companies are managing their business through projects. This is true for both non-project-driven and project-driven organizations. Virtually all activities in an organization can be treated as some sort of project. Therefore, it is only fitting that well-managed companies regard a project management methodology as a way to manage the entire business rather than just projects. Business processes and project management processes will be merged together as the project manager is viewed as the manager of part of a business rather than just the manager of a project.

Developing a standard project management methodology is not for every company. For companies with small or short-term projects, such formal systems may not be cost-effective or appropriate. However, for companies with large or ongoing projects, developing a workable project management system is mandatory.

For example, a company that manufactures home fixtures had several project development protocols in place. When it decided to begin using project management systematically, the complexity of the company's current methods became apparent. The company had multiple system development methodologies based on the type of project. This became awkward for employees who had to struggle with a different methodology for each project. The company then opted to create a general, all-purpose methodology for all projects. The new methodology had flexibility built into it. According to one spokesman for the company:

> Our project management approach, by design, is not linked to a specific systems development methodology. Because we believe that it is better to use a (standard) systems development methodology than to decide which one to use, we have begun development of a guideline systems development methodology specific for our organization. We have now developed prerequisites for project success. These include:

- A well-patterned methodology
- A clear set of objectives
- Well-understood expectations
- Thorough problem definition

During the late 1980s, merger mania hit the banking community. With the lowering of costs due to economies of scale and the resulting increased competitiveness, the banking community recognized the importance of using project management for mergers and acquisitions. The quicker the combined cultures became one, the less the impact on the corporation's bottom line.

The need for a good methodology became apparent, according to a spokesperson at one bank:

> The intent of this methodology is to make the process of managing projects more effective: from proposal to prioritization to approval through implementation. This methodology is not tailored to specific types or classifications of projects, such as system development efforts or hardware installations. Instead, it is a commonsense approach to assist in prioritizing and implementing successful efforts of any jurisdiction.

In 1996, the information services (IS) division of one bank formed an IS reengineering team to focus on developing and deploying processes and tools associated with project management and system development. The mission of the IS reengineering team was to improve performance of IS projects, resulting in increased productivity and improved cycle time, quality, and satisfaction of project customers.

According to a spokesperson at the bank, the process began as follows:

> Information from both current and previous methodologies used by the bank was reviewed, and the best practices of all these previous efforts were incorporated into this document. Regardless of the source, project methodology phases are somewhat standard fare. All projects follow the same steps, with the complexity, size, and type of project dictating to what extent the methodology must be followed. What this methodology emphasizes are project controls and the tie of deliverables and controls to accomplishing the goals.

To determine the weaknesses associated with past project management methodologies, the IS reengineering team conducted various focus groups. These focus groups concluded that the following had been lacking from previous methodologies:

- Management commitment
- A feedback mechanism for project managers to determine the updates and revisions needed to the methodology
- Adaptable methodologies for the organization
- A training curriculum for project managers on the methodology
- Focus on consistent and periodic communication on the methodology deployment progress
- Focus on the project management tools and techniques

Based on this feedback, the IS reengineering team successfully developed and deployed a project management and system development methodology. Beginning June 1996 through December 1996, the target audience of 300 project managers became aware and applied a project management methodology and standard tool (MS Project).

The bank did an outstanding job of creating a methodology that reflects guidelines rather than policies and provides procedures that can easily be adapted to any project in the bank. Up to 2017, the bank had continuously added flexibility to its project management approach, making it easier to manage all types of projects. Some of the selected components of the project management methodology are discussed next.

Organizing

With any project, you need to define what needs to be accomplished and decide how the project is going to achieve those objectives.

Each project begins with an idea, vision, or business opportunity, a starting point that must be tied to the organization's business objectives. The project charter is the foundation of the project and forms the contract with the parties involved. It includes a statement of business needs, an agreement of what the project is committed to deliver, an identification of project dependencies, the roles and responsibilities of the team members involved, and the standards for how project budget and project management should be approached. The project charter defines the boundaries of the project, and the project team has a great deal of flexibility as long as the members remain within the boundaries.

Planning

Once the project boundaries are defined, sufficient information must be gathered to support the goals and objectives and to limit risk and minimize issues. This component of project management should generate sufficient information to clearly establish the deliverables that need to be completed, define the specific tasks that will ensure completion of these deliverables, and outline the proper level of resources. Each deliverable affects whether each phase of the project will meet its goals, budget, quality, and schedule. For simplicity's sake, some projects take a four-phase approach:

1. *Proposal.* Project initiation and definition
2. *Planning.* Project planning and requirements definition
3. *Development.* Requirement development, testing, and training
4. *Implementation.* Rollout of developed requirements for daily operation

Each phase contains review points to help ensure that project expectations and quality deliverables are achieved. It is important to identify the reviewers for the project as early as possible to ensure the proper balance of involvement from subject matter experts and management.

Managing

Throughout the project, management and control of the process must be maintained. This is the opportunity for the project manager and team to evaluate the project, assess project performance, and control the development of the deliverables. During the project, the following areas should be managed and controlled:

- Evaluate daily progress of project tasks and deliverables by measuring budget, quality, and cycle time.
- Adjust day-to-day project assignments and deliverables in reaction to immediate variances, issues, and problems.
- Proactively resolve project issues and changes to control unnecessary scope creep.
- Aim for client satisfaction.
- Set up periodic and structured reviews of the deliverables.
- Establish a centralized project control file.

Two essential mechanisms for successfully managing projects are solid status-reporting procedures and issues and change management procedures. Status reporting is necessary for keeping the project on course and in good health. The status report should include the following:

- Major accomplishment to date
- Planned accomplishments for the next period
- Project progress summary:
  - Percentage of effort hours consumed
  - Percentage of budget costs consumed
  - Percentage of project schedule consumed
- Project cost summary (budget versus actual)
- Project issues and concerns
- Impact on project quality
- Management action items

Issues-and-change management protects project momentum while providing flexibility. Project issues are matters that require decisions to be made by the project manager, project team, or management. Management of project issues needs to be defined and properly communicated to the project team to ensure the appropriate level of issue tracking and monitoring. This same principle relates to change management because inevitably the scope of a project will be subject to some type of change. Any change management on the project that impacts the cost, schedule, deliverables, and dependent projects is reported to management. Reporting of issues and change management should be summarized in the status report, noting the number of open and closed items of each. This assists management in evaluating the project's health.

Simply having a project management methodology and using it does not lead to maturity and excellence in project management. There must exist a "need" for improving the system so it moves toward maturity. Project management systems can change as the organization changes. However, management must be committed to the change and have the vision to let project management systems evolve with the organization.

4.3 ENTERPRISE PROJECT MANAGEMENT METHODOLOGIES

Most companies today seem to recognize the need for one or more project management methodologies but either create the wrong methodologies or misuse those that have been created. Many times, companies rush into the development or purchase of a methodology without any understanding of the need for one other than the fact that their competitors have a methodology. According to Jason Charvat:

> Using project management methodologies is a business strategy allowing companies to maximize the project's value to the organization. The methodologies must evolve and be "tweaked" to accommodate a company's changing focus or direction. It is almost a mind-set, a way that reshapes entire organizational processes: sales and marketing, product design, planning, deployment, recruitment, finance, and operations support.

It presents a radical cultural shift for many organizations. As industries and companies change, so must their methodologies. If not, they're losing the point.[1]

Methodologies are a set of forms, guidelines, templates, and checklists that can be applied to a specific project or situation. It may not be possible to create a single enterprise-wide methodology that can be applied to each and every project. Some companies have been successful doing this, but there are still many companies that successfully maintain more than one methodology. Unless the project manager is capable of tailoring the EPM methodology to his or her needs, more than one methodology may be necessary.

There are several reasons why good intentions often go astray. At the executive level, methodologies can fail if the executives have a poor understanding of what a methodology is and believe that a methodology is:

- A quick fix
- A silver bullet
- A temporary solution
- A cookbook approach for project success[2]

At the working level, methodologies can also fail if they:

- Are abstract and high level
- Contain insufficient narratives to support these methodologies
- Are not functional or do not address crucial areas
- Ignore the industry standards and best practices
- Look impressive but lack real integration into the business
- Use nonstandard project conventions and terminology
- Compete for similar resources without addressing this problem
- Do not have any performance metrics
- Take too long to complete because of bureaucracy and administration[3]

Deciding on the type of methodology is not an easy task. There are many factors to consider, such as:

- The overall company strategy—how competitive are we as a company?
- The size of the project team and/or scope to be managed
- The priority of the project
- How critical the project is to the company
- How flexible the methodology and its components are[4]

Project management methodologies are created around the project management maturity level of the company and the corporate culture. If the company is reasonably

1. J. Charvat, *Project Management Methodologies* (Hoboken, NJ: Wiley, 2003), p. 2.

2. Ibid., p. 4.

3. Ibid., p. 5.

4. Ibid., p. 66.

mature in project management and has a culture that fosters cooperation, effective communications, teamwork, and trust, then a highly flexible methodology can be created based on guidelines, forms, checklists, and templates. Project managers can pick and choose the parts of the methodology that are appropriate for a particular client. Organizations that do not possess either of these two characteristics rely heavily on methodologies constructed with rigid policies and procedures, thus creating significant paperwork requirements with accompanying cost increases and removing the flexibility that the project manager needs for adapting the methodology to the needs of a specific client.

Charvat describes these two types as light methodologies and heavy methodologies.[5]

Light Methodologies

Ever-increasing technological complexities, project delays, and changing client requirements brought about a small revolution in the world of development methodologies. A totally new breed of methodology—which is agile and adaptive and involves the client every part of the way—is starting to emerge. Many of the heavyweight methodologists were resistant to the introduction of these "lightweight" or "agile" methodologies.[6] These methodologies use an informal communication style. Unlike heavyweight methodologies, lightweight projects have only a few rules, practices, and documents. Projects are designed and built on face-to-face discussions, meetings, and the flow of information to clients. The immediate difference of using light methodologies is that they are much less documentation-oriented, usually emphasizing a smaller amount of documentation for the project.

Heavy Methodologies

The traditional project management methodologies (i.e., the systems development life-cycle approach) are considered bureaucratic or "predictive" in nature and have resulted in many unsuccessful projects. These heavy methodologies are becoming less popular. These methodologies are so laborious that the whole pace of design, development, and deployment slows down—and nothing gets done. Project managers tend to predict every milestone because they want to foresee every technical detail (i.e., software code or engineering detail). This leads managers to start demanding many types of specifications, plans, reports, checkpoints, and schedules. Heavy methodologies attempt to plan a large part of a project in great detail over a long span of time. This works well until things start changing, and the project managers inherently try to resist change.

EPM methodologies can enhance the project planning process and provide some degree of standardization and consistency. Companies have come to realize that enterprise project management (EPM) methodologies work best if the methodology is based on templates rather than rigid policies and procedures. The International Institute for Learning has created a Unified Project Management Methodology (UPMM™)[7]

---

5. Ibid., pp. 102–104.

6. M. Fowler, *The New Methodology, Thought Works*, 2005. Available: https://martinfowler.com/articles/newMethodology.html.

7. Unified Project Management Methodology (UPMM™) is registered, copyrighted, and owned by International Institute for Learning, Inc., © 2022; reproduced by permission.

with templates categorized according to the Areas of Knowledge in the sixth edition of the *PMBOK® Guide*:

Communication
Project Charter
Project Procedures Document
Project Change Requests Log
Project Status Report
PM Quality Assurance Report
Procurement Management Summary
Project Issues Log
Project Management Plan
Project Performance Report

Cost
Project Schedule
Risk Response Plan and Register
Work Breakdown Structure (WBS)
Work Package
Cost Estimates Document
Project Budget
Project Budget Checklist

Human Resources
Project Charter
Work Breakdown Structure (WBS)
Communications Management Plan
Project Organization Chart
Project Team Directory
Responsibility Assignment Matrix (RAM)
Project Management Plan
Project Procedures Document
Kick-Off Meeting Checklist
Project Team Performance Assessment
Project Manager Performance Assessment

Integration
Project Procedures Overview
Project Proposal
Communications Management Plan
Procurement Plan
Project Budget
Project Procedures Document
Project Schedule
Responsibility Assignment Matrix (RAM)

Risk Response Plan and Register
Scope Statement
Work Breakdown Structure (WBS)
Project Management Plan
Project Change Requests Log
Project Issues Log
Project Management Plan Changes Log
Project Performance Report
Lessons Learned Document
Project Performance Feedback
Product Acceptance Document
Project Charter
Closing Process Assessment Checklist
Project Archives Report

Procurement
Project Charter
Scope Statement
Work Breakdown Structure (WBS)
Procurement Plan
Procurement Planning Checklist
Procurement Statement of Work (SOW)
Request for Proposal Document Outline
Project Change Requests Log
Contract Formation Checklist
Procurement Management Summary

Quality
Project Charter
Project Procedures Overview
Work Quality Plan
Project Management Plan
Work Breakdown Structure (WBS)
PM Quality Assurance Report
Lessons Learned Document
Project Performance Feedback
Project Team Performance Assessment
PM Process Improvement Document

Risk
Procurement Plan
Project Charter
Project Procedures Document
Work Breakdown Structure (WBS)
Risk Response Plan and Register

Scope
Project Scope Statement
Work Breakdown Structure (WBS)
Work Package
Project Charter

Time
Activity Duration Estimating Worksheet
Cost Estimates Document
Risk Response Plan and Register Medium
Work Breakdown Structure (WBS)
Work Package
Project Schedule
Project Schedule Review Checklist

Stakeholder Management
Project Charter
Change Control Plan
Schedule Change Request Form
Project Issues Log
Responsibility Assignment Matrix (RAM)

4.4 BENEFITS OF A STANDARD METHODOLOGY

For companies that understand the importance of a standard methodology, the benefits are numerous. These benefits can be classified as both short- and long-term benefits. Short-term benefits were described by one company as:

- Decreased cycle time and lower costs
- Realistic plans with greater possibilities of meeting time frames
- Better communications as to "what" is expected from groups and "when"
- Feedback: lessons learned

These short-term benefits focus on key performance indicators (KPIs) or, simply stated, the execution of project management. Long-term benefits seem to focus more on critical success factors and customer satisfaction. Long-term benefits of development and execution of a world-class methodology include:

- Faster time to market through better scope control
- Lower overall program risk
- Better risk management, which leads to better decision-making
- Greater customer satisfaction and trust, which lead to increased business and expanded responsibilities for the tier 1 suppliers

- Emphasis on customer satisfaction and value-added rather than internal competition between functional groups
- Customer treating the contractor as a partner rather than as a commodity
- Contractor assisting the customer during strategic planning activities

Perhaps the largest benefit of a world-class methodology is its acceptance and recognition by customers. If one of your critically important customers develops its own methodology, that customer could "force" you to accept it and use it in order to remain a supplier. But if you can show that your methodology is superior or equal to the customer's, your methodology will be accepted, and an atmosphere of trust will prevail.

One contractor recently found that a customer had so much faith in and respect for its methodology that the contractor was invited to participate in the customer's strategic planning activities. The contractor found itself treated as a partner rather than as a commodity or just another supplier. This resulted in sole-source procurement contracts for the contractor.

Developing a standard methodology that encompasses the majority of a company's projects and is accepted by the entire organization is a difficult undertaking. The hardest part might very well be making sure that the methodology supports both the corporate culture and the goals and objectives set forth by management. Methodologies that require changes to a corporate culture may not be well accepted by the organization. Nonsupportive cultures can destroy even seemingly good project management methodologies.

During the 1980s and 1990s, several consulting companies developed their own project management methodologies, most frequently for information systems projects, and then pressured clients into purchasing the methodology rather than helping them develop a methodology more suited to their own needs. Although there may have been some successes, there appeared to be significantly more failures than successes. A hospital purchased a $130,000 project management methodology with the belief that this would be the solution to its information system needs. Unfortunately, senior management made the purchasing decision without consulting the workers who would be using the system. In the end, the package was never used.

Another company purchased a similar package, discovering too late that the package was inflexible and the organization, specifically the corporate culture, would need to change to use the methodology effectively. The vendor later admitted that the best results would occur if no changes were made to the methodology.

These types of methodologies are extremely rigid and based on policies and procedures. The ability to custom design the methodology for specific projects and cultures was nonexistent, and eventually these methodologies fell by the wayside—but after the vendors made significant profits. Good methodologies must be flexible.

4.5 CRITICAL COMPONENTS

It is almost impossible to become a world-class company with regard to project management without having a world-class methodology. Years ago, perhaps only a few companies really had world-class methodologies. Today, because of the need for survival and stiffening competition, numerous companies have good methodologies.

The characteristics of a world-class methodology include:

- Usually, a maximum of six life-cycle phases
- Life-cycle phases overlap
- End-of-phase gate reviews
- Integration with other processes
- Continuous improvement (i.e., hear the voice of the customer)
- Customer-oriented (interface with customer's methodology)
- Companywide acceptance
- Use of templates (level 3 WBS)
- Critical path scheduling (level 3 WBS)
- Simplistic, standard bar chart reporting (standard software)
- Minimization of paperwork

Generally speaking, each life-cycle phase of a project management methodology requires paperwork, control points, and perhaps special administrative requirements. Having too few life-cycle phases is an invitation for disaster, and having too many life-cycle phases may drive up administrative and control costs. Most companies prefer a maximum of six life-cycle phases.

Historically, life-cycle phases were sequential in nature. However, because of the necessity for schedule compression, life-cycle phases today will overlap. The amount of overlap will be dependent on the magnitude of the risks the project manager will take. The more the overlap, the greater the risk. Mistakes made during overlapping activities are usually more costly to correct than mistakes during sequential activities. Overlapping life-cycle phases requires excellent up-front planning.

End-of-phase gate reviews are critical for control purposes and verification of interim milestones. With overlapping life-cycle phases, there are still gate reviews at the end of each phase, but they are supported by intermediate reviews during the life-cycle phases.

World-class project management methodologies are integrated with other management processes, such as change management, risk management, total quality management, and concurrent engineering. This produces a synergistic effect, which minimizes paperwork, minimizes the total number of resources committed to the project, and allows the organization to perform capacity planning to determine the maximum workload that the organization can support.

World-class methodologies are continuously enhanced through KPI reviews, lessons-learned updates, benchmarking, and customer recommendations. The methodology itself could become the channel of communication between customer and contractor. Effective methodologies foster customer trust and minimize customer interference in the project.

Project management methodologies must be easy for workers to use as well as cover most of the situations that can arise on a project. Perhaps the best way is to have the methodology placed in a manual that is user-friendly.

Excellent methodologies try to make it easier to plan and schedule projects. This is accomplished by using templates for the top three levels of the WBS. Simply stated, using WBS level 3 templates, standardized reporting with standardized terminology exists. The differences between projects will appear at the lower levels (i.e., levels 4–6) of the WBS. This also leads to a minimization of paperwork.

Today, companies seem to be promoting the use of the project charter concept as a component of a methodology, but not all companies create the project charter at the same point in the project life cycle, as shown in Figure 4–1. The three triangles in the figure show possible locations where the charter can be prepared:

- In the first triangle, the charter is prepared immediately after the feasibility study is completed. At this point, the charter contains the results of the feasibility study as well as documentation of any assumptions and constraints that were considered. The charter is then revisited and updated once this project is selected.
- In the second triangle, which seems to be the preferred method, the charter is prepared after the project is selected and the project manager has been assigned. The charter includes the authority granted to the project manager, but for this project only.
- In the third method, the charter is prepared after detailed planning is completed. The charter contains the detailed plan. Management will not sign the charter until after detailed planning is approved by senior management. Then, and only then, does the company officially sanction the project. Once management signs the charter, it becomes a legal agreement between the project manager and all involved line managers as to what deliverables will be met and when.

Although we identified critical components in this section, there are many companies that have methodologies that do not follow these critical components but are highly successful in project execution because the employees have made the methodology work successfully.

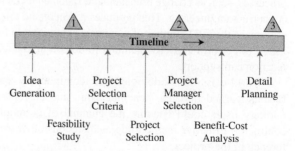

Figure 4–1. When to prepare the charter.

4.6 VALMET CUSTOMER PROJECT MANAGEMENT[8]

**Introduction to Valmet and
Valmet's Project Management**

Valmet is a leading global developer and supplier of process technologies, services, and automation to the pulp, paper, and energy industries. With our automation systems and flow control solutions, we serve an even wider base of process industries. Our 17,000 professionals working in more than 30 countries are committed to moving our customers' performance forward with our unique offering and way to serve. Valmet has over 220 years of industrial history and a strong track record of continuous improvement and renewal.

Projects in Valmet vary in size, complexity, and duration. There are typically many project teams and functions involved throughout the project lifecycle in multiple countries. To manage this complexity and to harmonize the work done in a project, Valmet project execution model (PEM) shown in Figure 4–2 was developed.

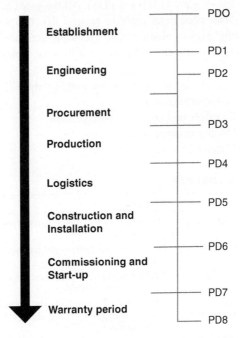

Figure 4–2. Project delivery process with nine gates used in Valmet PEM.

8. The section, "Valmet Customer Project Management", has been provided by Mr. Mikko Sillanpää, Vice President, Key Projects at Valmet. ©2022 by Valmet. All rights reserved. Reproduced with permission.

Valmet PEM

Valmet PEM is a PEM that standardizes project management in Valmet. Valmet PEM gives projects a structure and a common language that form the basis of successful and effective projects. The project model is used to ensure that a project is proceeding with the right quality and according to the schedule and the budget. Valmet PEM also enhances the transparency and predictability of the projects.

The project model describes the main phases, gates, and checklists that are common for all the projects. Valmet PEM is scalable and can be complemented with project-specific elements. Project gate is a checkpoint at which the completed and ongoing actions are reviewed and project's readiness to proceed is evaluated based on the facts in the checklists and the defined milestones.

Valmet PEM contains nine gates, and it is scalable by project categories. Mandatory gates are defined according to project category (A, B, C, and D). Large and/or complex projects are category A projects, and all nine gates are mandatory in these projects. Small and simple projects are in category D, and in these projects, there are only two mandatory gates (PD1 and PD8). All the gates are described in Table 4–1.

Gate checklists include around 10–30 questions that are used to ensure that required project planning and executing activities are proceeding and project can reach its targets. Gate checklists cover following topics:

1. Health, safety, environment, and sustainability management
2. Scope management
3. Schedule and progress management
4. Quality management
5. Financial management

TABLE 4–1. GATES IN VALMET PEM

Gate	Gate name	Purpose
PD0	Decision to set up project	Decision to start internal preparatory work to ensure prompt and efficient start for the project.
PD1	Decision to start delivery project	Decision to start delivery project: engineering, purchasing, production, and logistics.
PD2	Decision to start detail engineering	Decision to start the detail engineering and continue purchasing, manufacturing, and logistics activities.
PD3	Decision to start site operation preparations	Ensure that documentation is available to start installation contracting and that site planning can proceed. Review the status of purchasing, manufacturing, and testing plans.
PD4	Decision to start main shipments	Ensure that the equipment is ready for shipment and that the site is ready to receive the deliveries.
PD5	Decision to start construction/installation	Review the condition at site to ensure safe and efficient construction and installation work on schedule with required quality.
PD6	Decision to start commissioning and start-up	Review the condition at site to ensure safe and efficient commissioning work on schedule.
PD7	Decision to start performance testing	Ensure that the plans to fulfill performance guarantees are ready and that needed resources for the planned activities are available.
PD8	Decision to close project	Ensure that all project activities and costs are concluded.

It is the project manager's responsibility to follow-up and update the status of gate checklists. Checklists are reminding the project managers of the most important aspects of the project in each project phase, as it is otherwise very easy to lose sight of the big picture and focus on a specific topic only.

In Table 4–2, there is an example of some (but not all) of the questions in gate PD 5 "Decision to start construction/installation."

Gate decision dates for each gate are decided when project's main schedule baseline is created. Gate decisions and checklists are updated in the project portfolio management tool that is used for all Valmet's customer projects. Using a common tool has made it easy to create automated and visual report to follow-up on the status of the gate decisions, checklists, and upcoming gate decisions on project and on portfolio level.

Gate decisions available in Valmet PEM are as follows:

1. Go decision
   - There are no major deviations identified, and project can proceed
   - If there are minor deviations project can proceed, but there needs to be a plan of corrective actions with due dates for the minor deviations
2. No-Go decision
   - Major deviations to project plan and/or objectives are identified
   - Detailed action plan to manage the deviations is made and followed-up
   - Project manager needs to escalate deviations to Valmet management if no-go decision is made

TABLE 4–2. EXAMPLE OF A GATE CHECKLIST IN VALMET PEM

Questions	Answers	Notes
Project Management and Scope		
1 Have the customer payments been received according to plan?	Yes	75% of payments received. All payments are on time.
2 Has a change management procedure been agreed upon with the customer and third-party contractors?	Yes	Change management procedure is described in the customer contract.
QHSE, Sustainability, and Risks		
3 Is the site-specific HSE plan up to date?	No	HSE plan will be updated on 31 May 2021 after customer's updated HSE instructions are received.
4 Have personnel been informed about travel safety and the conditions on site?	Yes	Site induction material is available and shared with all persons traveling to the site.
Financials		
5 Have contract changes been agreed upon with the customer and third-party contractors?	Yes	No open items.
6 Are the project financials up to date (cost forecast, revenue recognition plan, invoicing, bank guarantees)?	Yes	Updated on 3 May 2021.
Planning and Scheduling		
7 Has the resource plan for the site been agreed upon with the customer?	Yes	Plan was reviewed in customer meeting on 15 May 2021.
8 Are construction/installation and commissioning schedules synchronized between involved parties?	Yes	Schedules were reviewed on 15 May 2021 with all parties working on site. Some changes were made.

Gate decision is made based on answers to checklist questions, other available information about the project, and overall understanding of the project status. Valmet management is involved in all no-go decisions.

On a project portfolio level, projects are reviewed monthly. In portfolio meetings projects' Valmet PEM gate decisions, internal project reports, risks, deviations, and financial statuses are reviewed to ensure that projects are able to reach their targets. Projects that have major deviations will be followed up in separate meetings focused on corrective actions.

Continuous improvement is important for all companies. Valmet PEM is promoting sharing of lessons learned between projects. Valmet PEM and its checklists are reviewed annually to keep them valuable and relevant for years to come.

Conclusion

PEM has helped Valmet to improve project management. During 2020 and 2021, while Coronavirus pandemic was impacting everyone's lives and work, Valmet was able to continue to execute hundreds of projects successfully around the world. Common processes and structures with project management did not make the work easy, but it did make it possible to succeed in a very challenging situation.

Valmet PEM is a simple and powerful system, but it is just a system. It is essential for project success to have competent, experienced people with positive and hardworking attitude who will execute the project based on the system and the project plans. Systems like Valmet PEM will help great people to do their best work as an effective team by giving individuals common targets, structure, language, and more.

4.7 PROJECT QUALITY GATES—STRUCTURED APPROACH TO ENSURE PROJECT SUCCESS[9]

Project quality is paramount in delivery of SAP projects; this fact is reflected in structured approach to quality management for SAP solution delivery—Quality Built In. The staple of the quality approach in SAP projects is the execution of formal project quality gates. The quality gates are defined in the SAP Activate methodology that is used for delivery of SAP implementation, scope enhancements, and upgrade projects. SAP customers and partners use SAP Activate to plan, manage, and deliver successful projects. Each project type has predetermined number of quality gates (Q-Gates) that are planned and run at key milestones of the project. Examples of quality gates for implementation of SAP S/4HANA Cloud are shown in Figure 4–3, and they occur at the end of Prepare, Explore, Realize and Deploy phases.

SAP believes that quality gates are essential for success of any project regardless of deployment strategy whether the project is managed as agile, traditional, or hybrid.

---

9. Material in this section has been provided by Jan Musil, Chief Product Owner, SAP Activate. ©2022 by SAP. All rights reserved. Reproduced by permission.

The quality gates are integrated not only into SAP Activate methodology but are also coded into SAP's delivery policies and internal systems.

SAP provides project team members with detailed quality gate checklist along with guide for structuring and running the quality gates during the project execution. Customers can access these SAP Activate assets in the Roadmap Viewer tool online at https://go.support.sap.com/roadmapviewer/. You can easily locate Quality Gates (see Figure 4–4) when you select filter for *Project Management* work stream in the filter

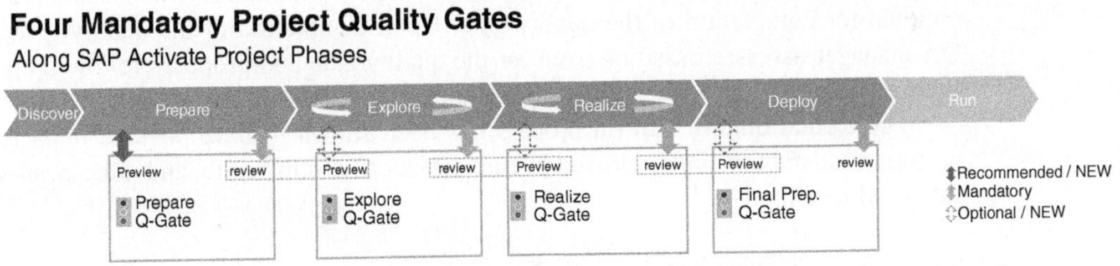

Four Mandatory Project Quality Gates
Along SAP Activate Project Phases

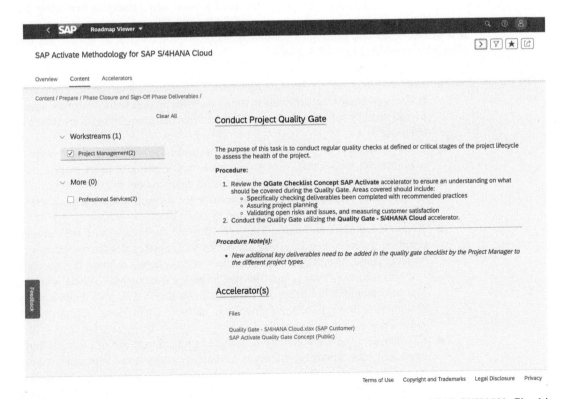

Figure 4–3. Project quality gates are defined in the SAP Activate methodology at critical stages in the project life cycle.

Figure 4–4. Example of SAP Activate project quality gate task for implementation of SAP S/4HANA Cloud in Roadmap Viewer.

selection and navigate the *Phase Closure and Sign-Off Phase Deliverables* in one of the phases that require quality gate (like Prepare, Explore, Realize of Deploy—we listed them in the text above).

The project quality gates in the SAP Activate methodology provide clear guidance to project managers, stakeholders, and project teams on how to structure and perform the quality gate review. During each quality gate, the quality assurance manager assesses completeness and quality of each deliverable produced in the project according to predefined quality gate checklist, which includes not only deliverable name but also acceptance criteria. Each deliverable in the checklist is marked as either mandatory or optional for completion of the quality gate. Upon completion of the quality gate, the QA manager assesses pass/fail score for the quality gate and proposes follow-up plans to take corrective actions to address deficiencies identified in this process.

The formal quality built-in process has been shown to have positive impacts on customer satisfaction and improved overall project portfolio health, and it has positively impacted revenue.

SAP Activate Methodology

SAP introduced SAP Activate methodology to market along with introduction of SAP S/4HANA. It replaced previously used ASAP methodology and introduced several new principles that allow for more efficient deployment and activation of SAP software, and it takes advantage of new technologies and principles for management of projects. Let us review the key principles embedded in SAP Activate:

- *Start with ready-to-run business processes*. Project teams leverage pre-configured business processes that are ready to run to show the process to the customer, assess fit, and identify delta requirements. The identified delta requirements are stored in the backlog.
- *Confirm solution fit*. While leveraging the ready-to-run business processes in requirements workshops, the project team steers business users to prioritize re-use of pre-delivered processes rather than development or heavy tailoring of the software. This approach helps accelerate the project, lower the total cost of ownership, and enables faster adoption of innovation.
- *Modular, scalable, and agile*. The SAP Activate methodology is built around common modules that cover all the various disciplines needed in successful project. These disciplines include project management, solution design, data management, technology, etc. The methodology is designed to scale with the size and complexity of the project and customer needs. This way it can support customers deploying software globally in multiple countries as well as organizations that are targeting smaller scope.
- *Cloud-ready*. The methodology is designed to maximize use of cloud technologies and take advantage of faster availability of systems with ready-to-run business processes. These technologies can be used not only by customers

deploying their software in the cloud but also customers targeting on-premises deployment. For example, the availability of sandbox environment can be achieved faster if cloud hosting is used.

- *Premium-engagement ready.* SAP Activate provides project teams with advice on where SAP Services can provide additional advisory and safeguarding services, such as SAP MaxAttention or SAP ActiveAttention. This advice is embedded into SAP Activate and provides companies with information about how to best leverage SAP Services and Support offerings in their implementation projects.

- *Quality built-in.* Planning and managing quality during the implementation project is one of the critical factors for delivering a successful implementation. SAP Activate provides structured approach quality management. It starts with quality planning in the Prepare phase, including development of quality management plans, then monitoring and controlling activities throughout the entire project, and finally executing quality gates in each SAP Activate phase as we outlined earlier in this chapter.

SAP Activate is structured into six phases that guide the project team through steps to deploy, activate, and configure the SAP software to fit customer needs. You can see the phases of SAP Activate methodology along with description of key activities in each phase in Figure 4–5.

Example: Deploy SAP S/4HANA Cloud with SAP Activate
Journey overview and key milestones

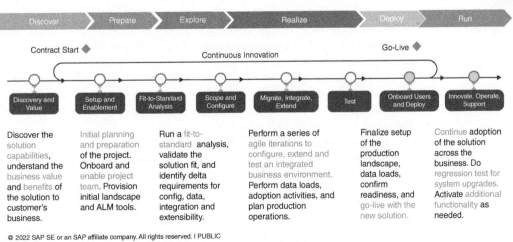

Figure 4–5. The six phases of SAP Activate with high-level explanation of work performed in each phase.

Let us review the phases and key activities in each of them:

- *Discover*. The objective of this phase is to assist in discovering the solution capabilities and understanding the business value and benefits of the solution to customer's business.
- *Prepare*. Project team performs initial planning and preparation of the project, including formal kickoff, establishment of project management plans, and governance. Next important step is to onboard and enable project team and provisioning of initial landscape and application lifecycle management (ALM) tools.
- *Explore*. Project team runs series of fit-to-standard analysis workshops to confirm the solution's fit and identify delta requirements for configuration, data migration, integration, analytics, extensibility, and user access management. These requirements are captured in the backlog.
- *Realize*. Project team performs a series of agile iterations to consume the backlog. During the sprints, the project team configures, extends, and tests an integrated business environment. Additionally, the project team performs data loads, adoption and organizational change management activities, and plans production environment operations.
- *Deploy*. The goal of this phase is to finalize setup of the production landscape, completion of data loads to production environment, confirmation of readiness for cutover, and execution of go-live activities for customer to use the new solution.
- *Run*. In this phase, customer will continue the adoption of the new solution across the business by adding new users. They will prepare and execute pre- and post-upgrade activities, like regression testing. And lastly, they will take advantage of SAP's innovation and activate additional functionality as needed for their business needs.

The SAP Activate methodology covers key aspects of SAP implementation from project management guidance structured around the PMI *PMBOK® Guide*[10] in the project management work stream. In addition to project management, the methodology provides detailed guidance for business process design, business value management, application life-cycle management, organizational change management, technical solution and architecture, data management, and other topics important for successful deployment and activation of SAP solutions.

SAP customers can access the SAP Activate methodology in Roadmap Viewer tool online at https://go.support.sap.com/roadmapviewer/#. Additionally, project teams will use SAP Activate methodology in SAP Cloud ALM environment that SAP provides to customers to plan, execute, and monitor the project. The environment provides capabilities for business project modeling, requirements management, testing, etc. that support the project team on their deployment and activation journey.

You can learn more about SAP Activate, the available pre-configured business content, methodology assets, and training on SAP Activate community at https://community

---

10. PMBOK is a registered mark of the Project Management Institute, Inc.

.sap.com/topics/activate. Alternatively, you can consider exploring details in the SAP PRESS publication *SAP Activate, The project manager's handbook for SAP S/4HANA, cloud and on-premise* by Sven Denecken, Jan Musil, Srivatsan Santhanam at https://www.sap-press.com/sap-activate_5463/.

4.8 TÉCNICAS REUNIDAS[11]

Open Book Estimate as a Successful Contract Alternative to Execute Projects in the Oil and Gas Sector

Introduction

As a result of the projected rate of energy demand growth, the oil and gas industry has a wide range of challenges and opportunities across different areas. For that reason, for a number of years, the sector has been developing new facilities, which in many cases are megaprojects.

The typical complete life cycle of a capital project in the oil and gas sector is focused on the overall stages that are shown in Figure 4–6. Understanding and managing these stages is crucial to the long-term success of the project.

Lump-sum turnkey (LSTK) and cost-plus contracting are both very prevalent types of contracts within projects in the oil and gas industry. The level of risk the client of a project is willing to accept, budget constraints, and the client's organization's core competencies determine which method is best for a project.

A large number of projects in this sector are performed under engineering, procurement, and construction (EPC)–LSTK contracts; The major experience of Técnicas Reunidas (TR) is based mainly on this type of project, which in general means managing the whole project and carrying out the detailed engineering (in some cases it is included the basic engineering or front-end engineering design [FEED] in the scope of work); procuring all the equipment and materials required; and then constructing, precommissioning, and starting up to deliver a functioning facility ready to operate. LSTK contracting tends to be the riskiest, and all risks are assumed by the EPC contractor.

The open book estimate (OBE) or open book cost estimate (OBCE) is an alternative way to execute EPC projects. With this type of contract, the final purpose of the work is to define the total price of the project in collaboration with the client; the global costs of the project are established in a transparent manner (open book).

The Open Book Estimate

The main purpose of the OBE methodology is to build up an accurate EPC price by applying some parameters previously agreed on (between client and the contractor), the base cost through an OBE, the development of an extended front-end engineering effort, and in some cases, the placement of purchase orders for selected long lead and critical items to ensure the overall schedule of the project. The OBE will fix the project base cost and will become the basis for determining the lump-sum EPC price for the project.

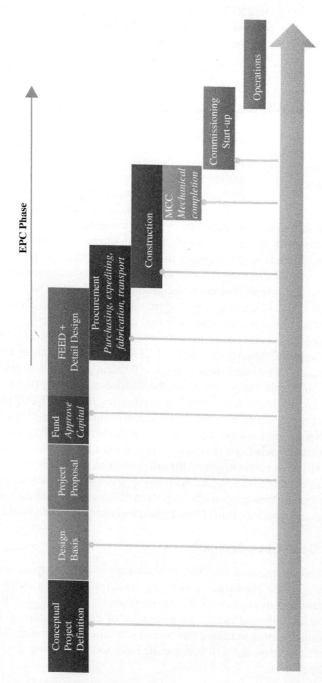

Figure 4–6. Typical life-cycle phases.

During an OBE phase, the contractor develops a FEED and/or part of the detailed engineering under a reimbursable basis or lump-sum price or alternatives, including a complete and open cost estimate of the plant. After an agreed period (usually between 6 and 12 months, mainly dependent on the accuracy grade, schedule, and other factors required by client) of engineering development and after the client and contractor agree on the base cost, the contract is changed or converted to an EPC LSTK contract by applying previously agreed-on multiplying factors.

Cost Estimate Methodology   Principal Cost Elements and Pricing Categories

The OBE usually is based on sufficient engineering development in accordance with the deliverables identified in OBE contract. These deliverables are developed as much as possible in the normal progress of the project. Required deliverables are prepared and submitted to the client prior to completion of the conversion phase.

The principal cost elements, as seen in Figure 4–7, that comprise the OBE are addressed below. The OBE cost estimate shall include the total scope of work:

1. Detailed engineering, procurement, and construction services
2. Supply of equipment, bulk materials, and spare parts
3. Transport to construction site
4. Customs clearance
5. Construction and erection at site
6. Provision of subcontractors' temporary construction facilities and services
7. Construction and precommissioning services
8. Commissioning and start-up services

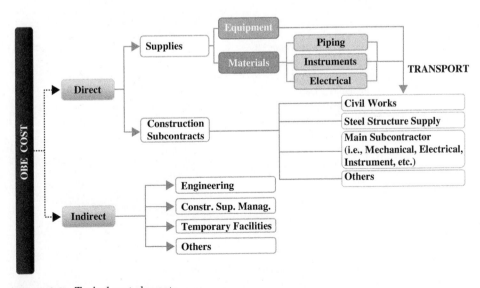

Figure 4–7. Typical cost elements.

9. Training services and vendor's assistance
10. Bonds, insurances, hedging charges
11. Other costs, including third-party inspection and contractor insurance
12. Others

Cost Base

- An OBE procedure is developed during the contract stage and implemented during the project's OBE phase. All details on how to prepare an OBE have to be agreed upon and included as an annex in the contract.
- Preagreement of allowances, growth, conditioning, technical design allowances, surplus, and cut and waste.
- Material take-offs (MTOs) based on PDS software, measured in process and instrument diagrams (P&ID), and plot plans and estimates. All details on procedures are to be agreed on before OBE contract signature.

In executing projects on convertible basis, TR develops the OBE in parallel with the normal project execution, ensuring that both activities can flow smoothly without interference. During the OBE phase, in certain cases and if agreed with the client, TR can advance the procurement of the main equipment and initiate negotiations for construction subcontractors. The execution of these activities in advance facilitates the fulfillment of project schedule requirements.

This OBE phase of the project is jointly developed between the client and TR. The OBE is fully transparent to client, and the conversion to LSTK is made once the risk/reward element is fixed.

In Figure 4–8, we see the main steps and activities that are developed to achieve OBE goals and to convert to next phase of the project.

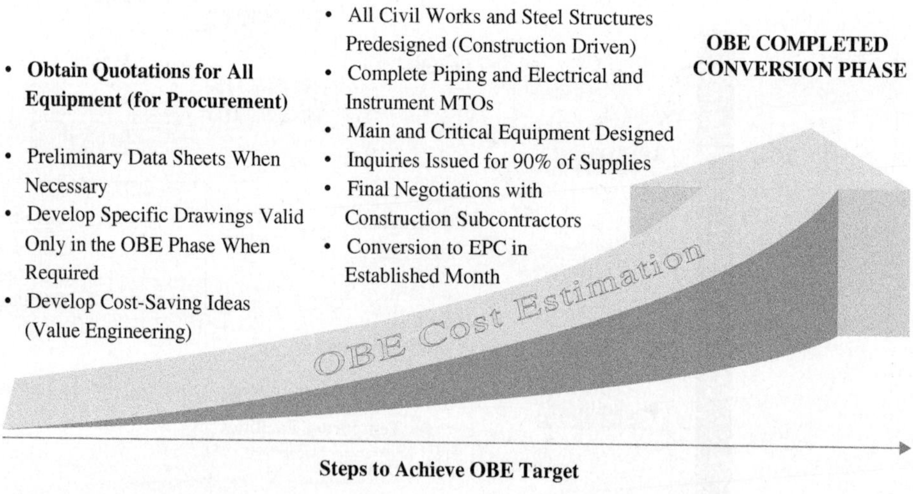

- **Obtain Quotations for All Equipment (for Procurement)**

- Preliminary Data Sheets When Necessary
- Develop Specific Drawings Valid Only in the OBE Phase When Required
- Develop Cost-Saving Ideas (Value Engineering)

- All Civil Works and Steel Structures Predesigned (Construction Driven)
- Complete Piping and Electrical and Instrument MTOs
- Main and Critical Equipment Designed
- Inquiries Issued for 90% of Supplies
- Final Negotiations with Construction Subcontractors
- Conversion to EPC in Established Month

OBE COMPLETED CONVERSION PHASE

OBE Cost Estimation

Steps to Achieve OBE Target

Figure 4–8. Steps to achieve OBE target.

During the OBE phase, cost-saving ideas are developed in order to adjust the final cost estimate; to do so, a special team of engineering specialists is appointed to work with both the engineering manager and the estimating manager, with the purpose to determine those areas where potential savings can be achieved by optimizing the design, without jeopardizing safety, quality, or schedule. Any of these changes that could lead to cost savings are carefully evaluated from a technical point of view. If the feasibility of the potential change is proven, the alternative solution, together with the cost-saving impact evaluation, will be forwarded to the clients for consideration.

The EPC contract price is the result of the base cost multiplied by fixed percentages for fees and markups related to equipment, bulk materials, construction, and ancillary costs agreed to by the client and contractor. During the conversion phase, this price is converted to lump-sum price and thereafter remains fixed during the EPC-LSTK phase.

Contracts

There are two typical models of contracts under this OBE alternative:

1. One contract, two parts: OBE and EPC. The price for the EPC part is to be included at conversion.
2. Two contracts, one OBE and the other EPC. Both may be signed at the beginning, or one at the beginning and the other at conversion.

The methodology of the OBE is included in the contract.
In the case of no conversion:

I. The contractual relationship disappears, and both client and contractor break their commitment. The client may break the contract if it is not interested. Consequences:
   - Six months for a new LSTK offer. Two to three months plus evaluating the offers
   - Repeat FEED with different contractor.
II. Contract provides mechanisms in the case of no agreement.
   - Continue contract on a service base (better LS contract).
   - Agree on partial conversion.
   - Other actions as per contract agreement.

Advantages

An OBE phase followed by an LSTK contract could optimize all project execution, especially in cost and schedule.

- In terms of cost, the client and contractor could together determine the project cost through an OBCE because an estimation methodology, conversion conditions such as multiplying factors are agreed upon, etc. Both the client and the contractor determine by mutual agreement the final price of the contract, sharing all information. This will generate a feeling of trust between both companies. This model results in an accurate cost because unnecessary contingencies are avoided.

● However, this model results in schedule advantages because the bidding period is shortened or replaced by a conversion negotiation phase; an EPC stage is shortened because all the work is developed during the extended feed and conversion stage. A representation of the schedule advantage is shown in Figure 4–9.

In summary, the advantages in cost and schedule are shown in Table 4–3.

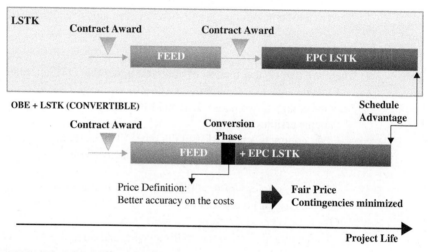

Figure 4–9. Typical schedule advantage.

TABLE 4–3. COST AND SCHEDULE ADVANTAGES OF THE OBE METHODOLOGY

Cost	Schedule Reduction
Develops an EPC estimate during 6–12 months. This provides much more accurate cost.	Short bidding period, as cost estimate does not need to be as detailed in an EPC-LSTK common bidding process.
Accurate prices based on real offers and an agreed conversion factor assure fairness to client and contractor.	Shortening of the overall project schedule: time for extended FEED and EPC bidding is shortened dramatically.
There is enough time to develop the project and to avoid unnecessary contingencies.	Contract award procedure is much easier and shorter.
Application of cost saving to match project cost to client's budget.	Some long-lead items and critical equipment could be awarded or negotiated.
Facilitates the possibility of funding, because of a more accurate estimation.	
Risks are reduced and better controlled to the benefit of both client and contractor.	

Close-out

The OBE has been demonstrated to be a successful contract alternative for executing projects in the oil and gas sector because it aligns clients and contractors with the project's goals. Both are motivated to pursue the best cost estimation or project target cost, and at the same time the schedule is optimized.

As mentioned in the introduction to this section, in the oil and gas sector, most current projects can be considered megaprojects, where there are many risks associated with a high workload from suppliers, contractors, subcontractors, and others. Through an OBE alternative, clients can better manage risks through a more cooperative and agreed-on approach, where the risks are reduced during an accurate estimate and then embraced rather than totally transferred to contractors. In this way, project outcomes can be improved.

TR has converted successfully 100 percent from OBE to EPC-LSTK projects.

Definitions

- *Client*. The owner of the oil and gas company.
- *Contractor*. Affiliated company responsible for performing the engineering, procurement, and construction services.
- *EPC*. Engineering, procurement, and construction. Type of contract typical of industrial plant construction sector, comprising the provision of engineering services, procurement of materials, and construction.
- *FEED*. Front-end engineering design. This refers to basic engineering conducted after completion of the conceptual design or feasibility study. At this stage, before the start of EPC, various studies take place to figure out technical issues and estimate rough investment cost.
- *LS contract*. Lump-sum contract. In a lump-sum contract, the contractor agrees to do a specific project for a fixed price.
- *LSTK contract*. lump-sum contract. In a lump-sum turnkey contract, all systems are delivered to the client ready for operations.
- *MTOS*. Material take-offs. This refers to piping, electrical, and instrumentation.
- *OBE*. Open book estimate or open book cost estimate (OBCE).
- *PDS*. Plant design system. Software used for designing industrial plants through a multidisciplinary engineering activity.
- *P&ID*. Process and instrument diagrams.
- *TR*. Técnicas Reunidas.

4.9 SONY CORPORATION AND EARNED VALUE MANAGEMENT[12]

Earned value management (EVM) is one of the most commonly used tools in project management. When it is used correctly, EVM becomes a best practice and allows

---

12. Section 4.9 © 2022 by Jun Makino and Koichi Nagachi of Sony Corporation.

managers and executives to have a much clearer picture of the true health of a project. EVM can also lead to significant continuous improvement efforts. Such was the case at Sony.

Sony suffered from some of the same problems that were common in other companies. Because project planning at Sony often did not have desired level of detail, Sony viewed itself as operating in a "negative cycle," as shown in Figure 4–10. Sony's challenge was to come up with effective and sustainable ways to break out of this negative cycle.

Sony's basic idea or assumption was that unless people recognize the need for change and want to get involved, nothing will happen, let alone further improvements. Sony realized that, at the beginning of the EVM implementation process, it might need to sacrifice accuracy of the information.

Sony sought the easiest possible or the most elementary way for project managers and team members to implement progress monitoring continuously.

Sony started by:

1. Using information on a list of final deliverables together with a completion date for each final deliverable. Team members did not have to make an extra effort to produce this level of information because it was being provided to them.
2. Selecting the fixed ratio method among several EVM methods, such as the weighted milestone method, percentage-complete method, and criteria achievement method, for reporting progress. The fixed-ratio method required the least effort from project managers and team members.

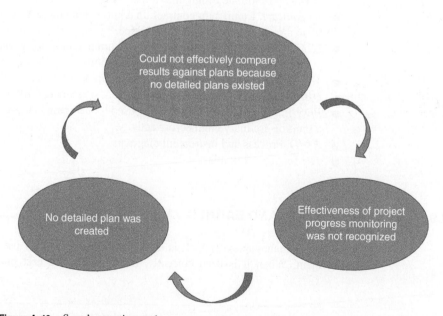

Figure 4–10. Sony's negative cycle.

Visualizing project progress and monitoring process using various graphs, such as Figures 4–11 and 4–12 for schedule performance index (SPI) reporting.

Shortly after people started to practice this elementary method of progress monitoring and reporting, we began to observe the following improvements:

1. Increased awareness by project managers and team members that forecasting, early alerts, and taking countermeasures earlier improve productivity (i.e., SPI).
2. Increased awareness by project managers and team members that reviewing and creating more detailed plans were critical to further improve productivity (i.e., SPI).

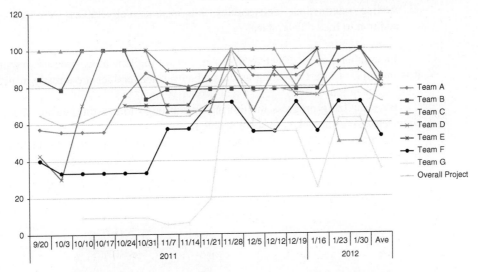

Figure 4–11. SPI (progress rate) transition by team.

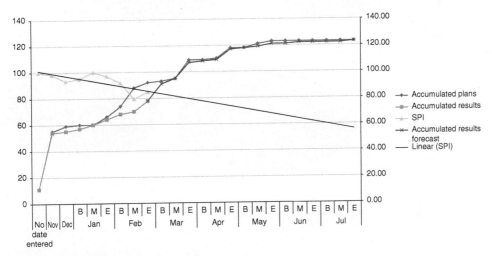

Figure 4–12. Overall project progress forecasting (linear approximation).

Visualization or performance certainly helped the project teams to easily understand how important and beneficial practicing project progress monitoring was.

Project managers and team members began to take initiatives to improve project progress monitoring and reporting, as shown in Figure 4–13. For example, team members noted that delays in progress were difficult to detect when data accuracy was poor or not detailed enough. Team members started to improve data accuracy by:

1. Dividing a month into three parts. Previously, data was given on a monthly basis.
2. Making changes to the fixed ratio method. Previously, the 1/100 rule had been applied but now the 20/80 rule is used.
3. Adding intermediate deliverables. Intermediate deliverables were reported in addition to final deliverables.

In summary, as an effective first step toward implementing progress monitoring, it is important to start by using as data whatever deliverables are already available in your organization.

By visualizing progress monitoring and through forecasting, ensure that correct countermeasures are taken to solve problems.

Accuracy will be improved once people become aware of the effectiveness of project progress monitoring.

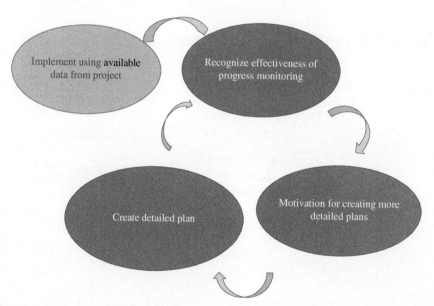

Figure 4–13. The progress improvement cycle.

FURTHER READING

Nagachi, K. 2006. "PM Techniques Applied in Nile Firmware Development: An Attempt to Visualize Progress by EVM." Paper presented at the PMI Tokyo Forum.

Nagachi, K. and Makino, J. 2012. "Practicing Three Earned Value Measurement Methods." Paper presented at the PMI Japan Forum.

Tominaga, A. August 20, 2003. *EVM: Earned Value Management for Japan*. Society of Project Management.

Yamato, S. and Nagachi, K. April 20, 2009. "IT Project Management by WBS/EVM," Soft Research Center Inc.

4.10 PROJECT MANAGEMENT TOOLS AND SOCIALIZED PROJECT MANAGEMENT

In the early years of project management, EVM was the only tool used by many companies. Customers such as the Department of Defense created standardized forms that every contractor was expected to complete for performance reporting. Some companies had additional tools, but these were for internal use only and not to be shared with the customers.

As project management evolved, companies created EPM methodologies that were composed of multiple tools displayed as forms, guidelines, templates, and checklists. The tools were designed to increase the chances of repeatable project success and were designed such that they could be used on multiple projects. Ideas for the additional tools often came from an analysis of best practices and lessons learned captured at the end of each project. Many of the new tools came from best practices learned from project mistakes, so that the mistakes would not be repeated on future projects. Project teams now could have as many as 50 different tools to be used. Some tools were used for:

- Defining project success since the definition could change from project to project
- Capturing best practices and lessons learned throughout the project life cycle rather than just at project completion
- Advances in project performance reporting techniques
- Capturing benefits and value throughout the life cycle of the project
- Measuring customer satisfaction throughout the life cycle of the project
- Handing off project work to other functional groups

As project management continued to evolve, companies moved away from co-located teams to distributed or virtual teams. Now additional tools were needed to help support the new forms of project communications that would be required. Project managers were now expected to communicate with everyone, including stakeholders, rather than just project team members. Some people referred to this as PM 2.0, where the emphasis was on social project management practices.

Advances in technology led to a growth in collaborative software, such as Facebook and Twitter, as well as collaborative communications, platforms such as company intranets. New project management tools, such as dashboard reporting systems, would be needed. Project management was undergoing a philosophical shift away from centralized command and control to socialized project management, and additional tools were needed for effective communications. These new tools are allowing for a more rigorous form of project management to occur, accompanied by more accurate performance reporting. The new tools also allow for decision-making based on facts and evidence rather than guesses.

4.11 ARTIFICIAL INTELLIGENCE AND PROJECT MANAGEMENT

It appears that the world of artificial intelligence (AI) is now entering the project management community of practice, and there is significant interest in this topic. Whether AI will cause an increase or decrease in project management tools is uncertain, but an impact is expected.

A common definition of AI is intelligence exhibited by machines.[13] From a project management perspective, could a machine eventually mimic the cognitive functions associated with the mind of a project manager such as decision-making and problem-solving? The principles of AI are already being used in speech recognition systems and search engines, such as Google Search and Siri. Self-driving cars use AI concepts as do military simulation exercises and content delivery networks. Computers can now defeat most people in strategy games such as chess. It is just a matter of time before we see AI techniques involved in project management.

The overall purpose of AI is to create computers and machinery that can function in an intelligent manner. Doing this requires the use of statistical methods, computational intelligence, and optimization techniques. The programming for such AI techniques requires not only an understanding of technology but also an understanding of psychology, linguistics, neuroscience, and many other knowledge areas.

The question regarding the use of AI is whether the mind of a project manager can be described so precisely that it can be simulated using the techniques just described. Perhaps there is no simple logic that will accomplish this in the near term, but there is hope because of faster computers, the use of cloud computing, and increases in machine learning technology. However, there are some applications of AI that could assist project managers in the near term:

- The growth in the use of competing constraints rather than the traditional triple constraints will make it more difficult to perform trade-off analyses. The use of AI concepts could make life easier for project managers.
- We tend to take it for granted that the assumptions and constraints given to us at the onset of the project will remain intact throughout the project's life cycle. Today, we know that this is not true and that all assumptions and constraints must be tracked throughout the life cycle. AI could help us in this area.

---

13. This definition and parts of this section have been adapted from Wikipedia contributors, "Artificial Intelligence," *Wikipedia, The Free Encyclopedia,* https://en.wikipedia.org/w/index.php?title=Artificial_intelligence&oldid=802537752.

- Executives quite often do not know when to intervene in a project. Many companies today are using crises dashboards. When an executive looks at the crises dashboard on the computer, the display identifies only those projects that may have issues, which metrics are out of the acceptable target range, and perhaps even the degree of criticality. AI practices could identify immediate actions that could be taken and thus shorten response time to out-of-tolerance situations.

- Management does not know how much additional work can be added to the queue without overburdening the labor force. For that reason, projects are often added to the queue with little regard for (1) resource availability, (2) skill level of the resources needed, and (3) the level of technology needed. AI practices could allow for the creation of a portfolio of projects that have the best chance to maximize the business value the firm will receive while considering effective resource management practices.

- Although some software algorithms already exist for project schedule optimization, practices still seem to be a manual activity using trial-and-error techniques. Effective AI practices could make schedule optimization significantly more effective by considering all present and future projects in the company rather than just individual projects.

Project managers are often pressured to make rapid decisions based on intuition rather than by step-by-step deduction used by computers. Nothing is simply true or false because we must make assumptions. Generally speaking, the more information we have available, the fewer the assumptions that must be made. With a sufficient database of information, AI tools could perform reasoning and problem-solving based on possibly incomplete or partial information. AI can visualize the future and provide us with choices that can maximize the value of the decision.

If AI practices are to be beneficial to the project management community of practice, then pockets of project management knowledge that existed in the past must be consolidated into a corporate-wide knowledge management system that includes all of the firm's intellectual property, as shown in Figure 4–14.

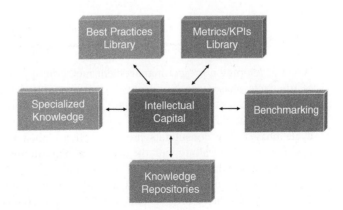

Figure 4–14. Components of intellectual property.

The more information available to the AI tools, the greater the value of the outcome. Therefore, the starting point must be a consolidation of project management intellectual property, and the AI tools must have access to this information. PMOs will most likely have this responsibility.

While all of this sounds workable, there are still some downside risks based on which area of knowledge in the *PMBOK® Guide* we apply the AI tools. As an example, using the Human Resources Knowledge Area, can AI measure and even demonstrate empathy in dealing with people? In the Integration Management Knowledge Area, can AI add in additional assumptions and constraints that were not included in the business case when the project was approved? In the Stakeholder Management Knowledge Area, can the AI tools identify the power and authority relationships of each stakeholder? And with regard to machine ethics, can an AI tool be made to follow or adhere to PMI's *Code of Ethics and Professional Responsibility* when making a decision?

While all of this seems challenging and futuristic to some, AI is closer than you think. Amazon, Google, Facebook, IBM, and Microsoft have established a nonprofit partnership to formulate best practices on AI technologies, advance the public's understanding, and serve as a platform for AI. In a joint statement, the companies stated: "This partnership on AI will conduct research, organize discussions, provide thought leadership, consult with relevant third parties, respond to questions from the public and media, and create educational material that advances the understanding of AI technologies, including machine perception, learning, and automated reasoning." Though not one of the original members in 2016, Apple joined other tech companies as a founding member of the Partnership on AI in January 2017. The corporate members will make financial and research contributions to the group, while engaging with the scientific community to bring academics onto the board.

Given the fact that Amazon, Google, Facebook, IBM, Microsoft, and Apple are all heavy users of project management, and some are considered to have world-class project practices, how long do you think it will be before they develop AI practices for their own project management community of practice? The implementation of AI practices in project management may very well be right around the corner.

4.12 LIFE-CYCLE PHASES

When developing a project management methodology, determining the best number of life-cycle phases can be difficult. As an example, let us consider IT. During the 1980s, with the explosion in software, many IT consulting companies came on the scene with the development of IT methodologies using systems development life-cycle phases. The consultants promised clients phenomenal results if clients purchased the package along with the accompanying training and consulting. Then, after spending hundreds of thousands of dollars, clients read the fine print that stated that the methodology must be used as is, and no customization of the methodology would take place. In other words, clients would have to change their company to fit the methodology rather than vice versa. Most of the IT consultancies that adopted this approach no longer exist.

For an individual company, agreeing on the number of life-cycle phases may be difficult at first. But when an agreement is finally reached, all employees should live by the same phases. However, for today's IT consulting companies, the concept of one-package-fits-all will not work. Whatever methodology they create must have flexibility in it so that client customization is possible. In doing so, it may be better to focus on processes rather than phases, or possibly a framework approach that combines the best features of each.

4.13 EXPANDING LIFE-CYCLE PHASES

Historically, we defined the first phase of a project as the initiation phase. This phase included bringing project managers on board, handing them a budget and a schedule, and telling them to begin project execution. Today, there is a preinitiation phase, which Russ Archibald and his colleagues refer to as the Project Incubation/Feasibility Phase.[14]

In this phase, we look at the benefits of the project, the value expected at completion, whether sufficient and qualified resources are available, and the relative importance of the project compared to other projects that may be in the queue. It is possible that the project may never reach the Initiation Phase.

In the past, project management was expected to commence at the initiation phase because it was during this phase that the project manager was assigned. Today, project managers are expected to possess a much greater understanding of the business as a whole, and companies have found it beneficial to bring project managers on board earlier than the initiation phase to assist in making business decisions rather than purely project decisions.

In the same context, we have traditionally viewed the last life-cycle phase as project closure, which includes the implementation of contractual closure, administrative closure, and financial closure. After closure, the project manager would be reassigned to another project.

Today, we are including a post-project evaluation phase. Some companies refer to this as a customer satisfaction management phase. In this phase, selected members of the project team and sales/marketing personnel, as well as members from the governance committee, meet with the client to see what changes can be made to the methodology or processes used to execute the project and what can be done differently on future projects for this client to further improve the working relationship among client, contractor, and stakeholders.

---

14. R. D. Archibald, I. Di Filippo, and D. Di Filippo, December 2012, "The Six-Phase Comprehensive Project Life Cycle Model Including the Project Incubation/Feasibility Phase and the Post-Project Evaluation Phase," *PM World Journal.*

4.14 CHURCHILL DOWNS INCORPORATED

Churchill Downs Incorporated has created a project management methodology that clearly reflects its organization. According to Chuck Millhollan, formerly director of program management at Churchill Downs:

> While we based our methodology on professional standards, we developed a graphic (and used terminology) understood by our industry to help with understanding and acceptance. For example, we have a structured investment request, approval, and prioritization process (see Figure 4–15). We used the analogy of bringing the Thoroughbred into the paddock prior to race and then into the starting gate. The project, or race, is not run until the Thoroughbred has entered the starting gate (approved business case and project prioritization).

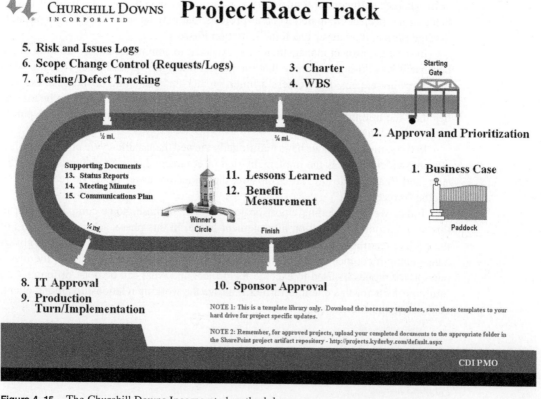

Figure 4–15. The Churchill Downs Incorporated methodology.

4.15 INDRA: THE NEED FOR A METHODOLOGY

As mentioned in Chapter 3, the quest for excellence in project management is almost always accompanied by the development of a project management methodology. Such was the case at Indra. Indra defines excellence in project management as follows: "Excellence in project management is achieved by being able to repeatedly reach the project targets, creating business opportunities, and improving the management process itself when managing the assigned projects." Enrique Sevilla Molina, PMP, formerly corporate PMO director, discusses the journey to excellence:

> A project management methodology was formally defined in the mid-'90s based upon the experience gained in our major international contracts. The main problems we faced were related to the definition of the scope, the limits of the methodology, and the adoption of the correct strategy to spread this knowledge throughout the company. To solve these issues, our management chose to hire an external consulting company to act as a dynamic factor that boosted and drove the cultural change.
>
> Yes, the process was carefully sponsored from the beginning by senior executives and closely followed up until its complete deployment in all areas of the company.
>
> The major milestones of the process have roughly been:

Project management strategy decision: mid-1990s
Methodology definition and documentation: mid- late 1990s
Tools definition and preparation: late 1990s
Training process start: 2000
Risk management at department level: 2002
PMP® certification[15] training start: 2004
Risk management process defined at corporate level: 2007
Program and portfolio management processes definition start: 2008

> A PM methodology was developed in the early '90s and formalized during that decade. It eventually was updated to cope with the company's and the industry's evolution. It is being used as a framework to develop and maintain the PMIS [project management information system], and to train the PMs throughout the company.
>
> It is based on the project life cycle and structured in two stages and six phases, as shown in Figure 4–16

Precontractual Stage
Phase 1. Initiation
Phase 2. Concept development, creation of offers and proposals
Phase 3. Offer negotiation

---

15. PMP is a registered mark of the Project Management Institute, Inc.

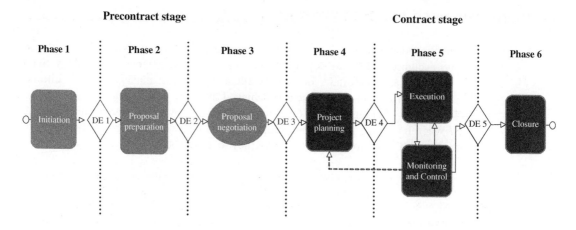

Figure 4–16. Project management life cycle.

Contractual Stage
Phase 4. Project planning
Phase 5. Execution, monitoring, and control
Phase 6. Closure

The precontractual and contractual stages are both part of the project and its life cycle. Most problems that appear during a project's life span originate during its definition and in the negotiation of its objectives, contents, and scope with the customers. A proper management of the precontractual stage is the best way to prevent problems later on.

At the end of each phase, there is a specific result that will allow a key decision to be made, focusing and directing the actions of the next phase and thereby reducing the initial risks and uncertainties of the project.

The decision on the stages and phases was a decision mainly based on the needs of our standard cycle of a project conception and development, and based on the most significant types of projects we were involved with.

Risk management processes are integrated into the methodology and into the corporate PM tools. An initial risk identification process is performed during the proposal phase, followed by a full risk management plan during the planning phase of the contract stage, and the subsequent monitoring processes during the execution phase of the project. QA [quality assurance] and change control processes are considered main support processes in the methodology.

4.16 IMPLEMENTING THE METHODOLOGY

Just because a methodology exists does not mean that it is a world-class methodology. Methodologies are nothing more than pieces of paper. What converts a standard methodology into a world-class one is the culture of the organization and the way the methodology is implemented.

The existence of a world-class methodology does not by itself constitute excellence in project management. Its corporate-wide acceptance and use do lead to excellence. It is through excellence in execution that an average methodology becomes a world-class methodology.

One company developed an outstanding methodology for project management. About one-third of the company used the methodology and recognized its true long-term benefits. The other two-thirds of the company would not support the methodology. The president eventually restructured the organization and mandated the use of the methodology.

The importance of execution cannot be overestimated. One characteristic of companies with world-class project management methodologies is that they have world-class managers throughout their organization.

Rapid development of a world-class methodology mandates an executive champion, not merely an executive sponsor. Executive sponsors generally are on an as-needed basis. Executive champions, in contrast, are hands-on executives who drive the development and implementation of the methodology from the top down. Most companies recognize the need for the executive champion. However, many companies fail to recognize that the executive champion position is lifelong. One Detroit company reassigned its executive champion after a few successes were realized using the methodology. As a result, no one was promoting continuous improvement in the methodology.

Good project management methodologies allow you to manage your customers and their expectations. If customers believe in your methodology, then they usually understand it when you tell them that no further scope changes are possible once you enter a specific life-cycle phase. One automotive subcontractor carried the concept of trust to its extreme by inviting customers to attend its end-of-phase review meetings. This fostered extreme trust between the customer and the contractor. The contractor did ask the customer to leave during the last 15 minutes of the end-of-phase review meetings, when project finances were being discussed.

Project management methodologies are an "organic" process, which implies that they are subject to changes and improvements. Typical areas of methodology improvement might include:

- Improved interfacing with suppliers
- Improved interfacing with customers
- Better explanation of subprocesses
- Clearer definition of milestones
- Clearer role delineation of senior management
- Recognition of the need for additional templates
- Recognition of the need for additional metrics

- Template development for steering committee involvement
- Enhancement of the project management guidebook
- Ways to educate customers on how the methodology works
- Ways of shortening baseline review meetings

4.17 IMPLEMENTATION BLUNDERS

Even though companies recognize the driving forces that indicate a need for project management improvement, the actual decision to make an investment to do it may not happen until some crisis occurs or a significant amount of red ink appears on the company's balance sheet. Recognizing a need is a lot easier than doing something about it because doing it requires time and money. Too often, executives procrastinate giving the go-ahead in hopes that a miracle will occur and project management improvements will not be necessary. And while they procrastinate, the situation often deteriorates further.

Delayed investment in project management capabilities is just one of many blunders. Another common blunder, which can occur in even the best companies, is the failure to treat project management as a profession. In some companies, project management is a part-time activity to be accomplished in addition to one's primary role. The career path opportunities come from the primary role, not through project management. In other companies, project management may be regarded merely as a specialized skill in the use of scheduling tools.

4.18 OVERCOMING DEVELOPMENT AND IMPLEMENTATION BARRIERS

Making the decision that the company needs a project management methodology is a lot easier than actually implementing it. Several barriers and problems surface well after the design and implementation team begins their quest. Typical problem areas include:

- Should we develop our own methodology or benchmark best practices from other companies and try to use their methodology in our company?
- Can we get the entire organization to agree on a singular methodology for all types of projects or must we have multiple methodologies?
- If we develop multiple methodologies, how easy or difficult will it be for continuous improvement efforts to take place?
- How should we handle a situation where only part of the company sees a benefit in using this methodology and the rest of the company wants to do its own thing?
- How do we convince employees that project management is a strategic competency and that the project management methodology is a process to support this strategic competency?
- For multinational companies, how do we get all worldwide organizations to use the same methodology? Must it be intranet based?

These are typical questions that plague companies during the methodology development process. These challenges can be overcome, and with great success, as illustrated by the companies identified in the next sections.

4.19 WÄRTSILÄ: RECOGNIZING THE NEED FOR SUPPORTING TOOLS[16] _____

Although we have always had a strong passion for engines at Wärtsilä, we are now much more than an engine company. Today, professional project management has become essential for our continuing success due to the bigger and more complex marine and power plant projects we deliver.

Excellent Project Management—A Prerequisite to Customer Satisfaction

The Wärtsilä Project Management Office (WPMO) was established in 2007 to develop a project management culture, processes, competencies, and tools that would guarantee our customers receive the satisfaction they deserve.

One of the first things we did was to conduct a detailed project management analysis in order to identify improvement areas. At that time, we did not have any software available for project and portfolio management. Therefore, one of the first actions the WPMO took was to initiate a global improvement program called Gateway to develop and implement a set of project and project portfolio management processes with a supporting application.

According to the program owner, Antti Kämi, the starting point for Gateway and the reason why Wärtsilä needed to improve project management even further was because "professional project management was seen as truly essential for our profitability, competitiveness, and for providing value to our customers."

Projects Divided into Three Categories

To achieve as many of the expected benefits as possible, it was decided that relevant parts of the unified processes of the new tool should be used in all divisions and in all three project categories in the company:

1. Customer delivery projects
2. Operational development projects
3. Product and solution development projects

Using this new approach meant that thousands of projects could be managed with the new tool, involving approximately 2,000 people in project management.

---

16. Section 4.19 provided by Wärtsilä Project Management Office (WPMO). Copyright to Wärtsilä Corporation 2022. Reproduced by permission.

Good Project Management Practices

Today we have unified business processes (gate models) in use throughout Wärtsilä with harmonized guidelines and terminology. Additionally, we maintain this resource through a professional training and certification path for project management.

As with all projects of this magnitude, there have been challenges to face on the way, especially when developing both the way of working and the software in parallel. The varying project management maturity levels within the company have also proven to be challenging. On the upside, a continuous and active dialogue around project management is now in place, experiences are openly exchanged between divisions and project categories, and the work gives a true feeling of "One Wärtsilä."

In several project management areas, we can already see improvements and benefits, especially in portfolio management and resource management.

Currently, we use the new application as a project database for our research and development portfolio planning. This enables the projects to be arranged in portfolios, which means that there is a more structured follow-up process. This in turn leads to better transparency and visibility in projects, easier and quicker responses to stakeholder inquiries, and more efficient project reporting on the whole.

First-class resource management is important today since information management resources are used in operational development projects throughout the company. Having a shared software tool ensures good resource availability, transparency for managing and monitoring the workload, as well as reliable facts for good planning.

Further benefits of a common project and portfolio management tool include the possibility to record and utilize lessons learned and the ability to collaborate and have information easily available for all project team members.

Tools Really Make a Difference in Project Management

In a nutshell, this is what Gateway at Wärtsilä is all about: to work out and apply a more effective way to plan and run projects and a common tool to help us gather, handle, and share project-related information. And by doing this, we ensure that both internal and external customers are satisfied.

4.20 GENERAL MOTORS POWERTRAIN GROUP

For companies with small or short-term projects, project management methodologies may not be cost-effective or appropriate. For companies with large projects, however, a workable methodology is mandatory. General Motors (GM) Powertrain Group is another example of a large company achieving excellence in project management. The company's business is based primarily on internal projects, although some contract projects are taken on for external customers. The size of the group's projects ranges from $100 million to $1.5 billion or greater. Based in Pontiac, Michigan, the GM Powertrain Group developed and implemented a four-phase project management methodology that has become the core process for its business. The company decided to go with project management in order to get its products out to the market faster. According to Michael Mutchler, former vice president and group executive:

> The primary expectation I have from a product-focused organization is effective execution. This comprehends disciplined and effective product program development, implementation, and day-to-day operations. Product teams were formed to create an environment in which leaders could gain a better understanding of market and customer needs; to foster systems thinking and cross-functional, interdependent behavior; and to enable all employees to understand their role in executing GM Powertrain strategies and delivering outstanding products. This organizational strategy is aimed at enabling a large organization to be responsive and to deliver quality products that customers want and can afford.

The program management process at GM Powertrain was based on common templates, checklists, and systems. Several elements that were common across all GM Powertrain programs during the 1990s are listed next.

- Charter and contract
- Program team organizational structure with defined roles and responsibilities
- Program plans, timing schedules, and logic networks
- Program-level and part-level tracking systems
- Four-phase product development process
- Change management process

Two critical elements of the GM Powertrain methodology were the program charter and program contract. The program charter defined the scope of the program with measurable objectives, including:

- Business purpose
- Strategic objective
- Results sought from the program
- Engineering and capital budget
- Program timing

The program contract specifies how the program will fulfill the charter. The contract became a shared understanding of what the program team will deliver and what the GM Powertrain staff will provide to the team in terms of resources, support, and so on.

Although the information here on GM Powertrain may appear somewhat dated, it shows that GM was several years ahead of most companies in the development of an EPM methodology. GM has made significant changes to its methodology since then. Many companies are just beginning to develop what GM accomplished more than a decade ago. Today, GM uses the above-mentioned methodology for new product development and has a second methodology for software projects.

4.21 INDRA: CLOSING THE PROJECT[17]

In a technological company like Indra, with projects being managed to develop, manufacture, and maintain complex hardware and software systems, an immature project closure can be, if not well treated, a cause of great efficiency losses.

Projects usually require a curve of effort with the peak at the beginning and half of a project life cycle (see Figure 4–17) or in other words, from the project manager's point of view, planning, monitoring, and control are the phases that require more attention.

During planning stage, project manager works toward clear goals. At the same time, planning depends on established commitments with either the sponsor or the customer. While in the monitoring and control stages, the project manager's attention is focused on coordinating team efforts to achieve project milestones, identifying variances to baselines, and protecting the project from changes, which really take most of the manager's time.

This is not the case at the end of the project: When commitments are fulfilled, most of the pressure on project management is removed. This occasionally means that the last of the milestones (project closure) is not properly achieved, as PM attention and effort have been reduced. A new assignment might be waiting for the project manager, so she is released to start the new responsibility without properly closing the previous one.

In the context of an organization like Indra, whose main business is delivering project results to its customers, we intend to organize our resources in the most efficient manner, responding effectively to all commitments to customers in a business improvement framework.

Project managers may have little motivation to perform a good project closure; they may consider it to be a simple and administrative task. Therefore, they might forget that if we do not pay attention to the opportunity to consolidate the efficiencies that were gained in the project, that benefit can be lost for the organization, particularly in the management of scope and resources (and scope and resources are the main values used to calculate productivity).

If we focus on scope management, if project closure is not well done, there is a risk that the customer might dilute, reopen, or reinterpret acceptance agreements of the deliverables. This happens if the project end is not well settled and if it is confused and mixed with the warranty period.

Consider this: After a new system is established, a customer's needs may change, and this may mean that the interpretation of the requirements must evolve also. No longer can the requirements be traced to the initial project scope and former conditions of validation. As time passes, the perceptions of the person at the client who performs the requirements validation on deliverables may change without formal project closure. Then the customer may try to relocate new needs back on the project instead of placing them in a project extension, as they should be.

When managing a project based on agile models, which are so popular today, it is especially advisable to pay particular attention to the efforts dedicated to customer requirements acceptance.

---

17. Section 4.21 © 2022 Indra. Reproduced by permission. All rights reserved. Material on Indra has been provided by Alfredo Vázquez Díaz, PMP, Director, Corporate Project Management Office.

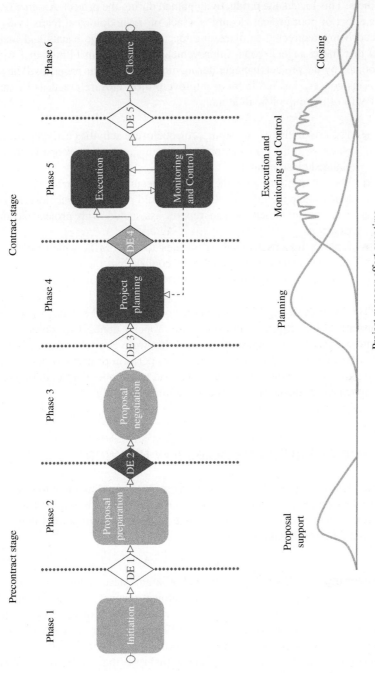

Figure 4–17. Indra's project management life cycle.

If we focus on management of resources, several organizational staffing roadblocks can occur; one of the most frequent is the failure to release resources from our project to others. This lowers the productivity gained during the project. Another potential negative effect of poor project closure is a lack of methodological focus, with the process instead being redirected to managing reactions to change events and incidents. This risks that the changes, scope, improvements, and responsibilities that were properly negotiated by the project manager during other stages of the project will be jeopardized.

This concern has led Indra to improve project closure practices by implementing the PMIS of a group of facilitators:

- The possibility of beginning project closure activities early, by overlapping this phase with the previous phase (e.g., in cases where scope is cut or there is a planned closure date)
- The use of information the PMIS has accumulated from the project over its lifetime to help identify situations that could prevent formal closure
- The use of indicators and reports associated with project closure, tracking its status
- Linking lessons learned to updates to the PMIS, which allows searches in previous experiential knowledge; could affect the closing process (and other processes) by requiring more time to capture lessons learned

The Spanish saying "Close the door [or] the cat will escape" shows the risks that organizations face if project closure is not properly done. If we do not close the door, the project escapes; in other words, risks that were controlled at one point ultimately may still occur. Project managers who do not perform project closure carefully may be adding great risks to the project, risks that had been under control during earlier phases when the project manager's effort and attention were high.

4.22 WHEN TRADITIONAL METHODOLOGIES MAY NOT WORK _____

While methodologies serve a viable purpose, traditional methodologies may not work well when projects become distressed and rapid action is required to save a failing project. In such a case, other factors may become more important than following the traditional life-cycle phases.

Insights about Recovering Troubled Projects and Programs   Today's projects and programs have become so complex that, on a daily basis, effective project recovery techniques may be required regardless of the country you are in or the business base of your company. We must be willing to face different or unforeseen situations that are not related to citizenship, the language we speak, or the experiences we all have. We are affected by a multitude of internal and external risks that can become real issues during project execution.

Following are some insights from Dr. Alexandre Sörensen Ghisolfi, who for many years has collaborated with the International Institute for Learning and faced these challenges in the global project management community. The following are some of his best practices and things to think about.

Project recovery can be source of many ideas and lessons learned. When projects require recovery, they are normally accompanied by conflicts, disagreements, and even fights.

When projects are in trouble, you will probably get a better understanding of who the people really are and whether they are truly committed to the organization. In other words, when things go well, we can easily see smiles and we often know the best side of people. When things go bad, a different individual usually emerges.

In this kind of environment, we learned that recovery usually requires a team of experts and effective project management leadership to be successful. Not all project managers have skills in recovery project management techniques. Trust in both the project manager and the solution is probably the most essential criteria for recovery to work. Furthermore, dedicated project management teams are usually essential.

When conditions indicate that failure may be imminent, we must be able to clearly change the cultural aspects where bad feelings can and will surface. In this way, we may be able to create a new culture conducive to recovery.

Recovery project management teams are composed of senior people, professional experts, young people with new ideas, new talent that recognizes that successful recovery may benefit their career goals and recovery project managers with leadership skills. They must all work together such that we can convert a bad-feeling environment to an environment where people believe we can still bring back the project and deliver what is required. If the team succeeds, this will, even more, make team members proud of their accomplishments. This can develop a great buy-in feeling that remains over a lifetime.

During recovery, we must consider two different environments:

1. Human behaviors
2. Application of technical expertise

Human Behaviors

Each project that is in trouble has very different scenarios and alternatives. The recovery process depends on your experience and ability to find solutions. It also depends on how well you can influence the different stakeholders to bring them to an agreed-on vision where they recognize that the game can be won. For successful solutions that are based on the different contributions of the team members, the project manager needs to know how to extract the best from them or influence them to achieve what is expected by the leader.

But before you start to identify and evaluate different alternatives, it would be good to consider some different aspects related to human behaviors that strongly influence the output. For simplicity's sake, we will not talk about politics, hierarchy, knowledge, and other aspects that influence human behaviors, but we suggest you at least clearly understand what kind of company and team you have. We can try to understand it through the study of organizational maturity.

Do you have a mature team? Is the company likewise mature in project management?

Some Best Practices to Always Try to Put in Place

- The organizational maturity of a company and/or team will directly affect the outcomes, so the more mature and professional team you have, the better the ability to recover failing programs. The best practice is to first deeply analyze company and team member maturity; having results at your disposal, you will be able to identify gaps, issues, and conditions that may require change. After the maturity analysis, and with the maturity reports in your hands, it will be necessary to prepare a recovery plan and show the project sponsors justification for why some important actions must take place urgently. When working in matrix organizations, we can face important difficulties related to resources that are not directly part of our project organizational structure and that may have different interests in the outcome of the project.

Additional Best Practices

- Remind people about the necessity for change. If required, let us remind people every day about our mission and daily tasks.
- Give training to people as appropriate and act as a role model for the attitudes that are necessary. It is probably the best way to improve team and organizational maturity.
- Empower team members; make our challenge their challenge.
- Make sure the entire team is deeply committed to the challenge of bringing back good results.

Ensure that the communication processes are effective. You can communicate less or even more, but at the end of the day, communication must be effective. Effective communication depends again on your company and team's maturity and on how much you are committed to the project. Truly committed team members will focus on communication. Communication must naturally flow.

When talking about processes and work, flexibility is important. But on other side, discipline to deliver key critical activities must be in place. Again, here the team's organizational structure, departments, and suppliers' interests can have a huge impact on good and bad things to happen.

Human behavior is probably more challenging than the application of the required technical expertise.

The study of organizational and team maturity can point us to a more secure way of proceeding. Since faster and better performance needs to be in place quickly, you cannot fail again; actions must be effective.

Application of Technical Expertise

On the technical side, when recovering projects, it maybe even more important to clearly know or define priorities.

You will probably put emphasis on the quality criteria (i.e., quality acceptance criteria) of the products you need to deliver; undoubtedly, you cannot sacrifice quality. The equilibrium of constraints will depend on your negotiation abilities as well as the contractual conditions you may have with your customers.

You may not be able to deliver everything that is required by the project because sacrifices must be made. Perhaps you will deliver results differently when compared to the project baselines. A recovery plan needs to be in place immediately.

What to sacrifice: Costs? Project costs? Product costs? Timing? Downgrade specifications? Change product/project delivery strategies? Communication downgrading? Documentation writing, presentations?

Many factors can significantly impact success and increase the issues on troubled projects. Here are some additional best practices you may try to put in place:

- Emphasis on risk management is a must-have condition. On troubled projects, risk management becomes even more important. When you encourage the team to perform a plan driven by risk management, the team is already defining priorities, which are the ones resulting from risk analysis. Risk management, as a holistic vision, can drive everything else around, such as scope, time, team organization, skills, communications, and others.
- Put the best people you have on the most difficult activities first.
- Emphasize critical activities to shorten their respective durations.
- Avoid bringing new people on board who lack sufficient experience; yet you could bring on board new people where you need to change the cultural aspects and/or interests in place.
- Avoid conflict of interest; we cannot lose time or waste resources solving unnecessary issues. Work even harder with your sponsor to gain their support.
- Adaptation of best practices available in the project scenario is also a key point. You must find out a way, often "out of the box," to make your team perform things that possibly never have been in practice before. Challenge your team members; ask them what they think and how we can start to work in different ways. In this way, you are developing the buy-in feeling.
- You will probably look for quick wins. You will soon observe that some of the best practices the team tried to apply have been useful and some other ones were not so well received. Replace or adopt best practices with no previous adoption and not necessarily applicable to your project by other ones where you can quickly have satisfactory results. Fast identification of what is working and what is not working is crucial to recover the time lost.

Successfully recovering bad projects can be an amazing experience, and when you have a team with the proper mindset in place, it can significantly contribute to increasing enterprise and team project management maturity. Great results can be achieved on future projects by preventing them from entering into critical situations that can lead to failure.

5 Integrated Processes

5.0 INTRODUCTION

Companies that have become extremely successful in project management have done so by performing strategic planning for project management. These companies are not happy with just matching the competition. Instead, they opt to exceed the performance of their competitors. To do this on a continuous basis requires processes and methodologies that promote continuous rather than sporadic success as well as continuous improvement efforts from best practices.

Figure 5–1 identifies the hexagon of excellence. The six components identified in the hexagon of excellence are the areas where the companies' excellence in project management exceeds their competitors. Each of these six areas is discussed in Chapters 5–10. We begin with integrated processes.

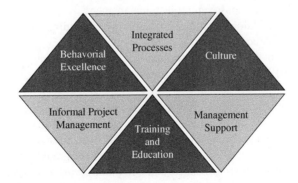

Figure 5–1. Six components of excellence.
Source: Reproduced from H. Kerzner, *In Search of Excellence in Project Management* (Hoboken, NJ: Wiley, 1998), p. 14.

5.1 UNDERSTANDING INTEGRATED MANAGEMENT PROCESSES

As we discussed in Chapter 1, since 1985 several new management processes (e.g., concurrent engineering) have supported the acceptance of project management. The most important complementary management processes and the years they appeared to be integrated into project management are listed next. It should be understood that many of these processes were introduced years before they were integrated into project management processes:

- 1985: Total quality management (TQM)
- 1990: Concurrent engineering
- 1992: Employee empowerment and self-directed teams
- 1993: Reengineering
- 1994: Life-cycle costing
- 1995: Change management
- 1996: Risk management
- 1997–1998: Project offices and centers of excellence
- 1999: Colocated teams
- 2000: Multinational teams
- 2001: Maturity models
- 2002: Strategic planning for project management
- 2003: Intranet status reporting
- 2004: Capacity-planning models
- 2005: Six Sigma integration with project management
- 2006: Virtual project management teams
- 2007: Lean/agile project management
- 2008: Knowledge/best practices libraries
- 2009: Project management business process certification
- 2010: Managing complex projects
- 2011: Governance by committees
- 2012: Competing constraints including a value component
- 2013: Advances in metrics management and dashboard reporting systems
- 2014: Value-driven project management
- 2015: Global project management including the management of cultural differences
- 2016–2017: Merger and acquisition project management growth
- 2018–2022: Corporate strategic planning and VUCA analysis

The integration of project management with these other management processes is key to achieving sustainable excellence. Not every company uses every process all the time. Companies choose the processes that work best for them. However, whichever processes are selected, they are combined and integrated into the project management methodology. Earlier we stated that companies with world-class methodologies try to employ a single, standard methodology based on integrated processes. This includes business processes as well as project management-related processes.

As each of these integrated processes undergoes continuous improvement efforts, so does the project management methodology that uses them. Best practices libraries and knowledge repositories contain best practices on the integrated processes as well as the overall project management methodology.

The ability to integrate processes is based on which processes the company decides to implement. For example, if a company implemented a stage-gate model for project management, the company might find it an easy task to integrate new processes, such as concurrent engineering. The only precondition would be that the new processes were not treated as independent functions but were designed from the onset to be part of a project management system already in place. The four-phase model used by the General Motors Powertrain Group and the PROPS model used at Ericsson Telecom AB readily allowed for the assimilation of additional business and management processes.

Earlier we stated that project managers today are viewed as managing part of a business rather than just a project. Therefore, project managers must understand the business and the processes to support the business as well as the processes to support the project. This chapter discusses some of the management processes listed and how the processes enhance project management. Then it looks at how some of the integrated management processes have succeeded using actual case studies.

5.2 EVOLUTION OF COMPLEMENTARY PROJECT MANAGEMENT PROCESSES ___

Since 1985, several new management processes have evolved in parallel with project management. Of these processes, TQM and concurrent engineering may be the most relevant. Agile and Scrum also have a significant impact and will be discussed later in this book (Chapter 15). Companies that reach excellence are the quickest to recognize the synergy among the many management options available today. Companies that reach maturity and excellence the quickest are those that recognize that certain processes feed on one another. As an example, consider the seven points listed below. Are these seven concepts part of a project management methodology?

1. Teamwork
2. Strategic integration
3. Continuous improvement
4. Respect for people
5. Customer focus
6. Management by fact
7. Structured problem solving

These seven concepts were actually the basis of Sprint's TQM process. They could just as easily have been facets of a project management methodology.

During the 1990s, Kodak taught a course entitled "Quality Leadership." The five principles of Kodak's quality leadership program included:

Customer Focus

"We will focus on our customers, both internal and external, whose inputs drive the design of products and services. The quality of our products and services is determined solely by these customers."

Management Leadership

"We will demonstrate, at all levels, visible leadership in managing by these principles."

Teamwork

"We will work together, combining our ideas and skills to improve the quality of our work. We will reinforce and reward quality improvement contributions."

Analytical Approach

"We will use statistical methods to control and improve our processes. Data-based analyses will direct our decisions."

Continuous Improvement

"We will actively pursue quality improvement through a continuous cycle that focuses on planning, implementing, and verifying of improvements in key processes."

Had we looked at just the left column, we could argue that these are the principles of project management as well.

Figure 5–2 shows what happens when an organization does not integrate its processes. The result is totally uncoupled processes. Companies with separate methodologies for each process may end up with duplication of effort, possible duplication of resources, and even duplication of facilities. Although there are several processes in Figure 5–2, we focus on project management, TQM, and concurrent engineering only.

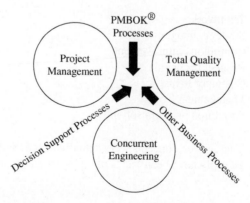

Figure 5–2. Totally uncoupled processes.

As companies begin recognizing the synergistic effects of putting several of these processes under a single methodology, the first two processes to become partially coupled are project management and TQM, as shown in Figure 5–3. As the benefits of synergy and integration become apparent, organizations choose to integrate all of these processes, as shown in Figure 5–4.

Excellent companies are able to recognize the need for new processes and integrate them quickly into existing management structures. During the early 1990s, integrating project management with TQM and concurrent engineering was emphasized. Since the middle 1990s, two other processes have become important in addition: risk management and change management. Neither of these processes is new; it is the emphasis that is new.

During the late 1990s, Steve Gregerson, formerly vice president for product development at Metzeler Automotive Profile System, described the integrated processes in its methodology:

> Our organization has developed a standard methodology based on global best practices within our organization and on customer requirements and expectations. This methodology also meets the requirements of ISO 9000. Our process incorporates seven gateways that require specific deliverables listed on a single sheet of paper. Some of these deliverables have a procedure and, in many cases, a defined format. These guidelines, checklists, forms, and procedures are the backbone of our project management structure and also serve to capture lessons learned for the next program. This methodology is incorporated into all aspects of our business systems, including risk management,

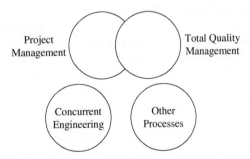

Figure 5–3. Partially integrated processes.

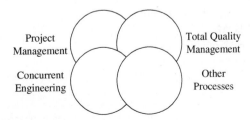

Figure 5–4. Totally integrated processes.

concurrent engineering, advanced quality planning, feasibility analysis, design review process, and so on.[1]

Another example of integrated processes was the methodology employed by Nortel. During the late 1990s, Bob Mansbridge, then vice president, supply chain management at Nortel Networks, stated:

> Nortel Networks project management is integrated with the supply chain. Project management's role in managing projects is now well understood as a series of integrated processes within the total supply chain pipeline. Total quality management (TQM) in Nortel Networks is defined by pipeline metrics. These metrics have resulted from customer and external views of "best-in-class" achievements. These metrics are layered and provide connected indicators to both the executive and the working levels. The project manager's role is to work with all areas of the supply chain and to optimize the results to the benefit of the project at hand. With a standard process implemented globally, including the monthly review of pipeline metrics by project management and business units, the implementation of "best practices" becomes more controlled, measurable, and meaningful.[2]

The importance of integrating risk management is finally being recognized. According to Frank T. Anbari, professor emeritus of project management, Drexel University:

> By definition, projects are risky endeavors. They aim to create new and unique products, services, and processes that did not exist in the past. Therefore, careful management of project risk is imperative to repeatable success. Quantitative methods play an important role in risk management. There is no substitute for profound knowledge of these tools.

Risk management has been a primary focus among healthcare organizations for decades, for obvious reasons, as well as among financial institutions and the legal profession. Today, in organizations of all kinds, risk management keeps us from pushing our problems downstream in the hope of finding an easy solution later on or of the problem simply going away by itself. Change management as a complement to project management is used to control the adverse effects of scope creep: increased costs (sometimes double or triple the original budget) and delayed schedules. With change management processes in place as part of the overall project management system, changes in the scope of the original project can be treated as separate projects or subprojects so that the objectives of the original project are not lost.

Today, almost all companies integrate five main management processes (see Figure 5–5):

1. Project management
2. TQM
3. Risk management
4. Concurrent engineering
5. Change management

---

1. H. Kerzner, Advanced Project Management: Best Practices on Implementation (Hoboken, NJ: Wiley, 2000), p. 188.

2. Personal communication with the author.

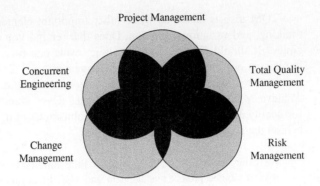

Project Management

Concurrent
Engineering

Total Quality
Management

Change
Management

Risk
Management

Figure 5–5. Integrated processes for twenty-first century.

Self-managed work teams, employee empowerment, reengineering, and life-cycle costing are also combined with project management in some companies. We briefly discuss these less widely used processes after we discuss the more commonly used ones.

5.3 TOTAL QUALITY MANAGEMENT

During the past three decades, the concept of TQM (and then Six Sigma) has revolutionized the operations and manufacturing functions of many companies. Companies have learned quickly that project management principles and systems can be used to support and administer TQM programs and vice versa. Ultimately excellent companies have completely integrated the two complementary systems.

The emphasis in TQM is on addressing quality issues in total systems. Quality, however, is never an end goal. TQM systems run continuously and concurrently in every area in which a company does business. Their goal is to bring to market products of better and better quality and not just of the same quality as last year or the year before.

TQM was founded on the principles advocated by W. Edwards Deming, Joseph M. Juran, and Phillip B. Crosby. Deming is famous for his role in turning postwar Japan into a dominant force in the world economy. TQM processes are based on Deming's simple plan-do-check-act cycle.

The cycle fits completely with project management principles. To fulfill the goals of any project, first you plan what you're going to do, then you do it. Next, you check on what you did. You fix what did not work, and then you execute what you set out to do. But the cycle doesn't end with the output. Deming's cycle works as a continuous-improvement system too. When the project is complete, you examine the lessons learned in its planning and execution. Then you incorporate those lessons into the process and begin the plan-do-check-act cycle all over again on a new project.

TQM also is based on three other important elements: customer focus, process thinking, and variation reduction. Does that remind you of project management principles? It should. The plan-do-check-act cycle can be used to identify, validate, and implement best practices in project management.

In the mid-1990s, during a live videoconference on the subject titled "How to Achieve Maturity in Project Management," Dave Kandt, then group vice president for quality and program management at Johnson Controls, commented on the reasons behind the company's astounding success:

> We came into project management a little differently than some companies. We have combined project management and TQC [total quality control] or total quality management. Our first design and development projects in the mid-1980s led us to believe that our functional departments were working pretty well separately, but we needed to have some systems to bring them together. And, of course, a lot of what project management is about is getting work to flow horizontally through the company. What we did first was to contact Dr. Norman Feigenbaum, who is the granddaddy of TQC in North America, who helped us establish some systems that linked together the whole company. Dr. Feigenbaum looked at quality in the broadest sense: quality of products, quality of systems, quality of deliverables, and, of course, the quality of projects and new product launches. A key part of these systems included project management systems that addressed product introduction and the product introduction process. Integral to this was project management training, which was required to deliver these systems.
>
> We began with our executive office, and once we had explained the principles and philosophies of project management to these people, we moved to the management of plants, engineering managers, analysts, purchasing people, and of course project managers. Only once the foundation was laid did we proceed with actual project management and with defining the role and responsibility so that the entire company would understand their role in project management once these people began to work. Just the understanding allowed us to move to a matrix organization and eventually to a stand-alone project management department. So how well did that work? Subsequently, since the mid-1980s, we have grown from two or three projects to roughly 50 in North America and Europe. We have grown from two or three project managers to 35. I don't believe it would have been possible to manage this growth or bring home this many projects without project management systems and procedures and people with understanding at the highest levels of the company.
>
> In the early 1990s, we found that we were having some success in Europe, and we won our first design and development project there. And with that project, we carried to Europe not only project managers and engineering managers who understood these principles but also the systems and training we incorporated in North America. So, we had a company-wide integrated approach to project management. What we've learned in these last 10 years that is the most important to us, I believe, is that you begin with the systems and the understanding of what you want the various people to do in the company across all functional barriers, then bring in project management training, and last implement project management.

Of course, the people we selected for project management were absolutely critical, and we selected the right people. You mentioned the importance of project managers understanding business, and the people that we put in these positions are very carefully chosen. Typically, they have a technical background, a marketing background, and a business and financial background. It is very hard to find these people, but we find that they have the necessary cross-functional understanding to be able to be successful in this business.

At Johnson Controls, project management and TQM were developed concurrently. Kandt was asked during the same videoconference whether companies must have a solid TQM culture in place before they attempt the development of a project management program. He said:

I don't think that is necessary. The reason why I say that is that companies like Johnson Controls are more the exception than the rule of implementing TQM and project management together. I know companies that were reasonably mature in project management and then ISO 9000 came along, and because they had project management in place in a reasonably mature fashion, it was an easier process for them to implement ISO 9000 and TQM. There is no question that having TQM in place at the same time or even first would make it a little easier, but what we've learned during the recession is that if you want to compete in Europe and you want to follow ISO 9000 guidelines, TQM must be implemented. And using project management as the vehicle for that implementation quite often works quite well.

There is also the question of whether successful project management can exist within the ISO 9000 environment. According to Kandt:

Not only is project management consistent with ISO 9000, a lot of the systems that ISO 9000 require are crucial to project management's success. If you don't have a good quality system, engineering change system, and other things that ISO requires, the project manager is going to struggle in trying to accomplish and execute that project. Further, I think it's interesting that companies that are working to install and deploy ISO 9000, if they are being successful, are probably utilizing project management techniques. Each of the different elements of ISO requires training, and sometimes the creation of systems inside the company that can all be scheduled, teams that can be assigned, deliverables that can be established, tracked, and monitored, and reports that go to senior management. That's exactly how we installed TQC at Johnson Controls, and I see ISO 9000 as having a very similar thrust and intent.

Total Quality Management   *While the principles of TQM still exist, the importance of Six Sigma concepts has grown. The remainder of Section 5.3 was provided by Eric Alan Johnson, Satellite Control Network Contract Deputy Program Director, AFSCN, and the winner of the 2006 Kerzner Project Manager of the Year Award, and by Jeffrey Neal, Blackbelt/Lean Expert and Lecturer, Quantitative Methods, University of Colorado, Colorado Springs.*

*   *   *

In addition to the TQM PDCA cycle, the continuous improvement DMAIC (define, measure, analyze, improve, and control) model can be used to improve the effectiveness of project management. This model has been successfully employed for Six Sigma and lean enterprise process improvement, but the basic tenets of its structured, data-enabled problem-solving methodology can also be employed to improve the success of project management.

By assessing data collected on both project successes and root causes of project failures, the DMAIC model can be used to improve and refine both the management of projects and the ultimate quality of products produced.

In the defined phase, specific project definitions and requirements are based on data gathered from the customer and on historical project performance. Gathering as much information as possible in these areas allows the project manager to concentrate on what is truly important to the customer while reviewing past performance in order to avoid the problems and continue to propagate the successes of past projects. In the defined stage, available data on the people, processes, and suppliers are reviewed to determine their ability to meet the cost, quality, and schedule requirements of the project. The defined phase, in short, should assess not only the requirements of the customer, but it should also assess the capability of your system to meet those requirements. Both of these assessments must be based on data gathered by a dedicated measurement system. Additionally, the defined stage should establish the metrics to be used during project execution to monitor and control project progress. These metrics will be continually evaluated during the measure and analysis phase. (These DMAIC phases are concurrent with the PMI phases of project management).

The next phase of the DMAIC model, measure data (the metrics identified in the defined stage) from the measurement system is continually reviewed during project execution to ensure that the project is being effectively managed. The same data metrics used in the defined stage should be updated with specific project data to determine how well the project is progressing. The continual assessment of project performance, based on data gathered during the execution phase, is the key to data-enabled project management.

During the continual measurement of the progress of the project, it is likely that some of these key metrics will indicate problems either occurring (present issues) or likely to occur (leading indicators). These issues must be addressed if the project is to execute on time and on budget to meet requirements. This is where the analysis aspect of the DMAIC model becomes a critical aspect of project management. The analysis of data is an entire field onto itself. Numerous books and articles have addressed the problem of how to assess data, but the main objective remains. The objective of data analysis is to turn data into usable information from which to base project decisions.

The methods of data analysis are specific to the data type and to the specific questions to be answered. The first step (after the data have been gathered) is to use descriptive techniques to get an overall picture of the data. This overall picture should include a measure of central tendency (i.e., mean) and a measure of variation such as standard deviation. Additionally, graphic tools such as histograms and Pareto charts are useful in summarizing and displaying information. Tests of significance and confidence interval development are useful in determining if the results of the analysis are statistically significant and for estimating the likelihood of obtaining a similar result.

In the continual monitoring of processes, control charts are commonly used tools to assess the state of stability of processes and to determine if the variation is significant enough to warrant additional investigation. In addition, control charts provide a basis for determining if the type of variation is special cause or common cause. This distinction is critical in the determination of the appropriate corrective actions that may need to be taken.

To provide a basis for the identification of potential root causes for project performance issues, tools such as failure modes and effects analysis and the fishbone (also known as the Ishikawa) diagram can be used to initiate and document the organized thought process needed to separate main causes of nonconformities from contributing causes.

If the data meets the statistical condition required, such tests as analysis of variance (ANOVA) and regression analysis can be extremely useful in quantifying and forecasting process and project performance. Because ANOVA (the general linear model) can be used to test for mean differences of two or more factors or levels, ANOVA can be used to identify important independent variables for various project-dependent variables. Various regression models (simple linear, multiple linear, and binary) can be used to quantify the different effects of independent variable on critical dependent variable that are key to project success.

In short, this phase uses the data to conduct an in-depth and exhaustive root cause investigation to find the critical issue that was responsible for the project execution problem and its effects on the project if left uncorrected.

The next phase involves the process correction and improvement that addressed the root cause identified in the previous phase. This is corrective action (fix the problem you are facing) and preventive action (make sure it or one like it doesn't come back). So, once the root cause has been identified, both corrective and preventive process improvement actions can be taken to address current project execution and to prevent the reoccurrence of that particular issue in future projects. To ensure that current projects do not fall victim to that problem recently identified and that future projects avoid the mistakes of the past, a control plan is implemented to monitor and control projects. The cycle is repeated for all project management issues.

The continuing monitoring of project status and metrics along with their continual analysis and correction is an ongoing process and constitutes the control phase of the project. During this phase, the key measurements instituted during the initiation phase are used to track project performance against requirements. When the root cause of each project problem is analyzed, this root cause and the subsequent corrective and preventive action are entered into a "lessons learned" database. This allows for consistent problem-resolution actions to be taken. The database is also then used to identify potential project risks and institute a priori mitigation actions.

Risk/Opportunity Management Using Six Sigma Tools and Probabilistic Models

Risk/opportunity management is one of the most critical, if not the most, tools in a project or program manager's toolbox—regardless of contract type. Typically, projects/programs focus on the potential impact and/or probability of a risk occurrence. While these are very critical factors in developing a good risk mitigation plan, the adroit ability of the project team to *detect* the risk will have the greatest impact on successful project execution. If you can't detect the risk, then your ability to manage it will always be reactive. The undetectable risk is a greater threat to execution, than the high-probability or high-impact factors. This is where using one of the Six Sigma tools failure modes and effects analysis (FMEA) can be very effective. The FMEA tool can help a project team evaluate risk identification. Focusing on risk identification will help the team think outside of the box in proposing, planning, or executing a successful project.

Example: If your project/program has a risk that has a significant probability of occurrence, then it is probably not really a risk—it is an issue/problem. If the impact is great and the probability is low, then you will keep an eye on this but not usually spend management reserve to mitigate. However, if the risk has a high impact or probability but has a low level of detectability, the results could be devastating.

The other side of managing a project/program is the lack of focus on opportunity identification and management. If a project team is only risk management focused, they may miss looking at the project's potential opportunities. Opportunities need to be evaluated with the same rigor as risks. The same level of focus in the areas of impact, probability, *and* the ability to *recognize* the opportunity must occur for a project team. The FMEA is also very useful for opportunity recognition and management. Sometimes undetectable risks will occur, but the ability to recognize and realize opportunities can counter these risk impacts. The use of opportunity recognition can have the greatest impacts on fixed-price projects where saving costs can increase the project's profit margin.

If a project has risk schedule, how can we quantify that risk? One method is through the use of probabilistic modeling. Probabilistic modeling of your schedule can help you forecast the likelihood of achieving all your milestones within your period of performance. If the risk of achieving your schedule is too high, you can use these models to perform what-if analysis until the risk factors can be brought to acceptable levels. This analysis should be done *before* the project is baselined or (ideally) during the proposal phase.

The key to successful implementation of this strategy is a relational database of information that will allow you to build the most realistic probabilistic model possible. This information must be gathered on a wide variety of projects so that information on projects of similar size and scope/complexity can be evaluated. It must be integrated with lessons learned from these other projects in order to build the best probabilistic model to mitigate your schedule risks. Always remember that a model is only as good as the information used to build it.

5.4 CONCURRENT ENGINEERING

The need to shorten product development time has always plagued U.S. companies. During favorable economic conditions, corporations have deployed massive amounts of resources to address the problem of long development times. During economic downturns, however, not only are resources scarce, but time becomes a critical constraint. Today, the principles of concurrent engineering have been almost universally adopted as the ideal solution to the problem.

Concurrent engineering requires performing the various steps and processes in managing a project in tandem rather than in sequence. This means that engineering, research and development, production, and marketing all are involved at the beginning of a project before any work has been done. That is not always easy, and it can create risks as the project is carried through. Superior project planning is needed to avoid increasing the level of risk later in the project. The most serious risks are delays in bringing product to market and cost when rework is needed as a result of poor planning. Improved planning is essential to project management, so it is no surprise that excellent companies integrate concurrent engineering and project management systems.

5.5 RISK MANAGEMENT

Risk management is an organized means of identifying and measuring risk and developing, selecting, and managing options for handling those risks. Throughout this book, we have emphasized that tomorrow's project managers will need superior business skills in assessing and managing risk. This includes both project risks and business risks. In the past, project managers were not equipped to quantify risks, respond to risks, develop contingency plans, or keep lessons-learned records. They were forced to go to senior managers for advice on what to do when risky situations developed. Now senior managers are empowering project managers to make risk-related decisions and doing that requires project managers to have solid business skills as well as technical knowledge.

Preparing a project plan is based on history. Simply stated: What have we learned from the past? Risk management encourages us to look at the future and anticipate what can go wrong and then develop contingency strategies to mitigate these risks.

We have performed risk management in the past, but only financial and scheduling risk management. To mitigate a financial risk, we increased the project's budget. To mitigate a scheduling risk, we added more time to the schedule. But in the 1990s, technical risks became critical. Simply adding into the plan more time and money is not the solution to mitigate technical risks. Technical risk management addresses two primary questions:

1. Can we develop the technology within the imposed constraints?
2. If we do develop the technology, what is the risk of obsolescence, and when might we expect it to occur?

To address these technical risks, effective risk management strategies are needed based on technical forecasting. On the surface, it might seem that making risk management an integral part of project planning should be relatively easy. Just identify and address risk factors before they get out of hand. Unfortunately, the reverse is likely to be the norm, at least for the foreseeable future.

For years, companies provided lip service to risk management and adopted the attitude toward risk that it is something we should simply learn to live with. Very little was published on how to develop a structure risk management process. The disaster with the space shuttle *Challenger* in January 1986 was a great awakening on the importance of effective risk management.[3]

Today risk management has become so important that companies are establishing separate internal risk management organizations. However, many companies have been using risk management functional units for years, and yet this concept has gone unnoticed. An overview of the program management methodology of the risk management department of an international manufacturer headquartered in Ohio follows. This department has been in operation for approximately 25 years.

The risk management department is part of the financial discipline of the company and ultimately reports to the treasurer, who reports to the chief financial officer. The overall objective of the department is to coordinate the protection of the company's assets. The primary means of meeting that objective is eliminating or reducing potential losses through loss prevention programs. The department works very closely with the internal environmental health and safety department. Additionally, it utilizes outside loss control experts to assist the company's divisions in loss prevention.

One method employed by the company to ensure the entire corporation's involvement in the risk management process is to hold its divisions responsible for any specific losses up to a designated self-insured retention level. If there is a significant loss, the division must absorb it and its impact on their bottom-line profit margin. This directly involves the divisions in both loss prevention and claims management. When a claim does occur, risk management maintains regular contact with division personnel to establish protocol on the claim and reserves and ultimate resolution.

The company does purchase insurance above designated retention levels. As with the direct claims, the insurance premiums are allocated to its divisions. These premiums are calculated based upon sales volume and claim loss history, with the most significant percentage being allocated to claim loss history.

Each of the company's locations must maintain a business continuity plan for its site. This plan is reviewed by risk management and is audited by the internal audit and environmental health and safety department.

Risk management is an integral part of the corporation's operations as evidenced by its involvement in the due diligence process for acquisitions or divestitures. It is involved at the onset of the process, not at the end, and provides a detailed written report of findings as well as an oral presentation to group management.

Customer service is part of the company's corporate charter. Customers served by risk management are the company's divisions. The department's management style with its customers is one of consensus building and not one of mandating. This is exemplified by the company's use of several worker's compensation third-party administrators (TPAs) in states where it is self-insured. Administratively, it would be much easier to utilize one nationwide TPA. However, using strong regional TPAs with offices in states where divisions operate provides knowledgeable assistance with specific state laws to the divisions. This approach has worked very well for this company that recognizes the need for the individual state expertise.

---

3. The case study "The Space Shuttle *Challenger* Disaster" appears in H. Kerzner, *Project Management Case Studies*, 6th ed. (Hoboken, NJ: Wiley, 2022), p. 363.

The importance of risk management is now apparent worldwide. The principles of risk management can be applied to all aspects of a business, not just projects. Once a company begins using risk management practices, it can always identify other applications for those processes.

For multinational companies that are project-driven, risk management takes on paramount importance. Not all companies, especially in undeveloped countries, have an understanding of risk management or its importance. These countries sometimes view risk management as an overmanagement expense on a project.

Consider the following scenario. As your organization gets better and better at project management, your customers begin giving you more and more work. You're now getting contracts for turnkey projects or complete-solution projects. Before, all you had to do was deliver the product on time and you were through. Now you are responsible for project installation and startup as well, sometimes even for ongoing customer service. Because the customers no longer use their own resources on the project, they worry less about how you're handling your project management system.

Alternatively, you could be working for third-world clients who have not yet developed their own systems. One hundred percent of the risk for such projects is yours, especially as projects grow more complex (see Figure 5–6). Welcome to the twenty-first century!

One subcontractor received a contract to install components in a customer's new plant. The construction of the plant would be completed by a specific date. After construction was completed, the contractor would install the equipment, perform testing, and then startup. The subcontractor would not be allowed to bill for products or services until after a successful startup. There was also a penalty clause for late delivery.

The contractor delivered the components to the customer on time, but the components were placed in a warehouse because plant construction had been delayed. The contractor now had a cash flow problem and potential penalty payments because of external dependencies that sat on the critical path. In other words, the contractor's schedule was being controlled by actions of others. Had the project manager performed business risk management rather than just technical risk management, these risks could have been reduced.

For the global project manager, risk management takes on a new dimension. What happens if the culture in the country with which you are working neither understands risk management nor has any risk management process? What happens if employees are afraid to surface bad news or identify potential problems? What happens if the project's constraints of time, cost, and quality/performance are meaningless to the local workers?

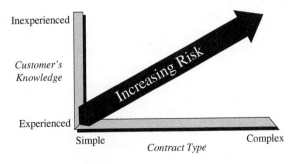

Figure 5–6. Future risks.

5.6 WÄRTSILÄ: THE NEED FOR PROACTIVE RISK MANAGEMENT _____

Proactive Project Risk Management in Wärtsilä Power Plant Projects

At Wärtsilä, project risk management has traditionally been much about identifying and planning. We have seen that this now needs to be expanded to cover reflection and proactive action taking in today's complex projects. Risks need to be tackled up front before they occur and potentially jeopardize project objectives. Here we briefly present what we have done in this respect.

How uncertainty and risk are handled in projects very much depends on experience. It can be said that many project managers deal with risk and uncertainty only as a result of what they are actually experiencing in their projects. Experienced project managers, however, can recognize risks far ahead before they become issues. Likewise, opportunities and positive uncertainties can be recognized more easily by experienced project managers. However, the recognition of opportunities is not only restricted to experience, since a willingness to take risks is also required. In many cases, a change of mindset is required from project managers to be able to accomplish this.

As large projects are becoming increasingly complex to manage today, it is essential that the project manager has enough experience to have an accurate perception of what is involved. Besides the project itself, it is of huge importance to know about the location, the customer, and the environment. Failure to have the knowledge or experience about these issues in advance will cause major problems, making the project more complex and challenging than necessary. In order to avoid such pitfalls, it is important to use the combined knowledge, experience, and creativity of the whole project team. Although risk management is the project manager's responsibility, it is not solely his or her task. The whole project team needs to share this responsibility.

This brings us to the importance of having a lessons-learned database with information that has been shared among the project teams. Such a database is an important resource for a new project manager or other team member joining the force. Likewise, when a project team accepts a new project type or a project in a totally new location, it is beneficial to be able to access the knowledge about similar cases. In light of this, a lessons-learned database is being implemented where all this knowledge and experience can be shared.

We have seen that knowledge and experience play an important role in managing risk, uncertainty, and other factors in projects. However, proactive risk management is not always easy to implement, since it depends on so many people's different perceptions of it. A lot of communication is needed in order to gain a common understanding of what the organization needs regarding risk and uncertainty as well as a clear understanding of the potential benefits they bring into the organization. Proactive risk management is not only about identifying, qualifying, and quantifying the risks; it is much more. The utilization of the risk management process is all about having the maturity to use the previously learned experience and knowledge to prevent risks from occurring in the first place, as well as the confidence to take the necessary actions to encourage positive opportunities to develop.

A project team needs a tool for project risk management where upcoming events, both foreseen and unforeseen, can be continuously followed. A risk management process tool does not need to be complex. The most important aspect is the way it is utilized in the organization. We see that in this case, the statement "The simpler the better" describes quite well what is needed.

The proactive risk management process taken into use at Wärtsilä consists of three different phases (see Figure 5–7). First, a project classification should be done to define the complexity of the project. Thereafter, the risk process itself will be managed as a continuous process throughout the whole project life cycle. In addition, lessons learned should be recorded on risks where the actions taken significantly differed from the planned response. At Wärtsilä we have implemented this entire process in one common project management tool used by all project management teams and management.

The classification process will provide important information for the risk identification steps. The intent of the process is to encourage project managers to think about the project and define where project complexity is situated and provide an input for the risk management process identification. It must describe the project from an objective point of view. One of the core added values that project classification brings to project management is defining needed resources for resource allocation.

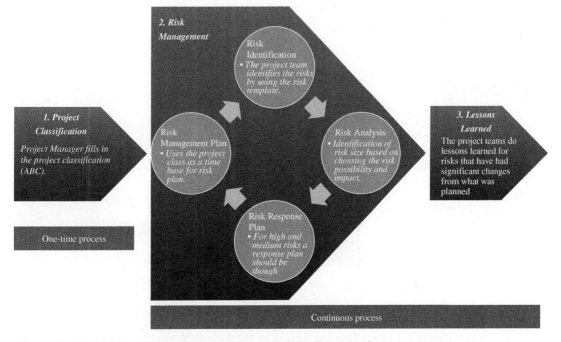

Figure 5–7. Proactive project risk management process in Wärtsilä power plants.

The risk management process basically also is one continuous process. All the same elements that were used in the classification process are implemented in this process. The traditional risk management process described in *PMBOK® Guide* (2008) has been used as the basis for the new risk management process.

In order for a proactive risk management process to be successful, it is vital that the project team makes full use of it. Ignorance of risks and uncertainties will cause large problems within project management when they materialize unexpectedly and become harmful issues.

A good communication system needs to be created in order for the project teams to implement a uniform risk management process. In addition, training should be given on how to use the risk management process in order to improve the understanding of how the proactive risk process can be utilized and to gain an appreciation as to why it is so important.

5.7 INDRA: WHEN A RISK BECOMES REALITY (ISSUE MANAGEMENT)[4]

In a company like Indra, with thousands of active projects geographically distributed, a solid and continuous risk management practice is vital. This is shown in Figure 5–8. Actions that might contain the impact that risks can have on the results of the project are planned and monitored throughout the project life cycle.

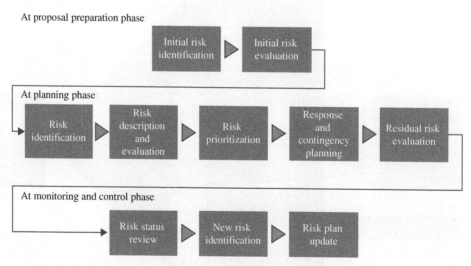

Figure 5–8. Indra's project risk management process.

4. Material in Section 5.7 has been provided by Alfredo Vázquez Díaz, PMP, Director, Corporate Project Management Office. © 2022 by Indra. Reproduced by permission. All rights reserved.

The percentage of projects with risk plans and risk registers is very high in Indra. However, we at the corporate project management office (PMO) noted that these high figures did not prevent some risks from happening; even worse, other unknown risks showed up as issues that project managers have not been able to identify in their plans.

However, an issue is not in the future; it is now already affecting project milestones or schedules, certain work breakdown structure elements, the committed budget, or the quality level of the project. For that reason, an issue usually demands immediate response, and it must be resolved as quickly and effectively as possible to avoid affecting other areas of the project.

We think that analyzing the relationship between risks and issues is essential for an integrated approach to risk management. By a premortem review of risks and their related issues, by sorting and classifying them, and by analyzing why those risks were not well addressed in earlier planning stages, we seek to understand what caused those risks and to determine earlier risk screening criteria.

In the second stage, we need to know the actual effects on the project of the risk and whether the solutions proposed to contain the effects have been effective. That will allow us to create a database and identify some lessons learned to help us to identify issue-prone risks early on and prevent them from appearing in future projects.

Not all issues are equal or affect the organization in the same way. Their impacts depend on the project size or volume, its internal or external visibility, project complexity, variances on initial economic forecast, time needed to put the project on track, or the issues' impact on the image of the company.

Having all those considerations up front, we have decided to focus our efforts on the projects with more critic issues. Those are considered projects that require close surveillance, and for those projects, key points are identifying the source of the issue, its immediate effects, and the follow-up of the action plan. To allow this we created a new functionality in our project management information system (PMIS) called issue registry. This is embedded within our PMIS issue management module.

The issue registry works like this: When a PMO or a user responsible for project control analyzes a project and detects it is having serious problems, the project manager is required to complete detailed information in the Issue Registry module. This can be triggered automatically through a red alert in the PMIS (see Figure 5–9).

The project manager must describe existing issues in the project, the action plans to cope with them, and the recovery target that must be obtained to get the project back on track. Once fulfilled, and to avoid this being a static snapshot of the project, the project manager is expected to update the information every reporting period from that moment on, indicating action plan updates and project status with regard to the initial targets.

What do we intend to achieve with the Issue Registry? We want to focus our attention and our efforts on:

- Detecting which problems and issues have emerged from previously identified risks and which have not.
- Getting a homogeneous classification and typification of projects that were seriously affected by issues.

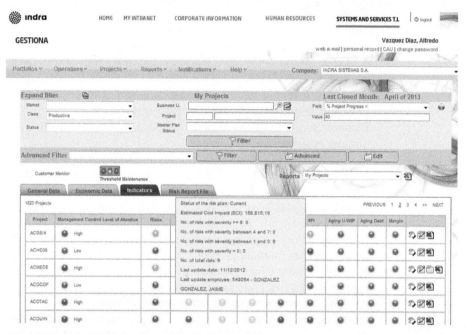

Figure 5–9. Indra's project risk monitoring in the PMIS.

- Tracking the effectiveness of the issue action plans.
- Automated and systematic reporting to business management on those issues and their projects from different perspectives (business unit, solution, type of project, geographic location, etc.).

Historically the efforts of our organization had been focused on the management of risks, leaving the management of issues and problems as a secondary process, not connected with risk management. Issue management was more dependent on the involvement and proactiveness of the project manager; for that reason, it was approached heterogeneously, usually in an internal project context and without close monitoring and follow-up by the organization, as it was the case with risk management.

The registration and follow-up of projects and the traceability between risks and issues can be made from the PMIS. To reverse actual dynamics, we will take a first step by learning from issues and problems that have already occurred; based on that analysis, in a second stage we will focus on prevention of recurrent issues—those that have been registered, diagnosed, and solved by other projects managers. This invaluable information will enable project managers to learn from others' experiences.

Inside the problem is the solution. Only by knowing the problem can we solve and avoid it.

5.8 THE FAILURE OF RISK MANAGEMENT

There are numerous reasons why risk management can fail. Typical reasons might include:

- The inability to
  - perform risk management effectively.
  - identify the risks.
  - measure the uncertainty of occurrence.
  - predict the impact, whether favorable or unfavorable.
- Having an insufficient budget for risk management work.
- Having team members who do not understand the importance of risk management.
- Fear that identification of the true risks could result in project cancellation.
- Fear that whoever identifies critical risks will get unfavorable recognition.
- Peer pressure from colleagues and superiors who want to see a project completed regardless of the risks.

All of these failures occur during project execution. We seem to understand these failures and can correct them with proper education and budget allocations for risk management activities. But perhaps the worst failures occur when people refuse to even consider risk management because of some preconceived notion about its usefulness or importance to the project. David Dunham discusses some of the reasons why people avoid risk management on new product development (NPD) projects.

Discussing risk in new product development certainly seems to be a difficult thing to do. Despite the fact that the high-risk nature of new product development is built into the corporate psyche, many corporations still take a fatalistic approach toward managing the risk. Reasons for not being anxious to dwell on risk differ depending on the chair in which you are sitting.

Program Manager
- Spending time on risk assessment and management is counter to the action culture of many corporations. "Risk management does not create an asset," to quote one executive.
- Management feels that the learning can/should be done in the market.

Project Manager
- There is a natural aversion among developers to focusing on the downside.
- Highlighting risk is counterintuitive for development teams who want to promote the opportunity when competing for NPD funding.[5]

---

5. D. J. Dunham, "Risk Management: The Program Manager's Perspective," in P. Belliveau, A. Griffin, and S. Somermeyer, *The PDMA Toolbook for New Product Development* (Hoboken, NJ: Wiley, 2002), p. 382.

5.9 DEFINING MATURITY USING RISK MANAGEMENT _____

For years, project management maturity was measured by how frequently we were able to meet a project's triple constraints of time, cost, and performance or scope. Today, we are beginning to measure maturity in components, such as the areas of knowledge in the *PMBOK® Guide*. Maturity is now measured in stages and components, such as how well we perform scope management, time management, risk management, and other areas of knowledge. Gregory Githens believes that the way we handle risk management can be an indicator of organizational maturity.

> Some firms have more capability to manage risk well, and these firms are the most consistent in their growth and profitability. Perhaps the simplest test for examining risk management maturity is to examine the level of authority given to the NPD program [project] manager: If authority is high, then the organization is probably positioning itself well to manage risks, but if authority is low, then the blinders may be on. Another test is the use of checklists: if ticking off a checklist is the sole company response to risk, then organizational maturity is low. Risk management provides an excellent lens by which to evaluate a firm's ability to integrate and balance strategic intent with operations.
>
> Many firms ignore risk management because they have not seen the need for it. They perceive their industry as stable and mostly focus on their competitive rivals and operational challenges. . . . By addressing risk at the project level, you encourage the organization to surface additional strategic concerns.
>
> Top NPD firms have a sophisticated capability for risk management, and they will "book" a project plan, pay attention to the details of product scope and project scope, use risk management tools such as computer simulations and principle-based negotiation, and document their plans and assumptions. These more mature firms are the ones that will consider risk in establishing project baselines and contracts. For example, Nortel uses a concept called "out of bounds" that provides the NPD program managers with the freedom to make trade-offs in time, performance, cost and other factors. Risk analysis and management is an important tool.
>
> Less mature firms typically establish a due date and pay attention to little else (and in my experience, this is the majority of firms). Firms that use the decision rule "Hit the launch date" default to passive acceptance—hiding the risk instead of managing it. Firefighting and crisis management characterize their organizational culture, and their strategic performance is inconsistent. These firms are like the mythological character Icarus: They fly high but come crashing down because they ignored easily recognizable risk events.[6]

---

6. G. D. Githens, "How to Assess and Manage Risk in NPD Programs: A Team-Based Risk Approach," in P. Belliveau, A. Griffin, and S. Somermeyer, *The PDMA Toolbook for New Product Development* (Hoboken, NJ: Wiley, 2002), p. 208.

5.10 BOEING AIRCRAFT COMPANY

As companies become successful in project management, risk management becomes a structured process that is performed continuously throughout the life cycle of the project. The two most common factors supporting the need for continuous risk management are how long the project lasts and how much money is at stake. For example, consider Boeing's aircraft projects. Designing and delivering a new plane might require 10 years and a financial investment of more than $15 billion.

From an academic perspective, Table 5–1 shows what the characteristics of risks might be at the Boeing Aircraft Company as considered by the author. (The table does not mean to imply that risks are mutually exclusive of each other, nor does it imply that these are the only risks.) New technologies can appease customers, but production risks increase because the learning curve is lengthened with new technology compared to accepted technology. The learning curve can be lengthened further when features are custom designed for individual customers. In addition, the loss of suppliers over the life of a plane can affect the level of technical and production risk. The relationships among these risks require the use of a risk management matrix and continued risk assessment.

5.11 CHANGE MANAGEMENT

Companies use change management to control both internally generated changes and customer-driven changes in the scope of projects. Most companies establish a configuration control board or change control board to regulate changes. For customer-driven

TABLE 5–1. RISK CATEGORIES AT BOEING

Type of Risk	Risk Description	Risk Mitigation Strategy
Financial	Up-front funding and payback period based on number of planes sold	Funding by life-cycle phases Continuous financial risk management Sharing risks with subcontractors Risk reevaluation based on sales commitments
Market	Forecasting customers' expectations on cost, configuration, and amenities based on a plane's 30- to 40-year life	Close customer contact and input Willingness to custom design per customer Development of a baseline design that allows for customization
Technical	Because of the long lifetime of a plane, must forecast technology and its impact on cost, safety, reliability, and maintainability	Structured change management process Use of proven technology rather than high-risk technology Parallel product improvement and new product development processes
Production	Coordination of manufacturing and assembly of a large number of subcontractors without impacting cost, schedule, quality, or safety	Close working relationships with subcontractors Structured change management process Lessons learned from other new airplane programs Use of learning curves

Note: The information in this section on how Boeing might characterize risks on a new airplane project is the author's opinion and not necessarily Boeing's official opinion.

changes, the customer participates as a member of the configuration control board. The configuration control board addresses these four questions at a minimum:

1. What is the cost of the change?
2. What is the impact of the change on project schedules?
3. What added value does the change represent for the customer or end user?
4. What are the risks?

The benefit of developing a change management process is that it allows you to manage your customer. When your customer initiates a change request, you must be able to predict immediately the impact of the change on schedule, safety, cost, and technical performance. This information must be transmitted to the customer immediately, especially if your methodology is such that no further changes are possible because of the life-cycle phase you have entered. Educating your customer as to how your methodology works is critical in getting customer buy-in for your recommendations during the scope change process.

Risk management and change management function together. Risks generate changes that, in turn, create new risks. For example, consider a company in which the project manager is given the responsibility for developing a new product. Management usually establishes a launch date even before the project is started. Management wants the income stream from the project to begin on a certain date to offset development costs. Project managers view executives as their customers during new project development, but executives view the stockholders who expect a revenue stream from the new product as their customers. When the launch date is not met, surprises result in heads rolling, usually executive heads first.

In a previous edition of this book, we stated that Asea Brown Boveri had developed excellent processes for risk management, so it is understandable that it also has structured change management processes. In companies excellent in project management, risk management and change management occur continuously throughout the project life cycle. The impact on product quality, cost, and timing is continuously updated and reported to management as quickly as possible. The goal is always to minimize the number and extent of surprises.

5.12 OTHER MANAGEMENT PROCESSES

Employee empowerment and self-directed work teams took the business world by storm during the early 1990s. With growing emphasis on customer satisfaction, it made sense to empower those closest to the customer—the order service people, nurses, clerks, and so on—to take action in resolving customers' complaints. A logical extension of employee empowerment is the self-managed work team. A self-directed work team is a group of employees with given day-to-day responsibility for managing themselves and the work they perform. This includes the responsibility for handling resources and solving problems.

Some call empowerment a basis for the next industrial revolution, and it is true that many internationally known corporations have established self-directed work teams. Such corporations include Lockheed-Martin, Honeywell, and Weyerhauser. Time will tell whether these concepts turn out to be a trend or only a fad.

Reengineering a corporation is another term for downsizing the organization with the (often unfortunate) belief that the same amount of work can be performed with fewer people, at lower cost, and in a shorter period of time. Because project management proposes getting more done in less time with fewer people, it seems only practical to implement project management as part of reengineering. It still is not certain that downsizing executed at the same time as the implementation of project management works, but project-driven organizations seem to consider it successful.

Life-cycle costing was first used in military organizations. Simply stated, life-cycle costing requires that decisions made during the research and development process be evaluated against the total life-cycle cost of the system. Life-cycle costs are the total cost of the organization for the ownership and acquisition of the product over its full life.

6 Culture

6.0 INTRODUCTION

Perhaps the most significant characteristic of companies that are excellent in project management is their culture. Successful implementation of project management creates an organization and cultures that can change rapidly because of the demands of each project and yet adapt quickly to a constantly changing dynamic environment, perhaps at the same time. Successful companies have to cope with change in real time and live with the potential disorder that comes with it. The situation can become more difficult if two companies with possibly diverse cultures must work together on a common project.

Change is inevitable in all organizations but perhaps more so in project-driven organizations. As such, excellent companies have come to the realization that competitive success can be achieved only if the organization has achieved a culture that promotes and sustains the necessary organizational behavior. Corporate cultures cannot be changed overnight. The time frame is normally years but can be reduced if executive support exists. Also, if as few as one executive refuses to support a potentially good project management culture, disaster can result.

In the early days of project management, a small aerospace company had to develop a project management culture in order to survive. The change was rapid. Unfortunately, the vice president for engineering refused to buy into the new culture. Prior to the acceptance of project management, the power base in the organization had been engineering. All decisions were either instigated or approved by engineering. How could the organization get the vice president to buy into the new culture?

The president realized the problem but was stymied for a practical solution. Getting rid of the vice president was one alternative, but not practical because of his previous successes and technical know-how. The corporation was awarded a two-year project that was strategically important to it. The vice president was then temporarily assigned as the project manager and removed from his position as vice president for engineering. At the completion of the project, the vice president was assigned to fill the newly created position of vice president of project management.

6.1 CREATION OF A CORPORATE CULTURE

Corporate cultures may take a long time to create and put into place but can be torn down overnight. Corporate cultures for project management are based on organizational behavior, not processes. Corporate cultures reflect the goals, beliefs, and aspirations of senior management. It may take years for the building blocks to be in place for a good culture to exist, but that culture can be torn down quickly through the personal whims of one executive who refuses to support project management.

Project management cultures can exist within any organizational structure. The speed at which the culture matures, however, may be based on the size of the company, the size and nature of the projects, and the type of customer, whether it is internal or external. Project management is a culture, not policies and procedures. As a result, it may not be possible to benchmark a project management culture. What works well in one company may not work equally well in another.

Good corporate cultures can also foster better relations with the customer, especially external clients. As an example, one company developed a culture of always being honest in reporting the results of testing accomplished for external customers. Customers, in turn, began treating the contractor as a partner and routinely shared proprietary information so that customers and the contractor could help each other.

Within the excellent companies, the process of project management evolves into a behavioral culture based on multiple-boss reporting. The significance of multiple-boss reporting cannot be overstated. There is a mistaken belief that project management can be benchmarked from one company to another. Benchmarking is the process of continuously comparing and measuring against an organization anywhere in the world in order to gain information that will help your organization improve its performance and competitive position. Competitive benchmarking is where organizational performance is benchmarked against the performance of competing organizations. Process benchmarking is the benchmarking of discrete processes against organizations with performance leadership in these processes.

Since a project management culture is a behavioral culture, benchmarking works best if we benchmark best practices, which are leadership, management, or operational methods that lead to superior performance. Because of the strong behavioral influence, it is almost impossible to transpose a project management culture from one company to another. As mentioned earlier, what works well in one company may not be appropriate or cost-effective in another company.

Strong cultures can form when project management is viewed as a profession and supported by senior management. A strong culture can also be viewed as a primary business differentiator. Strong cultures can focus on either a formal or an informal project management approach. However, with the formation of any culture, there are always some barriers that must be overcome.

According to a spokesperson from AT&T:

> Project management is supported from the perspective that the PM [project manager] is seen as a professional with specific job skills and responsibilities to perform as part of the project team. Does the PM get to pick and choose the team and have complete

control over budget allocation? No. This is not practical in a large company with many projects competing for funding and subject matter experts in various functional organizations.

A formal project charter naming an individual as a PM is not always done; however, being designated with the role of project manager confers the power that comes with that role. In our movement from informal to more formal, it usually started with project planning and time management, and scope management came in a little bit later.

In recent memory PM has been supported, but there were barriers. The biggest barrier has been in convincing management that they do not have to continue managing all the projects. They can manage the project managers and let the PMs manage the projects. One thing that helps this is to move the PMs so that they are in the same work group, rather than scattered throughout the teams across the company, and have them be supervised by a strong proponent of PM. Another thing that has helped has been the PMCOE's [project management center of excellence] execution of their mission to improve PM capabilities throughout the company, including impacting the corporate culture supporting PM.

Our success is attributable to a leadership view that led to creating a dedicated project management organization and culture that acknowledges the value of project management to the business. Our vision: Establish a global best-in-class project management discipline designed to maximize the customer experience and increase profitability for AT&T.

In good cultures, the role and responsibilities of the project manager are clearly identified. It is also supported by executive management and understood by everyone in the company. According to Enrique Sevilla Molina, formerly Corporate Project Management Office (PMO) director at Indra:

Based on the historical background of our company and the practices we set in place to manage our projects, we found out that the project manager role constitutes a key factor for project success. Our project management theory and practice has been built to provide full support to the project manager when making decisions and, consequently, to give him or her full responsibility for project definition and execution.

We believe that he or she is not just the one that runs the project or the one that handles the budget or the schedule but the one that "understands and looks at their projects as if they were running their own business," as our CEO used to say, with an integrated approach to his/her job.

Our culture sets the priority on supporting the project managers in their job, helping them in the decision-making processes, and providing them with the needed tools and training to do their job. This approach allows for a certain degree of a not-so-strict formal processes. This allows the project manager's responsibility and initiative to be displayed, but always under compliance with the framework and set of rules that allows for a solid accounting and results reporting.

We can say that project management has always been supported throughout the different stages of evolution of the company, and throughout the different business units, although some areas have been more reluctant in implementing changes in their established way of performing the job. One of the main barriers or drawbacks is the ability to use the same project management concepts for the different types of projects and products. It is still a major concern in our training programs to try to explain how the framework and the methodology is applied to projects with a high degree of definition in scope and to projects with a lesser degree of definition (fuzzy projects).

6.2 CORPORATE VALUES

An important part of the culture in excellent companies is an established set of values that all employees abide by. The values go beyond the normal "standard practice" manuals and morality and ethics in dealing with customers. Ensuring that company values and project management are congruent is vital to the success of any project. In order to ensure this congruence of values, it is important that company goals, objectives, and values be well understood by all members of the project team.

Many forms of value make up successful cultures. Figure 6–1 shows some of the types of values. Every company can have its own unique set of values that works well for it. Groups of values may not be interchangeable from company to company.

Figure 6–1. Types of values.

One of the more interesting characteristics of successful cultures is that productivity and cooperation tend to increase when employees socialize outside of work as well as at work.

Successful project management can flourish within any structure, no matter how terrible the structure looks on paper, but the culture within the organization must support the four basic values of project management:

1. Cooperation
2. Teamwork
3. Trust
4. Effective communication

Some companies prefer to add a fifth bullet, namely ethical conduct. This is largely due to PMI's *Code of Conduct and Professional Responsibility*.

6.3 TYPES OF CULTURES

There are different types of project management cultures, which vary according to the nature of the business, the amount of trust and cooperation, and the competitive environment. Typical types of cultures include:

- *Cooperative cultures.* These are based on trust and effective communication, not only internally but externally as well with stakeholders and clients.
- *Noncooperative cultures.* In these cultures, mistrust prevails. Employees worry more about themselves and their personal interests than what is best for the team, company, or customer.
- *Competitive cultures.* These cultures force project teams to compete with one another for valuable corporate resources. In these cultures, project managers often demand that employees demonstrate more loyalty to the project than to their line manager. This can be disastrous when employees are working on multiple projects at the same time and receive different instructions from the project and the functional manager.
- *Isolated cultures.* These occur when a large organization allows functional units to develop their own project management cultures. This could also result in a culture-within-a-culture environment within strategic business units. It can be disastrous when multiple isolated cultures must interface with one another.
- *Fragmented cultures.* Projects, where part of the team is geographically separated from the rest of the team, may lead to a fragmented culture. Virtual teams are often considered fragmented cultures. Fragmented cultures also occur on multinational projects, where the home office or corporate team may have a strong culture for project management, but the foreign team has no sustainable project management culture.

Cooperative cultures thrive on effective communications, trust, and cooperation. Decisions are made based on the best interest of all of the stakeholders. Executive sponsorship, whether individual or committee, is more passive than active, and very few problems ever go up to the executive levels for resolution. Projects are managed more informally than formally, with minimum documentation, and often meetings are held only as needed. This type of project management culture takes years to achieve and functions well during both favorable and unfavorable economic conditions.

Noncooperative cultures are reflections of senior management's inability to cooperate among themselves and possibly their inability to cooperate with the workforce. Respect is nonexistent. Noncooperative cultures can produce a good deliverable for the customer if the end justifies the means. However, this culture does not generate the number of project successes achievable with the cooperative culture.

Competitive cultures can be healthy in the short term, especially if an abundance of work exists. Long-term effects are usually not favorable. An electronics firm continuously bid on projects that required the cooperation of three departments. Management then implemented the unhealthy decision of allowing each of the three departments to bid on every job, thus creating internal competition as they bid against each other. One department would be awarded the contract, and the other two departments would be treated as subcontractors.

Management believed that this competitiveness was healthy. Unfortunately, the long-term results were disastrous. The three departments refused to talk to one another, and the sharing of information stopped. In order to get the job done for the price quoted, the departments began outsourcing small amounts of work rather than using the other departments, which were more expensive. As more and more work was being outsourced, layoffs occurred. Management then realized the disadvantages of a competitive culture.

The type of culture can be impacted by the industry and the size and nature of the business. According to Eric Alan Johnson and Jeffrey Alan Neal:[1]

> *Data-orientated culture*: The data-orientated culture (also known as the data-driven culture and knowledge-based management) is characterized by leadership and project managers basing critical business actions on the results of quantitative methods. These methods include various tools and techniques such as descriptive and inferential statistics, hypothesis testing, and modeling. This type of management culture is critically dependent on a consistent and accurate data collection system specifically designed to provide key performance measurements (metrics). A robust measurement system analysis program is needed to ensure the accuracy and ultimate usability of the data.
>
> This type of culture also employs visual management techniques to display key business and program objects to the entire work population. The intent of a visual management program is not only to display the progress and performance of the project, but to instill a sense of pride and ownership in the results with those who are ultimately responsible for project and program success . . . the employees themselves.

1. Eric Alan Johnson, Satellite Control Network Contract Deputy Program Director, AFSCN, was the winner of the 2006 Kerzner Project Manager of the Year Award; and Jeffrey Alan Neal, Blackbelt/Lean Expert and Lecturer, Quantitative Methods, University of Colorado, Colorado Springs.

Also critical to the success of this type of management culture is the training required to implement the more technical aspects of such a system. In order to accurately collect, assess, and enable accurate decision making the diverse types of data (both nominal and interval data), the organization needs specialists skilled in various data analysis and interpretation techniques.

6.4 CORPORATE CULTURES AT WORK

Cooperative cultures are based on trust, communication, cooperation, and teamwork. As a result, the structure of the organization may become unimportant. Restructuring a company simply to bring in project management may lead to disaster. Companies should be restructured for other reasons, such as getting closer to the customer.

Successful project management can occur within any structure, no matter how bad the structure appears on paper, if the culture within the organization promotes teamwork, cooperation, trust, and effective communication.

Boeing

In the early years of project management, aerospace and defense contractors set up customer-focused project offices for specific customers, such as the Air Force, Army, and Navy. One of the benefits of these project offices was the ability to create a specific working relationship and culture for that customer.

Developing a specific relationship or culture was justified because the projects often lasted for decades. It was like having a culture within a culture. When the projects disappeared and the project office was no longer needed, the culture within that project office might very well disappear as well.

Sometimes one large project can require a permanent culture change within a company. Such was the case at Boeing with the decision to design and build the Boeing 777 airplane.[2] The Boeing 777 project would require new technology and a radical change in the way people would be required to work together. The culture change would permeate all levels of management, from the highest levels down to the workers on the shop floor. Table 6–1 shows some of the changes that took place.[2] The intent of the table is to show that on large, long-term projects, cultural change may be necessary.

As project management matures and the project manager is given more and more responsibility, those managers may be given the responsibility for wage and salary administration. However, even excellent companies are still struggling with this new approach. The first problem is that project managers may not be on the management pay scale in the company but are being given the right to sign performance evaluations.

---

2. The Boeing 777 case study, "Phil Condit and the Boeing 777: From Design and Development to Production and Sales," appears in H. Kerzner, *Project Management Case Studies*, 5th ed. (Hoboken, NJ: Wiley, 2017), pp. 711–734.

TABLE 6–1. CHANGES DUE TO BOEING 777 NEW AIRPLANE PROJECT

Situation	Previous New Airplane Projects	Boeing 777
Executive communications	Secretive	Open
Communication flow	Vertical	Horizontal
Thinking process	Two dimensional	Three dimensional
Decision-making	Centralized	Decentralized
Empowerment	Managers	Down to factory workers
Project managers	Managers	Down to nonmanagers
Problem-solving	Individual	Team
Performance reviews (of managers)	One way	Three ways
Human resources problem focus	Weak	Strong
Meetings style	Secretive	Open
Customer involvement	Very low	Very high
Core values	End result/quality	Leadership/participation/customer satisfaction
Speed of decisions	Slow	Fast
Life-cycle costing	Minimal	Extensive
Design flexibility	Minimal	Extensive

Note: The information presented in this table is the author's interpretation of some of the changes that occurred, not necessarily Boeing's official opinion.

The second problem is determining what method of evaluation should be used for union employees. This is probably the most serious problem, and the jury is not yet in on what will and will not work. One reason why executives are a little reluctant to implement wage and salary administration that affects project management is because of union involvement, which dramatically changes the picture, especially if a person on a project team decides that a union worker is considered promotable when in fact his or her line manager says that promotion must be based on a union criterion." There is no black-and-white answer for the issue, and most companies have not even addressed the problem yet.

Midwest Corporation (Disguised Company)

The larger the company, the more difficult it is to establish a uniform project management culture across the entire company. Large companies have pockets of project management, each of which can mature at a different rate. A large Midwest corporation had one division that was outstanding in project management. The culture was strong, and everyone supported project management. This division won awards and recognition for its ability to manage projects successfully. Yet at the same time, a sister division was approximately five years behind the excellent division in project management maturity. During an audit of the sister division, the following problem areas were identified:

- Continuous process changes due to new technology.
- Not enough time allocated for effort.
- Too much outside interference (meetings, delays, etc.).

- Schedules laid out based on assumptions that eventually change during execution of the project.
- Imbalance of workforce.
- Differing objectives among groups.
- Use of a process that allows for no flexibility to "freelance."
- Inability to openly discuss issues without some people taking technical criticism as personal criticism.
- Lack of quality planning, scheduling, and progress tracking.
- No resource tracking.
- Inheriting someone else's project and finding little or no supporting documentation.
- Dealing with contract or agency management.
- Changing or expanding project expectations.
- Constantly changing deadlines.
- Last-minute requirements changes.
- People on projects having hidden agendas.
- Project scope unclear right from the beginning.
- Dependence on resources without having control over them.
- *Finger pointing.* "It is not my problem."
- No formal cost-estimating process.
- Lack of understanding of a work breakdown structure.
- Little or no customer focus.
- Duplication of efforts.
- Poor or lack of "voice of the customer" input on needs/wants.
- Limited abilities to support people.
- Lack of management direction.
- No product/project champion.
- Poorly run meetings.
- People not cooperating easily.
- People taking offense at being asked to do the job they are expected to do, while their managers seek only to develop a high-quality product.
- Some tasks lacking a known duration.
- People who want to be involved but do not have the skills needed to solve the problem.
- *Dependencies.* Making sure that when specifications change, other things that depend on them also change.
- Dealing with daily fires without jeopardizing the scheduled work.
- Overlapping assignments (three releases at once).
- Not having the right personnel assigned to the teams.
- Disappearance of management support.
- Work being started in days-from-due-date mode rather than in as-soon-as-possible mode.
- Turf protection by nonmanagement employees.
- Nonexistent risk management.

- Project scope creep (incremental changes that are viewed as small at the time but that add up to large increments).
- Ineffective communications with overseas activities.
- Vague/changing responsibilities (who is driving the bus?).

Large companies tend to favor pockets of project management rather than a company-wide culture. However, there are situations in which a company must develop a company-wide culture to remain competitive. Sometimes it is simply to remain a major competitor; other times it is to become a global company.

6.5 GEA AND HEINEKEN COLLABORATION: A LEARNING EXPERIENCE[3]

One of the most important aspects of the project management discipline is to be adapted to the specific characteristics of the project, to the work culture of the country where the project is developed as well as to the customer that owns the project.

GEA, as a world world-class technology supplier of turnkey plants in a wide range of process industries, and particularly one of the largest suppliers for the food and beverage industry, and Heineken, the world-leading brewery company, have been working together worldwide in close partnership for executing different types of projects at the Heineken plants according to Heineken needs. The last projects executed for the Heineken plants in Spain, and especially, the close collaboration between the Heineken and GEA local teams in Spain is a great example to identify from the cultural perspective the best practices learned and applied in PM to meet the strategical objectives of both companies.

Cultural aspects have really been key in the projects developed by GEA for Heineken in several plants of Spain. To be able to do it successfully, both teams had to work on an open-mind approach to combine the different PM Methodologies between both companies.

Project Management is a core competency of GEA. To enable all Project Managers to deliver projects consistently on time, within budget, and according to customers' expectations, GEA has developed Project Management methods, tools, and training. This is all covered on the GEA Project Process House (see Figure 6–2), which, for project execution, starting on the project handover, cover all aspects of projects development, linking the execution of the projects with the related processes of the company.

Along with these operative tools, GEA, conscious about the importance of the discipline, has also a supportive intranet portal for the guidance of the Project Management community (see Figure 6–3).

---

3. Material in this section has been provided by Martin Garcia Gil and Miguel Antonio Martinez Carrizo, Project Management Office, Liquid & Powder Technologies, GEA Process Engineering, Spain. ©2022 by GEA and Heineken. All rights reserved. Reproduced with permission.

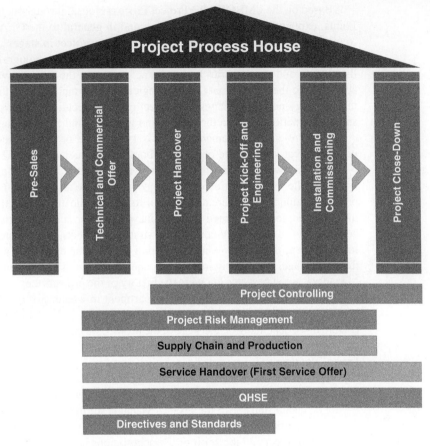

Figure 6–2. Project Process House.

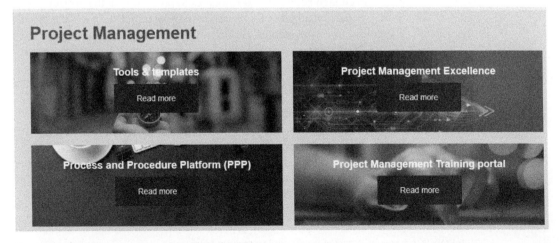

Figure 6–3. Project Management Portal (entry page).

Based on the PMBOK,[4] the Project Process House, throughout the Critical Process Points, explains to the Project Managers and in general to the GEA organization how to manage and execute the projects. It divides the project in stages with phases, which can be considered small projects on their own. It also has enough flexibility to adapt the process and its deliveries to the characteristics of the projects, which are classified according to their complexity and strategically importance during the Sales stage. The methodology not only secures the correct project execution and governance but also a smooth transition between the project and the Sales and Service stages.

This Project Management Model is key on the projects executed by GEA and provides great value added on the collaboration between GEA and Heineken in Spain.

It is also worth adding that the involvement of the Project Managers is not only limited to the pure execution of the projects. They are also involved in the Sales stage, providing support on different aspects of it, such as the creation of the contractual timing schedule, lessons learned from similar projects or customers, risk assessment, and contract revisions. In very complex or strategic projects, such some of the ones done for Heineken, the Project Manager is part of the four eye reviews performed during the Sales stage, particularly in the costs and timing fields. In the same way, the Project Managers, as the best role knowing the history of the projects they executed, collaborate by providing feedback to the Service department in whatever service issue.

Following the sequence of steps shown in the Critical Process Points diagram (Figure 6–4), depending on the stage of the project, we could make the following remarks:

Project Handover
- Project Handover:
  - To be sure that all the findings of the sales stage are captured, GEA Sales team transfer to the Project Manager all the relevant information gathered. Particularly important is to be sure that customer expectations about the project and the spirit of the negotiations is kept. The intention is to provide to the Project Manager enough information for understanding the targets of the project, not only for GEA but also for Heineken, empowering him/her also with enough authority and resources to proceed with the project.
  - Some tools are put in place for the Project Managers at the time of the Project Handover (for very complex projects even before). The most relevant ones are the Project Home Site (PHS), a SharePoint that is used by the project team as the repository not only the documents shared with Heineken but many others, technical and not technical, for internal use. Along with it, the project team has a tool for sharing large and very large files with Heineken in a secured way, called GEA Assist.
  - As part of the quality control of the process, the Phase Gate Review 3, PGR3, (Phase Gate Revies 1 and 2 pertain to the GEA Sales team) allows the Project Manager to check with the main stakeholders of the project whether the team is ready to face the next phase of the project, the Project Kick-Off as part of the Project Kick-Off and Engineering stage of the project.

---

4. PMBOK is a registered mark of the Project Management Institute, Inc.

PROJECT PROCESS HOUSE: Critical Process Points (mandatory) – Project Management, Version 3.0 (Sept 2021)

Depending on project category standardized/customized/tailor made/special and contract value

GE/\ Engineering for a better world.

Phase bars: **Project Sales** | **Project Handover** | **Project Kick-Off & Engineering** | **Project Management** — **Supply Chain** | **Installation & Commissioning** | **Service** | **Service Handover**

	Project Handover	Project Kick-Off³	Baseline Planning³	Engineering	Procurement	Manufacturing and Logistics	Installation	Commissioning	Project Close-Down
PGR	PGR3	PGR3	PGR4	(Interface Project manager with)			PGR5	PGR6 / PGR7	PGR7
All categories and contract value	• Project classification from Sales • Handover meeting • First Service Offer • Project governance	• Kick-off meetings¹	• Project Home Site • Allocated budgets for each project activities • Project schedule • Project Supply Chain plan • HSE activities • Mobility & Security • Project controlling	• Engineering detailed schedule • Design/sizing approval • Initiate Service contract development • Project controlling (monthly review) • Installation planning Step • FAT automation	• Use of preferred suppliers • Competitive tendering/RFQ • Lead time management • Manage warranty events with suppliers • POC recognition • Project controlling (monthly review) • Spare parts acquisition where applicable	• Schedule control • Quality control • Expediting • Shipping and customs	• HSE management • Installation detailed schedule • Quality control • Installation qualification • Punch-list • Project controlling (monthly review)	• HSE management • Commissioning detailed schedule • SAT • Punch-list • Project controlling (monthly review)	• Remaining activities, detailed budget and schedule • As-built documentation • Service handover (First Service Offer) • Handover to customer and customer feedback meeting • Manage warranty events • Final project review • Lessons learned
Standardized (>7Me)/Customized (>5Me)/Tailor Made/Special *(additional activities)*	• **PGR3:** Authorization to proceed	• Kick-off meetings (internal, customer, supply chain)	• Detailed WBS and activity list / Project deliverables • Communication plan • Risk analysis and response plan • **PGR4:** Project plans completed	• Process design review and approval • Acceptance procedure/technical KPIs • Change management⁴ • **DQ:** Design Qualification	• Project Procurement Manager • Procurement plan review meettings • Savings reporting	• ED Control • Change Management⁴ • FAT manufacturing	• **PGR5:** Ready for installation • ED Control • EVM² • Change Management⁴	• **PGR6:** Ready for commissioning • ED Control • EVM² • Change Management⁴ • Operational qualification • Performance qualification	• **PGR7:** Ready for close down • Close-out meeting • Close-out with suppliers

PGR Phase Gate Review (template on PPP)

Note: This document together with the project Scorecard is used to ensure compliance with our Processes. It shows minimum doesn't mean that e.g. PGRs or risk analysis should not be considered to some extent for the first group of projects. For this group of projects less formalism is acceptable.

1 Kick-off with project team can be combined with handover (only for category "standardized")
2 Earned Value Management or similar to monitor physical progress of site operations
3 Baseline planning can either happen before, during or after internal kick-off depending on project context
4 Full adherence to the new PPP "change management" process is expected

Figure 6–4. Critical Process Points.

Project Kick-Off and Engineering
- Project Kick-Off:
  - The interaction between Heineken project teams and project managers is intense during the sales phase to ensure the needs of both parties are included in the submitted quotation.
  - The aim to clarify the scope of the project as well as all the terms and conditions is a must for creating solid bases for the development of the project. Even issues that were relevant for the project future stages, such as OH&S, plant infrastructure, were discussed with Heineken at this point.
  - Frequent alignment meetings were held and they created a kind of excellent relationship between project managers and project team members of both companies. Of course, whenever possible, with a predetermined agenda and periodicity. From GEA perspective the exercise allowed a better understanding not only about the requirements of the project but also a good acquaintance of the most important stakeholders of the customer.
- Baseline Planning:
  - The contact between both project managers was also very deep. From the contractual dates all the process for creating the different management plans were done jointly, with a deep focus on the communication, risk assessment, resources allocation, schedule (of activities, procurement, logistics. . .), supply chain, quality expected, etc.
  - Passionate discussions were held until reaching a common ground that was, at the end, very beneficial for the project execution and for the relationship of the project team members. Definitely for both companies and their top management, from that point on, only one joint project team exists.
  - Although belonging to the other stages of the project we have to remark that the updates of the timing schedule based on the evolution of the projects in their different aspects were very useful for both parties and a way of facing the issues and risks of the project successfully.
  - It is important to mention that GEA system creates the PHS, a SharePoint that serves as the repository of the relevant documentation of the project that will be used even after the project is closed. Along with it, the GEA Assist, a tool for sharing project documentation, in this case to Heineken.
  - We cannot forget to mention that this phase starts the Project Controlling process that will be extended until the completion of the project. On it, the Project Manager not only takes control of the economics of the project (which is reviewed monthly with Finance) but also accountability on the management of the factors that could impact the economic result of the project.
  - Once all those activities are completed the GEA team performs with the main stakeholders the Phase Gate Review 4 (PGR4), where the status of the project is analyzed with the goal to see its preparation for, within the Project Kick-Off and Engineering stage, the next project phase, Engineering and Design.

- Engineering and Design:
  - It was also done jointly. GEA, working with the set of requirements of the project established by Heineken, developed the detailed engineering in all its aspects (P&IDs, 3D solids, layout, skids, hydraulic design, equipment, and component selection. . .).
  - Rounds of preliminary consultations, work development, and further validation were carried on one after another and in all disciplines (mechanical and process, electrical).
  - As consequence, the relatively large time consumed was later on recovered by reducing the scope variations to minor changes in the manufacturing of equipment, erection of the installation and its commissioning. It is remarkable that the open mind from both sides were really a factor that allowed a good engineering development.
  - Heineken showed its commitment to GEA by being open to share the key factors of the process for GEA to design and engineer. Transparency, mutual trust, and clear set of goals are very important to secure the success of this part of the project.
  - Contrary to the previous stages, where the number of deliverables were not so high, the design and engineering stage have a significant number of them, being then made storage in the PHS and shared with Heineken using the GEA Assist tool. Along with the update of the already existing documents (mainly timing schedule and risk assessment) we can mention the following ones:
    - P&IDs, electrical and entities drawings
    - Hydraulic, electrical, and structural calculations
    - 3D designs
    - Layout
  - Also, it is important to remark that this stage was also relevant regarding the procurement of components and manufacturing of the equipment. Heineken was consulted by GEA about the technical specifications for the equipment, and about the use of certified suppliers and contractors, respectively. Procurement dates were also agreed in order to secure the readiness of Heineken to store safely and securely all the material to be needed for the further erection of the installation in its plant.
  - As important part of the procurement, GEA provided to Heineken the recommended spare parts list that will be used later on to support the commissioning of the project. This practice ensures that the commissioning of the project is not affected by missing replacement parts.
  - At the end of this phase, design qualification should be achieved.

Supply Chain
- Procurement, Manufacturing, and Logistics:
  - With the information gathered during the design and detailed engineering the Supply Chain department, following the basis and strategy determined in the Procurement Management Plan created on the Baseline Planning phase of the project, start the procurement of material and components needed for the project.

- For the special equipment required usually provided by external or internal workshops, Supply Chain count with the support of technical specifications provided by the Engineering departments (Mechanical, Electrical, and Automation). This equipment is also part of the Procurement Management Plan and it is covered by the Quality Management Plan, not only in terms of Quality Planning and Assurance but specifically in the Quality Control, which is performed by GEA engineers. In many cases, Heineken pays a visit to workshops for doing a mechanical and electrical assessment and whenever is possible a FAT of the equipment.
- Once the components, material, and/or special equipment are ready, GEA's Logistic area of Supply Chain takes the lead for sending all of it to Heineken. The interaction here with its Logistic department is also key, being aligned both departments about the time and conditions on which the shipments have to be done. On international projects, the coordination has to be even more intense, not only for following the incoterms conditions agreed contractually and export control limitations, but also for facilitating the customs clearance and further transportation and storage in the plant.
- At the end of the Supply Chain stage the Project Manager, in the same way than previous phases, performs the PGR5, where the project team and project stakeholders check the readiness of the project for moving forward to the next stage of the project.

Installation and Commissioning
- Installation:
  - In this phase, the design done previously becomes something real.
  - The number of stakeholders on the customer side usually expands significantly. From having contact just with the Project Manager or the key engineers during the previous stages, now many other roles suddenly appear. From the relevant people from other contractors (civil works, utilities, OH&S, etc.) passing through other areas of the customer (plant management, maintenance, warehousing, etc.) to even government officials (for permits, regulations, formal authorizations, etc.) there is a long list of different people to deal with.
  - The projects developed with Heineken have been a good example of what has been described. The solid relationship developed during previous stages is now very important for smooth completion of the installation where clashes between the integrator and the customer come more frequently. In that regard, Heineken and GEA have been working without issues. The equipment and components were according to the approved P&ID. Their quality and standard were the correct ones. Infrastructures for mobilization, warehouse space, services, etc. were already available for the start of the operations in the plant.
  - Also, in this stage, the interaction with other contractors is very intense. Daily joint meetings for aligning the work to be done are key not only for better coordination but also for avoiding incidents and accidents. In that regard, the work done by Heineken has to be praised.

- Another relevant subject for the development of the project is the control of the subcontractors involved in the mechanical and electrical installation. The interaction between GEA, its subcontractors, and Heineken has been very intense and always managed from a proactive perspective, obtaining good results for the project development.
- In larger projects, subcontractors' advance is measured using the EVM techniques and registered using ED Controls tool, making it easy for project engineers to be in control of all the facts and events related to the erection of the installation.
- Since the installation is from the hazards' perspective one of the riskier phases of the project, OH&S matters became first priority to Heineken and GEA both. Representatives of both companies have to be absolutely aligned and in permanent communication for coordinating the works to be done in the plant.
- Once the installation is close to its completion, a common verification round should be done for assuring that all the issues are properly registered and with actions to be done for their resolution. Once finalized, installation qualification is achieved.
- At the end of the installation another Phase Gate Review, this time the number 6, is carried on for checking that the project is ready to move forward to the commissioning phase.
- Commissioning:
  - In this phase, it is important to start with the installation completely finished and verified. If that is not possible, a good register of pending points, including owners and dates of resolution, is compulsory.
  - More than ever during the project, in the commissioning the interaction between the different integrators and the customer is frequent and sometimes conflictive. GEA approach is to have the customer, not only its management but also at all the operational levels, involved as much as possible. In that regard, Heineken has been very receptive and cooperative in all the projects developed in its Spanish plants. The usual conflicts due to the interference of the project with the regular production of the plant were minimized thanks to the previous alignment done with the management of the production departments. Also, services and utilities for the areas subjected to the project works were provided in due time, which ease the start of the commissioning.
  - In this phase, the plant becomes alive creating a new set of hazards. The use of energies (mainly steam and hot water) as well as dangerous chemicals (caustic and acid) force Heineken and GEA personnel to be more aware of the risks present when running an installation very recently built.
  - System Acceptance Tests (SAT) were performed successfully by Heineken and assure that the consequent production was reliable, robust, and repeatable. Also, as happened during the installation phase, a list of pending issues was created for recording, analyzing, and fix all the issues that, although not impeding the regular production, were needed to leave in good condition.

- The documentation of the project is provided at the end of the commissioning. It would cover all maintenance requirements as well as the material list to be used for the engineers and technician of the plant. The training of the operators of the plant is done in this phase. Here the intervention of Heineken for providing suitable people as well as the proper facilities is key for securing the proper environment for doing a good training. The main goal is to have the people ready for maintaining and operating the plant once the project would be handover to the customer.
- The start of the product trials and particularly the commercial production release the provisional acceptance of the installation, allowing, among the achievement of the operational qualification, the handover of the asset to Heineken, and the start of the warranty period. As mentioned in the design and engineering stage, the existence of the spare lists would mitigate the impact of breakdowns of components and equipment.
- From that point on, all the warranty issues should be managed by GEA Service and Heineken Production and Maintenance departments. In that regard, it is very important to secure a smooth transition from the project execution stage to the service stage, and the PM of both teams work together to support any issue that will need to be solved.
- The final Phase Gate Review of the project, the number 7, is performed, leaving the assurance to Heineken and GEA that the project is ready for the last phase.
- Project Close Down:
  - It serves the project to provide a fine-tuning of the installation and secure a good ramp-up until reaching the full capacity of the plant (performance qualification). If there are issues pending from the commissioning, they are also fixed until leaving the punch list of the project is absolutely cleared. The final accomplishment of the project KPIs is a must and it is verified once again, but now at full production. The transition of the operations, in GEA from Project Execution to Service and in Heineken from Engineering to Production, is complete and final, opening the Service stage that, although out of the project life cycle, should be transitioned carefully too.
  - This phase is also the one for closing the project administratively. From GEA side to update the project repository with all the data about the plant and for Heineken for providing the final acceptance of the project and release the final payments according to the financial milestones.
  - It is also the time for preparing the internal session of lessons learned and the project evaluation, internally and by the customer. Both exercises (lessons learned and project evaluation) allow to list the good practices that should be acclaimed and the bad ones that should be rectified in future projects are an important part of the continuous improvement philosophy praised by GEA and the basis of the first and most important of its value, the excellence.

As a summary, we would remark that the *framework for the collaboration* between Heineken and GEA for the projects developed jointly according to this methodology and the *support of the management*, not only during the project execution but also on the Sales and Service stages, were crucial for their success and for the creation of a strong and long last partnership between GEA and Heineken.

Along with that, the teams from both companies learned project after project how to improve their collaboration and how to work in future projects. The success of this partnership was not only the good approach and basis settled from the beginning but also the aim for improving and learning from the obstacles found on the way we covered together.

Finally, we would add that, along with following a project management methodology and to be close to the customer, the other most important best practices shared between Heineken and GEA would be:

- *Stakeholders management*: Heineken Project Manager involved all departments that could be impacted on every meeting. That was key for the stakeholder's engagement. Even if their needs were considered on every meeting, when they were participating, their involvement and buy-in of them was much higher.
- *Scope management*:
  - *Collect requirements*. The importance of managing assumptions and validate the acceptance criteria is very relevant. Technical meetings to review the project scope provided open discussions and valuable insights that allowed the project teams to find significant improvements on the engineering. On the initial projects, some items were assumed, which created some discrepancies in the next project phase. Those meetings became more important project after project and reduced the risks of issues during assembly and commissioning.
  - *Scope validation*. During the projects, GEA understood which engineering tasks were really important for Heineken and provided an added value for them. The Engineering reviews and especially the 3D reviews were something critical for Heineken and those were settled as an important milestone on the project schedule.
  - *Scope lessons learned*. Different engineering design criteria were used on the initial projects. The result was that during the commissioning, those were controlled in a different way than expected by Heineken. For the next projects, before to define the design criteria, it was verified by all parties that it met Heineken expectations, even if both design criteria were initially valid.
- *Resource management*: The expectations of resource management on-site during the execution phases were different between Heineken and GEA. In terms of number of resources, accountability. . . . Some phases were under the initial expectation and others were over them. The reason was the lack of task management between the teams. This was aligned and improved in the next projects creating an open communication to review this matter.
- *Communication management*: Even if regular and frequent communications were defined between the teams to cover the needs, it was learned that the success factor for managing the project priorities was the close communication between the GEA and Heineken Project Managers.

Taking care of all the abovementioned best practices is, without any doubt, the best way to assure a good project execution as well as to secure the sustainability of the relationship between GEA and its customers.

6.6 INDRA: BUILDING A COHESIVE CULTURE[5]

At Indra, the project manager role constitutes a key factor for project success. This is because running projects is a core part of our business. As such, company policies and practices are oriented to provide full support to project managers and to give them full responsibility for the project definition and execution. In the words of our former chief executive (CEO): "Project managers must look at their projects as if they were running their own business."

This sentence distills the basis of the project management culture at Indra. It implies that a project manager must have an integrated approach to their job, not only focusing on main objectives tied to the triple constraints, taking care of schedule and cost baselines, but also having a business perspective and pushing to deliver results that will fulfill their business unit objectives (profitability, cost efficiency, development of resources, productivity, etc.) The project management foundations are shown in Figure 6–5.

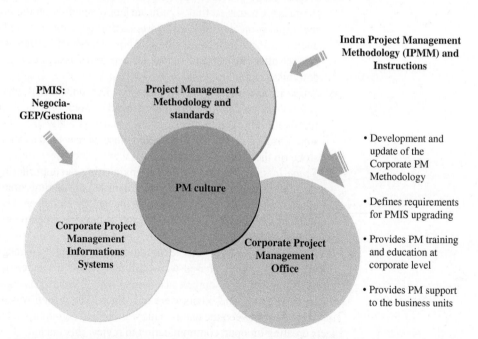

Figure 6–5. Project management foundations.

As of 2013, the corporate PMO provided support to around 3,300 project managers with clear directions, missions, strategies, methodologies, and a set of common tools and procedures to develop their jobs. We were responsible for developing and updating the Indra Project Management Methodology (IPMM), the Corporate Project Management Methodology. Based on that development, requirements for upgrading the company PMIS are defined and deployed. Ongoing support is provided both to business units and PM individuals, in terms of training and education, informal networking, and participation in different initiatives related to project management that are required by the business units. Our final objective is building and consolidating a strong and recognizable project management culture within Indra, whatever the performing unit, geography, or business sector. In 2005 we started an internal certification program as PMP® credential holders[6] for a small group of senior program and project managers and business unit managers. This certification program has been carried out yearly since then and has become one of the most sought-after training initiatives by project managers. Business managers carefully select the candidates that participate in the program.

In total, more than 950 professionals have been through the training process to become PMP® credential holders. We achieved the objective of counting on 500 certified PMP® credential holders by the end of 2012. As of May 2013, we had over 500 PMP® credential holders.

These figures would not mean nothing without a context. For us achieving these figures mean that an important proportion of the most experienced and talented professionals at Indra are well trained in project management best practices. Taking into account that our project management methodology, IPMM, is aligned with the *PMBOK® Guide*, then we could intuit that a certified PMP® credential holder could easily spread out the knowledge and experience in project management best practices in her area of influence, be this her program, project or business unit. This is a way that works when it comes to settle a strong project management culture in all branches within the company (see Figure 6–6).

We started in 2008 having PMP® credential holders collaborating as internal trainers by delivering content on the course "Project Management at Indra," created by the Corporate PMO. This course explains IPMM and project management information systems. Thanks to this initiative, we are training our people in the *PMBOK®* standard. At the same time, the experience of the trainer is used to provide a fitted project management context, using projects and services that Indra provides to its customers as training examples. In fact, this collaboration has been a success, having win-win result for all participants:

- PMP® credential holders contribute to create a better project management culture, spreading best practices within the company, and also getting professional development units to maintain their certification.
- Trainees connect directly with the content, without any interpretation that an external trainer could provide, as the teacher is a PMP® credential holder who

---

6. PMP is a registered mark of the Project Management Institute, Inc.

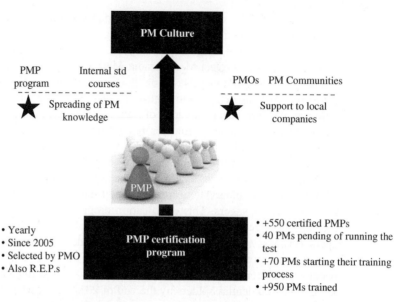

Figure 6–6. People: internal trainers.

knows well which issues must be handled when it comes to managing a project in our company.

- Human resources training departments also win, because they can invest money in other areas that could need external trainers.
- Corporate PMOs must supervise and support the consistency of the message being delivered in the training process.

In the end, it is Indra as whole that benefits, because this project management course content has been put into e-learning format, has been translated into English and Portuguese, and has been included as a mandatory content in the project management training paths of every Indra company, wherever in the world this might be (see Figure 6–7).

In addition to this, in 2010 the human resources department made available to all employees one platform accessed from the intranet aimed to let people connect, share, and learn from each other. This platform (named "Sharing Knowledge") has the look and feel of a social network and aims to support the informal exchange of knowledge and experiences between professionals. Its scope is corporate and local, and it helps to quickly and easily deliver content on best practices and methodologies, management, and technical issues and business information. It also has the possibility of creating groups and communities and even broadcasting digital content and courses.

For us, sharing knowledge has been a powerful tool to get our project managers into the loop and in touch with the Corporate PMO and also to keep building project management culture. We created PMPnet (see Figure 6–8) for certified professionals at Indra who want to be in touch, be updated with any interesting initiative or activity, or simply contribute with experiences and ideas.

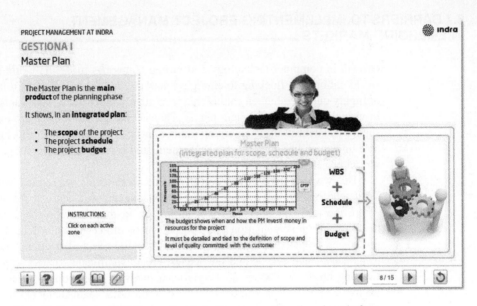

Figure 6–7. "Project Management at Indra" course on the e-learning platform.

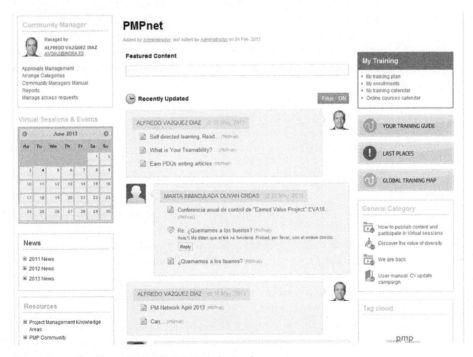

Figure 6–8. "PMPnet" in Sharing Knowledge tool.

6.7 BARRIERS TO IMPLEMENTING PROJECT MANAGEMENT IN EMERGING MARKETS

Growth in computer technology and virtual teams has made the world smaller. First-world nations are flocking to emerging market nations to get access to the abundance of highly qualified human capital that is relatively inexpensive and wants to participate in virtual project management teams. There is no question that there exists an ample supply of talent in these emerging market nations. These talented folks have a reasonable understanding of project management, and some consider it an honor to work on virtual project teams.

But working on virtual project management teams may come with headaches. While the relative acceptance of project management appears at the working levels where the team members operate, farther up in the hierarchy there might be resistance to the implementation and acceptance of project management. Because of the growth of project management worldwide, many executives openly pay lip service to its acceptance yet, behind the scenes, create significant barriers to prevent it from working properly. This creates significant hardships for those portions of the virtual teams in first-world nations that must rely on other team members for support. The ultimate result might be frustrations stemming from poor information flow, extremely long decision-making processes, poor cost control, and an abundance of external dependencies that elongate schedules beyond the buyer's contractual dates. Simply stated, there are strong cultural issues that need to be considered. In this section, we typically use the United States as an example of the first-world nations.

Barriers to effective project management implementation exist worldwide, not merely in emerging market nations. But in those nations, the barriers are more apparent. For simplicity's sake, the barriers can be classified into four categories:

1. Cultural barriers
2. Status and political barriers
3. Project management barriers
4. Other barriers

Culture

A culture is a set of beliefs that people follow. Every company could have its own culture. Some companies may even have multiple cultures. Some cultures are strong while others are weak. In some emerging market nations, there exist national cultures that can be so strong that they dictate corporate cultures. Numerous factors can influence the culture of an organization. Only those factors that can have an impact on the implementation and acceptance of project management are discussed here. They include:

- Bureaucratic centralization of authority in the hands of a few
- Lack of meaningful or real executive sponsorship
- Importance of the organizational hierarchy
- Improper legal laws
- The potential for corruption

Centralization of Authority

Many countries maintain a culture in which very few people have the authority to make decisions. Decision-making rests in the hands of a few, and it serves as a source of vast power. This factor exists in both privately held companies and governmental organizations. Project management advocates decentralization of authority and decision-making. In many countries, the seniormost level of management will never surrender authority, power, or right to make decisions to project managers. In these countries, an appointment to the senior levels of management is not necessarily based on performance. Instead, it is based on age, belonging to the right political party, and personal contacts within the government. The result can be executives who possess little knowledge of their own business or who lack leadership capacity.

Executive Sponsorship

Project sponsorship might exist somewhere in the company but most certainly not at the executive level. There are two reasons for this. First, senior managers know their limitations and may have absolutely no knowledge about the project. Therefore, they could be prone to making serious blunders that could become visible to the people who put them into these power positions. Second, and possibly most important, acting as an executive sponsor on a project that could fail could signal the end of the executive's political career. Therefore, sponsorship, if it exists at all, maybe at a low level in the organizational hierarchy, and at a level where people are expendable if the project fails. The result is that project managers end up with sponsors who either cannot or will not help them in times of trouble.

Organizational Hierarchy

In the United States, project managers generally have the right to talk to anyone in the company to get information relative to the project. The intent is to get work to flow horizontally as well as vertically. In some emerging market nations, project managers must follow the chain of command. The organizational hierarchy is sacred. Following the chain of command certainly elongates the decision-making process to the point where the project manager has no idea how long it will take to get access to needed information or for a decision to be made, even though a sponsor exists. No mature infrastructure is in place to support project management. The infrastructure exists to filter bad news from the executive level and to justify the existence of each functional manager.

In the United States, the "buck" stops at the sponsor. Sponsors have ultimate decision-making authority and are expected to assist project managers during a crisis. The role of the sponsor is clearly defined and may be described in detail in the enterprise project management methodology. But in some emerging market countries, even the sponsor might not be authorized to make a decision. Some decisions may need to go as high as a government minister. Simply stated, it is not always clear where and when the decision needs to be made and by whom it will be made. Also, in the United States, the project sponsor is responsible for reporting bad news. In some nations, the news may go as high as government ministers. Simply stated, you cannot be sure where project information will end up.

Improper But "Legal" Laws

Some laws in emerging market nations may be viewed by other nations as explicitly or implicitly condoning acts that would be illegal or elsewhere. Yet American project managers who partner with these nations must abide by those laws like any other. As an example, procurement contracts may be awarded not to the most qualified supplier or to the lowest bidder but to any bidder that resides in a city with a high unemployment level. As another example, some nations have laws that imply that bribes are an acceptable practice when awarding contracts. Some contracts might also be awarded to relatives and friends rather than the best-qualified supplier.

Potential for Corruption

Corruption can and does exist in some countries and plays havoc with project managers who focus on the triple constraints. Project managers traditionally lay out a plan to meet the objectives and the triple constraints. They also assume that everything will be done systematically and in an orderly manner, which assumes no corruption. But in some nations, potentially corrupt individuals or organizations will do everything possible to stop or slow down the project, until they can benefit personally.

Status and Politics

Status and politics are prevalent everywhere and can have a negative impact on project management. In some emerging market nations, status and politics actually sabotage project management and prevent it from working correctly. Factors that can affect project management include:

- Legal formalities and government constraints
- Insecurity at the executive levels
- Status consciousness
- Social obligations
- Internal politics
- Unemployment and poverty
- Attitude toward employees
- Inefficiencies
- Lack of dedication at all levels
- Misinformation or lack of information

Legal Formalities and Government Constraints

Here in the United States, we believe that employees who perform poorly can be removed from the project or even fired. But in some emerging market nations, employees have the legal right to hold a job even if their performance is substandard. There are laws that clearly state under what conditions a worker can be fired, if at all.

There are also laws on the use of overtime. Overtime may not be allowed because paying someone to work overtime could eventually end up creating a new social class. Therefore, overtime may not be used to maintain or accelerate a schedule that is in trouble.

Insecurity

Executives often feel more insecure than the managers beneath them because their positions may be the result of political appointments. As such, project managers may be seen by executives as the stars of the future and may be viewed as threats. Allowing project managers who are working on highly successful projects to be the ones to make presentations to the seniormost levels of management in the government could be seen as a threat. Conversely, if the project is in trouble, then the project manager may be forced to make the presentation.

Status Consciousness

Corporate officers in some cultures, particularly in some emerging market nations, are highly status conscious. They have a very real fear that the implementation of project management may cause them to lose status, yet they refuse to function as active project sponsors since this, too, may result in a loss of status if the project fails. Status often is accompanied by fringe benefits, such as a company cars and other special privileges.

Social Obligations

In emerging market nations, social obligations dictated by religious customs and beliefs or by politics may be more important than they would be in first-world nations. Social obligations are ways of maintaining alliances with those who have put an executive or a project manager in power. As such, project managers may be allowed to interface socially with certain groups but not others. This could also be an obstruction to project management implementation.

Internal Politics

Internal politics exist in every company in the world. Before executives consider throwing their support behind a new approach such as project management, they worry about whether they will become stronger or weaker, have more or less authority, and have a greater or lesser chance for advancement. This is one of the reasons why only a small percentage of emerging market companies have PMOs. Whichever executive gets control of the PMO could become more powerful than other executives. In the United States, we have solved this problem by allowing several executives to have their own PMO. But in emerging markets, this method is viewed as excessive headcount.

Unemployment and Government Constraints

Virtually all executives understand project management and the accompanying benefits, yet they remain silent rather than visibly showing their support. One of the benefits of project management implementation is that it can make organizations more efficient to the point where fewer resources are needed to perform the required work. This can be a threat to an executive because, unless additional business can be found, efficiency can result in downsizing the company, reducing the executive's power and authority, increasing the unemployment level, and possibly increasing poverty in the community. Therefore, the increased efficiencies of project management could be looked at unfavorably.

Attitude Toward Employees

In some nations, employees might be viewed as stepping stones to building an empire. Hiring three below-average workers to do the same work as two average workers is better for empire building, yet possibly at the expense of the project's budget and schedule. While it is true that finding adequate human resources may be difficult, sometimes companies simply do not put forth a good-faith effort in their search. Friends and family members may be hired first regardless of their qualifications. The problem is further complicated when people with project management expertise are sought.

Inefficiencies

Previously, we stated that companies might find it difficult to hire highly efficient people in project management. Not all people are efficient. Some people simply are not committed to their work even though they understand project management. Other people may get frustrated when they realize that they do not have the power, authority, or responsibility of their colleagues in first-world countries. Sometimes new hires who want to be efficient workers are pressured by the culture to remain inefficient or else the individual's colleagues will be identified as poor workers. Peer pressure exists and can prevent people from demonstrating their true potential.

Lack of Dedication

It is hard to get people motivated when they believe they cannot lose their jobs. Most people are simply not dedicated to the triple constraints. Some may even prefer to see schedules slip, believing it provides some degree of security for a longer period of time. There is also a lack of commitment to see the product through to project closure. As a project begins to wind down, employees will begin looking for a home on some other project. They may even leave their current project prematurely, before the work is finished, to guarantee employment elsewhere.

Misinformation

People working in emerging market countries may feel they must hide things from fellow workers and project managers, especially bad news, out of necessity, either to keep their jobs, their prestige, or to retain their power and authority. This creates a problem for project managers who rely on timely information, whether good or bad, in order to manage the project successfully. Delays in reporting could waste valuable time during which corrective action could have been taken.

Implementation of Project Management

While culture, status, and politics can create barriers for any new management philosophy, there are other barriers that are directly related to project management, including:

- Cost of project management implementation
- Risks of implementation failure

- Cost of training and training limitations
- Need for sophistication
- Lack of closure on projects
- Work ethic
- Poor planning

Cost of Implementation

There is a cost associated with the implementation of project management. The company must purchase hardware and software, create a project management methodology, and develop project performance reporting techniques. Doing this requires a significant financial expenditure, which the company might not be able to afford, and also requires tying up significant resources in implementation for an extended period of time. With limited resources, and because the better resources would be required for implementation and would be unavailable for ongoing work, companies shy away from project management even though they know the benefits.

Risk of Failure

Even if a company is willing to invest the time and money for project management implementation, there is a significant risk that the implementation will fail. Even if the implementation is successful, if projects begin to fail for any number of other reasons, blame will be placed on faulty implementation. Once executives have to explain the time and money expended for no real results, they may find their positions in the hierarchy insecure. This is why some executives refuse to accept or to visibly support project management.

Training Limitations

Implementation of project management is difficult without training programs for the workers. This creates three additional problems:

1. How much money must be allocated for training?
2. Who will provide the training, and what are the credentials of the trainers?
3. Should people be released from project work to attend training classes?

Training people in project management is time consuming and expensive. The costs of both implementation and training might prevent some executives from accepting project management.

Need for Sophistication

Project management requires sophistication, not only with the limited technology or tools that may be available but also with the ability of people to work together. This teamwork sophistication is generally lacking in emerging market countries. People may see no benefit in teamwork because others may be able to recognize their lack of competence and mistakes. They have not been trained to work properly in teams and are not rewarded for their contribution to the team.

Lack of Closure on Projects

Employees are often afraid to be attached to the project at closure when lessons learned, and best practices are captured. Lessons learned and best practices can be based on what we did well and what we did poorly. Employees may not want to see anything in writing that indicates that best practices were discovered from their mistakes.

Work Ethic

In some nations, the inability to fire people creates a relatively poor work ethic, which is contrary to effective project management practices. There is a lack of punctuality in coming to work and attending meetings. When people do show up at meetings, only good news is discussed in a group; bad news is discussed one on one. Communication skills are weak, as is report writing. There is a lack of accountability because accountability means explaining your actions if things go bad.

Poor Planning

Poor planning is paramount in emerging market nations. There exists a lack of commitment to the planning process. Because of a lack of standards, perhaps attributed to the poor work ethic, estimating duration, effort, and cost is very difficult. The ultimate result of poor planning is an elongation of the schedule. Workers who are unsure about their next assignment can view this as job security at least for the short term.

Other Barriers

Other barriers that are too numerous to mention, but some of the more important ones are listed next. These barriers are not necessarily universal in emerging market nations, and many of them can be overcome.

- Currency conversion inefficiencies
- Inability to receive timely payments
- Superstitious beliefs
- Laws against importing and exporting intellectual property
- Lack of tolerance for the religious beliefs of virtual team partners
- Risk of sanctions by partners' governments
- Use of poor or outdated technologies

Recommendations

Although we have painted a rather bleak picture, there are great opportunities in these emerging market nations, which have an abundance of talent that is yet to be fully harvested. The true capabilities of these workers are still unknown. Virtual project management teams might be the starting point for the full implementation of project management.

As project management begins to grow, senior officers will recognize and accept the benefits of project management, and see their business base increase. Partnerships and joint ventures using virtual teams will become more prevalent. The barriers that impede successful project management implementation will still exist, but project managers will begin to excel in how to live and work within the barriers and constraints imposed on the continually emerging virtual teams.

Greater opportunities are seen for the big emerging market economies. They are beginning to see more of the value of project management and have taken strides to expand its use. Some of the rapidly developing economies are very aggressive in providing the support needed for breaking many of the barriers just mentioned. As more success stories emerge, the various economies will strengthen, become more connected, and start to fully utilize project management for what it really is.

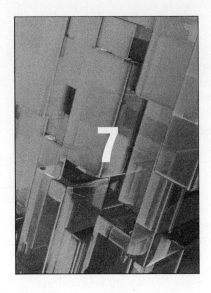

7 Management Support

As we saw in Chapter 6, senior managers are the architects of corporate culture. They are charged with making sure that their companies' cultures, once accepted, do not come apart. Visible management support is essential to maintaining a project management culture. And above all, the support must be continuous rather than sporadic.

This chapter examines the importance of management support in the creation and maintenance of project management cultures. Case studies illustrate the vital importance of employee empowerment and the project sponsor's role in the project management system.

7.1 VISIBLE SUPPORT FROM SENIOR MANAGERS

As project sponsors, senior managers provide support and encouragement to the project managers (PMs) and the rest of the project team. Companies excellent in project management have the following characteristics:

- Senior managers maintain a hands-off approach, but they are available when problems come up.
- Senior managers expect to be supplied with concise project status information, either in a written report format or using dashboards.
- Senior managers practice empowerment.

- Senior managers decentralize project authority and decision-making.
- Senior managers expect PMs and their teams to suggest both alternatives and recommendations for solving problems, not just identify the problems.

However, there is a fine line between effective sponsorship and overbearing sponsorship. Robert Hershock, former vice president at 3M, said it best during a videoconference on excellence in project management:

> Probably the most important thing is that they have to buy in from the top. There has to be leadership from the top, and the top has to be 100 percent supportive of this whole process. If you're a control freak, if you're someone who has high organizational skills and likes to dot all the *i*'s and cross all the *t*'s, this is going to be an uncomfortable process, because basically, it's a messy process; you have to have a lot of fault tolerance here. But what management has to do is project the confidence that it has in the teams. It has to set the strategy and the guidelines and then give the teams the empowerment that they need in order to finish their job. The best thing that management can do after training the team is get out of the way.

To ensure their visibility, senior managers need to believe in walk-the-halls management. In this way, every employee will come to recognize the sponsor and realize that it is appropriate to approach the sponsor with questions. Walk-the-halls management also means that executive sponsors keep their doors open. It is important that everyone, including line managers and their employees, feels supported by the sponsor. Keeping an open door can occasionally lead to problems if employees attempt to go around lower-level managers by seeking a higher level of authority. But such instances are infrequent, and the sponsor can easily deflect the problems back to the appropriate manager.

7.2 PROJECT SPONSORSHIP

Executive project sponsors provide guidance for PMs and project teams. They are also responsible for making sure that the line managers who lead functional departments fulfill their resource commitments to the projects underway. In addition, executive project sponsors maintain communication with customers and stakeholders.

Based upon the strategic importance of the project, the sponsor usually is an upper-level manager who, in addition to his or her regular responsibilities, provides ongoing guidance to assigned projects. An executive might take on sponsorship for several concurrent projects. Sometimes, on lower priority or maintenance projects, a middle-level manager may take on the project sponsor role. One organization I know of even prefers to assign middle managers instead of executives. The company believes this avoids the common problem of lack of line manager buy-in to projects (see Figure 7–1).

In some large, diversified corporations, senior managers do not have adequate time to invest in project sponsorship. In such cases, project sponsorship falls to the level below corporate senior management or to a committee or to special project management offices (PMOs).

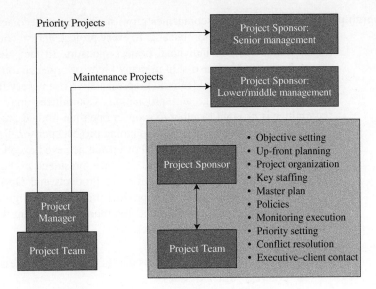

Figure 7–1. Roles of project sponsor.
Source: Reproduced from H. Kerzner, *In Search of Excellence in Project Management* (Hoboken, NJ: Wiley, 1998), p. 159/ John Wiley & Sons.

Some projects do not need project sponsors. Generally, sponsorship is required on large, complex projects involving a heavy commitment of resources. Large, complex projects also require a sponsor to integrate the activities of the functional lines, dispel disruptive conflicts, and maintain strong customer and stakeholder relations.

Consider one example of a project sponsor's support for a project. A PM who was handling a project in an organization in the federal government decided that another position would be needed on his team if the project were to meet its completion deadline. He had already identified a woman in the company who fit the qualifications he had outlined. But adding another full-time-equivalent position seemed impossible and was beyond his authority. The size of the government project office was constrained by a unit-manning document that dictated the number of positions available.

The PM went to the project's executive sponsor for help. The executive sponsor worked with the organization's human resources and personnel management department to add the position requested. Within 30 days, the addition of the new position was approved. Without the sponsor's intervention, it would have taken the organization's bureaucracy months to approve the position, too late to affect the deadline.

In another example, the president of a medium-size manufacturing company wanted to fill the role of sponsor on a special project. The PM decided to use the president of the division to the project's best advantage. He asked the president/sponsor to handle a critical situation. The president/sponsor flew to the company's headquarters and returned two days later with an authorization for a new tooling the PM needed. The company ended up saving time on the project, and the project was completed four months earlier than originally scheduled.

Sponsorship by Committee   As companies grow, it sometimes becomes impossible to assign a senior manager to every project, and so committees act in the place of individual project sponsors. In fact, the recent trend has been toward committee sponsorship in many kinds of organizations. A project sponsorship committee usually is made up of a representative from every function of the company: engineering, marketing, and production. Committees may be temporary when a committee is brought together to sponsor one time-limited project, or permanent when a standing committee takes on the ongoing project sponsorship of new projects.

For example, General Motors Powertrain achieved excellence in using committee sponsorship. Two key executives, the vice president of engineering and the vice president of operations, led the Office of Products and Operations, a group formed to oversee the management of all product programs. This group demonstrated visible executive-level program support and commitment to the entire organization. The roles and responsibilities of the group were to:

- Appoint the PM and team as part of the charter process
- Address strategic issues
- Approve the program contract and test for sufficiency
- Assure program execution through regularly scheduled reviews with program managers

Committee governance is now becoming commonplace. Companies that focus on maximizing the business value from a portfolio of strategic projects use committee governance rather than single-project sponsorship. The reason for this is that projects are now becoming larger and more complex to the point where a single individual may not be able to make all of the necessary decisions to support a PM.

Unfortunately, committee governance comes with some issues. Many of the people assigned to these governance committees may never have served on governance committees before and may be under the impression that committee governance for projects is the same as organizational governance. To make matters worse, some of the people may never have served as PMs and may not recognize how their decisions impact project management. Governance committee members can inflict a great deal of pain on the portfolio of projects if they lack a cursory understanding of project management. Organizations that are heavy users of agile and scrum practices use committee governance, and, in most organizations, committee members have a good understanding of their new roles and responsibilities.

Phases of Project Sponsorship   The role of the project sponsor changes over the life cycle of a project. During the planning and initiation phases, the sponsor plays an active role in the following activities:

- Helping the PM establish the objectives of the project
- Providing guidance to the PM during the organization and staffing phases

- Explaining to the PM what environmental or political factors might influence the project's execution
- Establishing the project's priority (working alone or with other company executives) and then informing the PM about the project's priority in the company and the reason that priority was assigned
- Providing guidance to the PM in establishing the policies and procedures for the project
- Functioning as the contact point for customers and clients

During the execution phase of a project, the sponsor must be very careful in deciding, which problems require his or her guidance. Trying to get involved with every problem that comes up on a project will result in micromanagement. It will also undermine the PM's authority and make it difficult for the executive to perform his or her regular responsibilities.

For short-term projects of two years or less, it is usually best that the project sponsor assignment is not changed over the duration of the project. For long-term projects of five years, more or less, different sponsors could be assigned for every phase of the project, if necessary. Choosing sponsors from among executives at the same corporate level works best, since sponsorship at the same level creates a "level" playing field, whereas favoritism can occur at different levels of sponsorship.

Project sponsors need not come from the functional area where the majority of the project work will be completed. Some companies even go so far as assigning sponsors from line functions that have no vested interest in the project. Theoretically, this system promotes impartial decision-making.

Customer Relations The role of executive project sponsors in customer relations depends on the type of organization (entirely project driven or partially project driven) and the type of customer (external or internal). Contractors working on large projects for external customers usually depend on executive project sponsors to keep the clients fully informed of progress of their projects. Customers with multimillion-dollar projects often keep an active eye on how their money is being spent. They are relieved to have an executive sponsor they can turn to for answers.

It is common practice for contractors heavily involved in competitive bidding for contracts to include both the PM's and the executive project sponsor's resumes in proposals. All things being equal, the resumes may give one contractor a competitive advantage over another.

Customers prefer to have a direct path of communication open to their contractors' executive managers. One contractor identified the functions of the executive project sponsor as:

- Actively participating in the preliminary sales effort and contract negotiations.
- Establishing and maintaining high-level client relationships.

- Assisting PMs in getting the project underway (planning, staffing, etc.).
- Maintaining current knowledge of major project activities.
- Handling major contractual matters.
- Interpreting company policies for PMs.
- Helping PMs identify and solve significant problems.
- Keeping general managers and client managers advised of significant problems with projects.

Decision-making

Imagine that project management is like car racing. A yellow flag is a warning to watch out for a problem. Yellow flags require action by the PM or the line manager. There is nothing wrong with informing an executive about a yellow-flag problem as long as the PM is not looking for the sponsor to solve the problem. Red flags, however, usually do require the sponsor's direct involvement. Red flags indicate problems that may affect the time, cost, and performance parameters of the project. So red flags need to be taken seriously, and decisions need to be made collaboratively by the PM and the project sponsor.

Serious problems sometimes result in serious conflicts. Disagreements between PMs and line managers are not unusual, and they require the thoughtful intervention of the executive project sponsor. First, the sponsor should make sure that the disagreement could not be solved without his or her help. Second, the sponsor needs to gather information from all sides and reflect on the alternatives being considered. Then the sponsor must decide whether he or she is qualified to settle the dispute. Often disputes are of a technical nature and require someone with the appropriate technical knowledge base to solve them. If the sponsor is unable to solve the problem, he or she will need to identify another source of authority that has the needed technical knowledge. Ultimately, a fair and appropriate solution can be shared by everyone involved. If there were no executive sponsor on the project, the disputing parties would be forced to go up the line of authority until they found a common superior to help them. Having executive project sponsors minimizes the number of people and the amount of time required to settle work disputes.

Strategic Planning

Executives are responsible for performing the company's strategic planning, and PMs are responsible for the operational planning on their assigned projects. Although the thought processes and time frames are different for the two types of planning, the strategic planning skills of executive sponsors can be useful to PMs. For projects that involve process or product development, sponsors can offer a special kind of market surveillance to identify new opportunities that might influence the long-term profitability of the organization. Furthermore, sponsors can gain a lot of strategically important knowledge from lower-level managers and employees. Who else knows better when the organization lacks the skill and knowledge base it needs to take on a new type of product? When the company needs to hire more technically skilled labor? What technical changes are likely to affect the industry?

7.3 EXCELLENCE IN PROJECT SPONSORSHIP

In excellent companies, the role of the sponsor is not to supervise the PM but to make sure that the best interests of both the customer and the company are recognized. However, as the next two examples reveal, it is seldom possible to make executive decisions that appease everyone.

Franklin Engineering (a pseudonym) had a reputation for developing high-quality, innovative products. Unfortunately, the company paid a high price for its reputation: a large research and development (R&D) budget. Fewer than 15 percent of the projects initiated by R&D led to the full commercialization of a product and the recovery of the research costs.

The company's senior managers decided to implement a policy that mandated that all R&D project sponsors periodically perform cost-benefit analyses on their projects. When a project's cost-benefit ratio failed to reach the levels prescribed in the policy, the project was canceled for the benefit of the whole company.

Initially, R&D personnel were unhappy to see their projects canceled, but they soon realized that early cancellation was better than investing large amounts in projects that were likely to fail. Eventually, PMs and team members came to agree that it made no sense to waste resources that could be better used on more successful projects. Within two years, the organization found itself working on more projects with a higher success rate but no addition to the R&D budget.

Another disguised case involves a California-based firm that designs and manufactures computer equipment. Let's call the company Design Solutions. The R&D group and the design group were loaded with talented individuals who believed that they could do the impossible and often did. These two powerful groups had little respect for the PMs and resented schedules because they thought schedules limited their creativity.

The company introduced two new products into the market barely ahead of the competition. The company had initially planned to introduce them a year earlier. The reason for the late releases: Projects had been delayed because of the project teams' desire to exceed the specifications required, not just meet them.

To help the company avoid similar delays in the future, the company decided to assign executive sponsors to every strategically important R&D project to make sure that the project teams adhered to standard management practices in the future. Some team members tried to hide their successes with the rationale that they could do better. But the sponsor threatened to dismiss those employees, and they eventually relented.

The lessons in both cases are clear. Executive sponsorship actually can improve existing project management systems to better serve the interests of the company and its customers.

7.4 WHEN SPONSORSHIP FAILS

Project management educators and practitioners promote the value that effective project sponsorship can bring to projects. Unfortunately, there are many instances where, despite starting out with good intentions, ineffective sponsorship occurs and leads to project disasters and even project failures. We will discuss several of these ineffective sponsorship situations and what can be done to improve project sponsorship practices going forward.

Defining Success

For more than 40 years, articles and books have appeared to extoll the successes in capturing project and organizational value that can be achieved from the effective implementation of project management practices. While many companies have achieved and maintained high levels of project management success, other companies have limited the continuous investment needed in project management practices to make the success sustainable (Chandler and Thomas 2015; Thomas and Mullaly 2008).

There are many definitions of success in a project management environment. The reason for the disparity is that most companies do not have a clear understanding of the factors that contribute to success (Bryde 2008). For simplicity's sake, project management success and the value it brings to an organization can be described in the following areas: (1) project success, (2) repeatable use of processes, tools, and techniques, (3) impact on the firm's business model, and (4) business results. These areas have been adapted from components in the model used by Thomas and Mullaly (2008).

- Project success has been traditionally defined by completing a deliverable within the triple constraints of time, cost, and scope followed by customer acceptance. The customer could be internal or external to the organization.
- Repeatable use of the project management processes, tools, and techniques is usually a characteristic of success when companies mandate a one-size-fits-all methodology approach for all of their traditional or operational projects.
- Business model success measures the amount of new business generated or an increase in market share because of successful use of project management. It can also measure the effectiveness of portfolio management practices and use of PMOs.
- Business results success is usually measured in financial terms obtained from revenue generated from completed projects.

There are other areas of success that could be considered, and many of them are industry related or dependent upon the type of project. Each of the areas of success can be broken down into critical success factors (CSFs), which are also most often industry specific.

The Project Sponsor/Project Manager Working Relationship

The birth of project sponsorship began in the early years of project management in the aerospace and defense industries. Most of the projects were highly technical and mandated that engineers, often those with advanced degrees, assume the lead role as PMs. Executive management was fearful that these highly technical PMs would make decisions that were reserved for the senior levels of management and restrictions had to be in place as to what decisions they were allowed to make.

The mistaken belief was that these PMs, because of their technical expertise, may be ineffective in making project business decisions. This was certainly not true, but senior management preferred to assign sponsors to handle all the business decisions on the projects and let the PMs handle the technical issues.

Many of the people assigned as project sponsors had a poor understanding of project management practices and sometimes the technology as well. As such, the sponsors and the PMs did not communicate as often as needed. The result was that project business decisions were being made without a full understanding of the technology and technical decisions were being made without an understanding of the impact on the customer and the business.

Result: Poor project decision-making

Customer Communications   Companies soon realized the abovementioned issue with the relationship between sponsors and PMs. Many companies considered assigning sponsors from the lower or middle levels of management rather than from the senior levels. While this approach was expected to increase collaboration and resolve some of the collaboration issues, government and military personnel did not see this as being in their best interest and exerted their influence.

Many government workers and military personnel believed that, because of their rank or title, their "equals" in the contractors' firms were at the executive levels. As such, even though lower-level individuals were assigned as sponsors, government and military personnel that controlled the funding for the projects communicated only with senior management, thus forcing them to remain as sponsors. Simply stated, sponsorship was often based upon the impact of two government rules:

- Rank has its privileges
- He/she who controls the "gold," rules! (i.e. makes the final decision)

Senior management succumbed to the pressure and remained as sponsors to appease the customers. Most of the time they functioned as "invisible" sponsors.

Result: Ineffective project sponsorship

Information is Power   In the early years of project management, senior management believed that allowing PMs to make business decisions was not only a risk but also diminished the role of senior management. Many executives believed then (and some still do) that information is power. Therefore, providing PMs with the necessary strategic or business information needed for business decisions would reduce their power base.

When information is power, project teams do not always have a line-of-sight to senior management and therefore make decisions that may not be aligned with strategic business objectives.

Result: Lack of alignment across project teams

Sponsorship Growth

It did not take long for the benefits of project management to appear. Companies began using project management for internal traditional or operational projects as well as projects for external clients. Now, there was a need for significantly more project sponsors.

Senior management recognized quickly that they could not function as project sponsors for all the projects. Sponsorship could be delegated to the middle or lower levels of management, but they would soon complain about the amount of time they would need to perform as sponsors and the fact that it could force them to reduce their efforts on other activities necessary to support daily activities.

Senior management made the decision that, for the internal projects, the business owners would assume the role of project sponsors. This created additional problems. The business owners had very limited knowledge about how project management should function. Many times, they did not understand the technology or the complexities in developing the technologies or creating product features. But what appeared as the worst situation was when business owners made project decisions based upon short-term profitability that could impact their year-end bonuses and sacrificed the long-term benefits and value the project could bring to the firm.

Result: Short-term project decision-making

Educating Sponsors

For decades, many of the people assigned as sponsors did not fully understand their role and had a very limited knowledge about project management. Some companies set up training programs to educate people on the role of a sponsor. Unfortunately, many of the project owners did not believe they needed to attend such a course even though most of the courses were less than two hours in duration. They felt that it was beneath their dignity to be told that they must be educated on how to properly function as a project sponsor given the fact that they were all in management positions already. These people believed that project sponsorship was the same as providing executive guidance.

Result: Understanding the role of project sponsor remained unsolved

Sponsor's Role in Project Staffing

PMs are at the mercy of functional managers for qualified staffing for the project. PMs may not know the skill sets needed from the functional groups. But when the PMs do know the skill set and the functional managers provide resources that the PMs consider inadequate, the PMs naturally expect the sponsors to intervene and assist them in obtaining the correct resources.

Many sponsors have shied away from participating in project staffing for fear of alienating functional managers whom they may have to work with in the future. As such, it was not uncommon for sponsors to avoid all responsibilities and participation in project staffing activities where they may have to usurp the authority of other managers. Project sponsors did not like the idea of telling functional managers in other functional units how to staff a given project especially since the sponsors did not know what other projects the functional units were responsible for staffing or the accompanying priorities.

Result: Projects are staffed with the wrong resources

Sponsorship Staffing with a Hidden Agenda

In the previous example, we showed that sponsors may not desire to participate in project staffing. At the other end of the spectrum, we have sponsors that may insist on project staffing participation, especially if the sponsor believes that the success of the project that they are sponsoring could have favorable implications on their career goals. This occurs when sponsors may have a hidden agenda related to this project.

Based upon the sponsor's rank and title, the sponsor may possess the authority to force functional units to staff a project with individuals handpicked by the project sponsor. This is often done with little regard for the impact of removing the workers from another project that desperately needed their skills.

Result: Project staffing is not done in the best interest of the company

Making Unrealistic Promises to the Customers

It is not uncommon for sponsors to handle communications with clients, especially with the senior levels of management in the clients' organizations. While this is an acceptable and often beneficial activity, it can create problems when the sponsor makes promises to the client as a way of appeasing the client or simply to look good in the eyes of the client.

As an example, during a discussion with the client, a sponsor promised the client that the company would perform additional testing to validate certain numbers in a report. The cost of the additional testing was more than $100,000. The sponsor told the PM to perform the additional testing within the original budget and that the sponsor would not be pleased if there were any cost overruns. The sponsor wanted this to be treated as a "no-cost scope change."

The project team was unable to hide the costs of the additional work and the profit of the project was reduced by $100,000. The sponsor reprimanded the team for not following his instructions even though they were unrealistic.

Result: No-cost scope changes rarely exist

Not Wanting to Hear Any Bad News

Sponsors exist to help project teams resolve problems and make the right decisions. Yet there are many sponsors that tell the teams that they do not want to hear any bad news. There are several reasons for this. The sponsor may not want to relay any bad news to the client and feels better not knowing about the issues. The sponsor may feel that bad news can be detrimental to his/her long-term goals. The sponsor may not wish to be involved in solving problems.

Perhaps the worst case of not wanting to hear bad news was identified as one of the causes for the Space Shuttle Challenger disaster where senior management expected lower-level managers to filter bad news from reaching the senior levels of management. There exists a valid argument that the filtering of bad news led to the death of seven astronauts.

Result: Filtering bad news can create very serious problems

Lessons Learned

What can be learned from the situations provided here? First and foremost, the success of a project is not entirely under the control of the PM. There can be numerous issues that are outside of the control of the PMs and require involvement and decisions by project sponsors. The role of a sponsor is quite complex and can be different in companies even in the same industry. Without a clear understanding of the roles and responsibilities of a sponsor, it is impossible to determine how sponsorship can and does contribute to project success.

Sponsors need to understand their role and the decisions they are expected to make. They should understand this before functioning as a sponsor, not by trial and error when performing as a sponsor. The PM in a telecom company became concerned that her sponsor was making decisions that she was unaware of and often disagreed with. She met with her sponsor. On a whiteboard, she drew a line down the center and listed many of the decisions that she expected would need to be made on the project. Then she looked at the sponsor and asked for clarification as to which decisions she was authorized to make, and which decisions must be made by the sponsor. The result of her meeting with the sponsor was a clarification of the lines of responsibility, which made the organization aware of the issue, and eventually led to the creation of a project sponsor's role template that became part of the firm's project management methodology.

The Need for Sponsorship Standards

Professional organizations have created standards for project management, but there do not appear to be any standards or guidelines for project sponsorship. The UK-based Association of Management (APM) defines a project sponsor as the individual/body, who is the primary risk taker, on whose behalf the project is undertaken and the US-based Project Management Institute (PMI) describes the sponsor as the person/group that provides the financial resources, in cash or in-kind, for the project. These two definitions characterize a project sponsor as being the primary risk taker or the resource provider, Bryde (2008).

Companies must understand the CSFs that lead to effective sponsorship and success. The following list, which is not in any specific order of importance, provides some guidance in understanding the role and responsibilities of a sponsor:

- Sponsors must understand that on some projects they may have to function as the primary communications link with the customer.
- PMs may not possess the authority to drive the projects to succeed without support from sponsors.
- We are now using project management on strategic as well as operational or traditional projects. Sponsors provide the knowledge and authority to make sure that project decisions are aligned with corporate strategy and strategic corporate objectives.
- Sponsors must be assigned during project selection activities to ensure that the best portfolios of projects are chosen and to get their buy-in. Sponsors should possess skills in conducting a SWOT analysis and application of established business models during project selection.
- More and more projects today are impacted by the enterprise environmental factors in the VUCA environment. Sponsors may possess a better understanding than PMs of how the company is impacted by the VUCA environment.
- The VUCA environment increases the risks that the company must face. Sponsors can provide guidance on how to best mitigate the risks.
- Sponsors are more than just business owners that fund projects. They possess the authority to ensure that the correct resources are assigned to the project.
- Sponsors must understand that status reporting is no longer based on just three metrics, namely time, cost, and scope. Sponsors must participate in selecting the proper mix of metrics such that the true project status can be determined quickly and that project sponsorship decisions will be based upon evidence and facts rather than guesses.
- Sponsors must provide PMs with the criteria (perhaps based on metrics selected) as to what will be defined as project success and project failure. Project failure criteria are essential so teams will know when to stop working on a project and wasting resources.
- Sponsors must understand how their decisions impact the outcome of projects and can lead to success or failure.
- Sponsors must recognize that the true success of a project rests in the benefits and value that come from the deliverables. It may be months or years after the project's deliverables have been produced before the real success of the project can be seen. Sponsors must therefore remain active as sponsors over the full life cycle of the project including benefits harvesting and sustainment of benefits and value. This is especially true for projects that lead to organizational changes in the firm's business model.
- PMs rely upon the PMs for guidance, leadership, networking, coaching, and mentorship during the execution of the project. Therefore, sponsors must possess more than just a cursory knowledge of how project management should work.

- Sponsors must be willing to attend periodic sponsorship courses to learn about sponsorship CSFs and best practices.
- Some companies have embarked upon committee sponsorship because one person may not possess all the necessary skills for sponsorship. Other companies have created specialized PMOs that have as their primary function the sponsorship of the portfolio of projects under their control. All members of the sponsorship committee as well as PMO leadership personnel must understand the role of a sponsor.

REFERENCES

Bryde, D. (2008). Perceptions on the Impact of Project Sponsorship Practices on Project Success. *International Journal of Project Management*, 26(8), pp. 800–809.

Chandler, D.E. and Thomas, J.L. (2015). Does Executive Sponsorship Matter for Realizing Project Management Value? *Project Management Journal*, 46(5), pp. 46–61.

Thomas, J. and Mullaly, M. (2008). *Researching the Value of Project Management*. Newtown Square, PA: Project Management Institute.

7.5 THE NEED FOR A PROJECT CANCELLATION CRITERIA

Not all projects will be successful. Executives must be willing to establish "exit criteria" that indicates when a project should be terminated. If the project is doomed to fail, then the earlier it is terminated, the sooner valuable resources can be reassigned to projects that demonstrate a higher likelihood of success. Without cancellation criteria, perhaps even identified in the project's business case, there is a risk that projects will linger on while squandering resources.

As an example, two vice presidents came up with ideas for pet projects and funded the projects internally using money from their functional areas. Both projects had budgets close to $2 million and schedules of approximately one year. These were somewhat high-risk projects because both required that a similar technical breakthrough should be made. No cancellation criteria were established for either project. There was no guarantee that the technical breakthrough could be made at all. And even if the technical breakthrough could be made, both executives estimated that the shelf life of both products would be about one year before becoming obsolete, but they believed they could easily recover their R&D costs.

These two projects were considered pet projects because they were established at the personal request of two senior managers and without any real business case. Had these two projects been required to go through the formal portfolio selection of projects process, neither one would have been approved. The budgets for these projects were way out of line for the value that the company would receive, and the return on investment would be below minimum levels even if the technical breakthrough could be made. The PMO, which is actively involved in the portfolio selection of projects process, also stated that it would never recommend approval of a project where the end result would have a shelf life of one year or less. Simply stated, these projects existed for the self-satisfaction of the two executives and to get them prestige from their colleagues.

Nevertheless, both executives found money for their projects and were willing to let them go forward without the standard approval process. Each executive was able to get an experienced PM from their group to manage their pet project.

At the first gate review meeting, both PMs stood up and recommended that their projects be canceled and that the resources be assigned to other, more promising projects. They both stated that the technical breakthrough needed could not be made in a timely manner. Under normal conditions, both of these PMs would have received medals for bravery in standing up and recommending that their project must be canceled. This certainly appeared as a recommendation in the best interest of the company.

But both executives were not willing to give up that easily. Canceling both projects would be humiliating for the executives who were sponsoring the projects. Instead, both executives stated that their project was to continue on to the next gate review meeting at which time a decision would be made regarding possible cancellation.

At the second gate review meeting, both PMs once again recommended that their projects be canceled. And as before, both executives asserted that the projects should continue to the next gate review meeting before a decision would be made.

As luck would have it, the necessary technical breakthrough was finally made, but six months late. That meant that the window of opportunity to sell the products and recover the R&D costs would be six months rather than one year. Unfortunately, the marketplace knew that these products might be obsolete in six months, and very few sales occurred of either product.

Both executives had to find a way to save face and avoid the humiliation of having to admit that they squandered a few million dollars on two useless R&D projects. This could very well impact their year-end bonuses. The solution that the executives found was to promote the PMs for creating the products and then blame marketing and sales for not finding customers.

Exit criteria should be established during the project approval process, and the criteria should be clearly visible in the project's business case. The criteria can be based on time to market, cost, selling price, quality, value, safety, or other constraints. If this is not done, the example just given can repeat itself again and again.

7.6 PROJECT GOVERNANCE

All projects have the potential of getting into trouble, but, in general, project management can work well as long as the project's requirements do not impose severe pressure on the PM and a project sponsor exists as an ally to assist the PM when trouble does appear. Unfortunately, in today's chaotic environment, this pressure appears to be increasing because:

- Companies are accepting high-risk and highly complex projects as a necessity for survival.
- Customers are demanding low-volume and high-quality products with some degree of customization.
- Project life cycles and new product development times are being compressed.
- Enterprise environmental factors are having a greater impact on project execution.
- Customers and stakeholders want to be more actively involved in the execution of projects.
- Companies are developing strategic partnerships with suppliers, and each supplier can be at a different level of project management maturity.
- Global competition has forced companies to accept projects from customers that are all at different levels of project management maturity and with different reporting requirements.

These pressures tend to slow down the decision-making processes at a time when stakeholders want the projects and processes to be accelerated. One person, while acting as the project sponsor, may have neither the time nor the capability to address all of these additional issues. The resulting project slowdown can occur because:

- The PM is expected to make decisions in areas where he/she has limited knowledge.
- The PM hesitates to accept full accountability and ownership of the projects.
- Excessive layers of management are superimposed on top of the project management organization.
- Risk management is pushed up to higher levels in the organization hierarchy, resulting in delayed decisions.
- The PM demonstrates questionable leadership ability on some nontraditional projects.

The problems resulting from these pressures may not be able to be resolved, at least easily and in a timely manner, by a single project sponsor. These problems can be resolved using effective project governance. Project governance is actually a framework by which decisions are made. Governance relates to decisions that define expectations, accountability, responsibility, the granting of power, or verifying performance. Governance relates to consistent management, cohesive policies, processes, and decision-making rights for a given area of responsibility. Governance enables efficient and effective decision-making to take place.

Every project can have different governance, even if each project uses the same enterprise project management methodology. The governance function can operate as a separate process or as part of project management leadership. Governance is not designed to replace project decision-making but to prevent undesirable decisions from being made.

Historically, governance was provided by a single project sponsor. Today, governance is a committee and can include representatives from each stakeholder's organization. Table 7–1 shows various governance approaches based on the type of project team. Committee membership can change from project to project and industry to industry. Membership may also vary based on the number of stakeholders and whether the project is for an internal or external client. On long-term projects, membership can change throughout the project.

Governance on projects and programs sometimes fails because people confuse project governance with corporate governance. The result is that members of the committee are not sure what their role should be. Some of the major differences include:

- *Alignment.* Corporate governance focuses on how well the portfolio of projects is aligned to and satisfies overall business objectives. Project governance focuses on ways to keep a project on track.
- *Direction.* Corporate governance provides strategic direction with a focus on how project success will satisfy corporate objectives. Project governance is more operation direction with decisions based on predefined parameters on project scope, time, cost, and functionality.
- *Dashboards.* Corporate governance dashboards are based on financial, marketing, and sales metrics. Project governance dashboards have operations metrics on time, cost, scope, quality, action items, risks, and deliverables.
- *Membership.* Corporate governance committees are composed of the senior-most levels of management. Project government membership may include some members from middle management.

Another reason why failure may occur is that members of the project or program governance group do not understand project or program management. This can lead to micromanagement by the governance committee. There is always the question of what decisions must be made by the governance committee and what decisions the PM can make. In general, the PM should have the authority for decisions related to actions necessary to maintain the baselines. Governance committees must have the authority to approve scope changes above a certain dollar value and to make decisions necessary to align the project to corporate objectives and strategy.

TABLE 7–1. TYPES OF PROJECT GOVERNANCE

Structure	Description	Governance
Dispersed locally	Team members can be full time or part time. They are still attached administratively to their functional area.	Usually, a single person acts as sponsor but may be an internal committee based on project complexity.
Dispersed geographically	This is a virtual team. Project managers may never see some team members. Team members can be full time or part time.	Usually, governance by committee and can include stakeholder membership.
Colocated	All team members are physically located in close proximity to the project manager. The project manager does not have any responsibility for wage and salary administration.	Usually, a single person acting as the sponsor.
Projectized	This is similar to a colocated team, but the project manager generally functions as a line manager and may have wage and salary responsibilities.	May be governance by committee, based on project size and number of strategic partners.

7.7 TOKIO MARINE: EXCELLENCE IN PROJECT GOVERNANCE[1] _____

**Executive Management Must
Establish IT Governance:
Tokio Marine Group**

Section 7.7 contributed by Yuichi (Rich) Inaba, CISA, and Hiroyuki Shibuya. Yuichi Inaba is a senior consultant specialist in the area of IT governance, IT risk management, and IT information security in the Tokio Marine and Nichido Systems Co. Ltd. (TMNS), a Tokio Marine Group company. Before transferring to TMNS, he had worked in the IT Planning Dept. of Tokio Marine Holdings Inc. and had engaged in establishing Tokio Marine Group's IT governance framework based on COBIT 4.1. His current responsibility is to implement and practice Tokio Marine Group's IT governance at TMNS. Inaba is a member of the ISACA Tokyo Chapter's Standards Committee and is currently engaged in translating COBIT 5 publications in Japanese.

Hiroyuki Shibuya is an executive officer in charge of IT at Tokio Marine Holdings Inc. From 2000 to 2005, he led the innovation project from the IT side, which has totally reconstructed the insurance product lines, their business processes, and the information systems of Tokio Marine and Nichido Fire Insurance Co. Ltd. To leverage his experience from this project as well as remediate other troubled development projects of group companies, he was named the general manager of the newly established IT planning department at Tokio Marine Holdings in July 2010. Since then, he has been leading the efforts to establish IT governance basic policies and standards to strengthen IT governance throughout the Tokio Marine Group.

* * *

Tokio Marine Group is a global corporate group engaged in a wide variety of insurance businesses. It consists of about 70 companies on five continents, including Tokio Marine and Nichido Fire Insurance (Japan), Philadelphia Insurance (US), Kiln (UK), and Tokio Marine Asia (Singapore).

In addition to Tokio Marine and Nichido Fire Insurance, which is the largest property and casualty insurance company in Japan, Tokio Marine Group has several other domestic companies in Japan, such as Tokio Marine and Nichido Life Insurance Co. Ltd, as well as service providers, such as Tokio Marine and Nichido Medical Service Co. Ltd. and Tokio Marine and Nachido Facilities Inc.

Implementing IT Governance at Tokio Marine Group

Tokio Marine Holdings, which is responsible for establishing the group's IT governance approach, observed that the executive management of Tokio Marine Group companies believes that IT is an essential infrastructure for business management, and it hoped to strengthen company management by utilizing IT. However, some directors and executives had a negative impression of IT—that IT is difficult to understand, costs too much, and results in frequent system troubles and system development failures.

---

1. Section 7.7 has been provided by Yuichi (Rich) Inaba, CISA, and Hiroyuki Shibuya, originally published in *COBIT® Focus* 1 (January 2013). © 2013 ISACA. All rights reserved. Reproduced with permission.

It is common for an organization's executive management to recognize the importance of system development but to put its development solely on the shoulders of the IT department. Other executives go even further, saying that the management or governance of IT is not anyone's business but the IT department's or CIO's. This line of thinking around IT is similar to the thought process that accounting is the job of the accounting department and handling personnel affairs is the role of the human resources department.

These are typical behaviors of organizations that fail to implement IT governance systems. Tokio Marine Holdings' executive management recognized that IT is not for IT's sake alone but is a tool to strengthen business.

Tokio Marine Holdings' management recognized that there were various types of system development failures (e.g. development delays for the service-in date, projects being over budget). Even more frequently, the organization was finding requirement gaps—for instance, where after building a system, the business people say, "This is not the system that we asked you to build" or "The system that you built is not easy to use. It is useless for the business."

Why the Requirement Gaps Occur

The process of system development is similar to that of a buildings construction. However, there is a distinct difference between the two: system development is not visible, whereas building construction is. Therefore, in system development, it is inevitable that there are recognition and communication gaps between business and IT.

Tokio Marine Group's Solution for System Development Success

To fill these gaps, business and IT must communicate enough to minimize the gaps between A and C in Figure 7–2 and maximize a common understanding of B. The road to success for system development is to improve the quality of communication between business and IT.

Such communication cannot be reached or maintained in a one-sided relationship. Ideal communication is enabled only with an equal partnership between business and IT with appropriate roles and responsibilities mutually allocated.

This is the core concept of Tokio Marine Group's Application Owner System.

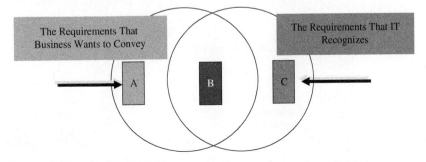

Figure 7–2. The requirements gap.

Implementing the Application Owner System

Tokio Marine Holdings decided to implement the Application Owner System as a core concept of the Group IT Governance System. Tokio Marine Holdings believes it is essential for the group's companies to succeed in system development and to achieve the group's growth in the current business environment.

The basic idea of the Application Owner System (Figure 7–3) is:

- Mutual cooperation between business and IT with proper check-and-balance functions, appropriately allocated responsibilities, and shared objectives.
- Close communication between business and IT, each taking their own respective roles and responsibilities into account.

Early Success in Tokio Marine and Nichido Fire Insurance

Tokio Marine and Nichido Fire Insurance Co. Ltd., the largest group company, implemented the Application Owner System in 2000. Implementation of the Application Owner System immediately reduced system troubles and problems by 80 percent (see Figure 7–4).

Mindset of IT

Tokio Marine's mindset is that only executive management can establish the enterprise's IT governance system. Thus, IT governance is the responsibility of executive management.

Furthermore, the organization is of the mindset that all employees, not only executive management, should understand the principle that strong IT systems cannot be

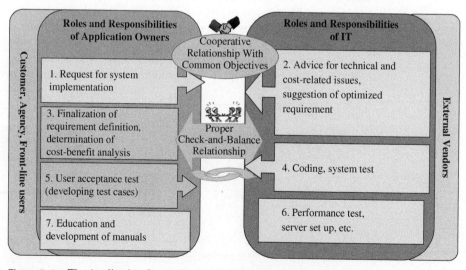

Figure 7–3. The Application Owner System in Tokio Marine Group.

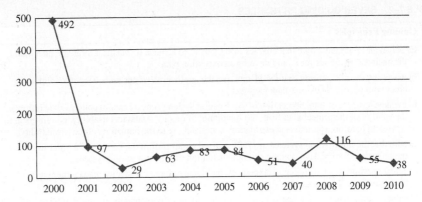

Figure 7–4. Number of system troubles.

realized by the IT department alone but require cooperation between business and IT. It is important that all employees recognize IT matters as their own, not as the matter of the IT function.

Establishing such a mindset within the enterprise is a role of the executive management.

Tokio Marine Group's IT Governance System

Characterized by the Application Owner System, Tokio Marine Holdings has introduced an IT governance framework, focused on the COBIT 4.1 framework, specifically the Plan and Organize (PO) domain.

The main goals of the IT Governance Framework are:

- *Establishing basic policies for IT governance.* Tokio Marine Holdings established the Basic Policies for IT Governance as the policies for the group's IT governance framework.
- *Establishing guiding principles for IT governance.* Tokio Marine Holdings defines seven principles as the guiding principles (see Table 7–2). These cover the five focus areas defined in the Board Briefing on IT Governance, particularly focusing on strategic alignment and value delivery. The seven principles are included in the Basic Policies for IT Governance. Tokio Marine Holdings thinks that the most important principle is the Application Owner System, which is stated as follows:
- In implementing the plan, it is important for the IT unit and the application owner units to cooperate with each other with proper checks and balance functions. Management shall clearly determine the appropriate sharing of roles between the IT unit and application owner units, secure human resources of adequate quality in both units and establish a management system to assure that each unit will execute the plan according to its responsibilities.

TABLE 7–2.   SEVEN GUIDING PRINCIPLES

No.	Guiding Principles	Focus Area
1	Establish an IT strategic plan that enables management to achieve its business strategic plan, build the business processes for it, and develop an execution plan.	Strategic Alignment
2	In executing the plan, ensure that the IT unit and the application owner units cooperate with each other with proper check-and-balance functions.	Strategic Alignment
3	In the development or implementation of information systems, ensure that management scrutinizes the validity of the project plan from the standpoint of quality assurance, usability, commitment to service-in date, appropriate cost estimation, and matching to the human resources availability.	Value Delivery
4	Ensure that the information systems are fully utilized by all staff in the company in order to achieve the objectives for the development or implementation of the information systems.	Value Delivery
5	Conduct appropriate IT resource management, including computer capacity management and human resources management.	Resource Management
6	Conduct appropriate risk management and information security management, and establish contingency plans for system faults in consideration of the accumulation of various risk factors in IT, such as high dependency of business processes on IT, centralization of important information, and threats from wider use of the internet.	Risk Management
7	Encourage the transparency of IT operations to be improved, and monitor their progress, which includes, for example, the progress of projects, the usage of IT resources, and utilization of information systems.	Performance Measurement

- *Establishing a governance and management system for Tokio Marine Group.* Tokio Marine Holdings defines the governance and management system to be implemented in the group companies. It covers five domains and consists of three major components: establishment of the organizational structure; establishment of policies and standards; and execution of the plan, do, check, and act (PCDA) cycle for improvement. The governance and management system required for Tokio Marine Group companies is detailed in the Group IT Governance Standard.
- *Establishing an IT governance standard* (the definition of Tokio Marine's priority processes). Tokio Marine Holdings has decided to utilize COBIT 4.1 to define the management system. However, the organization recognizes that it is difficult for a relatively small group of companies to implement mature processes for all COBIT 4.1 processes. To handle this concern, the organization focused on the minimal set and processes or more detailed control on objectives, which are essential for its group business in terms of IT governance and the most important controls for Tokio Marine Group.

In the IT Governance Standard, Tokio Marine Holdings defined the IT controls outlined in Figure 7–5 as priorities for the Tokio Marine Group. The priority IT controls are defined as five domains, 14 processes, and 39 control objectives, which are selected processes from the 210 control objectives of COBIT 4.1 (see Table 7–3).

The group companies are required to improve the priority controls to reach a maturity level 3, according to the COBIT Maturity Model, and report the progress of improvements of Tokio Marine Holdings.

TABLE 7–3. TOKIO MARINE GROUP'S PRIORITY CONTROLS

Domain Name	Id	Process Name
a. Planning and organization	a1	Annual IT planning
	a2	Definitions of roles and responsibilities of the IT unit and application owner units
	a3	Establishment of an IT steering committee
b. Projects management	b1	Management of development and implementation projects
c. Change management	c1	Change control
d. Operations management	d1	Incident/problem management
	d2	Vendor management
	d3	Security management
	d4	IT asset management
	d5	Computer capacity management
	d6	Disaster recovery and backup/restore
e. Monitoring performance and return on investment	e1	Annual IT review
	e2	Monitoring the IT steering committee
	e3	Monitoring project management, change management, and systems operation management

Figure 7–5. Tokio Marine group's priority controls.

Toward the Future

Since the establishment of the IT governance system for Tokio Marine Group, Tokio Marine Holdings has extensively communicated not only with CIOs but also with CEOs and executive management of the group companies to ensure that they understand, agree on, and take leadership for IT governance implementation.

Through these activities, the organization is confident that the core concept of IT governance has become better understood by management and good progress is being made as a result of the implementation of the Application Owner System in group companies. Tokio Marine Holdings will continue its evangelist mission to the group companies, realizing the benefit for the group business and giving value to stakeholders.

7.8 EMPOWERMENT OF PROJECT MANAGERS

One of the biggest problems with assigning executive sponsors to work beside line managers and PMs is the possibility that the lower-ranking managers will feel threatened with a loss of authority. This problem is real and must be dealt with at the executive level. Frank Jackson, formerly a senior manager at MCI, believes in the idea that information is power:

> We did an audit of the teams to see if we were really making the progress that we thought or were kidding ourselves, and we got a surprising result. When we looked at the audit, we found out that 50 percent of middle management's time was spent in filtering information up and down the organization. When we had a sponsor, the information went from the team to the sponsor to the operating committee, and this created a real crisis in our middle management area.
>
> MCI has found its solution to this problem. If there is anyone who believes that just going and dropping into a team approach environment is an easy way to move, it's definitely not. Even within the companies that I'm involved with, it's very difficult for managers to give up the authoritative responsibilities that they have had. You just have to move into it, and we've got a system where we communicate within MCI, which is MCI mail. It's an electronic mail system. What it has enabled us to do as a company is bypass levels of management. Sometimes you get bogged down in communications, but it allows you to communicate throughout the ranks without anyone holding back information.

Not only do executives have the ability to drive project management to success, but they also have the ability to create an environment that leads to project failure. According to Robert Hershock, former vice president at 3M:

> Most of the experiences that I had where projects failed, they failed because of management meddling. Either management wasn't 100 percent committed to the process, or management just bogged the whole process down with reports and a lot of other innuendos. The biggest failures I've seen anytime have been really because of management. Basically, there are two experiences where projects have failed to be successful. One is the management meddling where management cannot give up its decision-making capabilities, constantly going back to the team and saying you are doing this wrong or you are doing that wrong. The other side of it is when the team can't communicate its own objective. When it can't be focused, the scope continuously expands, and you get into project creep. The team just falls apart because it has lost its focus.

Project failure can often be a matter of false perceptions. Most executives believe that they have risen to the top of their organizations as solo performers. It is very difficult for them to change without feeling that they are giving up a tremendous amount of power, which traditionally is vested in the highest level of the company. To change this situation, it may be best to start small. As one executive observed:

> There are so many occasions where senior executives won't go to training and won't listen, but I think the proof is in the pudding. If you want to instill project management teams in your organizations, start small. If the company won't allow you to do it using

the Nike theory of just jumping in and doing it, start small and prove to them one step at a time that they can gain success. Hold the team accountable for results—it proves itself.

It is also important for us to remember that executives can have valid reasons for micromanaging. One executive commented on why project management might not be working as planned in his company:

> We, the executives, wanted to empower the project managers and they, in turn, would empower their team members to make decisions as they relate to their project or function. Unfortunately, I do not feel that we (the executives) totally support decentralization of decision-making due to political concerns that stem from the lack of confidence we have in our project managers, who are not proactive and who have not demonstrated leadership capabilities.

In most organizations, senior managers start at a point where they trust only their fellow managers. As the project management system improves and a project management culture develops, senior managers come to trust PMs, even though they do not occupy positions high on the organizational chart. Empowerment does not happen overnight. It takes time and, unfortunately, a lot of companies never make it to full PM empowerment.

7.9 MANAGEMENT SUPPORT AT WORK

Visible executive support is necessary for successful project management and the stability of a project management culture. But there is such a thing as too much visibility for senior managers. Take the following case example, for instance.

Midline Bank
Midline Bank (a pseudonym) is a medium-sized bank doing business in a large city in the Northwest. Executives at Midline realized that growth in the banking industry in the near future would be based on mergers and acquisitions and that Midline would need to take an aggressive stance to remain competitive. Financially, Midline was well prepared to acquire other small- and medium-sized banks to grow its organization.

The bank's information technology group was given the responsibility of developing an extensive and sophisticated software package to be used in evaluating the financial health of the banks targeted for acquisition. The software package required input from virtually every functional division of Midline. Coordination of the project was expected to be difficult.

Midline's culture was dominated by large, functional empires surrounded by impenetrable walls. The software project was the first in the bank's history to require cooperation and integration among the functional groups. A full-time PM was assigned to direct the project.

Unfortunately, Midline's executives, managers, and employees knew little about the principles of project management. The executives did, however, recognize the need for executive sponsorship. A steering committee of five executives was assigned to provide support and guidance for the PM, but none of the five understood project management. As a result, the steering committee interpreted its role as one of continuous daily direction of the project.

Each of the five executive sponsors asked for weekly personal briefings from the PM, and each sponsor gave conflicting directions. Each executive had his or her own agenda for the project.

By the end of the project's second month, chaos took over. The PM spent most of his time preparing status reports instead of managing the project. The executives changed the project's requirements frequently, and the organization had no change control process other than the steering committee's approval.

At the end of the fourth month, the PM resigned and sought employment outside the company. One of the executives from the steering committee then took over the PM's role, but only on a part-time basis. Ultimately, the project was taken over by two more PMs before it was completed, one year later than planned. The company learned a vital lesson: More sponsorship is not necessarily better than less.

Contractco

Another disguised case involves a Kentucky-based company I'll call Contractco. Contractco is in the business of nuclear fusion testing.

The company was in the process of bidding on a contract with the U.S. Department of Energy (DoE). The department required that the PM be identified as part of the company's proposal and that a list of the PM's duties and responsibilities be included. To impress the DoE, the company assigned both the executive vice president and the vice president of engineering as cosponsors.

The DoE questioned the idea of dual sponsorship. It was apparent to the DoE that the company did not understand the concept of project sponsorship, because the roles and responsibilities of the two sponsors appeared to overlap. The DoE also questioned the necessity of having the executive vice president serve as a sponsor.

The contract was eventually awarded to another company. Contractco learned that a company should never underestimate the customer's knowledge of project management or project sponsorship.

Health Care Associates

Health Care Associates (another pseudonym) provides health-care management services to both large and small companies in New England. The company partners with a chain of 23 hospitals in New England. More than 600 physicians are part of the professional team, and many of the physicians also serve as line managers at the company's branch offices. The physician managers maintain their own private clinical practices as well.

It was the company's practice to use boilerplate proposals prepared by the marketing department to solicit new business. If a client was seriously interested in Health Care Associates' services, a customized proposal based on the client's needs would be prepared. Typically, the custom-designed process took as long as six months or even a full year.

Health Care Associates wanted to speed up the custom-designed proposal process and decided to adopt project management processes to accomplish that goal. The company decided that it could get a step ahead of its competition if it assigned a physician manager as the project sponsor for every new proposal. The rationale was that the clients would be favorably impressed.

The pilot project for this approach was Sinco Energy (another pseudonym), a Boston-based company with 8,600 employees working in 12 cities in New England. Health Care Associates promised Sinco that the health-care package would be ready for implementation no later than six months from now.

The project was completed almost 60 days late and substantially over budget. Health Care Associates' senior managers privately interviewed each of the employees on the Sinco project to identify the cause of the project's failure. The employees had the following observations:

- Although the physicians had been given management training, they had a great deal of difficulty applying the principles of project management. As a result, the physicians ended up playing the role of invisible sponsors instead of actively participating in the project.
- Because they were practicing physicians, the physician sponsors were not fully committed to their role as project sponsors.
- Without strong sponsorship, there was no effective process in place to control scope creep.
- The physicians had not had authority over the line managers, who supplied the resources needed to complete a project successfully.

Health Care Associates' senior managers learned two lessons. First, not every manager is qualified to act as a project sponsor. Second, the project sponsors should be assigned on the basis of their ability to drive the project to success. Impressing the customer is not everything.

Indra

According to Enrique Sevilla Molina, PMP, formerly Corporate PMO Director at Indra:

Executive management is highly motivated to support project management development within the company. They regularly insist upon improving our training programs for project managers as well as focusing on the need that the best project management methods are in place.

Sometimes the success of a project constitutes a significant step in the development of new technology, in the launch in a new market, or for the establishment of a new partnership, and, in those cases, the managing directorate usually plays an especially active role as sponsors during the project or program execution. They participate with the customer in steering committees for the project or the program and help in the decision-making or risk management processes.

For a similar reason, due to the significance of a specific project but at a lower level, it is not uncommon to see middle-level management carefully watching its execution and providing, for instance, additional support to negotiate with the customer the resolution of a particular issue.

Getting middle-level management support has been accomplished using the same set of corporate tools for project management at all levels and for all kinds of projects in the company. No project is recognized if it is not in the corporate system, and to do that, the line managers must follow the same basic rules and methods, no matter if it is a recurring, a nonrecurring effort, or other type of project. A well-developed WBS [work breakdown structure], a complete foreseen schedule, a risk management plan, and a tailored set of earned value methods may be applied to any kind of project.

7.10 GETTING LINE MANAGEMENT SUPPORT

Management support is not restricted to senior management, as the previous section showed. Line management support is equally crucial for project management to work effectively. Line managers are usually more resistant to project management and often demand proof that project management provides value to the organization before they support the new processes. This problem was identified previously in the journey to excellence in project management. It also appeared at Motorola. According to a spokesperson at Motorola, getting line management support "was tough at first. It took years of having PMs provide value to the organization."

When organizations become mature in project management, sponsorship at the executive levels and middle management levels becomes minimal, and integrated project teams are formed where the integrated or core team is empowered to manage the project with minimal sponsorship other than for critical decisions. These integrated or core teams may or may not include line management. The concept of core teams became a best practice at Motorola:

Most project decisions and authority resides in the project core team. The core team is made up of middle- to low-level managers for the different functional areas (marketing, software, electrical, mechanical, manufacturing, system test, program management, quality, etc.) and has the project ownership responsibility. This core team is responsible for reviewing and approving product requirements and committing resources and scheduling dates. It also acts as the project change control board and can approve or reject project scope change requests. However, any ship acceptance date changes must be approved by senior management.

7.11 INITIATION CHAMPIONS AND EXIT CHAMPIONS

As project management evolved, so did the role of the executive in project management. Today, the executive plays three roles:

1. Project sponsor
2. Project (initiation) champion
3. Exit champion

The role of the executive in project management as a project sponsor has become reasonably mature. Most textbooks on project management have sections that are dedicated to the role of the project sponsor.[2] The role of the project champion, however, is just coming of age. Stephen Markham defines the role of the champion:

> Champions are informal leaders who emerge in a somewhat erratic fashion. Championing is a voluntary act by an individual to promote a particular project. In the act of championing, individuals rarely refer to themselves as champions; rather, they describe themselves as trying to do the right thing for the right company. A champion rarely makes a single decision to champion a project. Instead, he or she begins in a simple fashion and develops increasing enthusiasm for the project. A champion becomes passionate about a project and ultimately engages others based upon personal conviction that the project is the right thing for the entire organization. The champion affects the way other people think of the project by spreading positive information across the organization. Without official power or responsibility, a champion contributes to new product development by moving projects forward. Thus, champions are informal leaders who (1) adopt projects as their own in a personal way, (2) take on risk by promoting the projects beyond what is expected of people in their position, and (3) promote the project by getting other individuals to support it.[3]

With regard to new product development projects, champions are needed to overcome the obstacles in the "valley of death," as seen in Figure 7–6. The valley of death is the area in new product development where recognition of the idea/invention and efforts to commercialize the product come together. In this area, good projects often fall by the wayside and projects with less value often get added to the portfolio of projects. According to Markham:

> Many reasons exist for the Valley of Death. Technical personnel [left side of Figure 7–6] often do not understand the concerns of commercialization personnel [right side] and vice versa. The cultural gap between these two types of personnel manifests itself in the results prized by one side and devalued by the other. Networking and contact management may be important to salespeople but seen as shallow and self-aggrandizing by technical people. Technical people find value in discovery and pushing the frontiers of knowledge. Commercialization people need a product that will sell in the market and often considers the value of discovery as merely theoretical and therefore useless. Both technical and commercial people need help translating research findings into superior product offerings.[4]

---

2. See, for example, H. Kerzner, *Project Management: A Systems Approach to Planning, Scheduling and Controlling*, 12th ed. (Hoboken, NJ: Wiley, 2017), Chapter 10. Also, H. Kerzner and F. Saladis, *What Executives Need to Know About Project Management* (Hoboken, NJ: Wiley and International Institute for Learning, 2009).

3. S. K. Markham, "Product Champions: Crossing the Valley of Death," in P. Belliveau, A. Griffin, and S. Somermeyer (eds.), *The PDMA Toolbook for New Product Development,* vol. 1 (Hoboken, NJ: Wiley, 2002), p. 119.

4. Ibid., p. 120.

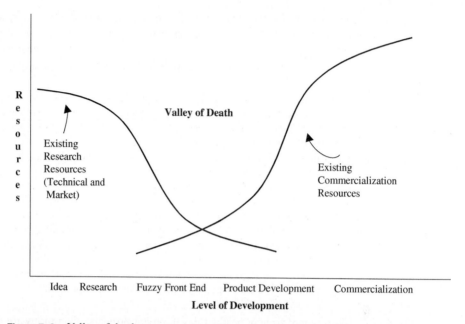

Figure 7–6. Valley of death.
Source: Adapted from S. K. Markham, "Product Champions: Crossing the Valley of Death," in P. Belliveau, A. Griffin, and S. Somermeyer (eds.), *The PDMA Toolbook for New Product Development* (Hoboken, NJ: Wiley, 2002), p. 119.

As seen in Figure 7–6, the valley of death seems to originate somewhere near the fuzzy front end. The fuzzy front end is

> the messy "getting started" period of product development, which comes before the formal and well-structured product development process when the product concept is still very fuzzy. It generally consists of the first three tasks (strategic planning, concept generation, and, especially, pretechnical evaluation) of the product development process. These activities are often chaotic, unpredictable, and unstructured. In comparison, the new product development process is typically structured and formal, with a prescribed set of activities, questions to be answered, and decisions to be made.[5]

Project champions are usually neither PMs nor project sponsors. The role of the champion is to sell the idea or concept until it finally becomes a project. The champion may not even understand project management and may not have the necessary skills to manage a project. Champions may reside much higher up in the organizational hierarchy than the PM.

Allowing the project champion to function as the project sponsor can be as bad as allowing them to function as the PM. When the project champion and the project sponsor are the same person, projects never get canceled. There is a tendency to prolong the pain of continuing with a project that should have been canceled.

5. P. Belliveau, A. Griffin, and S. Somermeyer, *The PDMA Toolbook for New Product Development* (Hoboken, NJ: Wiley, 2002), p. 444.

Some projects, especially very long-term projects where the champion is actively involved, often mandate the existence of a collective belief. The collective belief is a fervent, and perhaps blind, desire to achieve that can permeate the entire team, the project sponsor, and even the most senior levels of management. The collective belief can make a rational organization act in an irrational manner. This is particularly true if the project sponsor spearheads the collective belief.

When a collective belief exists, people are selected based on their support for that belief. Champions may prevent talented employees from working on the project unless they possess the same fervent belief as the champion does. Nonbelievers are pressured into supporting the collective belief, and team members are not allowed to challenge the results. As the collective belief grows, both advocates and nonbelievers are trampled. The pressure of the collective belief can outweigh the reality of the results.

There are several characteristics of the collective belief, which is why some large, high-technology projects are often difficult to kill:

- Inability or refusing to recognize failure
- Refusing to see the warning signs
- Seeing only what you want to see
- Fearful of exposing mistakes
- Viewing bad news as a personal failure
- Viewing failure as a sign of weakness
- Viewing failure as damage to your career
- Viewing failure as damage to your reputation

Project sponsors and project champions do everything possible to make their project successful. But what if the project champion and the project team and sponsor have blind faith in the success of the project? What happens if the strongly held convictions and the collective belief disregard the early warning signs of imminent danger? What happens if the collective belief drowns out dissent?

In such cases, an exit champion must be assigned. Sometimes the exit champion needs to have some direct involvement in the project in order to have credibility, but direct involvement is not always a necessity. Exit champions must be willing to put their reputations on the line and possibly face the likelihood of being cast out from the project team. According to Isabelle Royer:

> Sometimes it takes an individual, rather than growing evidence, to shake the collective belief of a project team. If the problem with unbridled enthusiasm starts as an unintended consequence of the legitimate work of a project champion, then what may be needed is a countervailing force—an exit champion. These people are more than devil's advocates. Instead of simply raising questions about a project, they seek objective evidence showing that problems in fact exist. This allows them to challenge—or, given the ambiguity of existing data, conceivably even to confirm—the viability of a project. They then take action based on the data.[6]

---

6. I. Royer, "Why Bad Projects Are So Hard to Kill," *Harvard Business Review* (February 2003): 11. Copyright © 2003 by the Harvard Business School Publishing Corporation. All rights reserved. D. Davis, "New Projects: Beware of False Economics," *Harvard Business Review* (March–April 1985): 100–101. Copyright © 1985 by the President and Fellows of Harvard College. All rights reserved.

The larger the project and the greater the financial risk to the firm, the higher up the exit champion should reside. If the project champion just happens to be the chief executive officer, then someone on the board of directors or even the entire board of directors should assume the role of the exit champion. Unfortunately, there are situations where the collective belief permeates the entire board of directors. In this case, the collective belief can force board members to shirk their responsibility for oversight.

Large projects incur large cost overruns and schedule slippages. Making the decision to cancel such a project, once it has started, is very difficult, according to David Davis:

> The difficulty of abandoning a project after several million dollars have been committed to it tends to prevent objective review and recosting. For this reason, ideally an independent management team—one not involved in the project's development—should do the recosting and, if possible, the entire review. . . . If the numbers do not hold up in the review and recosting, the company should abandon the project. The number of bad projects that make it to the operational stage serves as proof that their supporters often balk at this decision.
>
> . . . Senior managers need to create an environment that rewards honesty and courage and provides for more decision-making on the part of project managers. Companies must have an atmosphere that encourages projects to succeed, but executives must allow them to fail.

The longer the project, the greater the necessity for exit champions and project sponsors to make sure that the business plan has "exit ramps" such that the project can be terminated before massive resources are committed and consumed. Unfortunately, when a collective belief exists, exit ramps are purposefully omitted from the project and business plans. Another reason for having exit champions is so that the project closure process can occur as quickly as possible. As projects approach their completion, team members often are worried about their next assignment and try to stretch out the existing project until they are ready to leave. In this case, the role of the exit champion is to accelerate the closure process without impacting the integrity of the project.

Some organizations use members of a portfolio review board to function as exit champions. Portfolio review boards have the final say in project selection. They also have the final say as to whether or not a project should be terminated. Usually, one member of the board functions as the exit champion and makes the final presentation to the remainder of the board.

8 Training and Education

8.0 INTRODUCTION

Establishing project management training programs is one of the greatest challenges facing training directors because project management involves numerous complex and interrelated skills (qualitative/behavioral, organizational, and quantitative). In the early days of project management, project managers learned more from their own mistakes than from the experience of others. Today, companies that excel at project management are offering a corporate curriculum in project management. Effective training supports project management as a profession.

Some large corporations offer more internal courses related to project management than do most colleges and universities. Such companies treat education almost as a religion. Smaller companies have more modest internal training programs and usually send their people to publicly offered training programs.

This chapter discusses processes for identifying the need for training, selecting the students who need training, designing, and conducting the training, and measuring training's return on dollars invested.

8.1 TRAINING FOR MODERN PROJECT MANAGEMENT

During the early days of project management, in the late 1950s and throughout the 1960s, training courses concentrated on the advantages and disadvantages of various organizational forms (e.g., matrix, traditional, projectized, and functional). Executives learned quickly, however, that any organizational structure could be made to work effectively and efficiently when basic project management is applied. Project management skills based on trust, teamwork, cooperation, and communication can solve the worst structural problems.

Starting in the 1970s, emphasis turned away from organizational structures for project management. The old training programs were replaced with two basic programs:

- *Basic project management*, which stresses behavioral topics such as multiple reporting relationships, time management, leadership, conflict resolution, negotiation, team building, motivation, and basic management areas, such as planning and controlling.
- *Advanced project management*, which stresses scheduling techniques and software packages used for planning and controlling projects.

Today's project management training programs include courses on behavioral and quantitative subjects. The most important problem facing training managers is how to achieve a workable balance between the two parts of the coursework—behavioral and quantitative (see Figure 8–1). For publicly sponsored training programs, seminar leaders determine their own comfort levels in the "discretionary zone" between technical and behavioral subject matter. For in-house trainers, however, the balance must be preestablished by the training director on the basis of factors such as which students will be assigned to manage projects, types of projects, and average lengths of projects (see Table 8–1).

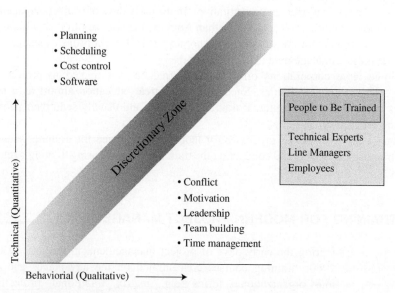

Figure 8–1. Types of project management training.
Source: Reproduced from H. Kerzner, *In Search of Excellence in Project Management* (Hoboken, NJ: Wiley, 1998), p. 174.

TABLE 8–1. EMPHASES IN VARIOUS TRAINING PROGRAMS

Type of Person Assigned for PM Training (PM Source)	Training Program Emphasis	
	Quantitative/Technology Skills	Behavior Skills
Training Needed to Function as a Project Manager		
Technical expert on short-term projects	High	Low
Technical expert on long-term projects	High	High
Line manager acting as a part-time project manager	High	Low
Line manager acting as a full-time project manager	High	Average to high
Employees experienced in cooperative operations	High	Average to high
Employees inexperienced in cooperative operations	High	Average to high
Training Needed for General Knowledge		
Any employees or managers	Average	Average

Source: Reproduced from H. Kerzner, *In Search of Excellence in Project Management* (Hoboken, NJ: Wiley, 1998), p. 175.

8.2 NEED FOR BUSINESS EDUCATION

The importance of the right balance is critical. We discussed the importance of determining the right balance between quantitative skills and behavioral skills. That balance is now changing because of how we view the role of a project manager. Today, we have a new breed of project managers. Years ago, virtually all project managers were engineers with advanced degrees. These people had a command of technology rather than merely an understanding of it. If the line manager believed that the project manager did in fact possess a command of technology, then the line manager would allow the assigned functional employees to take direction from the project manager. The result was that project managers were expected to manage people *and* provide technical direction. This meant that project managers would have to take responsibility for functional deliverables.

Most project managers today have an understanding of technology rather than a command of it. As a result, the accountability for the success of the project is now viewed as shared between the project manager and all affected line managers. With shared accountability, the line managers must have a good understanding of project management, which is why more line managers are now PMP® credential holders.[1] Project managers are now expected to manage deliverables as well as people. Management of the assigned resources is no longer just the responsibility of line managers.

Another important fact is that project managers are treated as though they are managing part of a business rather than simply a project, and they therefore are expected to make sound business decisions as well as project decisions. Project managers must understand business principles. In the future, project managers may be expected to

---

1. PMP is a registered mark of the Project Management Institute, Inc.

become externally certified by PMI and internally certified by their company on the organization's business processes.

Now, when designing training courses, we determine the correct balance between quantitative skills, behavioral skills, and business skills. Soft skills and business acumen are crucial elements for a flawless project execution, says Benny Nyberg, formerly group assistant vice president with responsibility for PM Methodologies and Talent Development at ABB:

> After implementing the ABB PM Process as a common high-level process throughout the company's project sales organizations as well as several product development organizations, one thing was very clear. In a technical company employing large numbers of highly skilled engineers, some of whom are promoted to project management, the technical aspects of project management such as planning, scheduling, and cost control are the least difficult to implement. Junior employees do need training in this area, but the real challenge for reaching operational excellence in project management, a flawless project execution, desirable project results, and a high level of customer satisfaction lies in identifying and developing project managers with the right business acumen. Project management is a management position requiring excellent commercial, communications, and leadership skills. A project manager must be very business minded, be able to communicate effectively with a variety of different stakeholders and possess the ability to lead and motivate people. For delivery projects, a precise understanding of the contract (i.e., terms and conditions, scope, and any promises made) is crucial for being able to deliver just that, meet customer expectations, and as such assure customer satisfaction and project success. Contract understanding is further a requisite for maximizing financial outcome and recognizing up-selling opportunities as they occur.
>
> The role of a project manager, especially for big contracts that take many months or several years to complete, is very close to the role of a key account manager. The following skills/abilities are among the most important for success: business mind, communication, negotiation, leadership, risk management, salesmanship.
>
> In order to identify and address training and other development activities required for the wide variety of competencies and skills, ABB have implemented a competency model. The model includes a definition of required competencies, questionnaires for self-assessment, interview questionnaires plus development guides, and last but not least a number of selectable training modules leading to appropriate level of certification.

8.3 SAP: IMPORTANCE OF A PROJECT MANAGEMENT CAREER PATH[2]

In SAP Professional Services, we have established a clear project management development path for anyone supporting customer projects and internal projects. The project management career path spans from entry-level position of project management associate through various levels of project managers all the way to the role of program manager or delivery executive (see Figure 8–2).

---

2. Material in this section has been provided by Jan Musil, Chief Product Owner, SAP Activate.

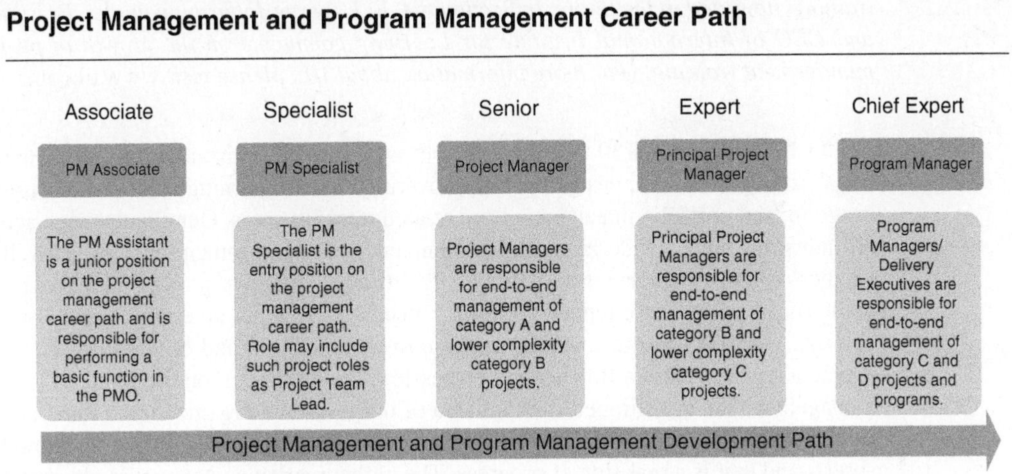

Project Management and Program Management Career Path

Associate	Specialist	Senior	Expert	Chief Expert
PM Associate	PM Specialist	Project Manager	Principal Project Manager	Program Manager
The PM Assistant is a junior position on the project management career path and is responsible for performing a basic function in the PMO.	The PM Specialist is the entry position on the project management career path. Role may include such project roles as Project Team Lead.	Project Managers are responsible for end-to-end management of category A and lower complexity category B projects.	Principal Project Managers are responsible for end-to-end management of category B and lower complexity category C projects.	Program Managers/ Delivery Executives are responsible for end-to-end management of category C and D projects and programs.

Project Management and Program Management Development Path

Figure 8–2. Project management and program management development path.

Each role has a clearly defined skill set profile, including precise definition of required job skills, and expected proficiency level for each skill that is required in the job. Each profile is also associated with the category of projects that the person at this career level is expected to manage—dimensions of this project category include project size, complexity, risk exposure, budget, and so on.

Each profile is tightly linked with a human resources job profile and respective professional development path that specifies recommended training classes for each skill in the profile. The training offerings for the PM practice specify target roles for the training, and the training learning objectives are tied to specific skills from the profile. The training catalog covers broad range of skills, from core project management skills based on PMI standards like PMI *PMBOK® Guide*[3] project management fundamentals; SAP-specific knowledge (like SAP solution knowledge, SAP Activate methodology, agile project delivery, and project management tools, etc.); and leadership and interpersonal skills.

Through the structured career path, project managers grow their knowledge, expertise, and experiences, which allows them to move along the PM career path. The PM career path is interlinked with other job profiles in SAP so that project managers can make career choices that allow them to move to different career tracks, such as product engineering, general management, and sales roles, or to remain in the PM career path.

8.4 INTERNATIONAL INSTITUTE FOR LEARNING

Given that an organization's strategy is delivered through projects, programs, and other major initiatives, there will be a continuing need for project management education and

---

3. PMBOK is a registered mark of the Project Management Institute, Inc.

training, now, and in the future. In Section 8.4, E. LaVerne Johnson, Founder, President, and CEO of International Institute for Learning comments on the growth of project management training. (For more information about IIL, please visit www.iil.com)

*   *   *

In IIL's more than 30-year history, we have worked with thousands of organizations around the world, designing and delivering custom learning solutions in project, program, and portfolio management and related learning areas. Our clients range across all industries and include large global companies and government entities, as well as smaller organizations trying to gain an edge through effective projects and programs. With IIL serving our clients' day-to-day needs as well as emerging trends on the horizon, we have been in a unique position to participate in and observe project management's growth into a full-fledged profession. We have been on the scene as project management has transitioned from an area of interest to an organizational imperative.

From our perspective, courses that were sufficient just a few years ago now fall short—and that is a real sign of progress. The global market and expanding importance of project management have dictated whole new families of courses, richer content, and a flexible range of delivery methods that allow participants to learn when, where, and how they need to—in face-to-face and live virtual classrooms, self-paced online interactive learning, or small group coaching. IIL takes pride in our Many Methods of Learning™ approach, which ensures that the education we provide serves a diversity of needs, styles, and interests.

In addition, we are now exploring opportunities to provide micro-learning, which can be accessed by learners at the point of need—on the job. This kind of learning requires a totally different design approach from traditional classroom or self-paced training because it needs to focus on those small steps that result in "just-in-time" knowledge or improved performance.

Evolutionary Years: Learning Trends

Training courses during the 1980s were mostly geared toward advancing the project manager's technical skills. The focus of training was on the basics: project planning and control, as well as the knowledge required to pass PMI's Project Management Professional (PMP)® Certification exam. In response, IIL launched training courses in project management fundamentals and established a comprehensive certification program that allowed individuals to prepare for and successfully pass PMI's PMP® exam. Companies then had access to a somewhat limited set of publications, classroom courses, and software products to help develop and assist their project managers in successfully executing their projects.

Revolutionary Years: Marketplace Trends

In recent years, an increasing number of companies, in virtually every industry sector, have recognized the importance of managing projects more effectively and analyzing the ways in which projects meet overall corporate goals.

Compared to previous years, a revolution has occurred in project management. This is evident in a number of trends.

- The volume of projects has steadily increased as more and more companies run their businesses through projects. Indeed, some leading organizations undertake hundreds of thousands of individual projects each year—some small and simple, others huge and complex.
- The ability to effectively manage projects has become critically important to business, and sound project management skills have become a competitive advantage for leading companies.
- As a result of this revolutionary growth, the status and value of project managers have grown in importance—having this know-how allows a company to complete projects faster, at lower cost, with greater customer satisfaction, and with more desirable project outcomes.
- Knowledge that was once deemed "nice to have" is now considered mandatory. A company's economic success and survival depend on its ability to determine which projects support its overall strategic objectives and to enable it to sequence them in ways to achieve that success.
- Today, project managers are not the only professionals who have, or who need to have, project management skills: team members and middle and upper management are developing expertise in the subject as well.
- The complexity and scope of project management methodologies have grown to include new skills and applications. For example, the accelerating adoption of agile and Scrum has provided new and novel approaches to project execution with positive results.
- Process development and improvement through direct hands-on support or knowledge management solutions have become a requirement for economic survival in these challenging times. For example, IIL's Unified Project Management® Methodology (UPMM™) Software Suite was developed to support consistency and quality in project, program, and portfolio management implementation.
- The number of planning and project management-related software applications continues to expand past the more common applications, such as Microsoft Project, to include applications for small projects and agile and Scrum projects as well (e.g., Trello, JIRA, VersionOne, and SmartSheet).
- Project management certification has become an even more valuable asset to an individual's career path. As a result, there are 1,036,367 active PMP-certified individuals and 314 chartered chapters across 214 countries and territories worldwide, as of July 31, 2020.
- For years, the project manager's skill set has remained mostly technical. Today, we are seeing project managers embrace additional knowledge areas such as leadership and interpersonal skills.
- Strategic planning for project management has gained importance. Organizations are now seeking systematic ways to better align their portfolio of projects with business objectives.

- More and more companies are establishing project and portfolio management offices.
- Approaches to project management within an organization remain relatively varied and non-standardized. Companies need to work toward a more mature and common methodology for repeatable and predictable success.
- Companies and their project management offices are placing a stronger emphasis on quality services, quality products, and improved processes using management models such as strategic management, benefits realization, knowledge management, and business relationship management.

Revolutionary Years: Learning Trends

In response to these trends, a greater variety of courses are now available to an increasing number of industries. New methods of learning have been introduced to meet the growing diversity of client needs. Here are some examples of how IIL has responded to these needs and established best practices in training and education for project and program managers, sponsors, and executives:

- In addition to the fundamentals, IIL offers a wide variety of specialized programs to further increase the depth of knowledge and skills of the project manager. Such courses include advanced concepts in risk management, innovation project management, requirements management, the design and development of a project office, and a whole curriculum focusing on agile and Scrum.
- Courses addressing the "softer" side of project management are designed to hone facilitation skills, interpersonal skills, leadership skills, and other non-technical areas.
- As organizations increase their level of project management maturity, there is a need for training in the effective use of enterprise project management software, such as Project Server and many others.
- More and more universities are either offering degrees in project management or including project management courses as part of specific degree requirements.
- The way we learn is changing. Working professionals today have less time to spend in the classroom than in years past. Accordingly, IIL offers live virtual training, where participants attend "class" from wherever they are. Additionally, many organizations prefer a blended approach to learning for their participants, which includes a variety of methods including face-to-face instruction, virtual delivery, and on-demand e-learning. Oftentimes, the participants attend this blended program in cohorts, thus providing a shared social experience that helps with knowledge retention and application.

A Look into the Crystal Ball: Trends and Learning Responses

It is always a challenge to try to predict the future, but there are some emerging trends that allow us to take a reasonable stab at this. For each of these trends, there will be the need to develop the appropriate learning responses. A key competitive factor for organizations will be their ability to select and successfully execute all types of projects, from those designed to develop new products and services, thus increasing revenue and profits, to those that help an organization improve internal performance and cut costs. To be sure, the ability to select a portfolio of projects that boost a company's competitive posture is of paramount importance in today's environment.

- Project management methodologies will blend with other proven business strategies and frameworks (such as Lean Six Sigma, Agile, Scrum, quality management, risk management, and business analysis). Training in these subjects will similarly become blended.
- Strategic and innovation projects will grow in scope and scale, and project managers are stepping up to fill roles in these areas. IIL is responding with new courses, coaching, and timely learning solutions.
- Project, program, and portfolio management will continue to grow in importance and become a strategic competitive advantage for organizations. Portfolio management software is now a requirement.
- Senior management will become more knowledgeable and involved in project management efforts. This will require project management training to meet their unique needs.
- Strategic planning for project management will become a way of life for leading organizations. The role of the project office/portfolio office will increase in importance and become more commonplace and vital in companies. Membership will include the highest levels of executive management. Senior management will take leadership of the company's project portfolio management efforts.
- IIL will continue to partner with organizations both large and small to offer external and internal certifications and learning solutions to support upskilling and reskilling.
- Demand for learning project management-related disciplines such as business analysis, business relationship management, and information technology service management, for example, remains strong. IIL will continue to partner with leading global strategic learning partners to offer world-renowned training and certifications in these areas.
- Executives will be increasingly involved in activities such as capacity planning, portfolio management, prioritization, process improvement, supply chain management, and strategic planning specifically for project management. More and more executives are earning project management certifications.
- The reward and recognition systems of organizations will change to stimulate and reinforce project management goals and objectives.
- The status of the certified project manager will grow significantly. The professional project manager will have a combination of technical, business, strategic planning, leadership, and communications skills.
- Project benchmarking and continuous project improvement will become essential for leading organizations. Current and new project management maturity models will help companies identify their strengths, weaknesses, and specific opportunities for improvement.

- The expanding importance of project, program, and portfolio management will require more individuals who are trained in project management. This in turn will necessitate the development of new and improved methods of training delivery. Online interactive video-based training will play an increasingly important role.
- We will see an order-of-magnitude increase in the number of organizations reaching higher levels of project management maturity.
- More colleges and universities will offer degree programs in project management and seek to align their courses with international standards and best practices.
- Project management will focus on providing the knowledge and best practices to support sustainable initiatives. Project sustainability in the global economy through values, leadership, and professional responsibility will be the mandate of all project, program, and portfolio managers and sponsors.

8.5 IDENTIFYING THE NEED FOR TRAINING

Identifying the need for training requires that line managers and senior managers recognize two critical factors: first, that training is one of the fastest ways to build project management knowledge in a company, and second, that training should be conducted for the benefit of the corporate bottom line through enhanced efficiency and effectiveness.

Identifying the need for training has become somewhat easier in the past 10 years because of published case studies on the benefits of project management training. The benefits can be classified according to quantitative and qualitative benefits. The quantitative results include:

- Shorter product development time
- Faster, higher-quality decisions
- Lower costs
- Higher profit margins
- Fewer people are needed
- Reduction in paperwork
- Improved quality and reliability
- Lower turnover of personnel
- Quicker "best practices" implementation

Qualitative results include:

- Better visibility and focus on results
- Better coordination
- Higher morale
- Accelerated development of managers

- Better control
- Better customer relations
- Better functional support
- Fewer conflicts requiring senior management involvement

Companies are finally realizing that the speed at which the benefits of project management can be achieved is accelerated through proper training.

8.6 SELECTING PARTICIPANTS

Selecting the people to be trained is critical. As we have seen in a number of case studies, it is usually a mistake to train only the project managers. A thorough understanding of project management and project management skills is needed throughout the organization if project management is to be successful. For example, one automobile subcontractor invested months in training its project managers. Six months later, projects were still coming in late and over budget. The executive vice president finally realized that project management was a team effort rather than an individual responsibility. After that revelation, training was provided for all of the employees who had anything to do with the projects. Virtually overnight, project results improved.

Dave Kandt, retired group vice president, quality, program management, and continuous improvement at Johnson Controls, explained how his company's training plan was laid out to achieve excellence in project management:

> We began with our executive office, and once we had explained the principles and philosophies of project management to these people, we moved to the managers of plants, engineering managers, cost analysts, purchasing people, and, of course, project managers. Only once the foundation was laid did we proceed with actual project management and with defining the roles and responsibilities so that the entire company would understand its role in project management once these people began to work. Just the understanding allowed us to move to a matrix organization and eventually to a stand-alone project management department.

8.7 FUNDAMENTALS OF PROJECT MANAGEMENT EDUCATION

Twenty years ago, we were somewhat limited as to availability of project management training and education. Emphasis surrounded on-the-job training in hopes that fewer mistakes would be made. Today, we have other types of programs, including:

- University courses and seminars
- In-house seminars and curriculums
- Vendor-provided corporate training

The means by which the learning takes place can be:

- Face-to-face (F2F)
- Virtual
- eLearning

With the quantity of literature available today, we have numerous ways to deliver the knowledge. Typical delivery systems include:

- Lectures
- Lectures with discussion
- Exams
- Illustrative case studies of external companies
- Working case studies on internal projects, custom-built scenarios, or standard scenarios built for general learning purposes
- Simulation and role-playing

Training managers are currently experimenting with "when to train." The most common choices include:

- *Just-in-time training*: This includes training employees immediately prior to assigning them to projects.
- *Exposure training*: This includes training employees on the core principles just to give them enough knowledge so that they will understand what is happening in project management within the firm.
- *Continuous learning*: This is training, first on basic topics, then on advanced topics, so that people continue to grow and mature in project management. Basic training may involve the principles of scheduling techniques, whereas advanced topics may include training on the use of a specific software package.
- *Self-confidence training*: This is similar to continuous learning but based on current state-of-the-art knowledge. This is to reinforce employees' belief that their skills are comparable to those in companies with excellent reputations for project management.

8.8 SOME CHANGES IN PROJECT MANAGEMENT EDUCATION

In the early years of project management, almost all project managers came from the engineering disciplines. The project managers were expected to possess a command of technology rather than just an understanding of it. When earned value management (EVM) principles were developed, project management coursework emphasized cost and schedule control. Seminars appeared in the marketplace entitled "Project Management," but the content for these two- or three-day courses was almost entirely PERT and EVM. In all but a few industries, project management was viewed as part-time work and an add-on to one's primary job. The need to fully understand the skills and competencies to be effective in project management was not considered to be that important.

Today, project management is a career-path position in almost all companies. Colleges and universities are now offering master's and doctorate degrees in project management. Typical coursework in such programs include:

Core Coursework
- Principles of Project Management
- Project Scheduling Techniques
- Project Estimating Techniques
- Project Financing Techniques
- Creativity and Brainstorming
- Problem-Solving and Decision-Making
- Global Project Management
- Managing Multiple Projects
- Project Management Leadership
- Managing Virtual Teams
- Project Portfolio Management

Electives
- Advanced Project Management
- Project Quality Management
- Project Procurement and Contracting
- Project Ethics and the Code of Professional Conduct
- Project Monitoring and Control Techniques
- Project Reporting Practices
- Stakeholder Relations Management
- Conducting Project Health Checks
- Managing Troubled Projects
- Capturing Best Practices
- Managing Cultural Differences

Some educational institutions also offer students specialized training for certification in various project management areas. The knowledge needed to pass the certification exams can come from specialized coursework or from the traditional core requirements and electives. A partial list of some certification programs related to project management might include:

- Project management
- Program management
- Project risk management
- Agile project management
- Business analyst

- Managing complex projects
- Other project management certifications
- Specialized or customized certifications

The greatest change in project management education appears to be in the softer skill requirements. This is understandable since projects require people to work together. Emphasis in the behavioral areas, as well as in some of the technical areas, is now being placed on:

- Problem-solving skills
- Decision-making skills
- Conceptualization skills
- Creativity/Brainstorming skills
- Process skills
- Coping with stress/pressure
- Leadership without authority
- Multiple boss reporting
- Counseling and facilitation
- Mentorship skills
- Negotiating skills
- Conflict resolution skills
- Presentation skills

In the future, social media skills may be added to the list as project performance information is transmitted over mobile devices.

8.9 DESIGNING COURSES AND CONDUCTING TRAINING

Many companies have come to realize that on-the-job training may be less effective than more formal training. On-the-job training virtually forces people to make mistakes as a learning experience, but what are they learning: how to make mistakes? It seems much more efficient to train people to do their jobs the right way from the start.

Project management has become a career path. More and more companies today allow or even require that their employees get project management certification. One company informed its employees that project management certification would be treated the same as a master's degree in the salary and career path structure. The cost of the training behind the certification process is only 5 to 10 percent of the cost of a typical master's degree in a business administration program. And certification promises a quicker return on investment (ROI) for the company. Project management certification can also be useful for employees without college degrees; it gives them the opportunity for a second career path within the company.

There is also the question of which is better: internally based or publicly held training programs. The answer depends on the nature of the individual company and how many employees need to be trained, how big the training budget is, and how deep the company's internal knowledge base is. If only a few employees at a time need training, it might be effective to send them to a publicly sponsored training course, but if large numbers of employees need training on an ongoing basis, designing and conducting a customized internal training program might be the way to go.

In general, custom-designed courses are the most effective. In excellent companies, course content surveys are conducted at all levels of management. For example, many years ago, the research and development group of Babcock and Wilcox in Alliance, Ohio, needed a project management training program for 200 engineers. The head of the training department knew that she was not qualified to select core content, so she sent questionnaires out to executive managers, line managers, and professionals in the organization. The information from the questionnaires was used to develop three separate courses for the audience. At Ford Motor Company, training was broken down into a two-hour session for executives, a three-day program for project personnel, and a half-day session for overhead personnel.

For internal training courses, choosing the right trainers and speakers is crucial. A company can use trainers currently on staff if they have a solid knowledge of project management, or the trainers can be trained by outside consultants who offer train-the-trainer programs. Either way, trainers from within the company must have not only the expertise the company needs but also the necessary facilitation skills to maximize training results. Most external educational providers, such as the International Institute for Learning, select trainers based on both technical knowledge and facilitation skills.

Some problems with using internal trainers include the following:

- Internal trainers may not be experienced in all areas of project management.
- Internal trainers may not have up-to-date knowledge of the project management techniques practiced by other companies.
- Internal trainers may have other responsibilities in the company and so may not have adequate time for preparation.
- Internal trainers may not be as dedicated to project management or as skillful as external trainers.

But the knowledge base of internal trainers can be augmented by outside trainers as necessary. In fact, most companies use external speakers and trainers for their internal educational offerings. The best way to select speakers is to seek out recommendations from training directors in other companies and teachers of university-level courses in project management. Another method is contacting speakers' bureaus, but the quality of the speaker's program may not be as high as needed. The most common method for finding speakers is reviewing the brochures of publicly sponsored seminars. Of course, the brochures were created as sales materials, and the best way to evaluate the seminars is to attend them.

After a potential speaker has been selected, the next step is to check his or her recommendations. Table 8–2 outlines many of the pitfalls involved in choosing speakers for internal training programs and how to avoid them.

TABLE 8–2.   COMMON PITFALLS IN HIRING EXTERNAL TRAINERS AND SPEAKERS

Warning Sign	Preventive Step
Speaker professes to be an expert in several different areas.	Verify speaker's credentials. Very few people are experts in several areas. Talk to other companies that have used the speaker.
Speaker's resume identifies several well-known and highly regarded client organizations.	See whether the speaker has done consulting for any of these companies more than once. Sometimes a speaker does a good job selling himself or herself the first time, but the company refuses to rehire the person again after the first presentation.
Speaker makes a very dramatic first impression and sells himself or herself well. Brief classroom observation confirms your impression.	Being a dynamic speaker does not guarantee that quality information will be presented. Some speakers are so dynamic that trainees do not realize until it is too late that "The trainer was nice but the information was marginal."
Speaker's resume shows 10–20 years or more experience as a project manager.	10–20 years of experience in a specific industry or company does not mean that the speaker's knowledge is transferable to your company's specific needs or industry. Ask the speaker what types of projects he or she has managed.
Marketing personnel from the speaker's company aggressively show the quality of their company rather than the quality of the speaker. The client list presented is the company's client list.	You are hiring the speaker, not the marketing representative. Ask to speak with or meet with the speaker personally and look at his or her client list rather than the parent company's client list.
Speaker promises to custom design materials to your company's needs.	Demand to see the speaker's custom-designed material at least two weeks before the training program. Also verify the quality and professionalism of graphs and other materials.

The final step is to evaluate the training materials and presentation the external trainer will use in the classes. The following questions can serve as a checklist:

- *Does the speaker use case studies?* If he or she does, are they illustrative case studies to demonstrate a point or technique, or are they working case studies, where a scenario is provided in which the learner has a chance to practice a skill or technique? Some companies find it best to develop their own case studies and ask the speaker to use those so that the cases will have relevance to the company's business.
- *Are role-playing and laboratory experiences planned?* They can be valuable aids to learning, but they can also limit class size.
- *Are homework and required reading a part of the class?* If so, can they be completed before the seminar?

8.10 MEASURING THE RETURN ON INVESTMENT IN EDUCATION

The last area of project management training is the determination of the value earned on the dollars invested in training. It is crucial to remember that training should not be performed unless there is a continuous return on dollars for the company. Keep in mind also that the speaker's fee is only part of the cost of training. The cost to the company of having employees away from their work during training must be included in the calculation. Some excellent companies hire outside consultants to determine ROI. The consultants base their evaluations on personal interviews, on-the-job assessments, and written surveys.

One company tests trainees before and after training to learn how much knowledge they really gained. Another company hires outside consultants to prepare and interpret posttraining surveys on the value of the specific training received.

The amount of training needed at any one company depends on two factors: whether the company is project-driven and whether it has practiced project management long enough to have developed a mature project management system. Figure 8–3 shows the amount of project management training offered (including refresher courses) against the number of years in project management. Project-driven organizations offer the most project management training courses, and organizations that have just started implementing project management offer the fewest. That is no surprise. Companies with more than 15 years of experience in applying project management principles show the most variance.

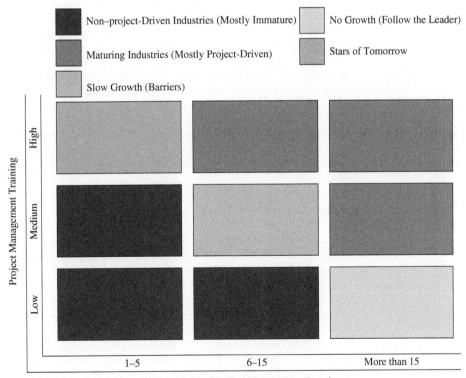

Figure 8–3. Amount of training by type of industry and year of project management experience.
Source: Reproduced from H. Kerzner, *In Search of Excellence in Project Management* (Hoboken, NJ: Wiley, 1998), p. 185.

8.11 PROJECT MANAGEMENT IS NOW A PROFESSION _____

For several years, project management was viewed as a part-time occupation, and, therefore, all training was designed for one's primary job description, whatever that may be, rather than project management. Because of this, there was no need to develop job descriptions for project and program managers. Today, these job descriptions exist, project management is viewed as a profession, and training programs are provided based upon the job descriptions. When asked if AT&T had job descriptions, a spokesperson for AT&T responded "yes" to both project and program management:

Project Manager

Provides end-to-end project management throughout the lifecycle of a project by directing the efforts of project team(s) using dotted-line authority to deliver a completed product and/or service. Has full accountability for managing larger low complexity to high complexity projects, or projects within programs, which may span multiple regions and/or multiple functions; multiple concurrent projects may be managed. Includes estimating, scheduling, coordinating, assigning resources, ensuring that project funding is secured, and assisting in recommending business solutions/alternatives for projects. Assesses, plans for, and manages project risks, issues, jeopardies, escalations and problem resolutions. Manages project scope, project budgeting, and cost reporting, and ensures completion of projects while meeting quality, schedule, and cost objectives using the organization's standard processes. Acts as project liaison between IT partners, client organizations and IT leadership. May assist in supplier management of existing vendors. May direct Associate Project Managers to provide support with project communications and tracking project progress. Does not include the management of extremely large and complex programs, with multiple sub-programs, requiring senior level oversight and extensive executive communications. Must spend 80% or more of time performing the project management duties described above.

Program Manager

Provides end-to-end project management and/or program management throughout the lifecycle of a project/program by directing the efforts of project/program team(s) using dotted-line authority to deliver a completed project and/or service. Has full accountability for managing concurrent high complexity projects and/or programs which may span multiple regions, functions and/or business units. Responsible for detailed planning including program/project structure and staffing, estimating, resource allocation and assignment, detailed scheduling, critical path analysis, consolidating project plans into an overall program plan, and negotiating any sequencing conflicts. Directs project and/or program activities utilizing the organization's standard processes to ensure the timely delivery of stated business benefits, comparing actuals to plans and adjusting plans as necessary. Assesses, plans for, and manages project/program risks including mitigation & contingency plans; manages issues, jeopardies, escalations and problem resolutions. Defines project/program scope and ensures changes to scope and deliverables are managed using the change control process. Manages large program or project budgets and cost reporting. Acts as liaison with client and IT leadership, providing communication and status regarding the progress of the project/program. May assist with RFP [request for proposal] development, evaluation, and supplier selection, as well as ongoing relationships with suppliers or consultants. Utilizes knowledge of business, industry and technology to incorporate business process improvements into the organization and/or to develop business strategies and functional/business/technical architectures. May direct the efforts of project managers when they manage a project or sub-program over which the Senior Project/Program Manager has authority. May include the management of extremely large and complex programs, with multiple sub-programs, requiring senior level oversight and extensive executive communications. Must spend 80% or more of time performing the project management duties described above.

The recognition of project management as a profession has spread worldwide. According to Enrique Sevilla Molina, PMP, formerly Corporate PMO Director at Indra:

> Project management is considered the result of a specific blend of knowledge and experience, gained through dedication to achieve success in the projects under the project manager responsibility.
>
> We have a set of management roles associated to the different levels of responsibilities and expertise to manage projects, programs and portfolios, and to develop business opportunities (i.e., Project Managers, Program Directors, etc.). For each role, a specific set of skills in a certain degree is defined, so performance and achievement may be assessed. The yearly evaluation of personal performance is done based on the job descriptions, the role maturity achieved so far, the expected performance for the role and the actual performance, so the evolution in personal development may also be assessed.

8.12 COMPETENCY MODELS

Thirty years ago, companies prepared job descriptions for project managers to explain roles and responsibilities. Unfortunately, the job descriptions were usually abbreviated and provided very little guidance on what was required for promotion or salary increases. Twenty years ago, we still emphasized the job description, but it was now supported by coursework, which was often mandatory. By the late 1990s, companies had begun emphasizing core competency models, which clearly depicted the skill levels needed to be effective as a project manager. Training programs were instituted to support the core competency models. Unfortunately, establishing a core competency model and the accompanying training is no easy task.

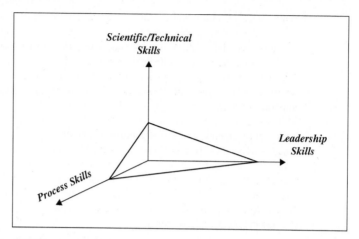

Figure 8–4. Eli Lilly competency model.

Several years ago, Eli Lilly developed perhaps one of the most comprehensive and effective competency models in industry.[4] The basis for the competency model is shown in Figure 8–4 A detailed listing of the skills needed along each axis was prepared, and training programs were established for skills needed.

4. A detailed description of Eli Lilly's Competency Model can be found in H. Kerzner, A. Zeitoun, & R. Vargas, (2022). *Project Management Next Generation: The Pillars for Organizational Excellence*, Hoboken, Wiley, pp. 304–313.

Informal Project Management

9.0 INTRODUCTION

Over the past 35 years, one of the most significant changes in project management has been the idea that informal project management does work. In the 1950s and 1960s, the aerospace, defense, and large construction industries were the primary users of project management tools and techniques. Because project management was a relatively new management process, customers of the contractors and subcontractors wanted evidence that the system worked. Documentation of the policies and procedures to be used became part of the written proposal. Formal project management, supported by hundreds of policies, procedures, and forms, became the norm. After all, why would a potential customer be willing to sign a $10 million contract for a project to be managed informally?

This chapter clarifies the difference between informal and formal project management, then discusses the four critical elements of informal project management.

9.1 INFORMAL VERSUS FORMAL PROJECT MANAGEMENT

Formal project management has always been expensive and time-consuming. In the early years, the time and resources spent on preparing written policies and procedures had a purpose: They placated the customer. As project management became established, formal documentation was created, mostly for the customer. Contractors began managing more informally, while the customer was still paying for formal project management documentation. Table 9–1 shows the major differences between formal and informal project management. As you can see, the most relevant difference is the amount of paperwork.

TABLE 9–1. FORMAL VERSUS INFORMAL PROJECT MANAGEMENT

Factor	Formal Project Management	Informal Project Management
Project manager's level	High	Low to middle
Project manager's authority	Documented	Implied
Paperwork	Exorbitant	Minimal

Paperwork is expensive. Even a routine handout for a team meeting can cost $500 to $2,000 per page to prepare. Executives at excellent companies know that paperwork is expensive. They encourage project teams to communicate without excessive amounts of paper. However, some people are still operating under the mistaken belief that ISO 9000 certification requires massive paperwork.

Figure 9–1 shows the changes in paperwork requirements in project management. The early 1980s marked the heyday for lovers of paper documentation. At that time, the average policies and procedures manual probably cost between $3 million and $5 million to prepare initially and $1 million to $2 million to update annually over the lifetime of the development project. Project managers were buried in forms to complete to the extent that they had very little time left for actually managing the projects. Customers began to complain about the high cost of subcontracting, and the paperwork boom started to fade.

Real cost savings did not materialize until the early 1990s, with the growth of concurrent engineering. Concurrent engineering shortened product development times by taking activities that had been done in series and performing them in parallel instead. This change increased the level of risk in each project, which required that project

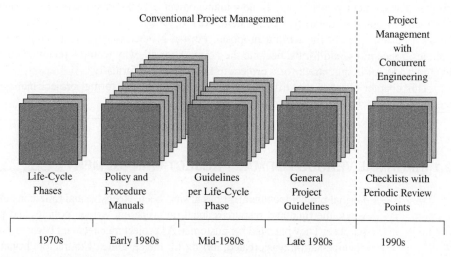

Figure 9–1. Evolution of policies, procedures, and guidelines.

management back away from some of its previous practices. Formal guidelines were replaced by less detailed and more generic checklists.

Policies and procedures represent formality. Checklists represent informality. But informality does not eliminate project paperwork altogether. It reduces paperwork requirements to minimally acceptable levels. Moving from formality to informality demands a change in organizational culture (see Figure 9–2). The four basic elements of an informal culture are these:

1. Trust
2. Communication
3. Cooperation
4. Teamwork

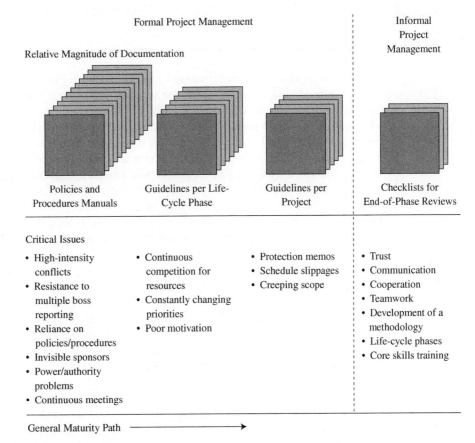

Figure 9–2. Evolution of paperwork and change of formality levels.

Large companies quite often cannot manage projects on an informal basis, although they want to. The larger the company, the greater the tendency for formal project management to take hold. A former vice president of IOC sales operations and customer service at Nortel Networks believes:

> The introduction of enterprise-wide project process and tools standards in Nortel Networks and the use of pipeline metrics (customer-defined, industry standard measures) provide a framework for formal project management. This is necessary given the complexity of telecom projects we undertake and the need for an integrated solution in a short time frame. The Nortel Networks project manager crosses many organizational boundaries to achieve the results demanded by customers in a dynamic environment.

Most companies manage either formally or informally. However, if your company is project-driven and has a very strong culture for project management, you may have to manage either formally or informally based on the needs of your customers.

9.2 TRUST

Trusting everyone involved in executing a project is critical. You wake up in the morning, get dressed, and climb into your car to go to work. On a typical morning, you operate the foot pedal for your brakes maybe 50 times. You have never met the people who designed, manufactured, or installed the brakes. Yet you still give no thought to whether the brakes will work when you need them. No one broadsides you on the way to work. You do not run over anyone. Then you arrive at work and push the button for the elevator. You have never met the people who designed, manufactured, installed, or inspected the elevator. But again, you feel perfectly comfortable riding the elevator up to your floor. By the time you get to your office at 8 a.m., you have trusted your life to uncounted numbers of people whom you have never even met. Still, you sit down in your office and refuse to trust the person in the next office to make a $50 decision.

Trust is the key to the successful implementation of informal project management. Without it, project managers and project sponsors would need all that paperwork just to make sure that everyone working on their projects was doing the work just as he or she had been instructed. Trust is also key in building a successful relationship between the contractor/subcontractor and the client. Let us look at an example.

Perhaps the best application of informal project management that I have seen occurred several years ago in the Heavy Vehicle Systems Group of Bendix Corporation. Bendix hired a consultant to conduct a three-day training program. The program was custom designed, and during the design phase, the consultant asked the vice president and general manager of the division whether he wanted to be trained in formal or informal project management. The vice president opted for informal project management. What was the reason for his decision? The culture of the division was already based on trust. Line managers were not hired solely based on technical expertise. Hiring and promotions were based on how well the new manager would communicate and cooperate with the other line managers and project managers in making decisions in the best interests of both the company and the project.

TABLE 9–2.   BENEFITS OF TRUST IN CUSTOMER–CONTRACTOR WORKING RELATIONSHIPS

Without Trust	With Trust
Continuous competitive bidding	Long-term contracts, repeat business, and sole-source contracts
Massive documentation	Minimal documentation
Excessive customer–contractor team meetings	Minimal number of team meetings
Team meetings with documentation	Team meetings without documentation
Sponsorship at executive levels	Sponsorship at middle-management levels

When the relationship between a customer and a contractor is based on trust, numerous benefits accrue to both parties. The benefits are apparent in companies such as Hewlett–Packard, Computer Associates, and various automobile subcontractors. Table 9–2 shows the benefits.

9.3 COMMUNICATION

In traditional, formal organizations, employees usually claim that communication is poor. Senior managers, however, usually think that communication in their company is just fine. Why the disparity? In most companies, executives are inundated with information communicated to them through frequent meetings and dozens of weekly status reports coming from every functional area of the business. The quality and frequency of information moving down the organizational chart are less consistent, especially in more formal companies. But whether it is a problem with the information flowing up to the executive level or down to the staff, the problem usually originates somewhere upstairs. Senior managers are the usual suspects when it comes to requiring reports and meetings. And many of those reports and meetings are unnecessary and redundant.

Most project managers prefer to communicate verbally and informally. The cost of formal communication can be high. Project communication includes dispensing information on decisions made, work authorizations, negotiations, and project reports. Project managers in excellent companies believe that they spend as much as 90 percent of their time on internal interpersonal communication with their teams. Figure 9–3 illustrates the communication channels used by a typical project manager. In project-driven organizations, project managers may spend most of their time communicating externally to customers and regulatory agencies.

Good project management methodologies promote not only informal project management but also effective communications laterally as well as vertically. The methodology itself functions as a channel of communication. A senior executive at a large financial institution commented on his organization's project management methodology, called Project Management Standards (PMS):

> The PMS guides the project manager through every step of the project. The PMS not only controls the reporting structure but also sets the guidelines for who should be involved in the project itself and the various levels of review. This creates an excellent

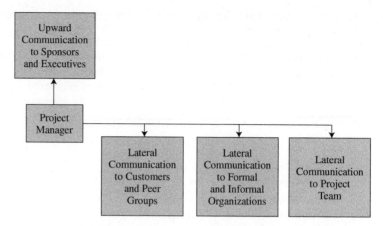

Figure 9–3. Internal and external communication channels for project management.

communication flow between the right people. The communication of a project is one of the most important factors for success. A great plan can only go so far if it is not communicated well.

Most companies believe that a good project management methodology will lead to effective communications, which will allow the firm to manage more informally than formally. The question, of course, is how long it will take to achieve effective communications. With all employees housed under a single roof, the timeframe can be short. For global projects, geographical dispersion and cultural differences may mandate decades before effective communication will occur. Even then, there is no guarantee that global projects will ever be managed informally.

Suzanne Zale, Hewlett–Packard operations director, emphasized:

> With any global project, communications become more complex. It will require much more planning up front. All constituents for buy-in need to be identified early on. In order to leverage existing subject matter, experts conversant with local culture, and suppliers, the need for virtual teams becomes more obvious. This increases the difficulty for effective communications.
>
> The mechanism for communication may also change drastically. Face-to-face conversations or meetings will become more difficult. We tend to rely heavily on electronic communications, such as video and telephone conferencing and electronic mail. The format for communications needs to be standardized and understood up front so that information can be sent out quickly. Communications will also take longer and require more effort because of cultural and time differences.

One of the implied assumptions for informal project management to exist is that employees understand their organizational structure and their roles and responsibilities within both the organizational and the project structures. Forms such as the linear responsibility chart and the responsibility assignment matrix are helpful. Communication tools are not used today with the same frequency as in the 1970s and 1980s.

For multinational projects, the organizational structure, roles, and responsibilities must be clearly delineated. Effective communication is of paramount importance and probably must be accomplished more formally than informally.

As Suzanne Zale stated:

> For any global project, the organizational structure must be clearly defined to minimize any potential misunderstandings. It is best to have a clear-cut definition of the organizational chart and roles and responsibilities. Any motivation incentives must also contemplate cultural differences. The drivers and values for different cultures can vary substantially.

The two major communication obstacles that must be overcome when a company truly wants to cultivate an informal culture are what I like to call hernia reports and forensic meetings. Hernia reports result from senior management's belief that what has not been written has not been said. Although there is some truth to such a belief, the written word comes with a high price tag. We need to consider more than just the time consumed in the preparation of reports and formal memos. There is all the time that recipients spend reading them, as well as all the support time taken up in processing, copying, distributing, and filing them.

Status reports written for management are too long if they need a staple or a paper clip. Project reports longer than five or 10 pages are often not even read. In companies that excel at project management, internal project reports answer these three questions as simply as possible:

1. Where are we today?
2. Where will we end up?
3. Are there any problems that require management's involvement?

All of these questions can usually be answered on one sheet of paper.

The second obstacle is the forensic team meeting. A forensic team meeting is a meeting scheduled to last 30 minutes that actually lasts for more than three hours. Forensic meetings are created when senior managers meddle in routine work activities. Even project managers fall into this trap when they present information to management that management should not be dealing with. Such situations are an invitation to disaster.

In the past, communication practices were directly related to the processes, tools, and techniques used in the project management methodology. Companies that utilized a one-size-fits-all methodology created standards for customer communications and often maintained a one-size-fits-all customer communications plan. Everything was based on information taken from the earned value measurement system.

The growth and acceptance of flexible methodologies changed how we view communications. Flexible methodologies such as Agile and Scrum have identified the importance of effective collaboration between everyone, including stakeholders. In addition, we now focus on multiple competing constraints rather than just time, cost, and scope. We are also allowing stakeholders some degree of freedom at the onset of the project to select the project metrics they need reported for the decisions they are expected to make. These changes can create unique project status reports and increase the frequency of communications.

9.4 COOPERATION

Cooperation is the willingness of individuals to work with others for the benefit of all. It includes the voluntary actions of a team working together toward a favorable result. In companies that excel at project management, cooperation is the norm and takes place without the formal intervention of those in authority. The team members know the right thing to do, and they do it.

In the average company (or the average group of any kind, for that matter), people learn to cooperate as they get to know each other. That takes time, something that is usually in short supply for project teams. But companies such as Ericsson Telecom AB, the General Motors Powertrain Group, and Hewlett–Packard create cultures that promote cooperation to the benefit of everyone.

9.5 TEAMWORK

Teamwork is the work performed by people acting together with a spirit of cooperation within the limits of coordination. Some people confuse teamwork with morale, but morale has more to do with attitudes toward work than it does with the work itself. Obviously, however, good morale is beneficial to teamwork.

In excellent companies, teamwork has these characteristics:

- Employees and managers share ideas with each other and establish high levels of innovation and creativity in work groups.
- Employees and managers trust each other and are loyal to each other and the company.
- Employees and managers are committed to the work they do and the promises they make.
- Employees and managers share information freely.
- Employees and managers are consistently open and honest with each other.

Making people feel that they are part of a team does not necessarily require a great deal of effort. Consider the situation at the Engineering and Construction Services Division of Dow Chemical Corporation several years ago. Dow Chemical had requested a trainer to develop a project management training course. The trainer interviewed several of the seminar participants before the training program to identify potential problem areas. The biggest problem appeared to be a lack of teamwork. This shortcoming was particularly evident in the drafting department. The drafting department personnel complained that too many changes were being made to the drawings. They simply could not understand the reasons behind all the changes.

The second problem identified, and perhaps the more critical one, was that project managers did not communicate with the drafting department once the drawings were complete. The drafting people had no idea of the status of the projects they were working on, and they did not feel as though they were part of the project team.

During the training program, one of the project managers, who was responsible for constructing a large chemical plant, was asked to explain why so many changes were being made to the drawings on his project. He said, "There are three reasons for the changes. First, the customers do not always know what they want up front. Second, once we have the preliminary drawings to work with, we build a plastic model of the plant. The model often shows us that equipment needs to be moved for maintenance or safety reasons. Third, sometimes we have to rush into construction well before we have final approval from the Environmental Protection Agency. When the agency finally gives its approval, that approval is often made contingent on making major structural changes to the work already complete." One veteran employee at Dow commented that in his 15 years with the company, no one had ever before explained the reasons behind drafting changes.

The solution to the problem of insufficient communication was also easy to repair once it was out in the open. Project managers promised to take monthly snapshots of the progress on building projects and share them with the drafting department. The drafting personnel were delighted and felt more like a part of the project team.

9.6 COLOR-CODED STATUS REPORTING

The use of colors for status reporting, whether it is for printed reports or intranet-based visual presentations, has grown significantly. Color-coded reports encourage informal project management to take place. Colors can reduce risks by alerting management quickly that a potential problem exists. One company prepared complex status reports, but color-coded the right-hand margins of each page designed for specific audiences and levels of management. One executive commented that he now reads only those pages that are color-coded for him specifically rather than having to search through the entire report. In another company, senior management discovered that color-coded intranet status reporting allowed senior management to review more information in a timely manner just by focusing on those colors that indicated potential problems. Colors can be used to indicate:

- Status has not been addressed.
- Status is addressed, but no problems exist.
- Project is on course.
- A potential problem might exist in the future.
- A problem definitely exists and is critical.
- No action is to be taken on this problem.
- Activity has been completed.
- Activity is still active, and completion date has passed.

9.7 CRISIS DASHBOARDS

Over the past several years, dashboards have become common for presenting project status information to the project team, clients, and stakeholders. The purpose of a dashboard is to convert raw data into meaningful information that can be easily understood and used for informed decision-making. The dashboard provides the viewer with situational awareness of what the information means now and what it might mean in the future if the existing trends continue. Dashboards function as communication tools that allow us to go to paperless project management, hold fewer meetings, and eliminate waste.

Projects in today's environment are significantly more complex than many of the projects managed in the past. With today's projects, governance is performed by a governance committee rather than just a single project sponsor. Each stakeholder or member of the governance committee may very well require different metrics and key performance indicators (KPIs). If each stakeholder wishes to view 20 to 30 metrics, the costs of metric measurement and reporting can be significant and can defeat the purpose of going paperless in project management.

The solution to effective communications with stakeholders and governance groups is to show them that they can most likely get all of the critical data they need for informed decision-making with 6 to 10 metrics, or KPIs, that can be displayed on one computer screen. This is not always the case, and drilling-down to other screens may be necessary. But, in general, one computer screenshot should be sufficient.

If an out-of-tolerance condition or crisis situation exists with any of the metrics or KPIs on the dashboard screen, then the situation should be readily apparent to the viewer. But what if the crisis occurs due to metrics that do not appear on the screen? In this case, the viewer will be immediately directed to a crisis dashboard that shows all of the metrics that are out of tolerance. The out-of-tolerance metrics will remain on the crisis dashboard until such time as the crisis or out-of-tolerance conditions are corrected. Each stakeholder will now see the regular screenshot and then be instructed to look at the crisis screenshot.

Defining a Crisis

A crisis can be defined as any event, whether expected or not, that can lead to an unstable or dangerous situation affecting the outcome of the project. Crises imply negative consequences that can harm the organization, its stakeholders, and the general public. The crisis can result in changes to the firm's business strategy, how it interfaces with the enterprise environmental factors, how its social consciousness is exhibited, and the way it maintains customer satisfaction. A crisis does not necessarily mean that the project will fail; nor does it mean that the project should be terminated. The crisis could simply be that the project's outcome will not be as expected.

Some crises may appear gradually and can be preceded by early warning signs. These crises can be referred to as smoldering crises. The intent of metrics and dashboards is to identify trends that could indicate that a crisis may be approaching and provide the project manager with sufficient time to develop contingency plans. The earlier you know about the impending crisis, the more options you may have as a remedy.

How do we determine whether the out-of-tolerance condition is just a problem or a crisis? The answer is in the potential damage that can occur. If any of the following can occur, then the situation will most likely be treated as a crisis:

- There is a significant threat to the outcome of the project.
- There is a significant threat to the organization as a whole, its stakeholders, and possibly the general public.
- There is a significant threat to the firm's business model and strategy.
- There is a significant threat to worker health and safety.
- There is a significant threat to consumers, such as with product tampering.
- There is a possibility for loss of life.
- There is the possibility of work delays because systems are being redesigned.
- There is the possibility of work delays due to necessary organizational changes.
- There is a significant chance that the firm's image or reputation will be damaged.
- There is a significant chance that the deterioration in customer satisfaction could result in current and future losses of significant revenue.

It is important to understand the difference between risk management and crisis management. According to Wikipedia:

> In contrast to risk management, which involves assessing potential threats and finding the best ways to avoid those threats, crisis management involves dealing with threats before, during, and after they have occurred. It is a discipline within the broader context of management consisting of skills and techniques required to identify, assess, understand, and cope with a serious situation, especially from the moment it first occurs to the point that recovery procedures start.[1]

Crises often require that immediate decisions be made. Effective decision-making requires information. If one metric appears to be in a crisis mode and shows up on the crisis dashboard, viewers may find it necessary to look at several other metrics that may not be in a crisis mode and may not appear on the crisis dashboard but are possible causes of the crisis. Looking at metrics on dashboards is a lot easier than reading reports.

The difference between a problem and a crisis is like beauty; it is in the eye of the beholder. What one stakeholder sees as a problem, another stakeholder may see it as a crisis. Table 9–3 shows how difficult it is to make the differentiation.

We can now draw the following conclusions about crisis dashboards:

- It is not always clear to the viewers what does or does not constitute a "crisis."
- Not all problems are crises.

---

1. Wikipedia contributors, "Crisis Management," *Wikipedia, The Free Encyclopedia*, https://en.wikipedia.org/wiki/Crisis_management

TABLE 9–3. DIFFERENTIATING BETWEEN A PROBLEM AND A CRISIS

Metric/KPI	Problem	Crisis
Time	The project will be late but still acceptable to the client.	The project will be late, and the client is considering cancellation.
Cost	Costs are being overrun, but the client can provide additional funding.	Costs are being overrun, and no additional funding is available. Cancellation is highly probable.
Quality	The customer is unhappy with the quality but can live with it.	If the quality of the deliverables is unacceptable, personal injury is possible, the client may cancel the contract, and no further work may come from this client.
Resources	The project is either understaffed or the resources assigned have marginal skills to do the job. A schedule delay is probably.	The quality of or lack of resources will cause a serious delay in the schedule, and the quality of workmanship may be so unacceptable that the project may be canceled.
Scope	There are numerous scope changes, which cause changes to the baselines. Delays and cost overruns are happening but are acceptable to the client for now.	The number of scope changes has led the client to believe that the planning is not correct and more scope changes will occur. The benefits of the project no longer outweigh the costs, and project termination is likely.
Action items	The client is unhappy with the amount of time taken to close out action items, but the impact on the project is small.	The client is unhappy with the amount of time taken to close out action items, and the impact on the project is significant. Governance decisions are being delayed because of the open action items, and the impact on the project may be severe.
Risks	Significant risk levels exist, but the team may be able to mitigate some of the risks.	The potential damage that can occur because of the severity of the risks is unacceptable to the client.
Assumptions and constraints	New assumptions and constraints have appeared and may adversely affect the project.	New assumptions and constraints have appeared such that significant project replanning will be necessary. The value of the project may no longer be there.
Enterprise environmental factors	The enterprise environmental factors have changed and may adversely affect the project.	The new enterprise environmental factors will greatly reduce the value and expected benefits of the project.

- Sometimes unfavorable trends are treated as a crisis and appear on crisis dashboards.
- The crisis dashboard may contain a mixture of crisis metrics and metrics that are treated as just problems.
- The metrics that appear on a traditional dashboard reporting system may have to be redrawn when placed on a crisis dashboard to make sure that they are easily understood.
- Crisis metrics generally imply that the situation must be monitored closely or that some decisions must be made.

9.8 THE RISKS OF USING INFORMAL PROJECT MANAGEMENT

Informal project management can save the company a great deal of money and time. However, there are always risks that must be considered. In one company, the customer and contractor agreed contractually to a cost-reimbursable contract that included monthly status reporting accompanied by detailed documentation of all issues, action items, and what the plans would be for solutions to the issues. The project was scheduled to be completed in two years.

During the first few months of the contract, both the customer and the contractor recognized the high cost of documentation. The decision was made verbally between the project managers in the customer's and contractor's organizations to manage informally rather than formally and to bypass much of the paperwork requirements.

Serious issues began to occur on the project. The customer's legal team became furious that the contractual document requirements were not available and believed that, had they been available, many of the serious issues would not have occurred. The customer's project manager was reprimanded, and the customer filed legal action against the contractor. The moral is that informal project management practices should not be a substitute for contractual requirements. Some projects are heavily data-driven rather than deliverables-driven.

It is generally assumed that communication between customers and contractors must be formal because of the potential for distrust when contracts are complex and involve millions of dollars. The use of on-site representatives, however, can change a potentially contentious relationship into one of trust and cooperation through informality.

As an example, a customer was apprehensive about relying solely on written reports to determine the status of a project. Two employees from the customer's organization were carefully chosen to be on-site representatives at the contractor's company to supervise the development of a highly technical new system. The working relationship between contractor's project management office and the customer's on-site representatives quickly developed into shared trust. Team meetings were held without the exchange of excessive documentation. And each party agreed to cooperate with the other. The contractor's project manager trusted the customer's representatives well enough to give them raw data from test results even before the contractor's engineers could formulate their own opinions on the data. The customer's representatives in turn promised that they would not relay the raw data to their company until the contractor's engineers were ready to share their results with their own executive sponsors.

The relationship on this project clearly indicates that informal project management can work between customers and contractors. Large construction contractors have had the same positive results in using informal project management and on-site representatives to rebuild trust and cooperation. Informality is not a replacement for formal project management activities. Rather, it simply means that some activities can be done more informally than formally. Formal and informal communications can exist simultaneously.

Informal project management practices generally work better on internal projects that are not impacted by customers, contractors, or contractual requirements. However, problems can still occur if the people wanting to manage informally have hidden agendas and try to hide the real status of a project or want to change the direction of the project without letting company leadership know.

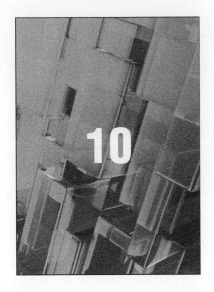

Behavioral Excellence

10.0 INTRODUCTION

Previously, we saw that companies that excel at project management strongly emphasize training for behavioral skills. In the past, it was thought that project failures were due primarily to poor planning, inaccurate estimating, inefficient scheduling, and lack of cost control. Today, excellent companies realize that project failures have more to do with behavioral shortcomings—poor employee morale, negative human relations, low productivity, and lack of commitment.

This chapter discusses these human factors in the context of situational leadership and conflict resolution. It also provides information on staffing issues in project management. Finally, the chapter offers advice on how to achieve behavioral excellence.

10.1 SITUATIONAL LEADERSHIP

As project management has begun to emphasize behavioral management over technical management, situational leadership, and other leadership styles as social leadership discussed in Chapter 1 has also received more attention. The average size of projects has grown, and so has the size of project teams. Process integration and effective interpersonal relations have also taken on more importance as project teams have gotten larger. Project managers now need to be able to talk with many different functions and departments. There is a contemporary project management proverb that goes something like this: "When researcher talks to researcher, there is 100 percent understanding. When researcher talks to manufacturing, there is 50 percent understanding. When researcher talks to sales, there is zero percent understanding. But the project manager talks to all of them."

Randy Coleman, former senior vice president of the Federal Reserve Bank of Cleveland, emphasizes the importance of tolerance:

> The single most important characteristic necessary in successful project management is tolerance: tolerance of external events and tolerance of people's personalities. Generally, there are two groups here at the Fed—lifers and drifters. You have to handle the two groups differently, but at the same time you have to treat them similarly. You have to bend somewhat for the independents (younger drifters) who have good creative ideas and whom you want to keep, particularly those who take risks. You have to acknowledge that you have some trade-offs to deal with.

A senior project manager in an international accounting firm states how his own leadership style has changed from a traditional to a situational leadership style since becoming a project manager:

> I used to think that there was a certain approach that was best for leadership, but experience has taught me that leadership and personality go together. What works for one person won't work for others. So, you must understand enough about the structure of projects and people and then adopt a leadership style that suits your personality so that it comes across as being natural and genuine. It's a blending of a person's experience and personality with his or her style of leadership.

Many companies start applying project management without understanding the fundamental behavioral differences between project management and line management. If we assume that the line manager is not also functioning as the project manager, here are the behavioral differences:

- Project managers have to deal with multiple reporting relationships. Line managers report up a single chain of command.
- Project managers have very little real authority. Line managers hold a great deal of authority by virtue of their titles.
- Project managers often provide no input into employee performance reviews. Line managers provide formal input into the performance reviews of their direct reports.
- Project managers are not always on the management compensation or career path ladder. Line managers always are.
- The project manager's position may be temporary. The line manager's position is usually permanent.
- Project managers sometimes are at a lower grade level than the project team members. Line managers usually are paid at a higher grade level than their subordinates.

Several years ago, a trainer was hired to conduct a three-day course on project management. During the customization process, the trainer was asked to emphasize planning, scheduling, and controlling and not to bother with the behavioral aspects of project management. At that time, the company offered a course on how to become a line supervisor that all the seminar participants had already taken. In the discussion that followed between the trainer and the course content designers, it became apparent that leadership, motivation, and conflict resolution were being taught from a superior-to-subordinate point of view in company's course. When the course content designers realized from the discussion that project managers provide leadership, motivation, and conflict resolution to employees who do not report directly to them, the trainer was allowed to include project management–related behavioral topics in the seminar.

Organizations must recognize the importance of behavioral factors in working relationships. When they do, they come to understand that project managers should be hired for their overall project management competency, not for their technical knowledge alone. Brian Vannoni, formerly site training manager and principal process engineer at GE Plastics, described his organization's approach to selecting project managers:

> The selection process for getting people involved as project managers is based primarily on their behavioral skills and their skills and abilities as leaders with regard to the other aspects of project management. Some of the professional and full-time project managers have taken senior engineers under their wing, coached and mentored them so that they learn and pick up the other aspects of project management. But the primary skills that we are looking for are, in fact, the leadership skills.

Project managers who have strong behavioral skills are more likely to involve their teams in decision-making, and shared decision-making is one of the hallmarks of successful project management. Today, project managers are more managers of people than they are managers of technology. According to Robert Hershock, former vice president at 3M:

> The trust, respect, and especially the communications are very, very important. But I think one thing that we have to keep in mind is that a team leader isn't managing technology; he or she is managing people. If you manage the people correctly, the people will manage the technology.

In addition, behaviorally oriented project managers are more likely to delegate responsibility to team members than technically strong project managers. In 1996, Frank Jackson, formerly a senior manager at MCI, said:

> Team leaders need to have a focus and a commitment to an ultimate objective. You definitely have to have accountability for your team and the outcome of your team. You've got to be able to share the decision making. You can't single out yourself as the exclusive holder of the right to make decisions. You have got to be able to share that. And lastly again, just to harp on it one more time, is communications. Clear and concise communication throughout the team and both up and down a chain of command is very, very important.

Some organizations prefer to have someone with strong behavioral skills act as the project manager, with technical expertise residing with the project engineer.

Other organizations have found the reverse to be effective. Rose Russett, formerly the program management process manager for General Motors Powertrain, stated:

> We usually appoint an individual with a technical background as the program manager and an individual with a business and/or systems background as the program administrator. This combination of skills seems to complement one another. The various line managers are ultimately responsible for the technical portions of the program, while the key responsibility of the program manager is to provide the integration of all functional deliverables to achieve the objectives of the program. With that in mind, it helps for the program manager to understand the technical issues, but they add their value not by solving specific technical problems but by leading the team through a process that will result in the best solutions for the overall program, not just for the specific functional area. The program administrator, with input from all team members, develops the program plans, identifies the critical path, and regularly communicates this information to the team throughout the life of the program. This information is used to assist with problem solving, decision making, and risk management.

Perhaps the biggest challenge over the next several years will be the need for a better understanding of effective project leadership. Project leadership may be horizontal, vertical, or team-centered and can vary from project to project and from life cycle phase to life cycle phase. Simply stated, project managers will need an "adaptable" leadership style based upon the needs of the project, skill level and expectations of the people assigned to the project team, and collaboration requirements with stakeholders. This is a necessity for situational leadership to work effectively.

Using an adaptable leadership style will force the project manager to focus on influence rather than formal authority. Project managers may also need to challenge the "status quo" on governance, culture, and channels of communication.

10.2 CULTURAL INTELLIGENCE

Advances in technology and the need for globalization have radically changed how companies work and formulate strategic plans. For a sustainable global growth to exist, companies must execute projects across boundaries that have cultures that are different than in the parent company's environment. The challenge facing most companies then becomes, "How do you manage projects effectively in a culturally diverse environment?" Many of these projects are taking place in the VUCA environment, with ill-defined requirements and high levels of risk and uncertainty. Many of the risks are related to the team's lack of knowledge about local cultures, religions, politics, and laws and how they can impact project performance expectations. The result can be extremely high levels of anxiety and stress that impact performance.

Cultural intelligence refers to the skills needed to relate and work effectively in culturally diverse environments. Most project teams have never been trained in these skills, but this is slowly changing. The most pressing issue is selecting the right project manager for such projects. Simply because an individual speaks the local language, it is not a guarantee he/she will be effective as a project manager. Also, superior technical knowledge alone may not be able to make a project interculturally successful.

Given that interacting competently across borders is a necessity, evaluating whether someone possesses these skills is still challenging for companies, even though companies are developing training programs on cultural intelligence. Exceptionally good candidates for multinational project management leadership positions often pass under the radar screen of HR personnel. Technical competency is important, but not the only skill needed. To maximize performance and minimize friction, other skills are needed related to how we must think, behave, and interface, such as:

- A recognition and understanding of the core cultural differences in values, beliefs, attitudes, and behaviors and the effect they can have on the way project management should work.
- Understand and acknowledge local values and behaviors that may or may not be acceptable on the project, as well as any cultural rules and non-verbal behaviors that must be followed for effective collaboration.
- Understand social interaction norms including gestures, facial expressions, tone of voice, and body language.
- Understand the actions that each team member can take to act and react appropriately to create a sense of belonging and produce effective results.
- Demonstrate cultural empathy, a high level of flexibility, and a tolerance for ambiguity.
- Understand how others see us and the differences.
- Recognize the possible existence of additional or new assumptions and constraints.

Effective project management leadership must understand and support all the components of cultural intelligence. Project managers must help the team see the benefits of a culturally diverse experience and view it as an enjoyable learning experience. However, the PM must also prepare the team for experiences that may be different than expectations.

10.3 EMOTIONAL INTELLIGENCE

Emotional intelligence and cultural intelligence have slowly become the main pillars for effective project leadership. Emotional intelligence is the ability to understand, manage, and express one's emotions in a healthy way. It also allows us to have a better understanding of the emotions displayed by others. We must be aware of our own emotions and learn to manage them before we can effectively manage our interaction and collaboration with others. Emotional intelligence becomes a critical skill for project managers on all types of projects, including those crossing cultural boundaries. Some of the characteristics of emotional intelligence include:

- Being aware of one's feelings and emotions
- Knowing how to interpret one's emotions

- Knowing how to regulate one's emotions
- Understanding how one's emotions may be interpreted by others
- Knowing how to use your emotions to help manage the emotions of others

Project managers with high emotional intelligence skills can recognize their own emotions and use this information to help guide the thinking and behavior of others. Effective use of emotional intelligence is often accompanied by a show of empathy and can give team members a sense of belonging and trust in the leadership of the project. Effective emotional intelligence can lead to better decision-making, improved collaboration with everyone, including stakeholders, and positive relationships with co-workers. This holds true even in situations of extreme team stress. Emotional intelligence can also lead to better working relationships with senior management and project governance personnel.

Emotional intelligence is just one important skill that PMs need in the future for motivating workers and creating a sense of belonging. However, project management performance is still being measured by the results of the project and the orders and decisions the PM had to make.

10.4 CONFLICT RESOLUTION

Opponents of project management claim that the primary reason why some companies avoid changing over to a project management culture is that they fear the conflicts that inevitably accompany change. Conflicts are a way of life in companies with project management cultures. Conflict can occur on any level of the organization, and it is usually the result of conflicting objectives. The project manager is a conflict manager. In many organizations, project managers continually fight fires and handle crises arising from interpersonal and interdepartmental conflicts. They are so busy handling conflicts that they delegate the day-to-day responsibility for running their projects to the project teams. Although this arrangement is not the most effective, it is sometimes necessary, especially after organizational restructuring or after a new project demanding new resources has been initiated.

The ability to handle conflicts requires an understanding of why conflicts occur. We can ask four questions, the answers to which are usually helpful in handling, and possibly preventing, conflicts in a project management environment:

1. Do the project's objectives conflict with the objectives of other projects currently in development?
2. Why do conflicts occur?
3. How can we resolve conflicts?
4. Is there anything we can do to anticipate and resolve conflicts before they become serious?

Although conflicts are inevitable, they can be planned for. For example, conflicts can easily develop in a team in which the members do not understand each other's roles and responsibilities. Responsibility charts can be drawn to map out graphically who is responsible for doing what on the project. With the ambiguity of roles and responsibilities gone, the conflict is resolved or future conflict is averted.

Resolution means collaboration, and collaboration means that people are willing to rely on each other. Without collaboration, mistrust prevails and progress documentation increases.

The most common types of conflict involve the following:

- Manpower resources
- Equipment and facilities
- Capital expenditures
- Costs
- Technical opinions and trade-offs
- Priorities
- Administrative procedures
- Schedules
- Responsibilities
- Personality clashes

Each of these types of conflict can vary in intensity over the life of the project. The relative intensity can vary as a function of:

- Getting closer to project constraints
- Having met only two constraints instead of three (e.g., time and performance but not cost)
- The project life cycle itself
- The individuals who are in conflict

Conflict can be meaningful if it results in beneficial outcomes. These meaningful conflicts should be allowed to continue as long as project constraints are not violated, and beneficial results accrue. An example of a meaningful conflict might be two technical specialists arguing that each has a better way of solving a problem. The beneficial result would be that each tries to find additional information to support his or her hypothesis.

Some conflicts are inevitable and occur over and over again. For example, consider a raw material and finished goods inventory. Manufacturing wants the largest possible inventory of raw materials on hand to avoid possible production shutdowns. Sales and marketing want the largest finished goods inventory so that the books look favorable and no cash flow problems are possible.

Consider five methods that project managers can use to resolve conflicts:

1. Confrontation
2. Compromise
3. Facilitation (or smoothing)

4. Force (or forcing)
5. Withdrawal

Confrontation is probably the most common method used by project managers to resolve conflict. Using confrontation, the project manager faces the conflict directly. With the help of the project manager, the parties in disagreement attempt to persuade one another that their solution to the problem is the most appropriate.

When confrontation does not work, the next approach that project managers usually try is compromise. In compromise, each of the parties in conflict agrees to trade-offs or makes concessions until a solution is arrived at that everyone involved can live with. This give-and-take approach can easily lead to a win-win solution to the conflict.

The third approach to conflict resolution is facilitation. Using facilitation skills, the project manager emphasizes areas of agreement and deemphasizes areas of disagreement. For example, suppose that a project manager said, "We've been arguing about five points, and so far, we've reached agreement on the first three. There's no reason why we can't agree on the last two points, is there?" Facilitation of a disagreement does not resolve the conflict. Facilitation downplays the emotional context in which conflicts occur.

Force is also a method of conflict resolution. A project manager uses force when he or she tries to resolve a disagreement by exerting his or her own opinion at the expense of the other people involved. Often, forcing a solution onto the parties in conflict results in a win-lose outcome. Calling in the project sponsor to resolve a conflict is another form of force that project managers sometimes use.

The least used and least effective mode of conflict resolution is withdrawal. A project director can simply withdraw from the conflict and leave the situation unresolved. When this method is used, the conflict does not go away and is likely to recur later. Personality conflicts might well be the most difficult conflicts to resolve. Personality conflicts can occur at any time, with anyone, and over anything. Furthermore, they can seem almost impossible to anticipate and plan for.

Let's look at how one company found a way to anticipate and avoid personality conflicts on one of its projects. Foster Defense Group (a pseudonym) was the government contracting branch of a Fortune 500 company. The company understood the potentially detrimental effects of personality clashes on its project teams, but it did not like the idea of getting the whole team together to air dirty laundry. The company found a better solution. The project manager put the names of the project team members on a list. Then he interviewed each team member one-on-one and asked each to identify who on the list the team member had had a personality conflict with in the past. The information remained confidential, and the project manager was able to avoid potential conflicts by separating clashing personalities.

If at all possible, the project manager should handle conflict resolution. When the project manager is unable to defuse the conflict, then and only then should the project sponsor be brought in to help solve the problem. Even then, the sponsor should not come in and force a resolution to the conflict. Instead, the sponsor should facilitate further discussion between the project managers and the team members in conflict.

10.5 STAFFING FOR EXCELLENCE

Project manager selection is always an executive-level decision. In excellent companies, however, executives go beyond simply selecting the project manager:

- Project managers are brought on board early in the life of the project to assist in outlining the project, setting its objectives, and even planning for marketing and sales. The project manager's role in customer relations becomes increasingly important.
- Executives assign project managers for the life of the project and through its termination. Sponsorship can change over the life cycle of the project, but the project manager usually does not change.
- Project management is given its own career ladder.
- Project managers given a role in customer relations are also expected to help sell future project management services long before the current project is complete.
- Executives realize that project scope changes are inevitable. The project manager is viewed as a manager of change.

Companies that are excellent in project management are prepared for crises. Both project managers and line managers are encouraged to bring problems to the surface as quickly as possible so that there is time for contingency planning and problem-solving. Replacing the project manager is no longer the first solution for problems on a project. Project managers are replaced only when they try to bury problems.

A defense contractor was behind schedule on a project, and the manufacturing team was asked to work extensive overtime to catch up. Two of the manufacturing people, both union employees, used the wrong lot of raw materials to produce a $65,000 piece of equipment needed for the project. The customer was unhappy because of the missed schedules and cost overruns that resulted from having to replace the useless equipment. An inquisition-like meeting was convened and attended by senior executives from both the customer and the contractor, the project manager, and the two manufacturing employees. When the customer's representative asked for an explanation of what had happened, the project manager stood up and said, "I take full responsibility for what happened. Expecting people to work extensive overtime leads to mistakes. I should have been more careful." The meeting was adjourned without anyone being blamed. When word spread through the company about what the project manager did to protect the two union employees, everyone pitched in to get the project back on schedule, even working uncompensated overtime.

Human behavior is also a consideration in assigning staff to project teams. Team members should not be assigned to a project solely based on their technical knowledge. It must be recognized that some people simply cannot work effectively in a team environment. For example, the director of research and development at a New England company had an employee, a 50-year-old engineer, who held two master's degrees in engineering disciplines. He had worked for the previous 20 years on mostly one-person projects. The director reluctantly assigned the engineer to a project team. After years of working alone, the engineer trusted no one's results but his own. He refused to work cooperatively with the other members of the team. He even went so far as to redo all the calculations passed on to him by other engineers on the team.

To solve the problem, the director assigned the engineer to another project on which he supervised two other engineers with less experience. Again, the older engineer tried to do all of the work by himself, even if it meant overtime for him and no work for the others.

Ultimately, the director had to admit that some people are not able to work cooperatively on team projects. The director went back to assigning the engineer to one-person projects on which the engineer's technical abilities would be useful.

Robert Hershock once observed:

> There are certain people whom you just don't want to put on teams. They are not team players, and they will be disruptive on teams. I think that we have to recognize that and make sure that those people are not part of a team or team members. If you need their expertise, you can bring them in as consultants to the team, but you never, never put people like that on the team.
>
> I think the other thing is that I would never, ever eliminate the possibility of anybody being a team member no matter what the management level is. I think if they are properly trained, these people at any level can be participators in a team concept.

In 1996, Frank Jackson believed that it was possible to find a team where any individual can contribute:

> People should not be singled out as not being team players. Everyone has got the ability to be on a team and to contribute to a team based on the skills and the personal experiences that they have had. If you move into the team environment, one other thing that is very important is that you do not hinder communications. Communications is the key to the success of any team and any objective that a team tries to achieve.

One of the critical arguments still being waged in the project management community is whether an employee (even a project manager) should have the right to refuse an assignment. At a public utility company, an open project manager position was posted, but nobody applied for the job. The company recognized that the employees probably did not understand what the position's responsibilities were. After more than 80 people were trained in the fundamentals of project management, there were numerous applications for the open position.

It's the kiss of death to the project to assign someone to a project manager's job if that person is not dedicated to the project management process and the accountability it demands.

10.6 VIRTUAL PROJECT TEAMS

Historically, project management was a face-to-face environment where team meetings involved all players meeting together in one room. Today, because of the size and complexity of projects, it is often impossible to find all team members located under one roof. Duarte and Snyder define seven types of virtual teams. These are shown in Table 10–1.

TABLE 10–1. TYPES OF VIRTUAL TEAMS

Type of Team	Description
Network	Team membership is diffuse and fluid; members come and go as needed. Team lacks clear boundaries within the organization.
Parallel	Team has clear boundaries and distinct membership. Team works in the short term to develop recommendations for an improvement in a process or system.
Project or product development	Team has fluid membership, clear boundaries, and a defined customer base, technical requirements, and output. Longer-term team task is nonroutine, and the team has decision-making authority.
Work or production	Team has distinct membership and clear boundaries. Members perform regular and outgoing work, usually in one functional area.
Service	Team has distinct membership and supports ongoing customer network activity.
Management	Team has distinct membership and works on a regular basis to lead corporate activities.
Action	Team deals with immediate action, usually in an emergency situation. Membership may be fluid or distinct.

Source: D. L. Duarte and N. Tennant Snyder, *Mastering Virtual Teams* (San Francisco: Jossey-Bass, 2001), p. 10. Reproduced by permission of John Wiley & Sons.

TABLE 10–2. TECHNOLOGY AND CULTURE

Cultural Factor	Technological Considerations
Power distance	Members from high-power-distance cultures may participate more freely with technologies that are asynchronous and allow anonymous input. These cultures sometimes use technology to indicate status differences between team members.
Uncertainty avoidance	People from cultures with high uncertainty avoidance may be slower adopters of technology. They may also prefer technology that is able to produce more permanent records of discussions and decisions.
Individualism–collectivism	Members from highly collectivistic cultures may prefer face-to-face interactions.
Masculinity–femininity	People from cultures with more "feminine" orientations are more prone to use technology in a nurturing way, especially during team startups.
Context	People from high-context cultures may prefer more information-rich technologies, as well as those that offer opportunities for the feeling of social presence. They may resist using technologies with low social presence to communicate with people they have never met. People from low-context cultures may prefer more asynchronous communications.

Source: D. L. Duarte and N. Tennant Snyder, *Mastering Virtual Teams* (San Francisco: Jossey-Bass, 2001), p. 60.

Culture and technology can have a major impact on the performance of virtual teams. Duarte and Snyder have identified some of these relationships in Table 10–2.

The importance of culture cannot be understated. Duarte and Snyder identify four important points to remember concerning the impact of culture on virtual teams. The four points are:

1. There are national cultures, organizational cultures, functional cultures, and team cultures. They can be sources of competitive advantages for virtual teams that know how to use cultural differences to create synergy. Team leaders and members who understand and are sensitive to cultural differences can create more robust outcomes than can members of homogeneous teams with members who think and act alike. Cultural differences can create distinctive advantages for teams if they are understood and used in positive ways.

2. The most important aspect of understanding and working with cultural differences is to create a team culture in which problems can be surfaced and differences can be discussed in a productive, respectful manner.

3. It is essential to distinguish between problems that result from cultural differences and problems that are performance based.

4. Business practices and business ethics vary in different parts of the world. Virtual teams need to clearly articulate approaches to these that every member understands and abides by.[1]

10.7 REWARDING PROJECT TEAMS

Today, most companies are using project teams. However, there still exist challenges in determining how to reward project teams for successful performance. Parker, McAdams, and Zielinski discuss the importance of how teams are rewarded:

> Some organizations are fond of saying, "We're all part of the team," but too often it is merely management-speak. This is especially common in conventional hierarchical organizations; they say the words but don't follow up with significant action. Their employees may read the articles and attend the conferences and come to believe that many companies have turned collaborative. Actually, though, few organizations today are genuinely team-based.
>
> Others who want to quibble point to how they reward or recognize teams with splashy bonuses or profit-sharing plans. But these do not by themselves represent a commitment to teams; they're more like a gift from a rich uncle. If top management believes that only money and a few recognition programs ("team of year" and that sort of thing) reinforce teamwork, they are wrong. These alone do not cause fundamental change in the way people and teams are managed.
>
> But in a few organizations, teaming is a key component of the corporate strategy, involvement with teams is second nature, and collaboration happens without great thought or fanfare. There are natural work groups (teams of people who do the same or similar work in the same location), permanent cross-functional teams, ad hoc project teams, process improvement teams, and real management teams. Involvement just happens.[2]

Why is it so difficult to reward project teams? To answer this question, we must understand what a team is and is not:

> Consider this statement: an organizational unit can act like a team, but a team is not necessarily an organizational unit, at least for describing reward plans. An organizational unit is just that, a group of employees organized into an identifiable business unit that appears on the organizational chart. They may behave in a spirit of teamwork, but for the

---

1. *Source*: D. L. Duarte and N. Tennant Snyder, *Mastering Virtual Teams* (San Francisco: Jossey-Bass, 2001), p. 10. Reproduced by permission of John Wiley & Sons.

2. G. Parker, J. McAdams, and D. Zielinski, *Rewarding Teams* (San Francisco: Jossey-Bass, 2000, p. 17). Reproduced by permission of John Wiley & Sons.

purposes of developing reward plans they are not a "team." The organizational unit may be a whole company, a strategic business unit, a division, a department, or a work group.

A "team" is a small group of people allied by a common project and sharing performance objectives. They generally have complementary skills or knowledge and an interdependence that requires that they work together to accomplish their project's objective. Team members hold themselves mutually accountable for their results. These teams are not found on an organization chart.

Incentives are difficult to apply because project teams may not appear on an organizational chart. Figure 10–1 shows the reinforcement model for employees. For project teams, the emphasis is on the three arrows on the right-hand side of Figure 10–1.

Project team incentives are important because team members expect appropriate rewards and recognition:

Project teams are usually, but not always, formed by management to tackle specific projects or challenges with a defined time frame—reviewing processes for efficiency or cost-savings recommendations, launching a new software product, or implementing enterprise resource planning systems are just a few examples. In other cases, teams self-form around specific issues or as part of continuous improvement initiatives such as team-based suggestion systems.

Project teams can have cross-functional membership or simply be a subset of an existing organizational unit. The person who sponsors the team—its "champion" typically creates an incentive plan with specific objective measures and an award schedule tied to achieving those measures. To qualify as an incentive, the plan must include pre-announced goals, with a "do this, get that" guarantee for teams. The incentive usually varies with the value added by the project.

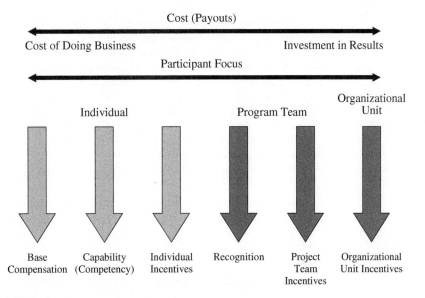

Figure 10–1. Reinforcement model.

Project team incentive plans usually have some combination of these basic measures:

Project milestones: Hit a milestone, on budget and on time, and all team members earn a defined amount. Although sound in theory, there are inherent problems in tying financial incentives to hitting milestones. Milestones often change for good reason (technological advances, market shifts, other developments) and you don't want the team and management to get into a negotiation on slipping dates to trigger the incentive. Unless milestones are set in stone and reaching them is simply a function of the team doing its normal, everyday job, it's generally best to use recognition-after-the-fact celebration of reaching milestones—rather than tying financial incentives to it.

Rewards need not always be time based, such that when the team hits a milestone by a certain date it earns a reward. If, for example, a product development team debugs a new piece of software on time, that's not necessarily a reason to reward it. But if it discovers and solves an unsuspected problem or writes better code before a delivery date, rewards are due.

- *Project completion*: All team members earn a defined amount when they complete the project on budget and on time (or to the team champion's quality standards).
- *Value added*: This award is a function of the value added by a project and depends largely on the ability of the organization to create and track objective measures. Examples include reduced turnaround time on customer requests, improved cycle times for product development, cost savings due to new process efficiencies, or incremental profit or market share created by the product or service developed or implemented by the project team.

One warning about project incentive plans: they can be very effective in helping teams stay focused, accomplish goals, and feel like they are rewarded for their hard work, but they tend to be exclusionary. Not everyone can be on a project team. Some employees (team members) will have an opportunity to earn an incentive that others (non-team members) do not. There is a lack of internal equity. One way to address this is to reward core team members with incentives for reaching team goals, and to recognize peripheral players who supported the team, either by offering advice, resources, or a pair of hands, or by covering for project team members back at their regular job.

Some projects are of such strategic importance that you can live with these internal equity problems and non–team members' grousing about exclusionary incentives. Bottom line, though, is this tool should be used cautiously.[3]

Some organizations focus only on cash awards. However, Parker et al. have concluded from their research that noncash awards can work equally well, if not better, than cash awards:

Many of our case organizations use non-cash awards because of their staying power. Everyone loves money, but cash payments can lose their motivational impact over time.

---

3. Ibid., pp. 38–39.

However, non-cash awards carry trophy value that has great staying power because each time you look at that television set or plaque you are reminded of what you or your team did to earn it. Each of the plans encourages awards that are coveted by the recipients and, therefore, will be memorable.

If you ask employees what they want, they will invariably say cash. But providing it can be difficult if the budget is small or the targeted earnings in an incentive plan are modest. If you pay out more often than annually and take taxes out, the net amount may look pretty small, even cheap. Non-cash awards tend to be more dependent on their symbolic value than their financial value.

Non-cash awards come in all forms: a simple thank-you, a letter of congratulations, time off with pay, a trophy, company merchandise, a plaque, gift certificates, special services, a dinner for two, a free lunch, a credit to a card issued by the company for purchases at local stores, specific items or merchandise, merchandise from an extensive catalogue, travel for business or a vacation with the family, and stock options. Only the creativity and imagination of the plan creators limit the choices.[4]

10.8 KEYS TO BEHAVIORAL EXCELLENCE

Project managers can take some distinguishing actions to ensure the successful completion of their projects. These include:

- Insisting on the right to select key project team
- Negotiating for key team members with proven track records in their fields
- Developing commitment and a sense of mission from the outset
- Seeking sufficient authority from the sponsor
- Coordinating and maintaining a good relationship with the client, parent, and team
- Seeking to enhance the public's image of the project
- Having key team members assist in decision-making and problem-solving
- Developing realistic cost, schedule, and performance estimates and goals
- Maintaining backup strategies (contingency plans) in anticipation of potential problems
- Providing a team structure that is appropriate yet flexible and flat
- Going beyond formal authority to maximize its influence over people and key decisions
- Employing a workable set of project planning and control tools
- Avoiding overreliance on one type of control tool
- Stressing the importance of meeting cost, schedule, and performance goals
- Giving priority to achieving the mission or function of the end item
- Keeping changes under control
- Seeking ways to assure job security for effective project team members

---

4. Ibid., pp. 190–191.

Earlier in this book, I claimed that a project cannot be successful unless it is recognized as a project and gains the support of top-level management. Top-level management must be willing to commit company resources and provide the necessary administrative support so that the project becomes part of the company's day-to-day routine of doing business. In addition, the parent organization must develop an atmosphere conducive to good working relationships among project managers, parent organization, and client organization.

Top-level management should take certain actions to ensure that the organization as a whole supports individual projects and project teams as well as the overall project management system. These actions include:

- Showing a willingness to coordinate efforts
- Demonstrating a willingness to maintain structural flexibility
- Showing a willingness to adapt to change
- Performing effective strategic planning
- Maintaining rapport
- Putting proper emphasis on past experience
- Providing external buffering
- Communicating promptly and accurately
- Exhibiting enthusiasm
- Recognizing that projects do, in fact, contribute to the capabilities of the whole company

Executive sponsors can take certain actions to make project success more likely, including:

- Selecting a project manager at an early point in the project who has a proven track record in behavioral skills and technical skills
- Developing clear and workable guidelines for the project manager
- Delegating sufficient authority to the project manager so that she or he can make decisions in conjunction with the project team members
- Demonstrating enthusiasm for and commitment to the project and the project team
- Developing and maintaining short and informal lines of communication
- Avoiding excessive pressure on the project manager to win contracts
- Avoiding arbitrarily slashing or ballooning the project team's cost estimate
- Avoiding buy-ins
- Developing close, not meddlesome, working relationships with the principal client contact and the project manager

The client organization can exert a great deal of influence on the behavioral aspects of a project by minimizing team meetings, rapidly responding to requests for information, and simply allowing the contractor to conduct business without interference. The positive actions of client organizations also include:

- Showing a willingness to coordinate efforts
- Maintaining rapport
- Establishing reasonable and specific goals and criteria for success
- Establishing procedures for making changes
- Communicating promptly and accurately
- Committing client resources as needed
- Minimizing red tape
- Providing sufficient authority to the client's representative, especially in decision-making

With these actions as the basic foundation, it should be possible to achieve behavioral success, which includes:

- Encouraging openness and honesty from the start from all participants
- Creating an atmosphere that encourages healthy competition but not cutthroat situations or liar's contests
- Planning for adequate funding to complete the entire project
- Developing a clear understanding of the relative importance of cost, schedule, and technical performance goals
- Developing short and informal lines of communication and a flat organizational structure
- Delegating sufficient authority to the principal client contact and allowing prompt approval or rejection of important project decisions
- Rejecting buy-ins
- Making prompt decisions regarding contract okays or go-aheads
- Developing close working relationships with project participants
- Avoiding arm's-length relationships
- Avoiding excessive reporting schemes
- Making prompt decisions on changes

Companies that are excellent in project management have gone beyond the standard actions just listed. Additional actions for excellence include the following:

- The outstanding project manager:
  - Understands and demonstrates competency as a project manager
  - Works creatively and innovatively in a nontraditional sense only when necessary; does not look for trouble
  - Demonstrates high levels of self-motivation from the start
  - Has a high level of integrity; goes above and beyond politics and gamesmanship

- Is dedicated to the company and not just the project; is never self-serving
- Demonstrates humility in leadership
- Demonstrates strong behavioral integration skills both internally and externally
- Thinks proactively rather than reactively
- Is willing to assume a great deal of risk and will spend the appropriate time needed to prepare contingency plans
- Knows when to handle complexity and when to cut through it; demonstrates tenaciousness and perseverance
- Is willing to help people realize their full potential; tries to bring out the best in people
- Communicates in a timely manner and with confidence rather than despair
- The project manager maintains high standards of performance for self and team, as shown by these approaches:
  - Stresses managerial, operational, and product integrity
  - Conforms to moral codes and acts ethically in dealing with people internally and externally
  - Never withholds information
  - Is quality conscious and cost conscious
  - Discourages politics and gamesmanship; stresses justice and equity
  - Strives for continuous improvement but in a cost-conscious manner
- The outstanding project manager organizes and executes the project in a sound and efficient manner by:
  - Informing employees at the project kickoff meeting how they will be evaluated
  - Preferring a flat project organizational structure over a bureaucratic one
  - Developing a project process for handling crises and emergencies quickly and effectively
  - Keeping the project team informed in a timely manner
  - Not requiring excessive reporting; creating an atmosphere of trust
  - Defining roles, responsibilities, and accountabilities up front
  - Establishing a change management process that involves the customer
- The outstanding project manager knows how to motivate:
  - Always uses two-way communication
  - Is empathetic toward the team and a good listener
  - Involves team members in decision-making; always seeks ideas and solutions; never judges an employee's idea hastily
  - Never dictates
  - Gives credit where credit is due
  - Provides constructive criticism rather than making personal attacks
  - Publicly acknowledges credit when credit is due but delivers criticism privately
  - Makes sure that team members know that they will be held accountable and responsible for their assignments

- Always maintains an open-door policy; is readily accessible, even for employees with personal problems
- Takes action quickly on employee grievances; is sensitive to employees' feelings and opinions
- Allows employees to meet the customers
- Tries to determine each team member's capabilities and aspirations; always looks for a good match; is concerned about what happens to the employees when the project is over
- Tries to act as a buffer between the team and administrative/operational problems
- The project manager is ultimately responsible for turning the team into a cohesive and productive group for an open and creative environment. If the project manager succeeds, the team will:
- Demonstrate innovation
- Exchange information freely
- Be willing to accept risk and invest in new ideas
- Have the necessary tools and processes to execute the project
- Dare to be different; is not satisfied with simply meeting the competition
- Understand the business and the economics of the project
- Try to make sound business decisions rather than just sound project decisions

10.9 PROACTIVE VERSUS REACTIVE MANAGEMENT

Perhaps one of the biggest behavioral challenges facing a project manager, especially a new project manager, is learning how to be proactive rather than reactive. Kerry R. Wills discusses this problem.[5]

* * *

Proactive Management Capacity Propensity

In today's world, project managers often get tapped to manage several engagements at once. This usually results in them having just enough time to react to the problems of the day that each project is facing. What they are not doing is spending the time to look ahead on each project to plan for upcoming work, thus resulting in more fires that need to be put out. There used to be an arcade game called "whack-a-mole" where the participant had a mallet and would hit each mole with it when one would pop up. Each time a mole was hit, a new mole would pop up. The cycle of spending time putting out fires and ignoring problems that cause more fires can be thought of as "project whack-a-mole."

It is my experience that proactive management is one of the most effective tools that project managers can use to ensure the success of their projects. However, it is a difficult situation to manage several projects while still having enough time to look ahead. I call this ability to spend time looking ahead the "Proactive Management Capacity Propensity" (PMCP). This article demonstrates the benefits of proactive management, defines the PMCP, and proposes ways of increasing the PMCP and thus the probability of success on the projects.

Proactive Management

Project management involves a lot of planning up front including work plans, budgets, resource allocations, and so on. The best statistics that I have seen on the accuracy of initial plans say there is a 30 percent positive or negative variance from the original plans at the end of a project. Therefore, once the plans have been made and the project has started, the project manager needs to constantly reassess the project to understand the impact of the 60 percent unknowns that will occur.

The dictionary defines proactive as "acting in advance to deal with an expected difficulty." By "acting in advance," a project manager has some influence over the control of the unknowns. However, without acting in advance, the impacts of the unknowns will be greater as the project manager will be reacting to the problem once it has snowballed.

When I drive to work in the morning, I have a plan and schedule. I leave my house, take certain roads, and get to work in 40 minutes. If I were to treat driving to work as a project (having a specific goal with a finite beginning and end), then I have two options to manage my commute: (see Figure 10–2).

By *proactively* managing my commute, I watch the news in the morning to see the weather and traffic. Although I had a plan, if there is construction on one of the roads that I normally take, then I can always change that plan and take a different route to ensure that my schedule gets met. If I know that there may be snow, then I can leave earlier and give myself more time to get to work. As I am driving, I look ahead on the

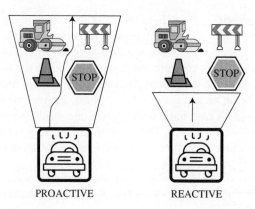

PROACTIVE   REACTIVE

Figure 10–2. Driving metaphor.

road to see what is coming up. There may be an accident or potholes that I will want to avoid, and this gives me time to switch lanes.

A *reactive* approach to my commute could be assuming that my original plan will work fully. As I get on the highway, if there is construction, then I have to sit in it because by the time I realize the impact, I have passed all of the exit ramps. This results in me missing my schedule goal. The same would happen if I walked outside and saw a foot of snow. I now have a chance to scope since I have the added activity of shoveling out my driveway and car. Also, if I am a reactive driver, then I will not see the pothole until I have driven over it (which may lead to a budget variance since I now need new axles).

Benefits

This metaphor demonstrates that reactive management is detrimental to projects because by the time you realize that there is a problem, it usually has a schedule, scope, or cost impact. There are several other benefits to proactive management:

- Proactively managing a plan allows the project manager to see what activities are coming up and start preparing for them. This could be something as minor as setting up conference rooms for meetings. I have seen situations where tasks were not completed on time because of something as minor as logistics.
- Understanding upcoming activities also allows for the proper resources to be in place. Oftentimes, projects require people from outside the project team, and lining them up is always a challenge. By preparing people in advance, there is a higher probability that they can be ready when needed.

The Relationship

The project manager should constantly be replanning. By looking at all upcoming activities as well as the current ones, it can give a gauge of the probability of success, which can be managed rather than waiting until the day before something is due to realize that the schedule cannot be met.

Proactive management also allows time to focus on quality. Reactive management usually is characterized by rushing to fix whatever "mole" has popped up as quickly as possible. This usually means a patch rather than the appropriate fix. By planning for the work appropriately, it can be addressed properly, which reduces the probability of rework.

As previously unidentified work arises, it can be planned for rather than assuming that "we can just take it on."

Proactive management is extremely influential over the probability of success of a project because it allows for replanning and the ability to address problems well before they have a significant impact.

I have observed a relationship between the amount of work that a project manager has and their ability to manage proactively. As project managers get more work and more concurrent projects, their ability to manage proactively goes down.

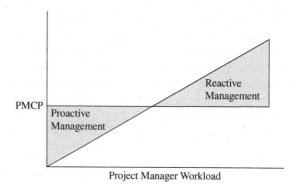

Figure 10–3. Proactivity graph.

The relationship between project manager workload and the ability to manage proactively is shown in Figure 10–3. As project managers have increased work, they have less capacity to be proactive to and wind up becoming more reactive.

Not all projects and project managers are equal. Some project managers can handle several projects well, and some projects require more focus than others. I have therefore labeled this factor the Project Management Capacity Propensity. That is, the sum of those qualities that allow a project manager to proactively manage projects.

There are several factors that make up the PMCP that I outline below.

Project manager skill sets have an impact on the PMCP. Having good time management and organization techniques can influence how much a PM can focus on looking ahead. A project manager who is efficient with their time has the ability to review more upcoming activities and plan for them.

Project manager's expertise in the project is also influential to the PMCP. If the PM is an expert in the business or the project, this may allow for quicker decisions since they will not need to seek out information or clarification (all of which takes away time).

The PMCP is also impacted by team composition. If the project manager is on a large project and has several team leads who manage plans, then they have an increased ability to focus on replanning and upcoming work. Also, having team members who are experts in their field will require less focus from the project manager.

Increasing the PMCP

The good news about the PMCP is that it can be increased.

Project managers can look for ways to increase their skill sets through training. There are several books and seminars on time management, prioritization, and organization. Attending these can build the effectiveness of the time spent by the PM on their activities.

The PM can also reevaluate the team composition. By getting stronger team leads or different team members, the PM can offload some of their work and spend more time focusing on proactive management.

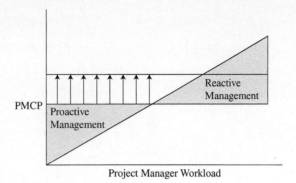

Figure 10–4. Increasing PMCP.

All of these items can increase the PMCP and result in an increased ability to manage proactively. Figure 10–4 shows how a PMCP increase raises the bar and allows for more proactive management with the same workload.

Conclusion

To proactively manage a project is to increase your probability of being successful. There is a direct correlation between the workload that a PM has and their ability to look ahead. Project managers do have control over certain aspects that can give them a greater ability to focus on proactive management. These items, the PMCP, can be increased through training and having the proper team.

Remember to keep your eyes on the road.

Measuring Return on Investment on Project Management Training Dollars

11

11.0 INTRODUCTION

For almost three decades, from the 1960s through the 1980s, the growth and acceptance of project management were restricted to the aerospace, defense, and heavy construction industries. In virtually all other industries, project management was nice to have but not a necessity. Very few project management training programs were offered in the public marketplace, and those that were offered covered the basics of project management with weak attempts to customize the material to a specific company. The concept of measuring the return on investment (ROI) on training, at least in project management courses, was nonexistent. More recently, however, several studies have quantified the benefits of project management, and there has been some pioneering work on the benefits of project management training.[1] There is still a great deal of effort needed, but at least we have recognized the need.

Today, our view of project management education has changed and so has our desire to evaluate ROI on project management training funds. There are several reasons for this:

- Executives realize that training is a basic necessity for companies to grow.
- Employees want training for professional growth and advancement opportunities.

---

1. W. Ibbs and J. Reginato, *Quantifying the Value of Project Management* (Newton Square, PA: Project Management Institute, 2002) W. Ibbs and Y.-H. Kwak, *The Benefits of Project Management* (Newton Square, PA: Project Management Institute, 1997). W. Ibbs, "*Measuring Project Management's Value: New Directions for Quantifying PM/ROI®*." Paper presented at the Proceedings of the PMI Research Conference, June 21–24, 2000, Paris, France. J. Knutson, "From Making Sense to Making Cents." A three-part series in *PM Network*: Part 1: Measuring Project Management ROI, vol. 13 no. 1 (January), 25–27; Part 2: Measuring Project Management, vol. 13 no. 2 (February), 23–24; Part 3: The Process, vol. 13 no. 7 (July), 17–19.

- Project management is now viewed as a profession and a strategic competency rather than a part-time occupation.
- The importance of becoming a PMP® credential holder[2] has been increasing.
- There are numerous university programs available worldwide leading to MS, MBA, and PhD degrees in project management.
- There are certificate programs in various project management concepts, such as risk management and program management.
- The pressure to maintain corporate profitability has increased, resulting in less money available for training. Yet more and more training funds are being requested by the workers who desire to become PMP® credential holders and then must accumulate 60 professional development units (PDUs) every three years to remain certified.
- Management realizes that a significant portion of training budgets must be allocated to project management education, but it should be allocated to those courses that provide the company with the greatest ROI. The concept of educational ROI is now upon us.

11.1 PROJECT MANAGEMENT BENEFITS

In the early years of project management, primarily in the aerospace and defense industries, studies were done to determine the benefits of project management. In an early study by Middleton,[3] the benefits discovered were:

- Better control of projects
- Better customer relations
- Shorter product development time
- Lower program costs
- Improved quality and reliability
- Higher profit margins
- Better control over program security
- Improved coordination among company divisions doing work on the project
- Higher morale and better mission orientation for employees working on the project
- Accelerated development of managers due to breadth of project responsibilities

These benefits were identified by Middleton through surveys and were subjective in nature. No attempt was made to quantify the benefits. At that time, there existed virtually no project management training programs. On-the-job training was the preferred method of learning project management, and most people learned from their own mistakes rather than from the mistakes of others.

Today, the benefits identified by Middleton still apply, and we have added other benefits to the list:

- Accomplishing more work in less time and with few resources
- More efficient and more effective performance

2. PMP is a registered mark of the Project Management Institute, Inc.

3. C. J. Middleton, "How to Set Up a Project Organization," *Harvard Business Review* (March–April 1967): 73–82.

- Increase in business due to customer satisfaction
- Potential for a long-term partnership relationship with customers
- Better control of scope changes

Executives wanted all of the benefits described here, and they wanted them yesterday. It is true that these benefits could be obtained just by using on-the-job training efforts, but this assumed that time was a luxury rather than a constraint. Furthermore, executives wanted workers to learn from the mistakes of others rather than their own. Also, executives wanted everyone in project management to look for continuous improvement efforts rather than just an occasional best practice.

Not every project management training program focuses on all of these benefits. Some courses focus on one particular benefit, while others might focus on a group of benefits. Deciding which benefits you desire is essential when selecting a training course. And if the benefits can be quantified after training is completed, then executives can maximize their ROI on project management training dollars by selecting the proper training organizations.

11.2 GROWTH OF ROI MODELING

In the past 15 years, the global expansion of ROI modeling has taken hold.[4] The American Society for Training and Development has performed studies on ROI modeling. Throughout the world, professional associations are conducting seminars, workshops, and conferences dedicated to ROI on training.

According to the 2001 Training's annual report, more than $66 billion was spent on training in 2001. It is therefore little wonder that management treats training with a business mindset, thus justifying the use of ROI measurement. But despite all of the worldwide commitment and documented successes, there is still a very real fear in many companies preventing the use of ROI modeling. Typical arguments are: "It doesn't apply to us"; "We cannot evaluate the benefits quantitatively"; "We don't need it"; "The results are meaningless"; "It costs too much." These fears create barriers to the implementation of ROI techniques, but most of these barriers are myths that can be overcome.

In most companies, human resources development (HRD) maintains the lead role in overcoming these fears and performing ROI studies. The cost of performing these studies on a continuous basis could be as much as 4 to 5 percent of the HRD budget. Some HRD organizations have trouble justifying this expense. And to make matters worse, HRD personnel may have a poor understanding of project management.

The salvation in overcoming these fears and designing proper project management training programs could very well be the project management office (PMO). Since the PMO has become the guardian of all project management intellectual property as well as designing project management training courses, it will most likely take the lead role

---

4. An excellent source is J. J. Phillips, *Return on Investment in Training and Performance Improvement Programs*, 2nd ed. (Burlington, MA: Butterworth-Heinemann, 2003). In the opinion of this author, this is by far one of the best, if not the best, text on this subject.

in calculating ROI on project management-related training courses. Members of the PMO might be required to become certified in educational ROI measurement the same way that they are certified as a PMP® credential holder or Six Sigma Black Belt.

Another reason for using the PMO is because of the enterprise project management (EPM) methodology. EPM is the integration of various processes, such as total quality management, concurrent engineering, continuous improvement, risk management, and scope change control, into one project management methodology that is utilized on a company-wide basis. Each of these processes has measurable output that previously may not have been tracked or reported, which has placed additional pressure on the PMO and project management education to develop the necessary metrics and measurements for success.

11.3 THE ROI MODEL

Any model used must provide a systematic approach to calculating ROI. It should be prepared on a life-cycle basis or in a step-by-step approach similar to an EPM methodology. Just as with EPM, there are essential criteria that must exist for any model to work effectively.

Because certain criteria are considered essential, any ROI methodology should meet the vast majority of, if not all, of these criteria. The bad news is that, generally, most ROI processes do not meet all of these criteria. A typical model is shown in Figure 11–1. The definitions of the levels in Figure 11–1 are shown in Table 11–1.

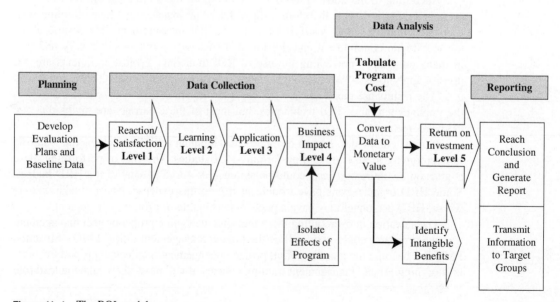

Figure 11–1. The ROI model.

TABLE 11–1. DEFINING LEVELS

Level	Description
1: Reaction/satisfaction	Measures the participants' reaction to the program and possibly creates an action plan for implementation of the ideas
2: Learning	Measures specific skills, knowledge, or attitude changes
3: Application	Measures changes in work habit or on-the-job performance as well as application and implementation of knowledge learned
4: Business impact	Measures the impact on the business as a result of implementation of changes
5: Return on investment	Compares monetary benefits with the cost of the training and expressed as a percentage

11.4 PLANNING LIFE-CYCLE PHASE

The first life-cycle phase in the ROI model is the development of evaluation plans and baseline data. The evaluation plan is similar to some of the *PMBOK® Guide*[5] knowledge areas that require a plan as part of the first process step in each knowledge area. The evaluation plan should identify:

- The objective(s) of the program
- The way(s) the objective(s) will be validated
- The target audience
- Assumptions and constraints
- The timing of the program

Objectives for the training program must be clearly defined before ROI modeling can be completed. Table 11–2 identifies typical objectives. The objectives must be clearly defined for each of the five levels of the model. Column 3 in the table would be representative of the objectives that a company might have when it registers a participant in a project management certificate program (PMCP) training course. In this example, the company funding the participant's training might expect the participant to become a PMP® credential holder and then assist the organization in developing an EPM methodology based on the *PMBOK® Guide,* with the expectation that this would lead to customer satisfaction and more business. Column 4 in Table 11–2 might be representative of a company that registers a participant in a course on best practices in project management. Some companies believe that if a seminar participant walks away from a training program with two good ideas for each day of the program and if these ideas can be implemented reasonably fast, then the seminar is considered a success. In this example, the objectives are to identify best practices in project management that other companies are doing and that can be implemented effectively in the participant's company.

---

5. PMBOK is a registered mark of the Project Management Institute, Inc.

TABLE 11–2. TYPICAL PROGRAM OBJECTIVES

		Objectives	
Level	Description	Typical PMCP Training	Typical Best Practices Training Course
1	Reaction/satisfaction	Understand principles of *PMBOK® Guide*	Understand that companies are documenting their best practices
2	Learning	Demonstrate skills or knowledge in domain groups and knowledge areas	Demonstrate how best practices benefit an organization
3	Application	Development of EPM processes based on the *PMBOK® Guide*	Develop a best practices library or ways to capture best practices
4	Business impact	Measurement of customer and user satisfaction with EPM	Determine the time and/or cost savings from a best practice
5	Return on investment	Amount of business or customer satisfaction generated from EPM	Measure ROI for each best practice implemented

There can be differences in training objectives, as seen through the eyes of management. As an example, looking at columns 3 and 4 in Table 11–2, objectives might be:

- Learn skills that can be applied immediately to the job. In this case, ROI can be measured quickly. This might be representative of the PMCP course in column 3.
- Learn about techniques and advancements. In this case, additional money must be spent to achieve these benefits. ROI measurement may not be meaningful until after the techniques have been implemented. This might be representative of the best practices course in column 4.
- A combination of the above.

11.5 DATA COLLECTION LIFE-CYCLE PHASE

In order to validate that each level's objectives for the training course were achieved, data must be collected and processed. Levels 1 to 4 in Figure 11–1 make up the data collection life-cycle phase.

To understand the data collection methods, we revisit the course on best practices in project management, which covers best practices implemented by various companies worldwide. The following assumptions will be made:

- Participants are attending the course to bring back to their company at least two ideas that can be implemented in their company within six months.
- Collecting PDUs is a secondary benefit.
- The course length is two days[6] Typical data collection approaches are shown in Table 11–3 and explained below for each level.

---

6. Some companies have one-day, two-day, and even week-long courses on best practices in project management.

TABLE 11–3. DATA COLLECTION

Level	Measures	Data Collection Methods and Instruments	Data Sources	Timing	Responsible Person
Reaction/ satisfaction	A 1–7 rating on end-of-course critique	Questionnaire	Participant (last day of program)	End of program	Instructor
Learning	Pretest, posttest, online courses, and case studies	In-class tests and skill practice sets	Instructor	Each day of course	Instructor
Application	Classroom discussion	Follow-up session or questionnaire	Participant and/or PMO	Three months after program[a]	PMO
Business impact	Measurement of EPM continuous improvement efforts	Benefit–cost monitoring by the PMO	PMO records	Six months after program	PMO
Return on investment	Benefit–cost ratios	PMO studies	PMO records	Six months after program	PMO

[a] Usually for in-house program only. For public seminars, this may be done by the PMO within a week after completion of training.

Level 1: Reaction and Satisfaction

Level 1 measures the participant's reaction to the program and possibly an action plan for implementation of the ideas. The measurement for level 1 is usually an end-of-course questionnaire where the participant rates the information presented, quality of instruction, instructional material, and other such topics on a scale of 1 to 7. All too often, the questionnaire is answered based on the instructor's presentation skills rather than the quality of the information. While this method is most common and often serves as an indication of customer satisfaction, hopefully leading to repeat business, it is not a guarantee that new skills or knowledge have been learned.

Level 2: Learning

This level measures specific skills, knowledge, or attitude changes learned during the course. Instructors use a variety of techniques for training, including:

- Lectures
- Lectures/discussions
- Exams
- Case studies (external firms)
- Case studies (internal projects)
- Simulation/role-playing
- Combinations

For each training technique, a measurement method must be established. Some trainers provide a pretest at the beginning of the course and a posttest at the end. The difference in scores is usually representative of the amount of learning that has taken place. This is usually accomplished through in-house training programs rather than public seminars. Care must be taken in the use of pretests and posttests. Sometimes a posttest is made relatively easy for the purpose of making it appear that learning has taken place. Out-of-class testing can also be accomplished using take-home case studies and online questions and exams.

Testing is necessary to validate that learning has taken place and knowledge has been absorbed. However, simply because learning has taken place, there is no guarantee that the information learned on best practices can or will be transferred to the company. The learning might simply confirm that the company is doing well and keeping up with its competitors.

Level 3: Application of Knowledge
This level measures changes in work habits or on-the-job performance as well as implementation of knowledge learned. Measurement at this level is normally done through follow-up sessions or follow-up questionnaires. However, for publicly offered courses with a large number of participants, it is impossible for the instructor to follow up with all participants. In such cases, the responsibility falls on the shoulders of the PMO. Participants may be required to prepare a short one- or two-page report on what they learned in the course and what best practices are applicable to the company. The report is submitted to the PMO, which might have the final decision on the implementation of the ideas. Based on the magnitude of the best practices ideas, the portfolio management of projects may be impacted. However, there is no guarantee at this point that there will be a positive impact on the business.

Level 4: Business Impact
This level measures the impact on the business as a result of implementation of the changes. Typical measurement areas are shown in Figure 11–2.

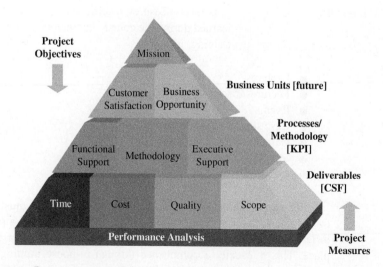

Figure 11–2. Postmortem pyramid.

The critical terms in Figure 11–2 are:

- *Critical success factor (CSF)*: This measures changes in the output of the project resulting from implementation of best practices. Hopefully, this will lead to improvements in time, cost, quality, and scope.
- *Key performance indicator (KPI)*: This measures changes in the use of the EPM system and support received from functional management and senior management.
- *Business unit impact*: This is measured by customer satisfaction as a result of the implementation of best practices and/or future business opportunities.

The measurement at level 4 is usually accomplished by the PMO. There are three reasons for this:

1. The information may be company-sensitive and not available to the instructor.
2. Since there may be a long time span between training and the implementation of best practices, the instructor may not be available for support.
3. The company may not want anyone outside of the company talking to its customers about customer satisfaction.

Although the implementation of best practices may have a favorable business impact, care must be taken that the implementation was cost-effective.

As shown in Figure 11–1, an important input into level 4 is to *isolate the effects of training*. It is often impossible to clearly identify the business impact that results directly from the training program. The problem is that people learn project management from multiple sources, including:

- Formal education
- Knowledge transfer from colleagues
- On-the-job experience
- Internal research on continuous improvements
- Benchmarking

Because of the difficulty in isolating specific knowledge, this step is often overlooked.

11.6 DATA ANALYSIS LIFE-CYCLE PHASE

In order to calculate the ROI, the business impact data from level 4 must be converted to a monetary value. The information can come from interviews with employees and managers, databases, subject matter experts, and historical data. Very rarely will all of the information needed come from one source.

Another input required for data analysis is the cost of the training program. Typical costs that should be considered include:

- Cost of course design and development
- Cost of materials
- Cost of the facilitator(s)
- Cost of facilities and meals during training
- Costs of travel, meals, and lodgings for each participant
- Fully burdened salaries of participants
- Administrative or overhead costs related to the training course or approach of participants to attend training
- Possible cost (loss of income) of not having the participants available for other work during the time of training

Not all benefits can be converted to monetary values. This is the reason for the "identify intangible benefits" box in Figure 11–1. Some business impact benefits that are easily converted to monetary values include:

- Shorter product development time
- Faster, higher-quality decisions
- Lower costs
- Higher profit margins
- Fewer resources needed
- Reduction in paperwork
- Improved quality and reliability
- Lower turnover of personnel
- Quicker implementation of best practices

Typical benefits that are intangible and cannot readily be converted to monetary value include:

- Better visibility and focus on results
- Better coordination
- Higher morale
- Accelerated development of managers
- Better project control
- Better customer relations
- Better functional support
- Fewer conflicts requiring some management support

Despite the fact that these benefits may be intangible, every attempt should be made to assign monetary values to these benefits.

Level 5: Return on Investment   Two formulas are required for completion of level 5. The first formula is the benefit–cost ratio (BCR), which can be formulated as

$$BCR = \frac{\text{Program benefits}}{\text{Program costs}}$$

The second formula is the ROI expressed as a percentage. The formula is based on "net" program benefits, which are the benefits minus the cost. Mathematically, we can describe it as

$$ROI = \frac{\text{Net Program benefits}}{\text{Program costs}} \times 100$$

To illustrate the usefulness of this level, we consider three examples, all based on the same training course. You attend a two-day seminar on best practices in project management. Your company's cost for attending the course is:

Registration fee	$475
Release time (16 hr at $100/hr)	1,600
Travel expenses	800
	$2,875

When the seminar is completed, you come away with three best practices to recommend to your company. Your company likes all three ideas and assigns you as the project manager to implement all three best practices. Additional funds must be spent to achieve the benefits desired.

Example 1

During the seminar, you discover that many companies have adopted the concept of paperless project management by implementing a traffic light status reporting system. Your company already has a web-based EPM system, but you have been preparing paper reports for status review meetings. Now every status review meeting will be conducted as paperless PowerPoint presentation displaying the web-based methodology with a traffic light display beside each work package in the work breakdown structure.

The cost of developing the traffic light system is:

Systems programming (240 hr at $100/hr)	$24,000
Project management (150 hr at $100/hr)	15,000
	$39,000

The benefits expressed by monetary terms are:

- Executive time in project review meeting (20 hr/project to 10 hr/project × 15 projects × 5 executives per meeting × $250/hr): $187,500
- Paperwork preparation time reduction (60 hr/project × 15 projects × $100/hr): $90,000

Total additional benefit is therefore $275,500:

$$BCR = \frac{\$275,000 - \$39,000}{\$2875} = 82$$

$$ROI = \frac{\$275,000 - \$39,000 - \$2875}{\$2875} = 8109$$

This means that for every dollar invested in the training program, there was a return of $8,109 in net benefits! In this example, it was assumed that workers were fully burdened at $100/hr and executives at $250/hr. The benefits were one-year measurements, and the cost of developing the traffic light system was not amortized but expensed against the yearly benefits.

Not all training programs generate benefits of this magnitude. Lear in Dearborn, Michigan, had a project management traffic light reporting system as part of its web-based EPM system. Lear had shown that in the same amount of time that it would review the status of one project using paper, it could review the status of all projects using traffic light reporting.

Example 2

During the training program, you discover that other companies are using templates for project approval and initiation. The templates are provided to you during the training program, and it takes very little effort to make the templates part of the EPM system and inform everyone about the update. The new templates will eliminate at least one meeting per week at a savings of $550:

$$Benefit = (\$500/meeting) \times (1\ meeting/week) \times 50\ week$$

$$= \$27,500$$

$$BCR = \frac{\$27,500}{\$2875} = 9.56$$

$$ROI = \frac{\$27,500 - \$2875}{\$2875} = 8.56$$

In this example, for each $1 invested in the best practices program, a net benefit of $8.56 was recognized.

Example 3

During the training program, you learn that companies are extending their EPM systems to become more compatible with systems utilized by their customers. This should foster better customer satisfaction. The cost of updating your EPM system to account for diversified customer report generators will be about $100,000.

After the report generator is installed, one of your customers, with whom you have four projects per year, informs you that it is so pleased with this change that it will now give you sole-source procurement contracts. This will result in a significant savings in procurement costs. Your company typically spends $30,000 preparing proposals:

$$BCR = \frac{(4 \text{ projects} \times \$30,000)}{\$2875} = 6.96$$

$$ROI(\%) = \frac{(4 \times \$30,000) - \$100,000 - \$2875}{\$2875} = 5.96$$

In this case, for every dollar invested in the best practices program, there was a net benefit of $5.96 received.

To date, there have been very few attempts to measure ROI specifically on project management education other than work done by a limited number of researchers. However, there have been some successes. In an insurance company, a $100 million project was undertaken. All employees were required to undergo project management training prior to working on the project. The project was completed 3 percent ($3 million) below budget.

Unsure of whether the $3 million savings was due to better project management education or poor initial estimating, the company performed a study on all projects where the employees were trained on project management prior to working on project teams. The result was an astounding 700 percent return on training dollars.

In another organization, the HRD people worked with project management to develop a computer-based project management training program. The initial results indicated a 900 percent ROI. The workers took the course on their own time rather than on company time. Perhaps this is an indication of the benefits of e-learning programs or virtual webinars. The e-learning programs and virtual webinars may produce a much higher ROI than traditional courses because the cost of the course is significantly reduced with the elimination of the cost of release time.

11.7 REPORTING LIFE-CYCLE PHASE

The final life-cycle phase in Figure 11–1 is reporting. The acceptance of the results could very well be based on how the report is prepared. The report must be self-explanatory to all target groups. If assumptions are made concerning costs or benefits, then they must be justified. If the ROI numbers are inflated to make a training program look better than it was, then people may be skeptical and refuse to accept the results of future ROI studies. All results should be factual and supported by realistic data.

11.8 EDUCATION AND ROI CHALLENGES

There are several challenges that organizations face that, if not recognized, could lower the ROI expectation.

Determining course content: A company hired a consultant from a well-known university to spend a week with senior management and discuss where each of their product lines should be positioned in 3, 5, and 10 years in order for the company to have a sustainable future. After the consultant left, the executives met to discuss what they had learned from the consultant. The conclusion was that the consultant told them "what" they should do, not "how" to do it. It quickly became apparent that the "how" to do it required the use of projects and project management. The company recognized the need for project management education and that deciding upon the course content was critical. HR was responsible for corporate training. The company sent a few HR personnel to attend publicly offered project management courses. When they returned, they were asked to customize the course content for trainers that would be hired to perform in-house training. HR training managers were the first people trained in the company in the hope that training could be customized correctly. Today, companies rely on PMOs for customization.

Not all courses are equal: The Commercial Products Division of a Fortune 500 Company recognized the need for project management training. The company had an Aerospace Division that had been using project management successfully for more than five years. The in-house project management trainers in the Aerospace Division were asked to conduct the same training for the Commercial Products Division. The training was poorly received because the types of projects in the Commercial Products Division were much shorter in length, significantly less costly, and less risky, and required different metrics for performance reporting. The company recognized the need for customized training.

Fear of the unknown: Management recognized the benefits of using project management and began training their workers on the basics of project management. The workers also recognized many of the benefits of the training and began implementing what they had learned. Wanting some of the advanced best practices implemented quickly, management hired a training to focus on future concepts in project management and what the workers might expect project management to look like a few years into the future. The workers became worried that they were being removed from their comfort zones and that the new performance reporting practices, which included new metrics, would appear as if senior management was now looking over the shoulders of the workers closely to make sure they were doing their work correctly. Some of the workers began sabotaging their projects. Management soon realized that the expectations from the learning process should be gradual rather than instantaneous.

When to evaluate ROI: At the end of each course, the HR Department would give the attendees an evaluation form to complete the day the training ended to determine if the trainer should be rehired and if the workers would be able to use the knowledge gained in the course. One of the speakers received high marks for being quite entertaining in classroom, and the participants stated that they believed they could use what was presented in class. The HR Department then changed the speaker evaluation process, asking the participants to complete the evaluation form between 60–90 days after the training was completed. This time, the evaluation forms showed that, even though the speaker was entertaining in the classroom, the workers did not learn as much as they expected and were not able to implement many of the practices that the speaker discussed. Speaker performance should be based upon ROI results from knowledge gained and used rather than from entertainment ability.

11.9 CONCLUSIONS

Because of the quantity and depth of available project management training programs, the concept of measuring ROI on training dollars can be expected to grow. Executives will recognize the benefits of this approach and its application to project management the same way they do for other training programs. Project management training organizations will be required to demonstrate expertise in ROI analysis. Eventually, PMI might even establish a special investigation group on ROI measurement.

12 The Project Management Office

12.0 INTRODUCTION

As companies begin to recognize the favorable effect that project management has on profitability, emphasis is placed on achieving professionalism in project management using the project office (PO) concept. The concept of a PO or project management office (PMO) could very well be the most important project management activity in this decade.

With this recognition of importance comes strategic planning for both project management and the PO. Maturity and excellence in project management do *not* occur simply by using project management over a prolonged period of time. Rather, they come through strategic planning for both project management and the PO.

General strategic planning involves the determination of where you wish to be in the future and then how you plan to get there. For PO strategic planning, it is often easier to decide which activities should be under the control of the PO than to determine how or when to do it. For each activity placed under the auspices of the PO, there may appear pockets of resistance that initially view removing this activity from its functional area as a threat to its power and authority. Typical activities assigned to a PO include:

- Standardization in estimating.
- Standardization in planning.
- Standardization in scheduling.
- Standardization in control.
- Standardization in reporting.
- Clarification of project management roles and responsibilities.
- Preparation of job descriptions for project managers.
- Preparation of archive data on lessons learned.

- Benchmarking continuously.
- Developing project management templates.
- Developing a project management methodology.
- Recommending and implementing changes and improvements to the existing methodology.
- Identifying project standards.
- Identifying best practices.
- Performing strategic planning for project management.
- Establishing a project management problem-solving hotline.
- Coordinating and/or conducting project management training programs.
- Transferring knowledge through coaching and mentorship.
- Developing a corporate resource capacity/utilization plan.
- Supporting portfolio management activities.
- Assessing risks.
- Planning for disaster recovery.
- Auditing the use of the project management methodology.
- Auditing the use of best practices.

In the first decade of the twenty-first century, the PO became commonplace in the corporate hierarchy. Although the majority of activities assigned to the PO had not changed, there was now a new mission for the PO:

- The PO now had the responsibility for maintaining all intellectual property related to project management and to actively support corporate strategic planning.
- The PO was now servicing the corporation, especially the strategic planning activities for project management, rather than focusing on a specific customer.
- The PO was transformed into a corporate center for control of project management intellectual property. This was a necessity as the magnitude of project management information grew almost exponentially throughout the organization.

During the past 20 years, the benefits to executive levels of management of using a PO have become apparent. They include:

- Standardization of operations.
- Company rather than silo decision-making.
- Better capacity planning (i.e., resource allocations).
- Quicker access to higher-quality information.
- Elimination or reduction of company silos.
- More efficient and effective operations.
- Less need for restructuring.
- Fewer meetings that rob executives of valuable time.
- More realistic prioritization of work.
- Development of future general managers.

All of these benefits are either directly or indirectly related to project management intellectual property. To maintain the project management intellectual property, the PO must maintain the vehicles for capturing the data and then for disseminating the data to the various stakeholders. These vehicles include the company project management intranet, project websites, project databases, and project management information systems. Since much of this information is necessary for both project management and corporate strategic planning, there must exist strategic planning for the PO.

The recognition of the importance of the PMO has now spread worldwide. Enrique Sevilla Molina, formerly Corporate PMO director for Indra, states:

> We have a PMO at corporate level and local PMOs at different levels throughout the company, performing a variety of functions. The PMO at corporate level provides directions on different project management issues, methodology clarifications, and tool use to local PMOs.
>
> Besides supporting the local PMOs and Project Managers as requested, the main functions of the corporate PMO include acting on the following areas:
>
> - Maintenance and development of the overall project management methodology, including the extensions for program and portfolio levels.
> - Definition of the training material and processes for the PMs.
> - Management of the PMP® certification[1] process and candidate training and preparation.
> - Definition of the requirements for the corporate PM tools.

The corporate PMO reports to the financial managing director.

Most PMOs are viewed as indirect labor and therefore are subject to downsizing or elimination when a corporation is under financial stress. To minimize the risk, the PMO should set up metrics to show that the office is adding value to the company. Typical metrics are listed next.

- Tangible measurements include:
  Customer satisfaction
  Projects at risk
  Projects in trouble
  The number of red lights that need recovery, and by how much-added effort
- Intangible elements may also exist, and these may not be able to be measured. They include:
  Early identification of problems
  Quality and timing of information

12.1 BOEING[2]

Not all companies use the term PMO. In some companies, it is also called a Community of Excellence (CoE) or a Community of Practice (CoP). Every company has its own unique goals and objectives for the PMO. As such, the responsibilities of the PMO can vary from company to company. The following information has been graciously provided by David Hunter – Boeing Program Management Specialist.

---

1. PMP is a registered mark of the Project Management Institute, Inc.

The Boeing Program Management Operations Council sponsors a Project Management Community of Excellence (PjMCoE) that exemplifies Project Management best practices and promotes the Project Management disciplines across The Boeing Company. The purpose of the PjMCoE is to provide a Boeing-wide, cross-functional forum for increased awareness of the skills, discipline, and profession of Project Management. Serving as a clearinghouse for ideas and information including industry, methodologies, tools, best practices, teams, and innovation.

A CoE within Boeing is a formally chartered group that functionally aligns with at least one business organization, has enterprise representation, and is committed to business engagement, knowledge sharing, and education across the entire Boeing Company.

The PjMCoE operates as a voluntary interest group with more than 6,800 active Boeing members (including 1,276 registered PMPs) and is one of the largest interest groups within Boeing. Membership is open to all Boeing employees (direct and contract), including non-US Boeing employees. The PjMCoE started in 1997 as an informal Project Management interest group that included only 75 members. The primary purpose is to act as a Boeing-wide forum for increased awareness of project management skills, discipline, and profession. The PjMCoE is foundational to project management success at Boeing and provides the following services to both its members and the business:

- Networking, collaboration, and support including *inSite,* a web tool that allows for sharing, learning, and replicating information and ideas across the company.
- An elite team of volunteer Subject Matter Experts (SMEs) and Boeing Designated Experts (BPEs) who collaborate to define and refine Project and Program Management Best Practices (PMBP).
- Project Management mentoring, coaching, and training.
- Assistance for managers seeking to hire or promote Project Managers.
- Opportunities to volunteer and support community service projects through Boeing's Global Corporate Citizenship team.

The PjMCoE has its own Steering Team with defined responsibilities that support CoE products and services and is representative across all Boeing operating groups and sites. The Steering Team is foundational to the achievement of the CoE charter, facilitation of regular meetings, oversees operation of an effective management information system, ensures all engagement requests receive a response, and notifies aligned organizations when contacted by project teams, programs, or functions with requests for support.

PjMCoE encourages the development of project management skills providing support to its members by offering the following services:

- Libraries containing Project Management related templates, for example, books, periodicals, methodologies, and presentations.
- SharePoint and web site with PjMCoE information and news.
- Learning and Development Project Management curriculum offered through the Learning and Development organization and internal Learning Together Program.

- Knowledge centers providing mentoring and coaching services and information on Project Management certifications and degrees.
- Annual PMP® exam prep course hosting live and virtual training sessions and lessons learned to assist in preparing for the PMP® exam. The first PMP® study group started in 2000 and has had a 96% pass rate for those who actually took the PMP® exam.
- Regular WebEx meetings hosting guest speakers and offering Professional Development Units used toward accreditation.
- Resources for current Project Managers including career advancement opportunities, skills assessment tools, and temporary positions.
- Annual Project Management Week attracting Boeing employees and contractors from around the world.

Many employees take advantage of these services to assist them in career development and project management skills. PjMCoE maintains a strong ongoing relationship with PMI and is a PMI Authorized Training Partner (ATP). The PjMCoE provides several training events for its members and supports the delivery of Boeing-specific project management classes. Another training is available for employees to obtain Professional Development Units toward the re-certification of their PMP® and other PM certifications. Life/Work Balance, MS Project, Milestones Professional, Risk, Issue and Opportunity Management, Leadership, Communication, and Virtual Team Management are some of the topics that have been presented.

12.2 KAUST IT PMO: BUILDING CAPABILITIES[3]

Background

In addition to it being a governance function, the Information Technology (IT) PMO at King Abdullah University of Science and Technology (KAUST) aims to build capabilities in the areas of project management, innovation, and digital transformation.

It is key to build awareness and knowledge of methodologies, best practices, and tools to maximize the value and impact of any transformation.

The IT PMO successfully launched a number of programs with participants from across the University sectors. More than 150 participants took part in the different specialized programs.

The selection and offering of these programs follow a thorough process of research and alignment with strategic priorities as well as the understanding of industry trends and key skills required in context of project management, innovation, and digital transformation. These programs are also delivered by experts and distinguished speakers from around the world.

---

3. Material in this section provided by Samara Barhamain, Program Manager, KAUST. ©2022 by KAUST. All rights reserved. Reproduced with permission.

In addition to the classroom virtual and physical learning programs, KAUST colleagues are also provided with a handpicked list of on-demand courses webinars from sound resources that are listed on the KAUST IT PMO online platform.

Building Capabilities continues to be a high pursuit of KAUST IT PMO and an effective tool for driving change based on sound practices.

Framework, Best Practices, and Local Context

The PMO framework was established in reference to PMI and PMP methodology and best practices, while also considering the local context of IT and Digital Transformation projects in KAUST.

The objective was to establish sound governance, increase the success rate and maximize the value of project investment. Another objective was to create visibility and situational awareness for sponsors and stakeholders on the status of project parameters including time, budget, and overall progress.

The established framework adopted a simplified version of the methodology in a way that provides governance yet minimizes the administrative overhead on project managers and project teams. For example, the project charter has a simplified approval and endorsement flow, and the project schedule captures a minimum of the key milestones of a project.

The framework along with its key tools and artifacts was established to standardize and simplify the process of adopting and following the PMO framework.

Awareness and Training for Adoption

The core PMO team is a small group of program/project managers that oversee large-scale and higher-complexity programs and projects. Other projects are managed by project managers reporting to the functional departments and overseen or supported by the core PMO team, mainly from a governance and framework compliance standpoint.

This consequently meant that a significant group of collogues in the functional departments were required to be trained on the use of the standard PMO methodology, artifacts, and tools.

The first learning program that was initiated by the PMO was aiming to provide an understanding of the methodology and its templates and tools to those who will be using them.

This setup also established a community of functional project managers and project professionals that come from a technical or business background and are required to wear the project management hat in their role.

Engaging Stakeholders and Project Team Members

Building capabilities to enable project success is key not only for project managers. The extended group of stakeholders and enablers for the success of projects include in addition to the core PMO and the functional project managers; the project teams, product managers, customer engagement managers, procurement officers, and business managers among others.

Therefore, it was important to involve this community and aim to elevate the awareness and knowledge level of all these key participants in project and program best practices and framework.

This group among others became a community that is participating and involved with the efforts to build capabilities, skills, and awareness related to project management, innovation, and digital transformation.

Some examples of the learning and development programs that were established and delivered between 2017 and 2022 to enable the adoption of best practices in Project Management given the evolving context of projects and programs include the following.

Agile Mindset   As a PMO that operates in an IT and Digital Transformation function, agile mindset and agile methodologies were important to incorporate into the framework of best practices.

The training program was not only targeted to project managers and core PMO team but also to other project team members and key stakeholders that enable agile product development, agile procurement, and continuous release.

The purpose was to create a Hybrid PMO where predictive and adaptive methods are used simultaneously to bring the best project results.

Following this program and the formal establishment of an agile flavor of the PMO framework, a number of projects in the e-services and mobile development domain were launched and managed using the best practices of agile methodologies.

Program Management   With the formation of a number of strategic initiatives of related projects, there was a need to manage these in a program structure.

The KAUST IT PMO invested in training and developing its core members as well as the key stakeholders on PMBP.

This led to the adoption of best practices in Program Management in running a number of high-impact programs.

Change Management   Project managers are by far change enablers. Change management is a key capability that is required for project managers and many others operating in the project environment.

In the search for the appropriate learning program, we opted for a program that uses more situational analysis and simulation of a real-life project scenario to practice and implement the change management methodology.

The outcome was a very engaging and enlightening experience with a practical toolkit of tools and techniques that the participants were able to implement and use in different projects.

Agile at Scale   With the increased adoption of agile methodologies and the increased complexity of the program structure, the context required a number of agile teams and release trains to operate simultaneously.

Following the search for best practices that would enable this transformation, Scaled Agile was selected as a framework of best practices that would facilitate the objectives of this transformation.

Following the learning program, the practice and adapted Scaled Agile Framework were established and the key events and artifacts were defined.

The training and certification allowed the teams to speak a common language and understand the purpose and value of different events, activities, and backlogs. It also enabled effective discussion between product and delivery teams around well-structured epics, features, and user stories. This resulted in an agile delivery cycle with better alignment and outcomes following sound practices.

Adaptive and Servant Leadership

Another learning program that we established and offered to the community of project professionals in KAUST was Adaptive Project Leadership in a Digital World.

The objective was to enable the team with the knowledge, tools, and techniques that support their project leadership in the context of digital transformation programs. The concept of Servant Leadership is also key to the culture change management that is required with implementing agile and hybrid projects.

Some examples of the topics of focus include: creating and executing a vision; coaching, facilitating, motivating others, influencing without authority, managing stakeholders, and managing conflict.

Network and Partnerships

The function and pursuit of building capabilities do not operate in silo. It is also not a one-party effort. In order for it to be successful, it is necessary to have strong leadership support, work with partners, and engage with the larger CoP on the national and international levels.

The engagement took several forms from actively participating in regional and international project management events to engaging with local PM organizations and chapters in order to share experiences and comprehend the evolving patterns, trends, and industry best practices. We actively discuss with different learning institutions and partners the strategy, priorities, and learning need in order to come up with solutions that are fit for our objectives.

12.3 PHILIPS HOSPITAL PATIENT (HPM) SERVICES AND SOLUTION DELIVERABILITY[4]

Achieving Solution Implementation and Services Excellence in Healthcare Business

Michael Bauer, Strategic Capability Portfolio Leader in Philips Hospital Patient Monitoring (HPM) Services and Solution Deliverability, and Mary Ellen Skeens, Solution Design and Delivery Innovation Leader at Philips Business Group HPM

Services and Solution Deliverability, describe why a Scalable Solution Design and Delivery Services approach is needed due to varying customer needs, a broad range of solution offerings—resulting in different project complexities. Both highlight as well how to combine the right mix of solution-specific and solution-independent capabilities for a successful implementation to achieve Customer Success.

In *Innovation Project Management: Methods, Case Studies, and Tools for Managing Innovation Projects*, Michael Bauer and Mary Ellen Skeens described healthcare driving Solution Innovation, understanding customer needs, and considering solution complexity, as well as enablers for achieving Solution Design and Delivery Service Excellence.[5]

In *Project Management Next Generation: The Pillars for Organizational Excellence*, Michael Bauer, and Mary Ellen Skeens described the dynamics of the shift to healthcare customer value-driven solution projects, as well as enablers for achieving Customer Success in Healthcare Solutions Business.[6]

In this section, we review key trends in healthcare driving Solution Innovation, translation of these into customer needs and solution complexity, as well as enablers for achieving **Solution Design and Delivery Service Excellence**, including:

- Scalable and role-specific Solution Design and Delivery Services framework.
- Customer Lifecycle and Customer Experience.
- Communities of Practice and Social Learning.
- Process Harmonization and Standardization.
- Continuous Improvement.

About Royal Philips   Royal Philips (NYSE: PHG, AEX: PHIA) is a leading health technology company focused on improving people's health and well-being and enabling better outcomes across the health continuum—from healthy living and prevention, to diagnosis, treatment, and home care. Philips leverages advanced technology and deep clinical and consumer insights to deliver integrated solutions. Headquartered in the Netherlands, the company is a leader in diagnostic imaging, image-guided therapy, patient monitoring, and health informatics, as well as in consumer health and home care. Philips generated 2021 sales of EUR 17.2 billion and employs approximately 78,000 employees with sales and services in more than 100 countries. News about Philips can be found at www.philips .com/newscenter.

The **HPM Business** is a software and solutions business encompassing patient monitoring and its capabilities. Reaching over 500 million people every year, HPM solutions are advanced intelligence platforms providing key insights and information to clinicians when and where they need it. The ultimate priority for the HPM Business

5. Kerzner, H. (2019) *Innovation Project Management: Methods, Case Studies, and Tools for Managing Innovation Projects*, New York, Wiley; p. 190–202.

6. Kerzner, H. (2022) *Project Management Next Generation: The Pillars for Organizational Excellence*, New York, Wiley; p. 449–465.

Group is to enable smart decision-making for caregivers, administrators, and patients such that workflows are improved, costs are controlled, efficiency is increased, and, importantly, better health outcomes are supported.

Mega Trends in Healthcare Toward Innovative Solutions[7]

The healthcare industry is quickly evolving. Digital technology and innovative solutions are shaping the industry to support individuals taking charge of their own health.

There are four key trends driving disruptive change in healthcare technology. They include:

1. The shift from volume to **value-based care**, due to global resource constraints. The World Health Organization estimates that 18 million more healthcare workers are needed to close the gap to meet the demands of the system in 2030.[8]
2. The growing population of **older patients** and increase in **chronic conditions** such as cardiovascular disease, cancer, and diabetes. The world's older population is forecasted to outpace the younger population over the next three decades.[9]
3. Patients are exerting **more control over healthcare decisions** and choosing which healthcare organizations they utilize as consumers. With access to digital healthcare tools and the incentive of reducing out-of-pocket expenses, patients are making more carefully informed decisions regarding care.[10]
4. The initiation of **healthcare digitalization**, triggering growing demand for **integrated solutions** over discrete products. Physicians can now leverage digital and artificial intelligence solutions to automate data collection and translate it into useful information to make evidence-based medical decisions.[11]

These trends have resulted in healthcare organizations striving to find solutions to reach the goals of improving clinical, patient, and financial outcomes while also addressing the well-being and engagement of healthcare employees.[12]

Philips has adopted a solution-oriented approach in delivering value to customers via integrated solution offerings. Philips defines a Solution as a combination of Philips (and third-party) systems, devices, software, consumables, and services, configured and delivered in a way that solves customer- (segment) specific needs.

---

7. See https://www.results.philips.com/publications/ar21

8. See Health workers density and distribution: https://www.who.int/data/gho/data/indicators/indicator-details/GHO/community-health-workers-density-(per-10-000-population)

9. He, W., Goodkind, D., Kowal, P., (March 2016) *An Aging World: 2015, International Population Reports,* pp. 6.

10. Cordina, J., Kumar, R., Martin, C.P., and Jones, E.P., (July 2018) *Healthcare consumerism 2018: An update on the journey,* pp. 4–6.

11. World Economic Forum and The Boston Consulting Group, (2017) *Value in Healthcare Laying the Foundation for Health System Transformation,* pp. 19.

12. Bodenheimer, T., Sinsky, C. (December 2014) *From Triple to Quadruple Aim: Care of the Patient Requires Care of the Provider,* Annals of Family Medicine, 12 (6), pp. 574–575.

Varying Customer Needs and Different Solution Complexities

Solutions address the customer need to cost-effectively maximize speed and consistency of clinical decisions, actions, and usage of patient information for reduced clinical variation and improved clinical performance within their IT ecosystem.

Designing and Delivering Solution Projects is a local activity performed at hospital organizations in every country, often in the local language. Philips operates with both local and centralized resources to support this. This global/local organizational design often leads to virtual working environments with specific requirements to efficiently drive the Solution Project Delivery. The requirements and maturity levels in each country, market, and hospital customer greatly vary. Each project in a hospital is unique and varies in duration (from weeks to years), in size (up to multimillion euros/dollars), and in complexity (from stand-alone solution for one clinician to regional distributed solution for thousands of users) (see Figure 12–1). The range of size and complexity for Solution Projects in healthcare is broad, it includes simple products, highly configurable systems, as well as software and services including clinical consulting. It is influenced by different customer situations, demands, and existing and new technologies. A Solution Design and Delivery Framework addresses customer needs and requirements, which vary from project to project:

- **From** Single Department **To** Multi-hospital deployment across country borders.
- **From** Standalone solutions in group practice or small departments **To** Complex solutions with different systems, software, and services fully integrated into the hospital infrastructure across multiple departments.
- **From** Simple clinical processes **To** Highly designed workflows.
- **From** "Greenfield" implementations across all modalities and applications **To** Customized solutions into an existing hospital environment.

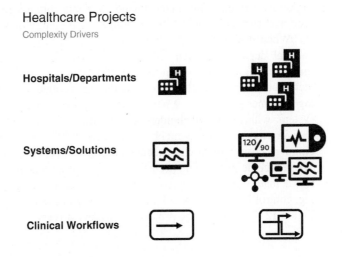

Healthcare Projects
Complexity Drivers

Hospitals/Departments

Systems/Solutions

Clinical Workflows

Figure 12–1. Healthcare Projects: Different Drivers Influence Complexity, © Koninklijke Philips N.V., [2022]. All rights reserved.

Healthcare Projects

Complexity Drivers

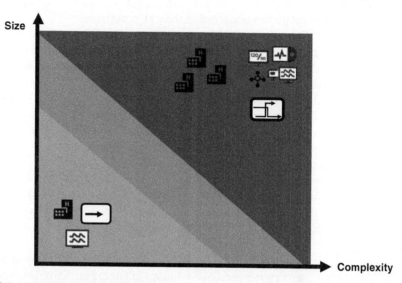

Figure 12–2. Healthcare Projects: Different Complexity Levels, © Koninklijke Philips N.V., [2022]. All rights reserved.

The variability in customer needs drives the Solution Commercialization process. Important elements considered include scalable solution requirements in product and service design, solution delivery readiness, and quality of execution in markets.

When designing and delivering low-complexity, single-solution projects in one hospital department on a simple, stand-alone network, the Project Manager will implement basic tasks within the five PMI[13] process groups. They include stakeholder identification, plan development, performing installation, controlling scope, and obtaining customer acceptance. When a high-complexity solution is delivered within a health system, with many stakeholders and a variety of solutions, the Solution Design and Delivery model becomes much more detailed. The Project Manager and the multidisciplinary Project Team will execute additional tasks from the five PMI process groups. These include performing a customer expectation analysis, developing a Stakeholder RACI matrix, performing a workflow analysis, performing solution integration testing, controlling risk, cost and labor budgets, and conducting Lessons Learned reviews. Figure 12–2 illustrates the project complexity drivers.

---

13. PMI stands for Project Management Institute, see www.pmi.org for more information.

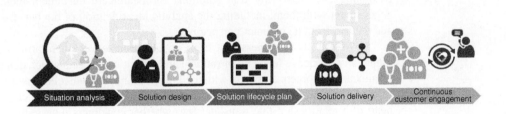

Figure 12–3. Solution Services along Customer Lifecycle, © Koninklijke Philips N.V., [2022]. All rights reserved.

Integrated Solutions and Services Offering Along the Customer Lifecycle

In 2015 Philips HPM Services and Solution Deliverability strategized to follow a fully integrated approach to how to offer, implement, and service solutions from a process and methodology perspective. This is gaining importance as the HPM portfolio transitions more and more into a Solutions and Services business, a holistic approach is key for scoping and designing, delivering, and servicing solutions at the customer along the whole **Customer Lifecycle**[14] (see Figure 12–3).

It starts with an intensive dialogue with the customer to have a complete **Clinical Discovery** and fully understand the customer needs, followed by the **Solution Design** phase during presales, where reference architectures and design guidelines help to shape a strong customer solution. This phase is essential for following Solution Delivery phases, builds the real foundation, and is documented in a Statement of Work (SOW). "Having a solid foundation is an essential element for delivering project excellence."[15] McKinsey emphasis the importance of technical and commercial capabilities as follows: "companies that invest in this capability are able to achieve win rates of 40 to 50 percent in new business and 80 to 90 percent in renewal business."[16] Afterward a multiyear **Solution Lifecycle Plan** is aligned with the customer before the **Solution Delivery** phase to implement the solution initially. Additional services are provided over the lifecycle to optimize customer value. **Continuous Customer Engagement** is key for full success and enabling the desired customer outcome (including continuous partnership and collaboration going forward).

14. Find more information about the Health Care Technology Life Cycle from the University of Vermont: its.uvm.edu/tsp

15. Source: Martin, M. G.: Delivering project excellence with the statement of work, Chapter 2, page 1. Management Concepts Inc., 2010.

16. Source: McKinsey & Company Podcast, Let's talk about sales growth, September 2016.

Customer Lifecycle and Customer Experience

Philips strives for long-term strategic partnerships with its customers and this goes beyond "just" delivering traditional project deliverables within the constraints of time, cost, and scope. Designing and delivering true solutions for healthcare customers requires a deep engagement with them, including the analytic identification of the benefits, outcomes, and value—from a customer perspective.

HPM has an awareness that each organization leaves an imprint with the customer, an experience made up of rational and emotional aspects which determines what Healthcare customers associate with the Philips brand, and what Philips means to them. This is especially pronounced in a services business. Customer experience is at the heart of a relationship that translates into whether customers repeatedly rely upon the organization's capabilities and embrace them as trusted advisors.[17] Therefore, another important aspect is how the organization actively and holistically "designs" the customer experience end-to-end (E2E) in terms of capabilities, tools, and processes. HPM strives to apply this customer experience-focused approach across the entire Customer Lifecycle from the point in time the customers share their vision and commission Philips with the realization, through Solution Design, Delivery, and Continuous Engagement and Improvement.

In this context, **Solutions Implementation and Services Excellence** are key strategic ingredients to ensure HPM reliably and repeatedly delivers the desired customer experience. Hence, building and sustaining Project Solution Implementation and Services Excellence and reaching a high level of project management maturity with solution implementation projects is an adamant ambition of vital importance for both, the customer and Philips.

According to the Project Management Institute's (PMI's) 2021 Pulse of the Profession survey,[18] only 73% of projects at organizations meet their original business goals and intent. For Philips HPM Services the ambitions for successful Solution Project Implementation are high and require high maturity in how Solution Projects are designed and delivered.

This ambition was the key strategic driver to implement a Global HPM Services and Solution Deliverability (SSD) function that has broad scope from a methodology, process, and tool perspective around the implementation of Solutions and Services.

The HPM Services and Solutions Deliverability Function and Strategic Capabilities

The focus on HPM Strategic E2E Capability Portfolio and the establishment of Solution Design and Delivery Innovation teams to realize it was a clear strategic decision and has full support from top management to drive Solutions Implementation and Services Excellence strategically.[19]

---

17. Sources for customer experience concepts: www.cxpa.org, www.temkingroup.com, www.beyondphilosophy.com

18. Source: PMI's Pulse of the Profession 2021, page 6.

19. See as well for importance of strategic support of Project Management: PMI's Pulse of the Profession™, March 2013, page 3.

Healthcare Projects

Solution needs a set of Solution Design and Delivery Services and Capabilities

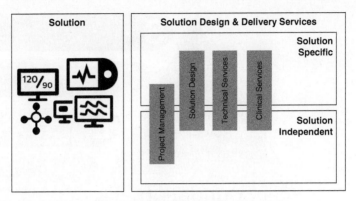

Figure 12–4. Solution Projects need Set of Services and Capabilities, © Koninklijke Philips N.V., [2022]. All rights reserved.

HPM considers the following important **Capabilities** with regard to **Project Solution Implementation and Services Excellence**:

- **Skills:** Well-educated, certified, skilled (hard, soft), and continuously trained Solution Architects, Customer Delivery Managers, Projects Managers and Team, and Services personnel with a professional mindset, appearance, and behavior. This also includes recruiting the best talent.[20]
- **Processes:** Highly efficient, standardized, lean, repeatable, and well-documented processes, which are continuously improved.
- **Tools:** Highly integrated and efficient tools, templates, and applications from the project acquisition until the end of the project.
- **Content**: Role-specific content (templates, training material, diagrams) around the Solution Design and Delivery, called Solution Standard Work (SSW).

Some of the solution-related **capabilities** are specific to the solution, some are more generic and independent of the solution (see Figure 12–4). For any solution, it is a combination of both. The solution-specific capabilities are directly linked to the solution Innovation. To be fully successful with selling, designing, and delivering solution projects these capabilities need to be prepared, designed, and deployed to executing organizations in the countries.

Solutions Implementation and Services Excellence is not seen as a static goal, the ambition is to continuously raise the bar for project implementation maturity as well as oversee overall competencies and project delivery capacity.

---

20. See as well for importance of Project Management talent management: PMI's Pulse of the Profession In Depth Study: Talent Management, March 2013.

Scalable Project Implementations

The SOLiD Design & Delivery Framework

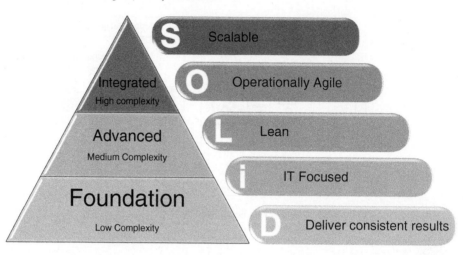

Figure 12–5. Scalable Solution Design and Delivery Framework, © Koninklijke Philips N.V., [2022]. All rights reserved.

HPM SSD enables the ambition of Solutions Implementation and Services Excellence where the following aspects need to be highlighted:

- **Solutions and Services Implementation Excellence matters**: key aspect to value and improving skills, processes, and tools.
- **Change Management**: identify, drive, and implement improvements and changes in the organization.
- **Standardization:** enable standardized and lean practices and processes across product domains and regions.[21]
- **Continuous learning**: train, review, and mentor as required.
- Facilitation of CoP **and Social Learning** for all the different professions—key aspect to enable to share, learn, leverage, network, and communicate.[22]

The SOLiD Framework   In close collaboration with the HPM Community around the globe, the HPM SSD team developed the SOLiD framework (see Figure 12–5). The SOLiD framework is HPM's approach

---

21. "High performing organizations are almost three times more likely than low-performing organizations (36 percent vs. 13 percent) to use standardized practices throughout the organization, and have better project outcomes as a result." Source: PMI's Pulse of the Profession™, March 2013, page 10.

22. See as well wenger-trayner.com/Intro-to-CoPs for more information about Community of Practice (CoP).

for designing and delivering customer-facing Solution Implementation Projects and Services. SOLiD is an abbreviation and stands for:

- **S**calable, which allows flexibility to meet the demands of our low, medium, and high-complexity projects.
- **O**perationally agile, meaning it is the first iteration and we will continue to build and improve via iterations over time.
- **L**ean only including the tasks that would add value to the Project and Services Team and even more important to hospital customers.
- **I**T-focused, including the structure, tools, and processes needed to successfully manage Projects and Services in an IT Solutions environment and lastly SOLiD will help to
- **D**eliver consistent results and bring business value by providing a standard and lean way of working.

The underpinnings of this framework are the process groups of initiating, planning, executing, monitoring/controlling, and closing as defined in the PMBOK (Project Management Body of Knowledge) by the PMI[23]. Each process group is then further broken down into more specific processes and procedures detailing how HPM manages the implementation of Solution Projects and Services. In 2019, Philips adopted SOLiD as the global, corporate standard and embedded the defined Solution design and delivery activities into the Philips Excellence Process Framework (PEPF).

Scalability in Project Implementations is key to allow the right, flexible, agile, and efficient approach per project but to leverage from a rich tool set. Solution Projects are defined by their level of complexity. Typical factors when defining complexity are total cost of the project, number of team members involved, number and size of deliverables, complexity of deliverables, and complexity of the customer environment and timeframes involved.

PMI defines a project as being different from other ongoing operations in an organization because, unlike operations, projects have a definite beginning and an end— they have a limited duration and bring value to the organization.

The SOLiD framework is designed to help offer guidance based on three complexity levels:

- **Foundation**: Designed for low complexity projects with basic Project Management tasks required, e.g., basic testing and simple SOW.
- **Advanced**: Incorporates the tasks in Foundation with additional activities/ processes to help better manage medium complexity projects, including Solution Design components.
- **Integrated**: Incorporates both Foundation and Advanced frameworks with additional activities needed to manage more intricate, high-complexity projects; usually, more technical integration and testing activities and different levels of risk and stakeholder management are needed.

---

23. See www.pmi.org/pmbok-guide-standards/foundational/pmbok for more information.

The SOLiD framework supports throughout the Customer Lifecycle (see Figure 12–6). The following diagram gives an overview with a focus on Solution Design and Delivery:

It shows how processes are structured and how the process frameworks overlap for smooth handoff between areas of responsibilities.

- **Solution design**: Tight teamwork between sales and project management is highly important in solution projects following a defined Sales Bid Management process.
- **Solution delivery**: Following the Solution Design a successful implementation, in line with what was scoped and following a lean and scalable Project Management approach.

Solution Delivery
The SOLiD Design & Delivery Framework for HPM

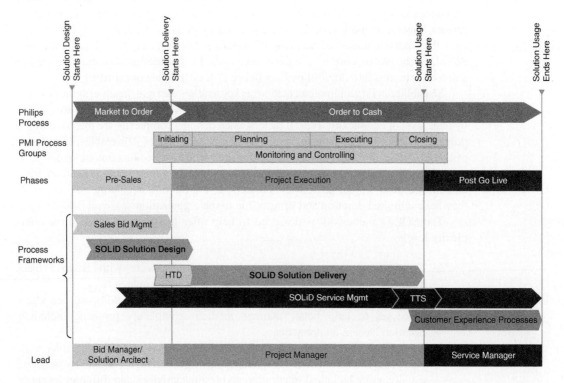

Figure 12–6. Scalable Solution Design and Delivery Framework across Processes, © Koninklijke Philips N.V., [2022]. All rights reserved.

- **Service management**: Using components from the ITIL[24] toolbox is an industry best practice to set up a highly efficient, state-of-the-art IT services concept.
- **Customer experience**: To make current customer experience visible at different project milestones and during the entire Customer Lifecycle customers are invited to key touch points to share their feedback with the HPM organization. The survey process continues beyond implementations and upgrades, with ongoing support and maintenance. Feedback received is evaluated under the aspect of congruence with the desired customer experience, identifying room for improvement, and also learning about strengths with regard to Solution Implementation and Services Excellence. This closed-loop process is a key characteristic of a continuously learning and improving organization.

An important takeaway is that the different processes are all interrelated and thus break the traditional silo approach. Communication and teamwork are some of the key aspects that prevailed during the definitions of these processes. Especially in a global organization like Philips HPM, it is important to implement, train and improve harmonized, standardized and lean processes. It is also important that everyone speaks the same Project Implementation and Services "language" and uses the same terms. This is one of the reasons why HPM Project Managers need to be trained on the SOLiD framework.

Value of Social Learning Within the Hospital Patient Monitoring Services and Solutions Delivery Business

In developing the solution design and delivery capabilities needed for customer success, it is imperative to ensure connections exist between project team members and their domain peers for the purpose of ongoing learning, knowledge sharing, continuous process improvement, and people development. In a cross-functional and global environment, it is important for the project team members to have access to a network of peers that share their experiences and lessons learned. Two specific approaches have been utilized in Philips: 1. CoPs and 2. Social Learning.

The basic model for a CoP includes three main parts—domain, community, and practice. First, the shared domain of interest people identify with is defined. The community then determines who should be included and what kind of relationships they should form. Lastly, within the practice, members determine what they want to do together and how they can make a difference in practice. These elements are essential for a CoP to thrive. There are key differences between CoPs and formal work groups, which should be recognized. The purpose of CoP's is to develop self-selecting members' capabilities and to exchange knowledge with one another whereas, a formal working groups' purpose is to deliver a product or service with everyone reporting to the group's manager. Both groups can be complimentary to each other and are essential for innovation to occur.[1]

---

24. ITIL (IT Infrastructure Library) developed by the Office of Government in Commerce (OGC) in the United Kingdom.

1. Wenger, E., Snyder, W. (2000) Harvard Business Review: *Communities of Practice: The Organizational Frontier.*, Retrieved from https://hbr.org/2000/01/communities-of-practice-the-organizational-frontier

The real value practitioners' gain from being part of a CoP is to help each other solve problems, reflect on practices to improve, keep up with change, cooperate on innovation, and find a voice to gain strategic influence.[2] As new communities emerge, within and across business groups, functions, and markets, even more knowledge is shared. In today's world, knowledge sharing is simply not enough. To truly transform and innovate faster, social learning spaces are necessary to pay attention to the data, engage uncertainty and move people to make a difference. Learning partnerships created to cross boundaries can turn into valuable learning assets. Value-creation frameworks provide a structure for capturing the flow of events or ideas from social learning spaces (virtual or in person), through data and stories. The information learned together in the space, new ideas, methods, and tools flow back into the real world through the small and big actions taken as a result. When the findings are then brought back to the community, that is referred to as creating learning loops. Activities to learn together include regular online meetings to exchange information on key topics, support in the preparation of face-to-face training and conferences, as well as to build new artifacts in creation of innovative content for a specific product domain or solution implementation and services processes. Regular interactions are the best way to get instant feedback from the community and for the community while also facilitating learning loops.

An example of how to develop learning loops is to encourage open collaboration, between boundaries, by bringing people together to share stories of both successes and failures and then to solicit feedback and allow time for questions and discussion. When people are open to share their struggles, it speeds up learning for others because they contribute to finding ways to help solve problems.[3] The results are impressive because not only does the person sharing get advice and new ideas but the community is also engaged in active problem-solving, which can be applied to their own challenges. Through active dialogue and problem-solving, important learning can thrive.

Customers are increasingly looking for solution propositions that will help them add value and address their business challenges. Communities of Practice create environments where exchanges can happen in real time between various functions that trigger new ideas leading to solutions customers really need. It is important for community moderators to be proactive in looking for relevant stories that can result in innovative successes when space is allowed for this to happen. Bringing failures forward in communities is an opportunity to apply the learning loop model while solving complex problems together. This is viewed as a very positive experience for members. Community moderators in Philips have witnessed that applauding failure can create powerful learnings in the right environment.

Key Takeaways to Achieve Solution Implementation and Services Excellence

Philips and HPM strive for solution excellence, which requires continuous improvement around all services and capabilities. Even though it is not an absolute objective per se, it is considered a proactive way to anticipate and fulfill the needs of customers with regards to solution project

---

2. See thesystemsthinker.com/communities-of-practice-learning-as-a-social-system/ for more information

3. Wenger, E., Trayner, B., & de Laat, M. Open Universiteit, Ruud de Moor Centrum. (2011). *Promoting and assessing value creation in communities and networks: a conceptual framework* (Rapport 18). Retrieved from website: http://www.bevtrayner.com/base/docs/Wenger_Trayner_DeLaat_Value_creation...

management, customer success, and outcomes realization. The following aspects are critical to build and improve solution design and delivery-related capabilities:

- **Scalable Project Implementation** enables success for different project complexities.
- Holistic and fully integrated approach for the **Customer Lifecycle** is needed and is key for scoping and designing, delivering, and servicing solutions for healthcare customers.
- **Solution Design and Delivery Services Excellence matters**—key aspect to value and improve capabilities (skills, processes, and tools). Some of the capabilities are solution specific, some capabilities are generic for any solution.
- **Process Harmonization and Standardization**[25] are highly important for the success of an organization operating globally and reducing complexity. Tight integration in the upstream processes (e.g., sales, bid management, analytics for outcome measurements) and downstream processes (e.g., entire lifecycle) are very important too. This includes Framework Integration as well.
- Solution Implementation and Service Excellence is not static objective. It requires **continuous improvement** around **people, processes, and tools**. Even though it is not an absolute objective per se, it is considered a **proactive way** to anticipate and fulfill the **needs of our customers with regard to Solutions and Services**.
- **Change management**—identify, drive, and implement improvements and changes in the organization.
- **Continuous learning**—train, review, and mentor all project team members as required.
- Facilitation of **CoP** and **Social Learning** for all the different professions—key aspect to enable sharing, learning, leverage, network, and communicate.

12.4 CHURCHILL DOWNS INCORPORATED: ESTABLISHING A PMO _____

Deciding to implement a PMO is easy. Being able to do it requires that certain obstacles be overcome. Chuck Millhollan, formerly director of program management at Churchill Downs Incorporated (CDI), discusses the chronology of events his organization went through and some of the obstacles that had to be overcome.

*   *   *

One of our primary barriers to implementing structured project, program, and portfolio management processes was "familiarity." The CDI PMO, chartered in April 2007, is the first in the thoroughbred horseracing industry. Our senior-most leadership understood the need for a structured, standardized approach to requesting, approving, and

---

25. "High performing organizations are almost three times more likely than low-performing organizations (36 percent vs. 13 percent) to use standardized practices throughout the organization and have better project outcomes as a result." Source: PMI's Pulse of the Profession (2013), The High Cost of Low Performance 2013 p. 10.

managing projects and maintaining the project portfolio; however, many of the organizational resources had never been exposed to formal project management concepts.

Our executives took an active role in the implementation process.

I would say this is one of the primary factors influencing the early success enjoyed by the CDI PMO. We chartered our PMO with clearly defined vision and mission statements and business objectives. Our CEO signed the charter, granting authority to the PMO to expend organizational resources related to managing capital projects.

- Our PMO was chartered in April 2007.
- We developed a threefold mission focused on the need identified by our senior leadership:
  1. Establish, facilitate, and manage the project portfolio selection and funding process.
  2. Create a foundation for consistent project success throughout the organization through development of a strong and pervasive project management discipline within CDI's project teams.
  3. Guide key projects to a successful conclusion by providing project management leadership, while improving the quality and repeatability of related processes.
- We defined the PMO's business objectives and linked progress to the PMO director's compensation plan. Objectives included:
  1. Develop and implement standards for project selection.
  2. Develop and implement a standardized project management methodology.
  3. Build project management professionalism among CDI staff.
  4. Manage the CDI project portfolio.
  5. Direct project management for key strategic initiatives.
  6. Ensure processes for benefits realization.
- We conducted training classes on project management, team building, critical thinking, and so on, to not only share our knowledge but also to build relationships with project team members and other stakeholders.
- The PMO facilitated a book club (also chartered with clearly defined objectives). This process received recognition throughout the organization and directly contributed to developing relationships between different departments. Our book club membership includes representatives from nine different departments, ranging from vice president-level members to individual contributors.

  Objective 1: Personal growth through completing chosen books and active involvement in discussions.

  Objective 2: Explore creative ideas and ways of addressing real-world business issues through practical application of concepts and shared learning as related to Churchill Downs and respective teams.

  Objective 3: Promote interaction among different functional areas within the Churchill team by active participation in book club discussions and sharing opportunities for addressing real-world work-related issues in a safe, confidential environment.

Objective 4: Share learning within respective teams through intradepartmental discussion and implementation of learning-related concepts.
- The primary driving factors behind CDI's decision to staff a PMO were challenges with defining and managing the scope of projects, effectively allocating resources among multiple projects, and bringing projects to a defined closure.

12.5 CHURCHILL DOWNS INCORPORATED: MANAGING SCOPE CHANGES[26]

Mature PMOs either participate directly in the scope changes above a certain dollar level or set up processes for controlling scope changes. Chuck Millhollan, formerly director of program management at CDI, identifies six steps necessary for scope definition and change control.

* * *

Step 1: Be Lean

Trying to introduce any type of structure or control in an organization or environment that has been absent of controls can present a significant challenge. Before a project management organization can address scope change control, it must implement a process to define scope. Getting organizational decision-makers to accept the project management precepts is not overly difficult, but changing organizational behavior to leverage these principles is another matter altogether. The more change we attempt to introduce into an environment, the more difficulty that environment has in adapting to, accepting, and embracing that change. To avoid the natural resistance to excessive change, a logical approach is to limit the scope of change and focus on immediate needs. Focus on the foundation and basics. Why have a complex, highly mature process if you are not consistently performing the basics well?

Step 2: Define Preliminary Scope

The immediate need for an organization without processes for capturing the business objectives associated with project requests is to define a structured approach for documenting, evaluating, and approving the preliminary scope of work. Note that *approving* the scope of work involves more than shaking heads, shaking hands, or a causal agreement on broad, subjective criteria. Approvals, in project management, imply documented endorsements, more simply, signatures that provide evidence of the agreement and a foundation to build upon. It is important to emphasize to stakeholders and sponsors unfamiliar with

---

26. Section 12.5 is from C. Millhollan, "Scope Change Control: Control Your Projects or Your Project Will Control You!" Paper presented at PMI® Global Congress 2008—North America, Denver, CO. Newtown Square, PA: Project Management Institute. Copyright © 2008 by Chuck Millhollan. Reproduced by permission of Chuck Millhollan.

our profession's structured approach to managing projects that accepting a preliminary scope of work does not mean that you are locked in for the remainder of execution. Nothing could be farther from the truth. Instead, you are protecting their interests by beginning to set boundaries upon which effective planning can begin. In other words, you are increasing the probability (remember the research) that the project will be successful.

Step 3: Develop Understanding of What Final Acceptance Means to Project Sponsor or Sponsors

How do we know when we have arrived at a destination? When traveling, we know our trip is complete when we reach our intended destination. Likewise, we know that a project is complete when we have delivered on the business objectives identified in the project charter, right? Well, yes . . . and then some. The "and them some" is the focus of scope change control. How does your organization define final sponsor acceptance? The recommended approach is to define sponsor acceptance for stakeholders using plain language. Sponsor acceptance is the formal recognition that the objectives defined in the original agreed-upon scope of work have been met, plus the objectives agreed upon in all of the formally approved change requests. This plain definition helps to avoid the differing perceptions around what was wanted versus what was documented.

Step 4: Define, Document, and Communicate a Structured Approach to Requesting, Evaluating, and Approving Change Requests

What is a change request? Some schools of thought suggest that changes are limited to requests for additional features, deliverables, or work. While this paper is focused on these types of change requests, or scope change requests, it is important to note that any change that has the potential to impact expectations should follow a formalized change request, approval, and communication process. Remember, aggressively managing expectations is our best opportunity to influence our stakeholder's perception of value. Scope, budget, schedules, and risks are typically interdependent and directly influence our stakeholder perceptions. Also, remember that the most effective change control processes include risk assessments that evaluate the potential risks of either approving or disapproving a change request.

Keep in mind that too much bureaucracy, too much analysis, or too much unnecessary paperwork will give stakeholders an incentive to circumvent your process. If you want your stakeholders to avoid, ignore, or completely bypass your process, including a great deal of administrivia. Administrivia is the new word for "trivial administrative process." (As the author, I reserve the right to add to the English language.) Remember, our profession's focus is on delivery and business results, not just adherence to a predefined process. Taking a lean approach to scope change request documentation can help influence acceptance of this sometimes painful, but vital, process for capturing change.

Process tip: Determine early (either as an enterprise standard or for your specific project) what the tripwires and associated levels of authority are for approving a requested change. What level of change can be approved internally? For example, a change with an impact of less than one-week schedule delay or budget impact of under $10,000 may be approved by the project manager. What needs to be escalated to the project sponsor, what needs to be reviewed by a change control board or governance council? Determining these decision points in advance can remove a great deal of the mystique around how to manage change.

Ensure that everyone understands the difference between the natural decomposition process and identifying new work that must be accomplished to deliver on a previously agreed upon business objective and work associated with new or modified deliverables. Remember that omissions and errors in planning may lead to schedule and budget changes, but are usually not scope changes.

Step 5: Document and Validate Full Scope of Work (Create Work Breakdown Structure) A great approach for defining all of the work required to complete a project is to start with the desired end state and associated expected benefits. What work is required to provide those benefits? What work is required to reach the approved end-state goals (or business objectives)? Plan to the level of detail necessary to effectively manage the work. Decomposing work packages beyond the level required for effective management is considered administrivia. Note that defining and communicating the processes for final sponsor acceptance and requesting changes both come before traditional decomposition. Why? Terrific question! The natural planning processes that we follow in breaking down business objectives into definable work packages can be a catalyst for change requests. We want to communicate up front that change is not free and that additional requests will need to be formally requested, documented, agreed upon, and approved before being included in the project scope of work.

Step 6: Manage Change Your foundation is laid, you have documented the preliminary scope, you have defined processes for sponsor acceptance, you have defined and documented scope change request processes, and you have developed your work breakdown structure, now the only thing left is to manage according to your policies and plan. Almost forgot . . . you have to manage the change requests that are guaranteed to come too! Scope change control protects the project manager, and the performing organization, from scope creep and contributes to managing stakeholder expectations.

A question that frequently comes up among practitioners is "What do I do when my leadership does not allow me to define, document, and manage change?" This is a real, practical question that deserves a response. The instinctive approach is to communicate the necessity for a structured approach to documenting and managing scope. As our peers will confess, this is not always sufficient to get the support we need to set organizational policy. We can attempt to implement these processes without formalization, or just "do it anyway." This can be an effective approach for demonstrating the value, but can also be perceived as a self-protective measure instead of a process used to increase the likelihood of project success. People can be leery of someone else documenting requests, justifications, etc. . . . for their needs. Ensure that you share the information and provide an explanation as to why this approach is designed to ensure you are managing to their expectations. In general, people have difficulty not accepting altruistic approaches to meeting their needs.

Learn from Other's Lessons: Leveraging experience, best practices, and lessons learned, the
A Real-World Application   Churchill Program Management Office began with the basics; they
                                     chartered their PMO. The threefold mission of the newly founded
PMO was to establish, facilitate, and manage the project portfolio selection and funding process; create a foundation for consistent project success throughout the organization through development of a strong and pervasive project management discipline; and to guide key projects to a successful conclusion by providing project management leadership while improving the quality and repeatability of related processes. Sounds fairly standard, right? The mission was then broken down into specific objectives and successful completion of these objectives was tied to the PMO director's compensation.

PMO Objectives:
1. Develop and implement standard processes for project requests, evaluation, and funding to ensure that approved projects were aligned with CDI's business goals and objectives.
2. Develop and implement a standardized project management methodology, to include policy, standards, guidelines, procedures, tools, and templates.
3. Build project management professionalism by providing mentorship, training, and guidance to project teams as they learn and adopt project management processes and best practices.
4. Manage the CDI project portfolio by ensuring required documentation is in place and that stakeholders are properly informed about the ongoing progress of the project portfolio through effective reporting of key performance indicators.
5. Direct project management for key strategic initiatives.
6. Ensure benefit realization by using processes for clearly defining business cases and the associated metrics for measuring project success. Facilitate post-implementation benefit measurement and reporting.

As related to change control, we wanted to ensure that the process was lean, that our stakeholders understood the importance of the process, and finally . . . arguably most important . . . communicated in a way that our stakeholders understood and could follow the change request processes. Here is a thought-provoking question for our practitioners: Why do we expect our stakeholders to learn and understand our vernacular? To aid in understanding and training, we developed visual tools documenting our overall project management processes in a language that they understood. For example, the project "race track" (see Figure 4.20 in Chapter 4) demonstrated to our leadership and project team members what we, in our profession, take for granted as universally understood; that projects have a defined start, a defined finish, and require certain documentation throughout the planning and execution processes to ensure everyone understands expectations and that we will realize the intended benefit from the investment.

For CDI, scope change control begins with the foundation of a completed investment request worksheet (or business case) and an agreed scope of work as outlined in a signed charter. The work is then decomposed to a level of detail required to control the effort and complete the work necessary to deliver on the requested and approved objectives as detailed in that charter and approved scope change requests. A scope change request consists of a simple-to-understand, fill-in-the-blank template and the process is facilitated by the project manager. More important, the scope change request form is used to document the business objectives for a change request, the metrics needed to ensure the change's benefits are realized, the impacts on schedule and costs, the funding source, and the necessary approvals required for including the request in the overall scope of work.

Some of the benefits that CDI has realized to date from this structured approach to documenting and controlling scope include:

1. Retroactively documenting scope for legacy projects, which resulted in canceling projects that were plagued with uncontrolled change to the point that the final product would no longer deliver the benefits presented in the business case.
2. Denying scope change requests based on factual return on investment and impact analysis.
3. Ensuring that requested scope changes would contribute to the business objectives approved by the investment council.
4. Empowering project team members to say "no" to informal change requests that may or may not provide a quantifiable benefit.
5. Demonstrating that seemingly great ideas might not stand up to structured impact analysis.

12.6 PROJECT MANAGEMENT OFFICE - BLITZSCALING AT NANOFORM[27]

Introduction

Nanoform Finland Plc is a publicly traded company based in Helsinki, Finland. A Physics research consortium based at Helsinki University, led by Professors Jouko Yliruusi and Edward Haeggström (one of the company's founders and CEO), invented a new technology in 2012, which was the platform to launch Nanoform in 2015. Nanoform specializes in nanotechnology for pharma and has the proprietary CESS® technology (Controlled Expansion of Supercritical Solutions), which brings hope to unmet needs in curing diseases and provides a more environmentally friendly process that is effective in the delivery of medicines to humans, aiming at reaching 1 billion patients.

The first trial in humans using the nanoformed ingredient piroxicam in late 2020 proved the concept that CESS® nanoparticles can have clear pharmacological benefits,

---

27. Material in this section has been provided by Miguel Cansado, PMO Director, Nanoform. © 2022 by Nanoform. All rights reserved. Reproduced by permission.

enabling the development of drug products with increased pharmaceutical efficacy in humans. Nanoform received the Finnish Medicines Agency (Fimea) GMP certificate in April 2020 and was listed on NASDAQ First North Premier Growth Market in June 2020.

The creation of intellectual property, the development of scientific know-how, the need for enhancing the agility of R&D activities, an effective change and portfolio management structure, and, in general, evolving as a company with advanced management systems and operational excellence have been crucial for enhancing the business and potential customer value of the company. Nanoform's values are quality, partnership, transparency, and ethics. The values are combined with three principles: "Yes, we can"; "The glass is half-full"; "Help your colleagues."

Nanoform has been strategically undertaking accelerated growth, a term the author R. Hoffman calls **_Blitzscaling_**, in which speed is prioritized over efficiency. The early hiring of professionals with 15 to 20 years of pharma industry experience and fostering the expansion of teams of young scientists, most of them PhDs, has been vital to accelerate the growth of the company in the 2019–2022 period, and the approach to the first stage of excellence in industrial operations, which requires a PMO on par with this lightning-fast growth. Currently, 26% of Nanoformers and 57% of the PMO team have a PhD. At the same time, Nanoform's amazing cultural diversity, with 30 nationalities in a 140+ headcount, is enriching and powerful.

Nanoform's accelerated growth faced a major roadblock in the form of the COVID-19 pandemic, which forced us to revise our stakeholder and communication management practices, executing extensive and mandatorily remote work by means of virtual teams (even in companies such as Nanoform, in which a hybrid system already prevailed). Many companies had to engage in remote work for months or even years, had to be even more agile in dealing with the unexpected, more selective on which projects to prioritize, and faced delays caused by supply chain disruptions. Our teams were under increased pressure to balance personal and professional life. In order to facilitate a new way of working, we revisited risk management and issue management, we became more agile, flexible, and proactive in the face of uncertainty. There was a rising technology dependency on project management.

We will describe the evolution of Nanoform's PMO, with a progressive adoption of project, program, and portfolio management best practices. This endeavor includes the stage-appropriate technical project management competencies, power skills, and business acumen of the project managers, suitable to the level of maturity of the company and the types of projects the company is undertaking. The way in which power skills are nurtured and developed early on for all project managers will be briefly explained. A perspective on the inception, growth, and adaptation of the PMO, in line with Nanoform's business growth, will be outlined. The PMO itself can be a vital organism with a gradually increasing role in linking strategy to execution, delivering benefits and value, and improving the customer experience.

Nanoform Company Background 2019–2022

Between 2015 and early 2019, Nanoform's headcount grew from 5 to ca. 20 people. Its workforce mainly comprised scientists, the majority holding a PhD. The company was at this stage applying the technology to internal projects and few clients were contracting out services to Nanoform, which at the time did not have a dedicated project management group, nor a sales and business development team.

In late 2018, there was a major turning point: a decision to hire senior management, including VP-level, and senior managers. These senior executives and the CEO developed the plan to build expertise into a PMO (reporting to Business Operations), and a PMO director was hired. All the senior managers and executives brought competencies and experience in pharma and project management. The company's senior management structure established in 2019–2020 had the required human capital and potential to take the company headcount five times upwards, enabling them to effectively manage a 200 to 250-strong company in 2025.

Regarding the PMO, by the end of 2019, it had two people; one year later, it had four project managers (PMs); by the end of 2021, six PMs. The headcount of the group is foreseen to potentially reach fifteen team members by the end of 2025, supporting 100 projects during that year. The creation of a Project Management toolkit, the adoption, and deployment of best practices throughout the organization, the learning by a blend of on-the-job training, formal training, and individual learning (including emotional intelligence [EI] skills), have created a strong momentum for the people and for the company. At the same time, high-level project management training was deployed to the whole company to raise awareness and place the scientists and technical experts into a project-driven mindset, with its dynamics, uncertainty, and focus on novelty and delivery of value and benefits.

Some key numbers on Nanoform's expansion are shown in Table 12–1.

The Business Case for a PMO at Nanoform

According to Mintzberg, strategy formation is "not just about values and vision, competencies and capabilities, but also about [. . .] commitment, organizational learning and punctuated equilibrium, industrial organization and social revolution." Strategy is a pattern, a consistency in behavior over time. Strategies can be intended and deliberate, but others are emergent, and may include elements of design, entrepreneurship, planning, positioning, cognition, learning, culture definition, power balances, configuration, and adaptation to the environment. In a disruptive company like Nanoform, the PMO must be able to link strategy

TABLE 12–1. KEY NUMBERS ON NANOFORM 2019–2025; F = FORECAST.

Year	2019	2020	2021	2022 F	2025 F
Headcount	40	60	125	142	200–250
PMO headcount	2	4	6	7	15
# projects signed/year	2	12	22	--	100
# operating lines	5	7	10	13	35
# GMP lines	0	1	1	3	7–14
#GMP projects signed	0	1	1	NaN	--
Revenue M EUR	0.05	0.69	2.0	NaN	--

and execution, having in mind all these dimensions of strategy. With these turbulent business conditions, it is a requirement to have a clear strategy, a rapid delivery, an adaptive mindset, and continuous improvement, all steered by strong leadership.

The need for a nimble and effective PMO, able to provide value and benefits to the stakeholders, using standardized methodologies and tools, with appropriate project and program management and skilled leadership, as well as the ability to further hire and develop human capital, capable of connecting and communicating across the whole organization, in which the key stakeholders can be involved, committed, and motivated, is of paramount importance.

Duggal states that we are now living in a different type of world, a **DANCE** world:

- **D**ynamic and changing
- **A**mbiguous and uncertain
- **N**onlinear and unpredictable
- **C**omplex
- **E**mergent nature of projects.

This ecosystem creates increased levels of entropy and requires new skill sets and new approaches, more organic than mechanic, for managing projects and the portfolio. And organic means adaptive: Schwartz claims that a business is a type of **Complex-Adaptive System (CAS)**, a self-organizing system with nonlinearities and complex interactions, one in which leadership can influence, but not control, outcomes." A business organization is presented as a Darwinian universe "in which leaders and managers can set parameters that determine what ideas and behaviors are most likely to survive." The role of leadership in the CAS is then "to influence the evolutionary process so that the organization delivers in its goals." The art of leading a business organization is therefore "the art of communicating an interpretation of business value so that teams and individuals in the organization will self-select the behaviors that will produce the mission of the organization."

A great deal of artistry is needed from each PM and of the PMO as a whole. The multitude of skills a PM must possess is well illustrated by the recently renewed PMI® Talent Triangle, which mentions **Ways of Working, Business Acumen, and Power Skills** as the key clusters of competencies for a PM. According to another author, Cahill, the five crucial capabilities in successful project delivery are **relationship building, collaborative leadership, strategic thinking, creative problem-solving**, and **commercial awareness**. These capabilities are at the core of what a Nanoform PM must be. Again, the PM as a *jongleur*, who is walking on a tightrope, or as a member of a rafting team in turbulent waters, comes to mind. Being comfortable with uncertainty is a must.

This PM artistry is crucial in managing nonlinear work, in which there is creation, intuition, and emotional thinking. Divergent thinking, individually and as a team, must happen during the innovation phase. But success in moving project work forward lies upon choices, criteria, and selection of work to be done, and work not to be done. Design thinking supports the creative process, but also supports making these crucial choices.

At the same time, the PMO facilitates the conversion of nonlinear to linear work, by applying standardized process management processes, facilitating faster iterations and effective learning, increasing the cross-functionality of the knowledge and learning processes, toggling back and forth between the two types of thinking and working processes, and by reducing variability based on experience gained. The outcomes become more and more predictable and hence less uncertain.

The 21st-century PMO (in which the world of 2022 is very different than the one of the early 2000s) must be capable of connecting the dots (and the people), constantly aiming at simplicity, being rigorous without having rigidity (Cf. Duggal). Nanoform's PMO has been built based on a strategic vision, in which the company CEO recognized the importance of establishing a PMO when the company headcount was ca. 20 people, and the annual number of projects and turnover was small.

The PMO at Nanoform has the following responsibilities:

- Aligning and continuously refining the project-support structure and mindset to the company vision and strategy, influencing excellence in operations, guiding decision-making, and supporting priority projects.
- Hiring, training, and developing the project managers, developing their competencies, and defining their career paths.
- Assigning PMs to new business projects, based on seniority, availability, and experience.
- The PM is the key point of contact for the customers, partners, and key internal stakeholders.
- Creation and improvement of standards, metrics, tools, and methodologies.
- To facilitate and deploy the use of current best practices and ways of working, adaptive and well-suited for the needs of different projects.
- Drivers of change, within and across the project teams.
- Guardians of value delivery and benefits realization for clients and partners, as well as for internal stakeholders.

A continued recruitment of project managers, their development on ways of working, by means of formal training (technical competencies, such as project management essentials and advanced-level formal training, agile, lean, and hybrid project management training), and also communication skills and EI and coaching (power skills), and business acumen (by continuous interactions with clients and partners, and sales and BD colleagues), is ongoing and endows them with the necessary knowledge and experience. Preparation for certifications, e.g., PMP®, is a path to maturing these competencies. Strengthening the emotional core of the PMs is crucial early on, therefore EI training, including 1:1 coaching and group coaching, is typically scheduled within the first 3 months at the company. EI skills are summarized in Figure 12–7.

Fostering on-the-job training and self-learning, the sharing of best practices between all PMs, the experience gained in a fast-paced environment, with ca. 40 projects carried out in less than 3 years, and the awareness and training that are given to the

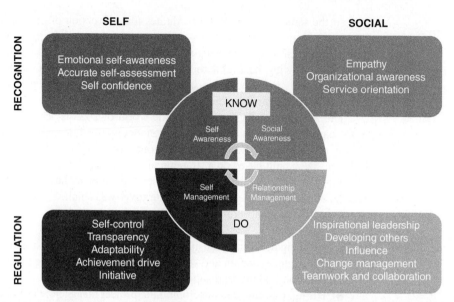

Figure 12–7. EI Skills (according to Daniel Goleman). © IIL, *Leverage the Power & Possibility of Emotional Intelligence!*, workshop by Jane Morgan, 2021.

whole company on the importance of project and program management have been vital to support the company expansion. Nanoform aims to achieve a higher level of business agility, that is, measuring, learning, and adapting rapidly to deliver business outcomes as fast and as sense worthy as possible.

The PMO of the current business environment must be prepared to transform into a **Next-Generation PMO,** in opposition to a traditional PMO, which is a rigid and bureaucratic one. The evolution of Nanoform's PMO advances forward and closer to a Next-Gen PMO, as in the right column in Table 12–2.

A comprehensive and adaptable toolkit was defined in the last 3 months of 2019 as shown in Table 12–3, upon the PMO creation. The tools and methodologies have been adapted over time. Each new PMO team member contributes to the creation or improvement of tools, as well as to share experiences and the most effective ways of working. In addition, the priority competencies and emotional skills are being constantly refined. Being comfortable with uncertainty, more than trendy buzzwords, are essential for all PMs in a company that is growing exponentially, in which projects are executed and managed faster and faster. There is speed but, in addition, there is also precision and accuracy.

For every customer or partner project, Nanoform uses all these templates and documentation, from initiation to closing, with a client satisfaction online survey and a Net Promoter Score (NPS) at the end of each project. The PMs support mainly external projects, with clients or partners, but some internal strategic projects have PM support as well (e.g., new facilities expansion and transversal company programs that are high priority). The PMO positions itself as the link between strategy and execution.

TABLE 12–2.   TRADITIONAL VS NEXT GEN PMO, © JACK DUGGAL, PROJECTIZE GROUP

Foundational/Traditional PMO	Next-Generation PMO
Mechanical of factory mindset (machine-oriented)	Organic mindset (knowledge-oriented)
Focus mostly on execution and delivery	Focus on strategic decision support and business value
Focus on science of management (technical expertise)	Focus on art and craft of management (organizational savvy)
Emphasis on monitoring and control	Emphasis on support and collaboration
Provides tools similar to a precise "map" to follow	Provides tools similar to a "compass" that shows the direction
Standard (heavy) methods and practices	Adaptable, flexible and simple methods and practices
Process-driven: Focus on process and methodology compliance	Customer and business-driven: Focus on value, experience, and impact
Emphasis on rules; follow rules	Based on guiding principles; follow rules and improvise if needed
Focus on the "WHAT" (task management)	Focus on "WHO" (stakeholder/relationship management)
Focus on efficiency: Manage inputs and outputs	Focus on effectiveness and experience: Ownership of results and outcomes
Measure and track compliance: certification and delivery	Measure and track benefits, value and experience
Heavy management and governance	Balanced management, governance, and leadership
Traditional score	Next-generation score

TABLE 12–3.   TOOLKIT USED BY NANOFORM PMO

Process group/knowledge area (as per PMBOK® 6)	Tools and methodologies	Applications
Initiation	Scope definition template, work order template Budgeting templates	Business case definition, scope definition, Project budgets
Planning	Project communication plan, project baseline plan, agile plan; kick-off template	Shared plan built by the team Buy-in from all Iteration, flexibility, teamwork, rigor, and discipline
Communication Regular project updates Documentation management and sharing	Project update templates Meeting agenda and minutes template % Level of completion and EVM methodology Action items register Risk register Brainstorming and Q-storming templates, problem-solving, design-thinking methodologies. Project shared portals (e.g., Box® and SharePoint®) Organizational governance: management team meetings	Rigorous and factual updates Culture of connection and collaboration Open feedback culture Cross-departmental/matrix communication, knowledge sharing, and continuous learning Constant alignment between sales, BD, PM, and operational area Sharing of data and reports between Nanoform and its clients and partners
Monitoring and controlling	Power BI, SAP S/HANA, and other systems Automation tools Daily stand-up meeting (operational) Weekly scheduling meeting (tactical)	Appropriate metrics to foster the right behaviors, effectively monitoring and controlling the projects
Closing	Closing template, project review NPS survey at the end of each project Quarterly and annual NPS reviews	Measure customer value and benefits

The company, as a living organism, is renewing itself and reinventing its DNA, and the PMO is solidifying and reaching the 1st stage of consolidation of best practices used, people development, and collective learning. Conway's law states that "any organization that designs a system will produce a design whose structure is a copy of the organization's communication structure." Nanoform is committed to design and evolve a system that supports the organization as adaptive networks of teams organized around specific goals. This vision and commitment are the north star for PMO excellence, which consists of a small, cross-functional, cross-hierarchy team with key representatives who work collaboratively across the organization.

Key Challenges Encountered at Nanoform, 2019–2022

As in most start-ups launched from a university group, the early Nanoform years (2015–2018) were mostly research-based, and in many ways exploratory. The development and scalability of the generated innovation were not being exploited in full since the use of the technology was still being intensified. Project leaders were the technical leads, who despite being scientifically advanced (PhDs in most cases) lacked any preparation or specific project management experience, or formal project management training. In 2019, running internal mock-up projects was an early way to train and develop the teams, not only in planning, and defining the scope and schedule, but also leading the execution forward while reporting to the partners, clients, and internal stakeholders.

Some level of resistance to change is a natural human characteristic, and this has been overcome by a constant communication reinforcement of the vision to the employees, an open sharing of information and context in a session open to all Nanoformers on a weekly basis (Town Hall). With this transparency, Nanoform was able to nurture the culture, identify gaps and hire the right talent and experience to all areas, keeping and building a cohesive team spirit when the company began to grow exponentially, and making sure everyone does meaningful work, with a clear purpose. Building synergies and coalitions with the functional areas, catalyzing collaboration and communication, creativity, and critical thinking, has been instrumental to the PMO being seen as a business partner, not as a bureaucratic department or a roadblock. Amidst the complexity, the elements of a department of simplicity are powerful and pervasive for a PMO. The COVID-19 context also made us become better and more pragmatic in decision-making, and less resistant to change.

At Nanoform, the communication is made most effective by a continuous process of discussion in the realm of the core project teams, a joint decision-making process, a no-blaming culture, and a collective acceptance of the decisions made. Trust and empathy are developed all the time, and frequent communication is encouraged and proactively occurring with our colleagues, both face-to-face and via platforms such as MS Teams. Colocated teams, in which office work was enabled most of the time due to the disciplined way in which Finland managed the pandemic, made it less restrictive to work at the office and enabled building cohesive relationships and closely connected teams. The value of communication platforms such as MS Teams should not be underestimated, especially since project teams at Nanoform are generally young professionals and scientists that enthusiastically embrace these communication methods.

Virtual communication also allows for collaboration across the world and has enabled project teams at Nanoform to work successfully with customers in different time zones. This again shows how adaptable Nanoform had to be to thrive during the pandemic and support the continued growth of the company. The project management tools only provide a structure, a common framework, but above all, frequent communication, be it face-to-face or virtual, is powerful at Nanoform.

During the COVID-19 lockdown periods, communication with clients, partners, and new business leads was kept alive by means of video communication, but above all Nanoform was able to deliver innovative and vibrant virtual plant tours, in which a multi-camera experience was provided to all stakeholders (bound by a confidentiality agreement), in an almost real tour and interaction with Nanoform's team.

Key PMO Achievements at Nanoform, 2019–2022

The first key achievement was to strengthen the PMO group, by defining its vision, mission, and objectives, and by collectively listening to the ecosystem, building and defining a standardized methodology and a toolkit, and fostering the right human competencies, with all the items mentioned above. The ways of working and toolkit were explained and communicated on a high level to all Nanoform stakeholders. Its simplified adoption was deployed to all functional areas. The toolkit and project management ways of working were adopted by all Nanoform project teams involved in external projects: 100% of such projects use the full toolkit and the project management methodologies, ways of working, and communication plan, from A to Z. All internal projects use a basic toolkit (project charter, project plans), even without a dedicated PM.

A portfolio dashboard was established in 2021 and it is being fine-tuned in 2022 (Figure 12–8). Its goal is to measure portfolio health (on a weekly basis) and take action to maximize value for the clients and partners.

All operational areas hired scientists and managers who were trained to become even more proficient, but also to be highly skilled at communicating with clients, not only with on-the-job training but also with outsourced communication training. They are also becoming more effective in teamwork and integration. The individual NPSs received from clients and partners show an increasing trend over these three years.

Six project managers were hired over 3 years, which required a significant investment on recruitment, reviewing ca. 150 CVs and carrying out more than 30 first screening interviews, after which selected PMs went through a 4-stage interview process, all the way up to the VP level and the CEO. Hiring PM professionals with a strong

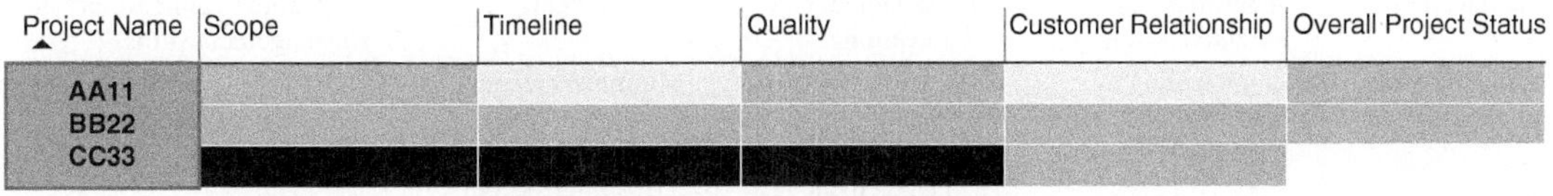

Project Name	Scope	Timeline	Quality	Customer Relationship	Overall Project Status
AA11					
BB22					
CC33					

Figure 12–8. An example of a traffic-lights portfolio health dashboard developed at Nanoform (dashboard is fictitious).

scientific background, pharma experience, ideally with CDMO (Contract Development and Manufacturing Organization) experience, and demonstrated project management competencies, is a tough but rewarding endeavor. The alignment of each new candidate to Nanoform's culture, the PMO culture, and team spirit has multiple checks since the candidates have exposure to an interview with at least two of the current PMs and are invited to visit the company offices before a contract proposal is offered. The interviews ensure alignment with the company culture, identifying the right levels of resilience, the right reasons to want to join Nanoform, no fear of change, and comfort in uncertainty.

Proactive and collective learning of the new PMs is allowing them to become autonomous in leading client meetings within 2–3 months after starting at Nanoform. A key success factor is the clear definition of roles, responsibilities, and expectations, and the conditions created to enable all the PMs to share experiences and observe meetings chaired by all other colleagues, with an increased dynamics during their induction period (collaboration is maximized). The competencies are acquired and developed by a balanced mix of on-the-job training, formal training, and individual learning (70:20:10 model). A career path has been defined: Project Manager – Senior Project Manager – Program Manager – Senior Program Manager.

PMO in the Growth Phase, Aiming for Maturity

The PMI® study "PMO Maturity" defines the elements of PMO maturity across five key dimensions: **governance, integration and alignment, processes, technology** and **data, and people**. "Successful execution against these principles allows the Top 10 Percent to have greater influence and impact and to deliver more successful project outcomes."

From inception to maturity, a PMO can travel across five phases, as shown in Table 12–4.

For the PMO to be successful, critical factors are shown in Table 12–5. Failure factors that must be avoided are also indicated.

TABLE 12–4. FIVE PHASES OF PM LIFE CYCLE

Embryonic	Executive Management Acceptance	Line Management Acceptance	Growth	Maturity
Recognize need	Get visible executive support	Get line management support	Recognize use of life-cycle phases	Develop a management cost/schedule control system
Recognize benefits	Achieve executive understanding of project management	Achieve line management commitment	Develop a project management methodology	Integrate cost and schedule control
Recognize applications	Establish project sponsorship at executive levels	Provide line management education	Make the commitment to planning	Develop an educational program to enhance project management skills
Recognize what must be done	Become willing to change way of doing business	Become willing to release employees for project management training	Minimize creeping scope Select a project tracking system	

Source: Adapted from Kerzner (2018).

TABLE 12–5. CRITICAL FACTORS IN THE PM LIFE CYCLE

Critical Success Factors	Critical Failure Factors
Executive Management Acceptance Phase	
Consider employee recommendations	Refuse to consider ideas of associates
Recognize that change is necessary	Unwilling to admit that change may be necessary
Understand the executive role in project management	Believe that project management control belongs at the executive levels
Line Management Acceptance Phase	
Willing to place company interest before personal interest	Reluctant to share information
Willing to accept accountability	Refuse to accept accountability
Willing to see associates advance	Not willing to see associates advance
Growth phase	
Recognize the need for a corporate-wide methodology	View a standard methodology as a threat rather than as a benefit
Support uniform status monitoring/reporting	Fail to understand the benefits of project management
Recognize the importance of effective planning	Provide only lip service to planning
Maturity Phase	
Recognize that cost and schedule are inseparable	Believe that project status can be determined from schedule alone
Track actual costs	See no need to track actual costs
Develop project management training	Believe that growth and success in project management are the same

Source: Adapted from Kerzner (2018).

Nanoform's PMO is in a stage of growth, aiming at maturity within the next three years.

Conclusions and Next Steps   Moving forward, the Nanoform PMO will aim at getting even closer to strategy definition and with an increased maturity, further deploying the best-standardized methodologies (focusing more on efficient value creation for customers, risk management, benefits measurement, and value for the stakeholders), with productive and synergistic interactions with all functional areas, strengthening two-way communication and fostering cross-functional resource planning, allocation, and prioritization. The most effective IT support and development of project management applications and systems will enable an integrative planning and monitoring of the projects' execution (e.g., with SAP S/HANA), and automation will be adopted.

In a world beyond the COVID-19 pandemic, project management will remain the key to success for many businesses. Well-defined decision models and portfolio management practices will reduce uncertainty, better manage volatility and render the organization more effective, agile, and adaptive. It is recognized that the customers care most about the solutions to their problems, through the timely delivery of value (hence the VMO concept, Value-Management Office), represented by product features and fixes or service innovations. This is the aim for future growth and enhancement of PMO practices at Nanoform. Leadership development at the PMO level will keep on developing. A company-wide program/project management culture and capabilities will be applied transversally, since practically all Nanoform employees are positioning themselves to some degree as project managers, and they do make business as well as project decisions.

Nanoform's vision and aspiration are to be a cutting-edge technology company that can touch the lives of 1 billion humans, that can double the number of new drugs entering the market annually, that will also be capable of being operationally excellent, while growing at the speed of light, with an adaptable and agile PMO that aims at reaching maturity by 2025, with the right momentum to fully meet the 2030 project trends and the lightning-fast pace of the pharma and healthcare services, in which only the most agile and evolutionary companies will thrive. This requires accelerated cycles of learning and transformation and exceptional adaptability of all Nanoformers, to grow individually as well as collectively.

ACKNOWLEDGMENTS

I would like to express my appreciation to: Dr. Edward Haeggström, Nanoform CEO, for his challenging questions and invaluable comments; Dr Gonçalo Andrade, Nanoform CBO, for his business review and leadership; Albert Haeggström, Nanoform CFO, Peter Hänninen, Nanoform General Counsel, for reviewing this document, and Dr. Helene Wahl, Senior Project manager.

FURTHER READING

Augustine, S., Cuellar, R., and Scheere, A. (2021). *From PMO to VMO. Managing for Value Delivery*. San Francisco: Berrett-Koehler Publishers.

Cahill, J. (2022). Guide your PMO's Evolution with this First-of-its Kind PMO Maturity Index. July 14, 2022 In: https://www.linkedin.com/pulse/guide-your-pmos-evolution-first-of-its-kind-/?trackingId=cAGdJK%2FqtoQjEssv3ZVhWA%3D%3D (viewed on July 15th, 2022).

Duggal, J. (2018). *The DNA of Strategy Execution. Next generation project management and PMO*. London: John Wiley & Sons.

Hoffman, R. and Yeh, C. (2018). *Blitzscaling -The Lightning-Fast Path to Building Massively Valuable Companies*. New York: Currency Books.

Kerzner, H. (2018). *Project management Best Practices. 4th Ed*. New York: IIL & Wiley.

Kotter, J.P. (2012). *Leading Change*. Cambridge, MA: Harvard Business Review Press.

Morgan, J. (2021). *Leverage the Power & Possibility of Emotional Intelligence!*. IIL Workshop.

Mintzberg, H., Ahlstrand, B., and Lampel, J. (2009). *Strategy Safari. Your complete guide through the Wilds of Strategic Management. 2nd Ed*. Harlow, UK: Prentice-Hall.

Project Management Institute (PMI®) (2017). PMBOK® 6th Ed. Newtown Square, PA.

Schwartz, M. (2016). *The Art of Business Value* Portland: OR, IT Revolution Press.

12.7 TYPES OF PROJECT OFFICES

Three types of project offices (POs) are commonly used in companies.

1. *Functional PO*. This type of PO is utilized in one functional area or division of an organization, such as information systems. The major responsibility of this type of PO is to manage a critical resource pool, that is, resource management. Many companies maintain an IT PMO, which may or may not have the responsibility for actually managing projects.
2. *Customer group PO*. This type of PO is for better customer management and customer communications. Common customers or projects are clustered together for better management and customer relations. Multiple customer group POs can exist at the same time and may end up functioning as a temporary organization. In effect, this acts like a company within a company and has the responsibility for managing projects.
3. *Corporate (or strategic) PO*. This type of PO services the entire company and focuses on corporate and strategic issues rather than functional issues. If this PMO does manage projects, it is usually projects involving cost reduction efforts.

As will be discussed later, it is not uncommon for more than one type of PMO to exist at the same time. For example, American Greetings maintained a functional PMO in IT and a corporate PMO at the same time. As another example, consider the following comments provided by a spokesperson for AT&T:

[A] client program management office (CPMO) represents an organization (e.g., business unit, segment) managing an assigned set of portfolio projects and interfaces with:

- Client sponsors and client project managers for their assigned projects.
- Their assigned department portfolio management office (DPMO).
- Their assigned portfolio administration office (PAO) representative.
- CPO-resource alignment (RA) organization factories.

The DPMO supports its client organization's Executive Officer, representing their entire department Portfolio. It serves as the primary point of contact between the assigned CPMOs within their client organization and the PAO for management of the overall departmental portfolio in the following areas:

- Annual portfolio planning.
- Capital and expense funding within portfolio capital and expense targets.
- In plan list change management and business case addendums.
- Departmental portfolio project prioritization.

The PMO is led by an executive director who is a peer to the line project management executive directors. All executive directors report to the vice president—project management.

The functions of the PMO include: Define, document, implement, and continually improve project management processes, tools, management information, and training requirements to ensure excellence in the customer experience. The PMO establishes and maintains:

- Effective and efficient project management processes and procedures across the project portfolio.
- Systems and tools focused on improving efficiency of project manager's daily activities while meeting external and internal customer needs.
- Management of information that measures customer experience, project performance, and organizational performance.
- Training/certification curriculum supporting organizational goals.

12.8 PROJECT AUDITS AND THE PMO

In recent years, the need for a structured independent review of various parts of a business, including projects, has become more evident. Part of this can be attributed to the Sarbanes–Oxley law compliance requirements. These audits are now part of the responsibility of the PMO.

These independent reviews are audits that focus on either discovery or decision-making. They also can focus on determining the health of a project. The audits can be scheduled or random and can be performed by in-house personnel or external examiners.

There are several types of audits. Some common types include:

- *Performance audits*. These audits are used to appraise the progress and performance of a given project. The project manager, project sponsor, or an executive steering committee can conduct this audit.
- *Compliance audits*. These audits are usually performed by the PMO to validate that the project is using the project management methodology properly. Usually, the PMO has the authority to perform the audit, but it may not have the authority to enforce compliance.
- *Quality audits*. These audits ensure that the planned project quality is being met and that all laws and regulations are being followed. The quality assurance group performs this audit.
- *Exit audits*. These audits are usually for projects that are in trouble and may need to be terminated. Personnel external to the project, such as an exit champion or an executive steering committee, conduct the audits.

- *Best practices audits.* These audits can be conducted at the end of each life-cycle phase or at the end of the project. Some companies have found that project managers may not be the best individuals to perform the audit. In such situations, the company may have professional facilitators trained in conducting best practice reviews.

12.9 PMO OF THE YEAR AWARD

Some people contend that the most significant change in project management in the first decade of the twenty-first century has been the implementation of the PMO concept. Therefore, it is no big surprise that the Center for Business Practices initiated the PMO of the Year award.

Award Criteria

The PMO of the Year Award is presented to the PMO that best illustrates—through an essay and other documentation—their project management improvement strategies, best practices, and lessons learned. Additional support documentation—such as charts, graphs, spreadsheets, brochures, and so on—cannot exceed five documents. While providing additional documentation is encouraged, each eligible PMO must clearly demonstrate its best practices and lessons learned in the awards essay. Judges review the essays to consider how the applicant's PMO links project management to their organization's business strategies and plays a role in developing an organizational project management culture. The essays are judged on validity, merit, accuracy, and consistency in addition to the applicant PMO's contribution to project and organizational success.

Types of best practices judges look for include:

- Practices for integrating PMO strategies to manage projects successfully.
- Improvements in project management processes, methodologies, or practices leading to more efficient and/or effective delivery of the organization's projects.
- Innovative approaches to improving the organization's project management capability.
- Practices that are distinctive, innovative, or original in the application of project management.
- Practices that promote an enterprise-wide use of project management standards.
- Practices that encourage the use of performance measurement results to aid decision-making.
- Practices that enhance the capability of project managers.

Best practice outcomes include:

- Evidence of realized business benefits—customer satisfaction, productivity, budget performance, schedule performance, quality, ROI, employee satisfaction, portfolio performance, and strategic alignment.

- Effective use of resources.
- Improved organizational project management maturity.
- Executive commitment to a project management culture expressed in policies and other documentation.
- A PMO that exhibits an organizational business results focus.
- Effective use of project management knowledge and lessons learned.
- Individual performance objectives and potential rewards linked to measurement of project success.
- Project management functions applied consistently across the organization.

The essay comprises three sections. Incomplete submissions are disqualified.

Completing the Essay

Section 1: Background of the PMO. In no more than 1,000 words, the applicants must describe their PMO, including background information on its scope, vision and mission, and organizational structure. In addition, they described:

- How long the PMO has been in place.
- Their role within the PMO.
- How the PMO's operation is funded.
- How the PMO is structured (staff, roles and responsibilities, enterprise-wide, departmental, etc.).
- How the PMO uses project management standards to optimize its practices.

Section 2: PMO Innovations and Best Practices. In no more than 1,500 words, the applicants must address the challenges their organization encountered prior to implementing the new PMO practices and how they overcame those challenges. They should describe clearly and concisely the practices implemented and their effect on project and organizational success.

Section 3: Impact of the PMO and Future Plans. In no more than 500 words, the applicants must describe the overall impact of the PMO over a sustained period (e.g., customer satisfaction, productivity, reduced cycle time, growth, building or changing organizational culture, etc.). If available, the applicants should provide quantitative data to illustrate the areas in which the PMO had the greatest business impact. Finally, they briefly described their PMO's plans for 2009 and how those plans will potentially impact their organization.

13 Six Sigma and the Project Management Office

13.0 INTRODUCTION

In Chapter 12, we discussed the importance of the project management office (PMO) for strategic planning and continuous improvements. In some companies, the PMO was established specifically for the supervision and management of Six Sigma projects. Six Sigma teams throughout the organization would gather data and make recommendations to the PMO for Six Sigma projects. The Six Sigma project manager, and possibly the team, would be permanently assigned to the PMO.

Unfortunately, not all companies have the luxury of maintaining a large PMO where the Six Sigma teams and other supporting personnel are permanently assigned to it. It is the author's belief that the majority of PMOs have no more than four or five people permanently assigned. Six Sigma teams, including the project manager, may end up reporting "dotted" to the PMO and administratively "solid" elsewhere in the organization. The PMO's responsibility within these organizations is primarily for the evaluation, acceptance, and prioritization of projects. The PMO may also be empowered to reject recommended solutions to Six Sigma projects.

For the remainder of this chapter, we focus on organizations that maintain small PMO staff. The people assigned to the PMO may possess a reasonable knowledge concerning Six Sigma but maybe neither Green nor Black Belts in Six Sigma. These PMOs can and do still manage selected Six Sigma projects but perhaps not the traditional type of Six Sigma projects taught in the classroom.

13.1 PROJECT MANAGEMENT—SIX SIGMA RELATIONSHIP

Is there a relationship between project management and Six Sigma? The answer is definitely yes. The problem is how to exchange the benefits such that the benefits of Six

Sigma can be integrated into project management and, likewise, the benefits of project management can be integrated into Six Sigma. Some companies have already recognized this important relationship, especially the input of Six Sigma principles to project management.

Today, there is a common belief that the majority of traditional, manufacturing-oriented Six Sigma failures occur because of the lack of project management; no one is managing the Six Sigma projects as projects. Project management provides Six Sigma with structured processes as well as faster and better execution of improvements.

From a project management perspective, problems with Six Sigma Black Belts include:

- Inability to apply project management principles to planning Six Sigma projects.
- Inability to apply project management principles to the execution of Six Sigma projects.
- Heavy reliance on statistics and minimum reliance on business processes.
- Inability to recognize that project management is value added.

If these problem areas are not resolved, then Six Sigma failures can be expected as a result of:

- Everyone plans but very few execute improvements effectively.
- There are too many projects in the queue and poor prioritization efforts.
- Six Sigma stays in manufacturing and is not aligned with overall business goals.
- Black Belts do not realize that executing improvements are projects within a project.

Six Sigma people are project managers and, as such, must understand the principles of project management, including statements of work, scheduling techniques, and so on. The best Six Sigma people know project management and are good project managers; Black Belts are project managers.

13.2 INVOLVING THE PMO

The traditional PMO exists for business process improvements and supports the entire organization, including Six Sigma Black Belts, through use of the enterprise project management (EPM) methodology. Project managers, including Black Belts, focus heavily on customer value-added activities, whether it is an internal or an external customer. The PMO focuses on corporate value-added activities.

The PMO can also assist with the alignment of Six Sigma projects with strategy. This includes the following:

- Continuous reprioritization may be detrimental. Important tasks may be sacrificed and motivation may suffer.

- Hedging priorities to appease everyone may result in significant work being prolonged or canceled.
- A cultural change may be required during alignment.
- Projects and strategies maybe working toward cross-purposes.
- Strategy starts at the top, whereas projects originate at the middle of the organization.
- Employees can recognize projects but may not be able to articulate strategy. Selecting the proper mix of projects during portfolio management of projects cannot be accomplished effectively without knowing the strategy. This may result in misinterpretation.
- "Chunking" breaks a large project into smaller ones to better support strategy. Chunking makes project revitalization or rejection easier.

The PMO can also assist in solving some of the problems associated with capturing Six Sigma best practices, such as:

- Introducing a best practice can raise the bar too soon and pressure existing projects to possibly implement a best practice that may not be appropriate at that time.
- Employees and managers are unaware of the existence of the best practices and do not participate in their identification.
- Knowledge transfer across the organization is nonexistent and weak at best.
- Falling prey to the superstitious belief that most of the best practices come from failures rather than from successes.

Simply stated, the marriage of project management with Six Sigma allows us to manage better from a higher level.

13.3 TRADITIONAL VERSUS NONTRADITIONAL SIX SIGMA

In the traditional view of Six Sigma, projects fall into two categories: manufacturing and transactional. Each category of Six Sigma is multifaceted and includes a management strategy, metric, and process improvement methodology. This is shown in Figure 13–1. Manufacturing Six Sigma processes utilize machines to produce products, whereas transactional Six Sigma processes utilize people and/or computers to produce services. The process improvement methodology facet of Six Sigma addresses both categories. The only difference is what tools you will use. In manufacturing, where we utilize repetitive processes that make products, we are more likely to use advanced statistical tools. In transactional Six Sigma, we might focus more on graphical analysis and creative tools/techniques.

The traditional view of a Six Sigma project has a heavy focus on continuous improvement to a repetitive process or activity associated with manufacturing. This traditional view includes metrics, possibly advanced statistics, rigor, and a strong desire

Figure 13–1. Six Sigma categories (traditional view).

to reduce variability. Most of these Six Sigma projects fit better for implementation in manufacturing than in the PMO. Six Sigma teams manage these manufacturing-related projects.

Not all companies perform manufacturing and not all companies support the PMO concept. Companies without manufacturing needs might focus more on the transactional Six Sigma category. Companies without a PMO rely heavily on the Six Sigma teams for the management of both categories of projects.

Those companies that do support a PMO must ask themselves these three questions:

1. Should the PMO be involved in Six Sigma projects?
2. If so, what type of project is appropriate for the PMO to manage, even if the organization has manufacturing capability?
3. Do we have sufficient resources assigned to the PMO to become actively involved in Six Sigma project management?

PMOs that are actively involved in most of the activities described in Chapter 12 do not have the time or resources required to support all Six Sigma projects. In such a case, the PMO must be selective as to which projects to support. The projects selected are commonly referred to as nontraditional projects that focus more on project management-related activities than manufacturing.

Figure 13–2 shows the nontraditional view of Six Sigma. In this view, operational Six Sigma includes manufacturing activities and all other activities from Figure 13–1, and transactional Six Sigma now contains primarily those activities to support project management.

In the nontraditional view, the PMO can still manage both traditional and nontraditional Six Sigma projects. However, there are some nontraditional Six Sigma projects

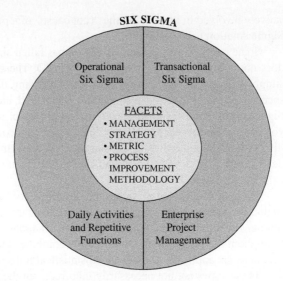

Figure 13–2. Six Sigma categories (nontraditional view).

that are more appropriate for management by the PMO. Some of the projects currently assigned to PMOs include enhancements to the EPM methodology, enhancements to the PMO tool set, efficiency improvements, and cost avoidance/reduction efforts. Another project assigned to the PMO involves process improvements to reduce the launch time of a new product and improve customer management. Experts in Six Sigma might view these as nontraditional types of projects. There is also some concern as to whether these are really Six Sigma projects or just a renaming of a continuous improvement project to be managed by a PMO. Since several companies now refer to these as Six Sigma projects, the author will continue this usage.

Strategic planning for Six Sigma project management is not accomplished just once. Instead, like any other strategic planning function, it is a cycle of continuous improvements. The improvements can be small or large, measured quantitatively or qualitatively, and designed for either internal or external customers.

There almost always exists a multitude of ideas for continuous improvements. The biggest challenge lies in effective project selection and then assigning the right players. Both challenges can be overcome by assigning Six Sigma project management best practices to the PMO. It may even be beneficial to have Six Sigma specialists with Green Belts or Black Belts assigned to the PMO.

13.4 UNDERSTANDING SIX SIGMA

Six Sigma is not about manufacturing widgets. It is about a focus on processes. And since the PMO is the guardian of the project management processes, it is only fitting that the PMO has some involvement in Six Sigma. The PMO may, however, be more actively involved in identifying the "root cause" of a problem than in managing the Six Sigma solution to the problem.

Some people contend that Six Sigma has fallen short of expectations and certainly does not apply to activities assigned to a PMO. These people argue that Six Sigma is simply a mystique that some believe can solve any problem. In truth, Six Sigma can succeed or fail, but the organizational intent and understanding must be there. Six Sigma gets you closer to the customer, improves productivity, and determines where you can get the biggest returns. Six Sigma is about process improvement, usually repetitive processes, and reducing the margin for human and/or machine error. Error can be determined only if you understand the critical requirements of either the internal or external customer.

There is a multitude of views and definitions of Six Sigma. Some people view Six Sigma as merely the renaming of total quality management (TQM) programs as Six Sigma. Others view Six Sigma as the rigorous application of advanced statistical tools throughout the organization. A third view combines the first two views by defining Six Sigma as the application of advanced statistical tools to TQM efforts.

These views are not necessarily incorrect, but they are incomplete. From a project management perspective, Six Sigma can be viewed as simply obtaining better customer satisfaction through continuous process improvement efforts. The customer could be external to the organization or internal. The word "satisfaction" can have a different meaning for external or internal customers. External customers expect products and services that are high quality and reasonably priced. Internal customers may define satisfaction in financial terms, such as profit margins. Internal customers may also focus on such items as cycle time reduction, safety requirements, and environmental requirements. If these requirements are met in the most efficient way without any non-value-added costs (e.g., fines, rework, overtime), then profit margins will increase.

Disconnects can occur between the two definitions of satisfaction. Profits can always be increased by lowering quality. This could jeopardize future business with the client. Making improvements to the methodology to satisfy a particular customer may seem feasible but may have a detrimental effect on other customers.

The traditional view of Six Sigma focused heavily on manufacturing operations using quantitative measurements and metrics. Six Sigma tool sets were created specifically for this purpose. Six Sigma activities can be defined as operational Six Sigma and transactional Six Sigma. Operational Six Sigma would encompass the traditional view and focus on manufacturing and measurement. Operational Six Sigma focuses more on processes, such as the EPM methodology, with emphasis on continuous improvements in the use of the accompanying forms, guidelines, checklists, and templates. Some people argue that transactional Six Sigma is merely a subset of operational Six Sigma. While this argument has merit, project management and specifically the PMO spend the majority of their time involved in transactional rather than operational Six Sigma.

The ultimate goal of Six Sigma is customer satisfaction, but the process by which the goal is achieved can differ, depending on whether we are discussing operational or transactional Six.

Table 13–1 identifies some common goals of Six Sigma. The left-hand column lists the traditional goals that fall more under operational Six Sigma, whereas the right-hand column indicates how the PMO plans on achieving the goals.

The goals for Six Sigma can be established at either the executive levels or the working levels. The goals may or may not be able to be completed with the execution of just one project. This is indicated in Table 13–2.

Six Sigma initiatives for project management are designed not to replace ongoing initiatives but to focus on those activities that may have a critical-to-quality and critical-to-customer-satisfaction impact in both the long and the short term.

Operational Six Sigma goals emphasize reducing the margin for human error. But transactional Six Sigma activities managed by the PMO may involve human issues, such as aligning personal goals to project goals, developing an equitable reward system for project teams, and project career path opportunities. Fixing people's problems is part of transactional Six Sigma but not necessarily of operational Six Sigma.

TABLE 13–1. GOALS OF SIX SIGMA

Goal	Method of Achievement
Understand and meet customer requirements (do so through defect prevention and reduction instead of inspection)	Improvements to forms, guidelines, checklists, and templates for understanding customer requirements
Improve productivity	Improve efficiency in execution of the project management methodology
Generate higher net income by lowering operating costs	Generate higher net income by streamlining the project management methodology without sacrificing quality or performance
Reduce rework	Develop guidelines to better understand requirements and minimize scope changes
Create a predictable, consistent process	Continuous improvement of the processes

Source: Adapted from *The Fundamentals of Six Sigma* (New York: International Institute for Learning, 2008).

TABLE 13–2. GOALS VERSUS FOCUS AREAS

Executive Goals	PMO Focus Areas
Provide effective status reporting	Identification of executive needs
	Effective utilization of information
	"Traffic light" status reporting
Reduce the time for planning projects	Sharing information between planning documents
	Effective use of software
	Use of templates, checklists, and forms
	Templates for customer status reporting
	Customer satisfaction surveys
	Extensions of the EPM methodology into the customer's organization

13.5 SIX SIGMA MYTHS[1]

Ten myths of Six Sigma are given in Table 13–3. These myths have been known for some time but have become quite evident when the PMO takes responsibility for project management transactional Six Sigma initiatives.

Works Only in Manufacturing   Much of the initial success in applying Six Sigma was based on manufacturing applications; however, recent publications have addressed other applications of Six Sigma. Breyfogle[2] includes many transactional/service applications. In GE's 1997 annual report, chief executive Jack Welch proudly states that Six Sigma "focuses on moving every process that touches our customers—every product and *service* (emphasis added)—toward near-perfect quality."

Ignores Customer in Search of Profits   This statement is not myth but rather a misinterpretation. Projects worthy of Six Sigma investments should (1) be of primary concern to the customer and (2) have the potential for significantly improving the bottom line. Both criteria must be met. The customer is driving this boat. In today's competitive environment, there is no surer way of going out of business than to ignore the customer in a blind search for profits.

Creates Parallel Organization   An objective of Six Sigma is to eliminate every ounce of organizational waste that can be found and then reinvest a small percentage of those savings to continue priming the pump for improvements. With the large amount of downsizing that has taken place throughout the world during the past two decades, there is no room or inclination to waste money through the duplication of functions. Many functions are understaffed as it is. Six Sigma is about nurturing any function that adds significant value to the customer while adding significant revenue to the bottom line.

TABLE 13–3. TEN MYTHS OF SIX SIGMA

1. Works only in manufacturing
2. Ignores the customer in search of bottom-line benefits
3. Creates a parallel organization
4. Requires massive training
5. Is an add-on effort
6. Requires large teams
7. Creates bureaucracy
8. Is just another quality program
9. Requires complicated, difficult statistics
10. Is not cost-effective

---

1. Section 13.5 has been adapted from F. W. Breyfogle III, J. M. Cupello, and B. Meadows, *Managing Six Sigma* (Hoboken, NJ: Wiley, 2001), pp. 6–8.

2. F. W. Breyfogle III, *Implementing Six Sigma: Smarter Solutions Using Statistical Methods* (Hoboken, NJ: Wiley, 1999).

Requires Massive Training   Peter B. Vaill states:

> Valuable innovations are the positive result of this age [we live in], but the cost is likely to be continuing system disturbances owing to members' nonstop tinkering. Permanent white water conditions are regularly taking us all out of our comfort zones and asking things of us that we never imagined would be required. It is well for us to pause and think carefully about the idea of being continually catapulted back into the beginner mode, for that is the real meaning of being a continual learner. We do not need competency skills for this life. We need incompetency skills, the skills of being effective beginners.

Is an Add-On Effort   This is simply the myth "creates a parallel organization" in disguise. Same question, same response.

Requires Large Teams   There are many books and articles in the business literature declaring that teams have to be small if they are to be effective. If teams are too large, the thinking goes, a combinational explosion occurs in the number of possible communication channels between team members, and hence no one knows what the other person is doing.

Creates Bureaucracy   A dictionary definition of bureaucracy is "rigid adherence to administrative routine." The only thing rigid about wisely applied Six Sigma methodology is its relentless insistence that the customer needs to be addressed.

Is Just Another Quality Program   After decades of poor performance of untold quality programs, at the time of its development,[3] Six Sigma represented "an entirely new way to manage an organization."[4]

Requires Complicated, Difficult Statistics   There is no question that a number of advanced statistical tools are extremely valuable in identifying and solving process problems. We believe that practitioners need to possess an analytical background and understand the wise use of these tools but do not need to understand all the mathematics behind the statistical techniques. The wise application of statistical techniques can be accomplished through the use of statistical analysis software.

---

3. J. Micklethwait and A. Wooldridge, *The Witch Doctors of the Management Gurus* (New York: Random House, 1997).

4. T. Pyzdek, Six Sigma Is Primarily a Management Program, *Quality Digest* (1999): 26.

Is Not Cost-effective   If Six Sigma is implemented wisely, organiza-
                                          tions can obtain a very high rate of return on their
                                          investment within the first year.

13.6 USE OF ASSESSMENTS

One of the responsibilities that can be assigned to a PMO is the portfolio management
of projects. Ideas for potential projects can originate anywhere in the organization.
However, ideas specifically designated as transactional Six Sigma projects may need to
be searched out by the PMO.

One way to determine potential projects is through an assessment. An assessment
is a set of guidelines or procedures that allows an organization to make decisions about
improvements, resource allocations, and even priorities. Assessments are ways to:

- Examine, define, and possibly measure performance opportunities.
- Identify knowledge and skills necessary for achieving organizational goals and
  objectives.
- Examine and solve performance gap issues.
- Track improvements for validation purposes.

A gap is the difference between what currently exists and what it should be. Gaps
can be in cost, time, quality, performance, or efficiency. Assessments allow us to pinpoint the gap and determine the knowledge, skills, and abilities necessary to compress
the gap. For project management gaps, the assessments can be heavily biased toward
transactional rather than operational issues, and this could easily result in behavior
modification projects.

Several factors must be considered prior to performing an assessment. These
factors might include:

- Amount of executive-level support and sponsorship.
- Amount of line management support.
- Focus on broad-based applications.
- Determining whom to assess.
- Bias of the participants.
- Reality of the answers.
- Willingness to accept the results.
- Impact on internal politics.

The purpose of the assessment is to identify ways to improve global business
practices first and functional business practices second. Because the target audience is
usually global, there must exist unified support and understanding for the assessment
process and recognition that it is in the best interests of the entire organization. Politics,
power, and authority issues must be put aside for the betterment of the organization.

Assessments can take place at any level of the organization. Such assessments can be:

- Global organizational assessments.
- Business unit organizational assessments.
- Process assessments.
- Individual or job assessments.
- Customer feedback assessments (satisfaction and improvements).

There are several tools available for assessments. A typical list might include:

- Interviews
- Focus groups
- Observations
- Process maps

Assessments for Six Sigma project management should not be performed unless the organization believes that opportunities exist. The amount of time and effort expended can be significant, as shown in Figure 13–3.

The advantages of assessment can lead to significant improvements in customer satisfaction and profitability. However, there are disadvantages, such as:

- Can be costly to implement.
- Can be labor-intensive.
- Difficulty determining which project management activities can benefit from assessments.
- May not provide any meaningful benefits.
- Inability to measure a return on investment from assessments.

Assessments can have a life of their own. There are typical life-cycle phases for assessments. These life-cycle phases may not be aligned with the life-cycle phases of

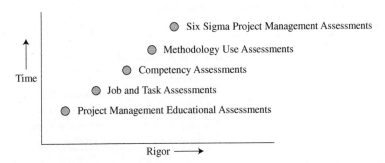

Figure 13–3. Time and effort expended.

the EPM methodologies and may be accomplished more informally than formally. Typical assessment life-cycle phases include:

● Gap or problem recognition.
● Development of the appropriate assessment tool set.
● Conducting the assessment/investigation.
● Data analyses.
● Implementation of the changes necessary.
● Review for possible inclusion in the best practices library.

Determining the best tool set can be difficult. The most common element of a tool set is a focus on questions. Types of questions include:

● Open-ended
  ● Sequential segments
  ● Length
  ● Complexity
  ● Time needed to respond
● Close-ended
  ● Multiple choice
  ● Forced choices (yes–no, true–false)
  ● Scales

Table 13–4 illustrates how scales can be set up. The left-hand column solicits a qualitative response and may be subjective whereas the right-hand column would be a quantitative response and more subjective.

It is vitally important that the assessment instrument undergo pilot testing. Pilot testing:

● Validates understanding of the instructions.
● Checks ease of response.
● Verifies time to respond.
● Verifies space to respond.
● Can analyze bad questions.

TABLE 13–4. SCALES

Strongly agree	Under 20%
Agree	Between 20% and 40%
Undecided	Between 40% and 60%
Disagree	Between 60% and 80%
Strongly disagree	Over 80%

13.7 PROJECT SELECTION

Six Sigma project management focuses on continuous improvements to the EPM methodology. Identifying potential projects for the portfolio is significantly easier than getting them accomplished. There are two primary reasons for this:

1. Typical PMOs may have no more than three or four employees. Based on the activities assigned to the PMO, the employees may be limited as to how much time they can allocate to Six Sigma project management activities.
2. If functional resources are required, then the resources may be assigned first to those activities that are mandatory for the ongoing business of the firm.

The conflict between ongoing business and continuous improvements occurs frequently. Figure 13–4 illustrates this point. The ideal Six Sigma project management activity would yield high customer satisfaction, high-cost reduction opportunities, and significant support for the ongoing business. Unfortunately, what is in the best interests of the PMO may not be in the best, near-term interests of the ongoing business.

All ideas, no matter how good or how bad, are stored in the "idea bank." The ideas can originate from anywhere in the organization, namely:

- Executives
- Corporate Six Sigma champions
- Project Six Sigma champions
- Master Black Belts
- Black Belts
- Green Belts
- Team members

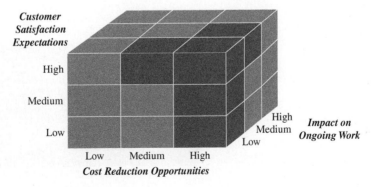

Figure 13–4. Project selection cube.

If the PMO is actively involved in the portfolio management of projects, then the PMO must perform feasibility studies and cost-benefit analyses on projects together with prioritization recommendations. Typical opportunities can be determined using Figure 13–5. In this figure, ΔX represents the amount of money (or additional money) being spent. This is the input to the evaluation process. The output is the improvement, ΔY, which is the benefits received or cost savings. Consider the next example.

Convex Corporation

Convex Corporation identified a possible Six Sigma project involving the streamlining of internal status reporting. The intent was to eliminate as much paper as possible from the bulky status reports and replace it with color-coded "traffic light" reporting using the company intranet. The PMO used the following data:

- Burdened hour at the executive level: $240
- Typical number of project status review meetings per project: 8
- Duration per meeting: 2 hours
- Number of executives per meeting: 5
- Number of projects requiring executive review: 20

Using the information provided, the PMO calculated the total cost of executives as:

$$(8\,\text{meetings}) \times (5\,\text{executives}) \times (2\,\text{hours/meeting}) \times (\$240/\text{hr}) \times (20\,\text{projects}) = \$384,000$$

Convex assigned one systems programmer (burdened at $100/hr) for four weeks. The cost of adding traffic light reporting to the intranet methodology was $16,000.

Six months after implementation, the number of meetings had been reduced to five per project for an average of 30 minutes each. The executives were now focusing on only those elements of the project that were color coded as a potential problem. On a yearly basis, the cost for the meetings on the 20 projects was now about $60,000. In the first year alone, the company identified savings of $324,000 for one investment of $16,000.

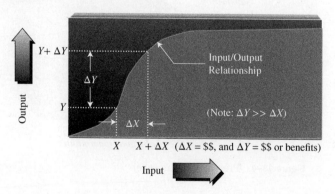

Figure 13–5. Six Sigma quantitative evaluation.

13.8 TYPICAL PMO SIX SIGMA PROJECTS

Projects assigned to the PMO can be operational or transactional but mainly the latter. Typical projects might include:

- *Enhanced status reporting*. This project could utilize traffic light reporting designed to make it easier for customers to analyze performance. This could be intranet based. The intent is to achieve paperless project management. The colors could be assigned based on problems; present or future risks; or title, level, and rank of the audience.
- *Use of forms*. The forms should be user-friendly and easy to complete. Minimal input by the user should be required, and the data inputted into one form should service multiple forms if necessary. Nonessential data should be eliminated. The forms should be cross-listed to the best practices library.
- *Use of checklists/templates*. These documents should be comprehensive yet easy to understand. They should be user-friendly and easy to update. The forms should be flexible such that they can be adapted to all situations.
- *Criteria for success/failure*. There must exist established criteria for what constitutes success or failure on a project. There must also exist a process that allows for continuous measurement against these criteria as well as a means by which success (or failure) can be redefined.
- *Team empowerment*. This project looks at the use of integrated project teams, the selection of team members, and the criteria to be used for evaluating team performance. This project is designed to make it easier for senior management to empower teams.
- *Alignment of goals*. Most people have personal goals that may not be aligned with the goals of the business. This includes project versus company goals, project versus functional goals, project versus individual goals, project versus professional goals, and other such alignments. The greater the alignment between goals, the greater the opportunity for increased efficiency and effectiveness.
- *Measuring team performance*. This project focuses on ways to uniformly apply critical success factors and key performance indicators to team performance metrics. This also includes the alignment of performance with goals and rewards with goals. This project may interface with the wage and salary administration program by requiring two-way and three-way performance reviews.
- *Competency models*. Project management job descriptions are being replaced with competency models. A competency criterion must be established, including goal alignment and measurement.
- *Financial review accuracy*. This type of project looks for ways of including the most accurate data into project financial reviews. This could include transferring data from various information systems, such as earned value measurement and cost accounting.
- *Test failure resolution*. Some PMOs maintain a failure reporting information system that interfaces with failure modes and effects analysis. Unfortunately, failures are identified, but there may be no resolution to the failures. This project attempts to alleviate this problem.

● *Preparing transitional checklists.* This type of project is designed to focus on transition or readiness of one functional area to accept responsibility. As an example, it may be possible to develop a checklist for evaluating the risks or readiness of transitioning the project from engineering to manufacturing. The ideal situation would be to develop one checklist for all projects.

This list is by no means comprehensive. However, it does identify typical projects managed by the PMO. Three conclusions can be reached by analyzing this list:

1. The projects can be both transactional and operational.
2. The majority of the projects focus on improvements to the methodology.
3. Having people with Six Sigma experience (i.e., Green, Brown, or Black Belts) would be helpful.

When a PMO takes the initiative in Six Sigma project management, it may develop a Six Sigma toolbox exclusively for that PMO. These tools most likely will not include the advanced statistical tools that are used by Black Belts in manufacturing but may be more process-oriented tools or assessment tools.

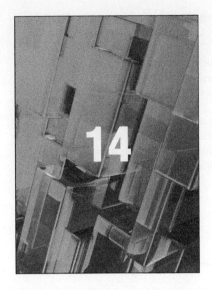

14 Project Portfolio Management

14.0 INTRODUCTION

Your company is currently working on several projects and has a waiting list of an additional 20 projects that it would like to complete. If available funding will support only a few more projects, how does a company decide which of the 20 projects to work on next? This is the project portfolio management process. It is important to understand the difference between project management and project portfolio management. Debra Stouffer and Sue Rachlin have made this distinction for information technology (IT) projects:

> An IT portfolio is composed of a set or collection of initiatives or projects. Project management is an ongoing process that focuses on the extent to which a specific initiative establishes, maintains, and achieves its intended objectives within cost, schedule, technical, and performance baselines.
>
> Portfolio management focuses attention at a more aggregate level. Its primary objective is to identify, select, finance, monitor, and maintain the appropriate mix of projects and initiatives necessary to achieve organizational goals and objectives.
>
> Portfolio management involves the consideration of the aggregate costs, risks, and returns of all projects within the portfolio, as well as the various trade-offs among them. Of course, the portfolio manager is also concerned about the "health" and well-being of each project that is included within the IT portfolio. After all, portfolio decisions, such as whether to fund a new project or continue to finance an ongoing one, are based on information provided at the project level.
>
> Portfolio management of projects helps determine the right mix of projects and the right investment level to make in each of them. The outcome is a better balance between ongoing and new strategic initiatives. Portfolio management is not a series of project-specific calculations such as ROI [return on investment], NPV [net present value], IRR [internal rate of return], payback period, and cash flow and then making the appropriate adjustment to account for risk. Instead, it is a decision-making process for what is in the best interest of the entire organization.

471

Portfolio management decisions are not made in a vacuum. The decision is usually related to other projects and several factors, such as available funding and resource allocations. In addition, the project must be a good fit with other projects within the portfolio and with the strategic plan.

The selection of projects could be based upon the completion of other projects that would release resources needed for the new projects. Also, the projects selected may be constrained by the completion date of other projects that require deliverables necessary to initiate new projects. In any event, some form of a project portfolio management process is needed.[1]

14.1 INVOLVEMENT OF SENIOR MANAGEMENT, STAKEHOLDERS, AND THE PMO

The successful management of a project portfolio requires strong leadership by individuals who recognize the benefits that can be accrued from portfolio management. The commitment by senior management is critical. Stouffer and Rachlin comment on the role of senior management in an IT environment in government agencies:

> Portfolio management requires a business and an enterprise-wide perspective. However, IT investment decisions must be made both at the project level and the portfolio level. Senior government officials, portfolio and project managers, and other decision makers must routinely ask two sets of questions.
>
> First, at the project level, is there sufficient confidence that new or ongoing activities that seek funding will achieve their intended objectives within reasonable and acceptable cost, schedule, technical, and performance parameters?
>
> Second, at the portfolio level, given an acceptable response to the first question, is the investment in one project or a mix of projects desirable relative to another project or a mix of projects?
>
> Having received answers to these questions, the organization's senior officials, portfolio managers, and other decision makers then must use the information to determine the size, scope, and composition of the IT investment portfolio. The conditions under which the portfolio can be changed must be clearly defined and communicated. Proposed changes to the portfolio should be reviewed and approved by an appropriate decision making authority, such as an investment review board, and considered from an organization-wide perspective.[2]

Senior management is ultimately responsible for clearly defining and communicating the goals and objectives of the project portfolio as well as the criteria and

---

1. D. Stouffer and S. Rachlin, "A Summary of First Practices and Lessons Learned in Information Technology Portfolio Management," prepared by the Chief Information Officer Council, Washington, DC, March 2002, p. 7.

2. Stouffer and Rachlin, "A Summary of First Practices and Lessons Learned in Information Technology Portfolio Management," p. 8.

conditions considered for the portfolio selection of projects. According to Stouffer and Rachlin, senior management must:

- Adequately define and broadly communicate the goals and objectives of the IT portfolio.
- Clearly articulate the organization's and management's expectations about the type of benefits being sought and the rates of returns to be achieved.
- Identify and define the type of risks that can affect the performance of the IT portfolio, what the organization is doing to avoid and address risk, and its tolerance for ongoing exposure.
- Establish, achieve consensus, and consistently apply a set of criteria that will be used among competing IT projects and initiatives.[3]

Senior management must also collect and analyze data in order to assess the performance of the portfolio and determine whether or not adjustments are necessary. This must be done periodically such that critical resources are not being wasted on projects that should be canceled. Stouffer and Rachlin provide insight on this through their interviews:

> According to Gopal Kapur, President of the Center for Project Management, organizations should focus on their IT portfolio assessments and control meetings on critical project vital signs. Examples of these vital signs include the sponsor's commitment and time, status of the critical path, milestone hit rate, deliverables hit rate, actual cost versus estimated cost, actual resources versus planned resources, and high-probability, high-impact events. Using a red, yellow, or green report card approach, as well as defined metrics, an organization can establish a consistent method for determining if projects are having an adverse impact on the IT portfolio, are failing and need to be shut down.
>
> Specific criteria and data to be collected and analyzed may include the following:

- Standard financial measures, such as return on investment, cost benefit analysis, earned value (focusing on actuals versus plan, where available), increased profitability, cost avoidance, or payback. Every organization participating in the interviews included one or more of these financial measures.
- Strategic alignment (defined as mission support), also included by almost all organizations.
- Client (customer) impact, as defined in performance measures.
- Technology impact (as measured by contribution to, or impact on, some form of defined architecture).
- Initial project and (in some cases) operations and schedules, as noted by almost all organizations.
- Risks, risk avoidance (and sometimes risk mitigation specifics), as noted by almost all participants.
- Basic project management techniques and measures.

---

3. Ibid., p. 13.

- And finally, data sources and data collection mechanisms also are important. Many organizations interviewed prefer to extract information from existing systems; sources include accounting, financial, and project management systems.[4]

One of the best practices identified by Stouffer and Rachlin for IT projects was careful consideration of both internal and external stakeholders:

Expanding business involvement in portfolio management often includes the following:

- Recognizing that the business programs are critical stakeholders, and improving that relationship throughout the life cycle
- Establishing service level agreements that are tied to accountability (rewards and punishment)
- Shifting the responsibilities to the business programs and involving them on key decision-making groups

In many organizations, mechanisms are in place to enable the creation, participation and "buy-in" of stakeholder coalitions. These mechanisms are essential to ensure the decision-making process is more inclusive and representative. By getting stakeholder buy-in early in the portfolio management process, it is easier to ensure consistent practices and acceptance of decisions across an organization. Stakeholder participation and buy-in can also provide sustainability to portfolio management processes when there are changes in leadership.

Stakeholder coalitions have been built in many different ways depending on the organization, the process and the issue at hand. By including representatives from each major organizational component who are responsible for prioritizing the many competing initiatives being proposed across the organization, all perspectives are included. The approach, combined with the objectivity brought to the process by using predefined criteria and a decision support system, ensures that everyone has a stake in the process and the process is fair.

Similarly, the membership of the top decision-making body is comprised of senior executives from across the enterprise. All major projects, or those requiring a funding source, must be voted upon and approved by this decision-making body. The value of getting stakeholder participation at this senior level is that this body works toward supporting the organization's overall mission and priorities rather than parochial interests.[5]

More and more companies today are relying heavily upon the project management office (PMO) for support with portfolio management. Typical support activities include capacity planning, resource utilization, business case analysis, and project prioritization. The role of the PMO in this regard is to support senior management, not to replace them. Portfolio management will almost always remain a prime responsibility for senior management, but recommendations and support by the PMO can make the job

---

4. Ibid., p. 18.
5. Ibid., pp. 22–23.

of the executive a little easier. In this role, the PMO may function as more of a facilitator. Chuck Millhollan, formerly director of program management at Churchill Downs Incorporated, describes portfolio management in his organization:

> Our PMO is responsible for the portfolio management process and facilitates portfolio reviews by our "Investment Council." We have purposefully separated the processes for requesting and evaluating projects (having projects approved in principle) and authorizing work (entry into the active portfolio).

When asked to describe the PMO's relationship to portfolio management, Millhollan commented:

> Investment Council: The Investment Council is comprised of senior (voting) members (CEO, COO, CFO, EVPs) and representatives from each business unit. There are regularly scheduled monthly meetings, facilitated by the PMO, to review and approve new requests and review the active portfolio. The Investment Council's goals and objectives include:
>
> 1. Prioritize and allocate capital to projects.
> 2. Approve/disapprove requested projects based on the merit of the associated Business Case.
> 3. Act individually and collectively as vocal and visible project champions throughout their representative organizations.
> 4. As necessary, take an active role in approving project deliverables, helping resolve issues and policy decisions, and providing project-related direction and guidance.
>
> Request, Evaluation, and Approval: We use an "Investment Request Worksheet" to standardize the format in which projects (called investment requests) are presented to the Investment Council. Elements include request description, success criteria and associated metrics, a description of the current and future state, alignment to strategic goals, preliminary risk assessment, identification of dependent projects, preliminary resource availability and constraint assessment and a payback analysis for ROI and Cost-Out initiatives.
>
> Work Authorization: If projects were approved during the annual operational planning processes and are capital investments that generate ROI or result in a Cost-Out, come back to the Investment Council for work authorization and addition to the portfolio of active projects. This can be done concurrently with request, evaluation, and approval for projects that are initiated mid-planning cycle.
>
> Portfolio Maintenance: We use a biweekly project status reporting process and only include projects that the Investment Council has identified as requiring portfolio review and/or oversight. The portfolio reports are provided biweekly and presented monthly during the Investment Council meetings.

When the PMO supports or facilitates the portfolio management process, the PMO becomes an active player in the strategic planning process and supports senior management by making sure that the projects in the queue are aligned with strategic objectives. The role might be support or monitoring and control. Enrique Sevilla Molina, PMP, formerly corporate PMO director at Indra, discusses portfolio management in his organization:

> Portfolio management is strongly oriented to monitor and control the portfolio performance, and to review its alignment with the strategic planning. A careful analysis of trends and forecasts is also periodically performed, so the portfolio composition may be assessed and reoriented if required.
>
> Once the strategic targets for the portfolio have been defined and allocated through the different levels in the organization, the main loop of the process includes reporting, reviewing and taking actions on portfolio performance, problems, risks, forecasts and new contracts planning. A set of alerts, semaphores, and indicators have been defined and automated in order to focus the attention on the main issues related with the portfolio management. Those projects or proposals marked as requiring specific attention are carefully followed by the management team, and a specific status reporting is provided for those.
>
> One of the key tools used for the portfolio management process is our Projects Monitor. It is a web-based tool that provides a full view of the status of any predefined set of projects (or portfolio), including general data, performance data, indicators, and semaphores. It has also the capability to produce different kinds of reports, at single project level, at portfolio level, or a specialized risk report for the selected portfolio.
>
> Besides the corporate PMO, major business units throughout the company use local PMOs in their portfolio management process. Some of them are in charge of risk status reporting for the major projects or programs in the portfolio. Others are in charge of an initial definition of the risk level for the projects and operations in order to provide an early detection of potential risk areas. And others play a significant role in providing the specific support to the portfolio managers when reporting the status to upper level management.
>
> Our corporate level PMO defines the portfolio management processes in order to be consistent with the project management level and, in consequence, the requirements for the implementation of those processes in the company tools and information systems.

Some companies perform portfolio management without PMO involvement. This is quite common when portfolio management might include a large amount of capital spending projects. According to a spokesperson at AT&T:

> Our PMO is not part of portfolio management. We maintain a Portfolio Administration Office (PAO), which approves major capital spending projects and programs through an annual planning process. The PAO utilizes change control for any modifications to the list of approved projects. Each project manager must track the details of their project and update information in the Portfolio Administration Tool (PAT). The Corporate Program Office uses data in PAT to monitor the health and well being of the projects. Individual projects are audited to ensure adherence to processes and reports are prepared to track progress and status.

14.2 PROJECT SELECTION OBSTACLES[6]

Portfolio management decision-makers frequently have much less information to evaluate candidate projects than they would wish. Uncertainties often surround the likelihood of success for a project, the ultimate market value of the project, and its total cost to completion. This lack of an adequate information base often leads to another difficulty: the lack of a systematic approach to project selection and evaluation. Consensus criteria and methods for assessing each candidate project against these criteria are essential for rational decision-making. Although most companies have established organizational goals and objectives, they are usually not detailed enough to be used as criteria for project portfolio management decision-making. However, they are an essential starting point.

Portfolio management decisions are often confounded by several behavioral and organizational factors. Departmental loyalties, conflicts in desires, differences in perspectives, and an unwillingness to openly share information can stymie the project selection, approval, and evaluation processes. Much project evaluation data and information is necessarily subjective in nature. Thus, the willingness of the parties to openly share and put trust in each other's opinions becomes an important factor.

The risk-taking climate or culture of an organization can also have a decisive bearing on the project selection process. If the climate is risk adverse, high-risk projects may never surface. Attitudes within the organization toward ideas and the volume of ideas being generated will influence the quality of the projects selected. In general, the greater the number of creative ideas generated, the greater the chances of selecting high-quality projects.

14.3 ROLE OF THE PROJECT MANAGER IN PROJECT SELECTION

In the past, project managers rarely participated during an organization's project selection process even though their expertise can be of value in decision-making. Project selection may occur for an individual project or part of a project portfolio. Portfolio managers may participate in the selection process, but usually not the project manager who will eventually be assigned.

There are textbooks dedicated to project selection practices. Unfortunately, most project management books, including the *PMBOK® Guide*, focus on traditional project management practices and do not discuss the different project selection techniques, which leads people to believe that the PM's involvement is unnecessary. However, the same books do discuss the financial measures that might define the project success factors.

---

6. Section 14.4 is adapted from W. Souder, *Project Selection and Economic Appraisal* (New York: Van Nostrand Reinhold, 1984), pp. 2–3.

Introduction to Project Selection

During the initiation stage of a project, executives and managers make decisions on what projects are needed for the business to grow and survive. Most often, business cases and statements of work (SOWs) are prepared, and the projects are prioritized and added to the queue. Then project managers are assigned, usually based upon availability, and often pressured to undertake something that may be unrealistic or even impossible. For fear of punishment, project managers undertake the assignments, and then we wonder why projects may not deliver the expected results or even fail.

The root cause of the problem is the timing when project selection decision-makers bring project managers onboard. While it is true that strategic planning is done by senior management rather than having others doing it for them, early assignment of project managers can significantly improve project selection efforts. Experienced project managers often possess skills that executives may lack such as estimating project risks due to uncertainties, success and failure factors of past projects, resource competencies needed, selecting potential contractors and partners, and identification of information needs for project decision-making.

Execution of a business strategy is accomplished by the selection of the correct projects. Project selection decision-makers may be doing a disservice to their company when they believe that "information is power" and project managers do not have a "need to know" until after the project is approved. Simply stating that a project is aligned with a strategic business objective does not mean that this is the correct project to be executed. Without participation in the project selection process, project managers must make both technical and business decisions based on just the information in the business case and SOW, both of which may be incomplete. This can result in suboptimal project decisions where project teams have only partial information and not privileged to the full thought process and variables that went into the selection of the project. This can then limit the project manager's ability to be innovative.

The path to a company's sustainable business future will require the execution of more strategic and innovative projects than in the past. The pathway for these nontraditional projects goes from idea generation to idea development and then to market entry. In the idea development stage, on some projects, SOWs and business cases may be replaced by just an idea or a short statement defining strategic goals and objectives. This can create severe migraine headaches for decision-making during project execution unless project managers are active participants in the idea generation stage.

The Fuzzy Front End

Most project management courses identify the first life cycle phase of a project as initiation. Project managers are trained to validate their understanding of the requirements for the project even though they did not participate in establishing them. The beginning of the initiation phase is called the fuzzy front end (FFE) and is generally where ideas are approved for development. Typical divisions of smaller life cycle phases that may be part of the FFE include:

- Identifying needs, ideas, or potential projects.
- Choosing the selection criteria for each project considered.

- Comparing projects using multiple methods both quantitative and qualitative.
- Analyzing the findings.
- Selecting a project.

Each of these FFE life cycle phases struggles with imprecise information, uncertainties, and risks. This is the reason for the fuzziness. There are templates and mathematical models to assist decision-makers in each life cycle phase, and they are often used differently in each company. Unfortunately, PMs were usually not trained in their use. Today, PMs are being trained in the use of these tools and are being brought on board projects to participate in project selection activities.

The expression, FFE, was seldom used in the early years of project management which resided in project-driven organizations that survived on competitive bidding practices. Sales and marketing personnel often made the decision on whether to bid on a job and made sure that the contractual SOW was reasonably well-defined. Sometimes, project managers were asked to participate in preparing the bidding package, but generally, project managers were not assigned until the contract was awarded and the requirements were well-understood.

The growth of the FFE also appeared in non-project-driven organizations where the projects were designed for the creation of new products or services and improvements in the way that the firm conducts business. Most of these projects started out as just ideas. Detailed requirements appeared as the projects progressed through project execution. Unlike traditional projects designed for a single customer, these nontraditional projects focused upon the problems and issues of all customers as well as future business opportunities.

The challenge with the FFE during project selection activities is in minimizing risks and maximizing opportunities given the number of project possibilities and the limited number of available resources. The situation is further complicated because much of the needed information on the attributes of each project considered is incomplete or imprecise, thus creating a great deal of risk and uncertainty. There is also uncertainty in determining the importance of each project attribute and the impact or relationship between attributes.

Even though there are tools and techniques for project selection, decisions are often based upon intuition rather than facts or partial information when dealing with threats and opportunities. Bringing project managers onboard during FFE activities, accompanied by effective use of project selection models, can make the difference between organizational success and failure.

Understanding Project Attributes

Even though every organization may have unique characteristics, there are some general attributes common to most project selection models. The attributes may appear as forms, guidelines, templates, or checklists and identify factors that should be considered for the evaluation and comparison of each project.

Attribute #1: Core Values and Management's Vision

Most companies have core values for products and services such as a specific customer age group, a specific price range, and perhaps a desire to stay within a certain geographical market segment. The core values, which are senior management's vision of the organization's future, establish broad boundaries for project selection. If management's vision for the project involves new opportunities that fall outside of the core values, project managers must be made aware of this because of the impact on project execution decisions.

Attribute #2: Alignment to Strategic Business Objectives

The projects selected must be aligned with strategic business objectives. For simplicity's sake, strategic projects can be defined under two categories:

- Business growth projects that focus on expanding the products and services of the business.
- Business model improvement projects focus on improvements in the efficiency and effectiveness of the organization's business model that describes how the firm will interact with its customers, suppliers, and distributors.

Typical strategic objectives within the two generic categories in this attribute are shown in Table 14–1. There are different paths and different decisions to be made in each category and possibly for each strategic objective.

Management must ensure that these business objectives are visible to project teams. Project managers must understand the rationale for establishing a specific strategy to maintain continuous alignment of project decisions to support strategic objectives. Even though the project manager may participate during the project selection process, there must still exist a line-of-sight between the project team and senior management during project execution in case business objectives change or the alignment of the project to the business objectives has issues.

The growth in metrics measurement techniques has allowed project teams to establish business, intangible, and strategic metrics for project measurement and reporting. The metrics can be used to measure business value created and alignment with strategic business objectives. Sometimes, small changes in the creation of strategic business

TABLE 14–1. GENERIC ATTRIBUTE CATEGORIES

Business Growth Projects	Business Model Improvement Projects
Revenue growth	Products can be sold by current sales force
Market share growth	Use similar marketing and distribution channels
Image and reputation	Purchased by existing customer base
Brand awareness	Reduction in time-to-market and product cycle times
Business importance	Cost reduction opportunities
Strategic partnerships	Improvements in customer satisfaction

goals and objectives make it easier for project teams to report alignment and creation of business value.

By participating in FFE activities, project managers have a better understanding of the strategic information that is important to senior management and the board of directors. The PM can then select the appropriate strategic metrics that focus on business benefits and value being created and report this information to the senior levels of management and possibly even the board of directors.

The information presented in Table 14–1 can be considered as suitability criteria for selecting projects. Approving projects that fall outside of the suitability criteria can delay time-to-market, increase the cost associated with a product launch, and incur additional costs associated with product redesign efforts. Project managers must understand the relationship between a project's deliverables and the suitability criteria.

Attribute #3: Core Competencies

The selection of projects should be aligned with the firm's core competencies, which can include use of existing technologies and can be manufactured using existing production facilities. Experienced project managers usually have a better understanding of the firm's core competencies than senior managers and can make this information available during FFE activities.

Attribute #4: Resource Requirements

Perhaps the most significant contribution project managers can make to the project selection processes is in the determination of the resources needed for the undertaking. Resources include people, time, money, and possible facilities. Executives often select and prioritize projects without considering the resources needed. Project managers have a good grasp on human resource needs such as full-time or part-time, worker ramp-up and ramp-down time, additional training needs if necessary, and risks associated with the untimely removal of highly skilled workers to satisfy the needs of other projects. The result can be a stalled project and a financial drain.

Attribute #5: Using a Knowledge Management System

When working on strategic or innovation projects, having the right resources with the necessary competencies may not be enough. The organization may also need access to an information warehouse or knowledge management system.

Project managers may have expertise in working with the organization's knowledge management systems on previous projects. By knowing the contents of the system, PMs can provide advice on the need to hire contractors, proprietary information restrictions, control of intellectual property, and knowledge needed about customers' purchasing decisions that may impact project selection.

Attribute #6: Establishing Failure Criteria

Another important reason for project managers to participate in FFE activities is to work with senior management in deciding upon the project failure criteria. Every organization has its own tolerance threshold for risk and the threshold limit of unacceptable risk can be different for each project. Without an established failure criterion, it can be hard for a project team to know when to pull the plug and minimize the potential damage. Some examples of failure criteria include:

- Insurmountable obstacles (business or technical).
- Inadequate know-how and/or lack of qualified resources.
- Legal/regulatory or product liability uncertainties.
- Too small a market or market share for the product; dependence on a limited customer base.
- Unacceptable dependence on some suppliers and/or specialized raw materials.
- Unwillingness to accept joint ventures and/or licensing agreements.
- Costs are excessive, and the selling price of the deliverable will be noncompetitive.
- Return on investment will occur too late.
- Competition is too stiff and not worth the risks.
- The product life cycle is too short.
- The technology is not like that used in our other products and services.
- The product cannot be supported by our existing production facilities.
- The product cannot be sold by our existing sales force and is not fit with our marketing and distribution channels.
- The product/services will not be purchased by our existing customer base.

Project managers are now recognizing the skills needed to work in the FFE environment. Decision-making for portfolio project selection activities requires an understanding of how to apply fuzzy logic when dealing with uncertainty.

The Growth of Fuzzy Logic   There are several project selection methods. Some methods use complex mathematical concepts involving hard numbers to turn real-world problems into solvable equations. Other models that use fuzzy logic principles attempt to reduce uncertainty, such as in project selection activities [Demircan Keskin (2020), Kuchta (2001), Mohagheghi et al. (2019), Tolga (2008), and Zolfaghari et al. (2021)].

The fuzzy logic concept, which was introduced in 1965, allows us to make decisions based upon vague or imprecise information by providing weights to each attribute element. In fuzzification, the value of each variable or attribute element is represented by a real number between 0 and 1. The number 0 may represent a completely false situation unlikely to occur and not applicable whereas the number 1 may mean completely true. Numbers between 0 and 1 represent partial truths to help us make project selection decisions based upon imprecise or nonnumerical information.

Fuzzy logic application can also be integrated with other mathematical formulations such as:

- Linear programming
- Nonlinear programing

- Dynamic programming
- Integer programming
- Multiple objective programming

Fuzzy logic mathematics allows us to manage project selection uncertainties and make decisions that would normally be just a best guess. Examples, where fuzzy logic has been shown to be of value, include:

- Resource dependencies attributed to multiple projects.
- Minimizing scheduling issues when starting and stopping projects intermittently due to scarcity of resources.
- Determining the maximum number of resources needed for all projects.
- Erratic supply dates for raw materials.
- The interaction between project selection attributes.

Nonfinancial Factors   There are nonfinancial reasons for selecting some projects. Examples might include supporting an organization's corporate social responsibility vision on environmental impact, usage and depletion of scarce natural resources, customer impact, maintaining a certain market share, keeping a product line open and workers employed, and maintaining the correct image and reputation. The nonfinancial factors are usually not quantified and are often difficult to compare against other projects with financial factors but may have significant strategic importance.

REFERENCES

Demircan Keskin, F. (2020). A Two-Stage Fuzzy Approach for Industry 4.0 Project Portfolio Selection Within Criteria and Project Interdependencies Context. *Journal of Multi-Criteria Decision Analysis*, 27, 65–83.

Kuchta, D. (2001). A Fuzzy Model for R&D Project Selection with Benefit, Outcome and Resource Interactions. *Engineering Economist*, 46(3), 164–180.

Mohagheghi, V., Mousavi, S. M., Antucheviciene, J., and Mojtahedi, M. (2019). Project Portfolio Selection Problems: A Review of Models, Uncertainty Approaches, Solution Techniques, and Case Studies. *Technological and Economic Development of Economy*, 25(6), 1380–1412.

Tolga, A. (2008). Fuzzy Multicriteria R&D Project Selection with a Real Options Validation Model. *Journal of Intelligent and Fuzzy Systems*, 19, 359–371.

Zolfaghari, S., Mousavi, S. M., and Antucheviciene, J. (2021). A Type-2 Fuzzy Optimization Model for Project Portfolio Selection and Scheduling by Incorporating Project Interdependency and Splitting. *Technological and Economic Development of Economy*, 27(2), 493–510.

14.4 IDENTIFICATION OF PROJECTS

For more than 20 years, companies have used the concept of a PMO as the organization's knowledge repository of project management practices, guardian of the best practices library, and control point for continuous improvements in project management. In some organizations, all project managers are solid line reporting to the PMO.

As companies began embarking on more strategic and innovation projects, senior management recognized that they could not commit the amount of time necessary for project selection activities without sacrificing their other duties. The solution appeared as the establishing of a portfolio PMO with the responsibility of selecting and managing a group of projects. The PMO became the organization's lifeblood for identifying ideas and turning them into successful projects. The portfolio PMO relied heavily upon participation of project managers for the identification, selection, and execution of the projects within the portfolio.

The overall project portfolio management process is a four-step approach, as shown in Figure 14–1. The first step is the identification of the ideas for projects and needs to help support the business. The identification can be done through brainstorming sessions, market research, customer research, supplier research, and literature searches. All ideas, regardless of merit, should be listed.

Because the number of potential ideas can be large, some sort of classification system is needed. There are three common methods of classification. The first method is to place the projects into two major categories, such as survival and growth. The sources and types of funds for these two categories can and will be different. The second method comes from typical research and development (R&D) strategic planning models, as shown in Figure 14–2. Using this approach, projects to develop new products or services are classified as either offensive or defensive projects. Offensive projects

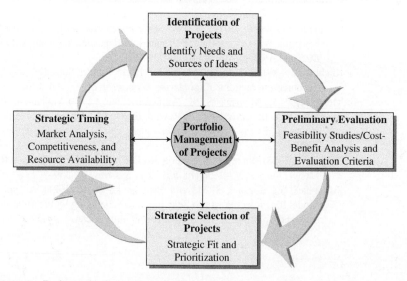

Figure 14–1. Project selection process.

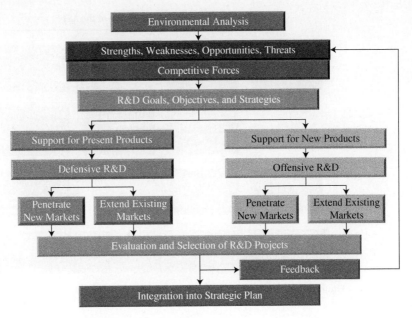

Figure 14–2. R&D strategic planning process.

are designed to capture new markets or expand market share within existing markets. Offensive projects mandate the continuous development of new products and services.

Defensive projects are designed to extend the life of existing products or services. This could include add-ons or enhancements geared toward keeping present customers or finding new customers for existing products or services. Defensive projects are usually easier to manage than offensive projects and have a higher probability of success.

Another method for classifying projects is:

- Radical technical breakthrough projects.
- Next-generation projects.
- New family members.
- Add-ons and enhancement projects.

Radical technological breakthrough projects are the most difficult to manage because of the need for innovation. Figure 14–3 shows a typical model for innovation. Innovation projects, if successful, can lead to profits that are many times larger than the original development costs. Unsuccessful innovation projects can lead to equally dramatic losses, which is one of the reasons why senior management must exercise due caution in approving innovation projects. Care must be taken to identify and screen out inferior candidate projects before committing significant resources to them.

There is no question that innovation projects are the most costly and difficult to manage. Some companies mistakenly believe that the solution is to minimize or limit the total number of ideas for new projects or to limit the number of ideas in each category. This could be a costly mistake.

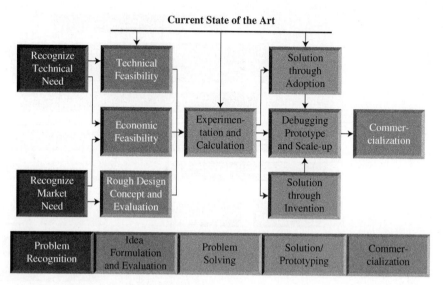

Figure 14–3. Modeling the innovation process.

In a study of the new-product activities of several hundred companies in all industries, Booz, Allen, and Hamilton[7] defined the new-product evolution process as the time it takes to bring a product to commercial existence. This process began with company objectives, which included fields of product interest, goals, and growth plans, and ended with, hopefully, a successful product. The more specifically these objectives were defined, the greater guidance would be given to the new-product program. This process was broken down into six manageable, fairly clear sequential stages:

1. *Exploration*. The search for product ideas to meet company objectives.
2. *Screening*. A quick analysis to determine which ideas were pertinent and merit more detailed study.
3. *Business analysis*. The expansion of the idea, through creative analysis, into a concrete business recommendation, including product features, financial analysis, risk analysis, market assessment, and a program for the product.
4. *Development*. Turning the idea-on-paper into a product-in-hand, demonstrable and producible. This stage focuses on R&D and the inventive capability of the firm. When unanticipated problems arise, new solutions and trade-offs are sought. In many instances, the obstacles are so great that a solution cannot be found, and work is terminated or deferred.
5. *Testing*. The technical and commercial experiments are necessary to verify earlier technical and business judgments.
6. *Commercialization*. Launching the product in full-scale production and sale; committing the company's reputation and resources.

---

7. *Management of New Products* (McLean, VA: Booz, Allen and Hamilton, 1984), pp. 180–181.

In the Booz, Allen & Hamilton study, the new-product process was characterized by a decay curve for ideas, as shown in Figure 14–4. This showed a progressive rejection of ideas or projects by stages in the product evolution process. Although the rate of rejection varied between industries and companies, the general shape of the decay curve is typical. It generally takes close to 60 ideas to yield just one successful new product.

The process of new-product evolution involves a series of management decisions. Each stage is progressively more expensive, as measured in expenditures of both time and money. Figure 14–5 shows the rate at which expense dollars are spent as

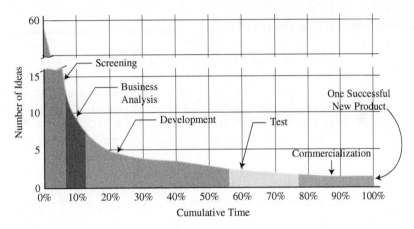

Figure 14–4. Mortality of new product ideas.

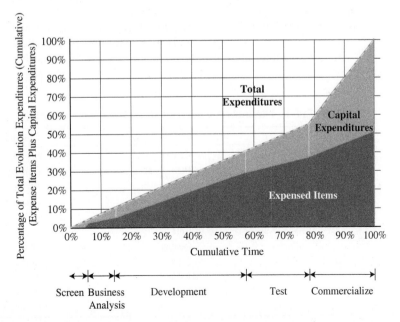

Figure 14–5. Cumulative expenditures and time.

time accumulates for the average project within a sample of leading companies. This information was based on an all-industry average and is, therefore, useful in understanding the typical industrial new-product process. It is significant to note that the majority of capital expenditures are concentrated in the last three stages of evolution. It is, therefore, very important to do a better job of screening for business and financial analysis. This will help eliminate ideas of limited potential before they reach the more expensive stages of evolution.

14.5 PRELIMINARY EVALUATION

As shown in Figure 14–1, the second step in project selection is preliminary evaluation. From a financial perspective, preliminary evaluation is basically a two-part process. First, the organization will conduct a feasibility study to determine whether the project can be done. Second, the organization will perform a cost-benefit analysis to see whether the company should do it (see Table 14–2).

The purpose of the feasibility study is to validate that the idea or project meets feasibility of cost, technological, safety, marketability, and ease of execution requirements. It is possible for the company to use outside consultants or subject matter experts to assist in both feasibility studies and benefit-to-cost analyses. A project manager may not be assigned until after the feasibility study is completed because the project manager may not have sufficient business or technical knowledge to contribute prior to this point in time.

If the project is deemed feasible and a good fit with the strategic plan, the project is prioritized for development along with other approved projects. Once feasibility is determined, a cost-benefit analysis is performed to validate that the project will, if executed correctly, provide the required financial and nonfinancial benefits. Cost-benefit analyses require that significantly more information is scrutinized than is usually available during a feasibility study. Such analyses can be expensive.

TABLE 14–2.   FEASIBILITY STUDIES AND COST-BENEFIT ANALYSES

	Feasibility Study	**Cost-Benefit Analysis**
Basic question	Can we do it?	Should we do it?
Life cycle phase	Preconceptual	Conceptual
Project manager selected	Not yet	Perhaps
Analysis	**Qualitative**	**Quantitative**
	Technical	NPV
	Cost	Discounted cash flow (DCF)
	Quality	IRR
	Safety	ROI
	Legal	Assumptions
	Economical	Reality
Decision criteria	Strategic fit	Benefits > cost

Estimating benefits and costs in a timely manner is very difficult. Benefits are often defined as:

- Tangible benefits, for which dollars may be reasonably quantified and measured.
- Intangible benefits, which may be quantified in units other than dollars or may be identified and described subjectively.

Costs are significantly more difficult to quantify, at least in a timely and inexpensive manner. The minimum costs that must be determined are those that are used specifically for comparison to the benefits. These include:

- The current operating costs or the cost of operating in today's circumstances.
- Future period costs that are expected and can be planned for.
- Intangible costs that may be difficult to quantify. These costs are often omitted if quantification would contribute little to the decision-making process.

There must be careful documentation of all known constraints and assumptions that were made in developing the costs and the benefits. Unrealistic or unrecognized assumptions are often the cause of unrealistic benefits. The go or no-go decision to continue with a project could very well rest on the validity of the assumptions.

14.6 STRATEGIC SELECTION OF PROJECTS

From Figure 14–1, the third step in the project selection process is the strategic selection of projects, which includes the determination of a strategic fit and prioritization. It is at this point where senior management's involvement is critical because of the impact that the projects can have on the strategic plan.

Strategic planning and the strategic selection of projects are similar in that both deal with the future profits and growth of the organization. Without a continuous stream of new products or services, the company's strategic planning options may be limited. Today advances in technology and growing competitive pressure are forcing companies to develop new and innovative products while the life cycle of existing products appears to be decreasing at an alarming rate. Yet, at the same time, executives may keep research groups in a vacuum and fail to take advantage of the potential profit contribution of R&D strategic planning and project selection.

There are three primary reasons that corporations work on internal projects:

1. To produce new products or services for profitable growth.
2. To produce profitable improvements to existing products and services (i.e., cost reduction efforts).
3. To produce scientific knowledge that assists in identifying new opportunities or in "fighting fires."

Successful project selection is targeted, but targeting requires a good information system, and this, unfortunately, is the weakest link in most companies. Information systems are needed for optimum targeting efforts, and this includes assessing customer and market needs, economic evaluation, and project selection.

Assessing customer and market needs involves opportunity-seeking and commercial intelligence functions. Most companies delegate these responsibilities to the marketing group, and this may result in a detrimental effort because marketing groups appear to be overwhelmed with today's products and near-term profitability. They simply do not have the time or resources to adequately analyze other activities that have long-term implications. Also, marketing groups may not have technically trained personnel who can communicate effectively with the R&D groups of the customers and suppliers.

Most organizations have established project selection criteria, which may be subjective, objective, quantitative, qualitative, or simply a seat-of-the-pants guess. The selection criteria are most often based on suitability criteria, such as:

- Similar in technology.
- Similar marketing methods used.
- Similar distribution channels used.
- Can be sold by current sales force.
- Will be purchased by the same customers as current products.
- Fits the company philosophy or image.
- Uses existing know-how or expertise.
- Fits current production facilities.
- Both research and marketing personnel enthusiastic.
- Fits the company long-range plan.
- Fits current profit goals.

In any event, there should be a valid reason for selecting the project. Executives responsible for selection and prioritization often seek input from other executives and managers before moving forward. One way to seek input in a quick and reasonable manner is to transform the suitability criteria shown above into rating models or fuzzy logic models discussed previously. Typical rating models are shown in Figures 14–6–14–8. These models can be used for both strategic selection and prioritization.

Prioritization is a difficult process. Factors such as cash flow, near-term profitability, and stakeholder expectations must be considered. Also considered are a host of environmental forces, such as consumer needs, competitive behavior, existing or forecasted technology, and government policy.

Being highly conservative during project selection and prioritization could be a road map to disaster. Companies with highly sophisticated industrial products must pursue an aggressive approach to project selection or risk obsolescence. This also mandates the support of a strong technical base.

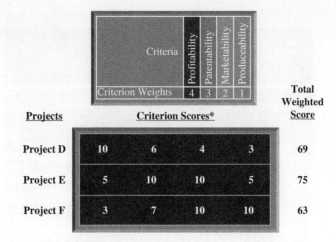

Figure 14–6. Scoring model.

Source: W. Souder, *Project Selection and Economic Appraisal* (New York: Van Nostrand Reinhold, 1984), pp. 66–69.

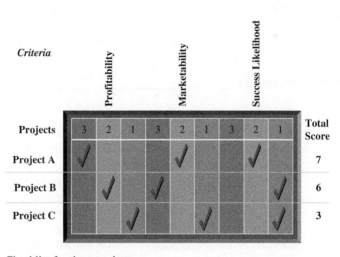

Figure 14–7. Checklist for three projects.

Source: W. Souder, *Project Selection and Economic Appraisal* (New York: Van Nostrand Reinhold, 1984), pp. 66–69.

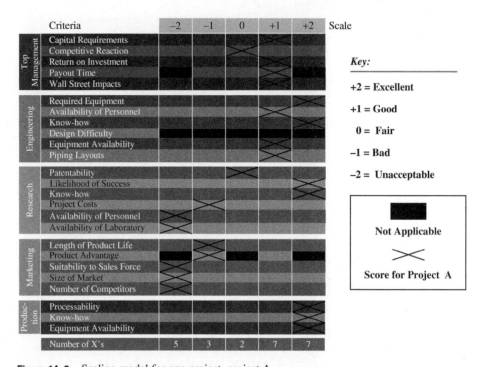

Figure 14–8. Scaling model for one project, project A.
Source: W. Souder, *Project Selection and Economic Appraisal* (New York: Van Nostrand Reinhold, 1984), pp. 66–69.

14.7 STRATEGIC TIMING

Many organizations make the fatal mistake of taking on too many projects without regard for the limited availability of resources. As a result, highly skilled labor is assigned to more than one project, creating schedule slippages, lower productivity, less-than-anticipated profits, and never-ending project conflicts.

The selection and prioritization of projects must be made based on the availability of qualified resources. Planning models are available to help with the strategic timing of resources. These models are often referred to as *aggregate planning models*.

Another issue with strategic timing is the determination of which projects require the best resources. Some companies use a risk–reward cube, where the resources are assigned based on the relationship between risk and reward. The problem with this approach is that the time required to achieve the benefits (i.e., payback period) is not considered.

Aggregate planning models allow an organization to identify the overcommitment of resources. This could mean that high-priority projects may need to be shifted in time or possibly be eliminated from the queue because of the unavailability of qualified resources. It is a pity that companies also waste time considering projects for which they know that the organization lacks the appropriate talent.

Another key component of timing is the organization's tolerance level for risk. Here the focus is on the risk level of the portfolio rather than the risk level of an individual project. Decision-makers who understand risk management can then assign resources effectively such that the portfolio risk is mitigated or avoided.

14.8 ANALYZING THE PORTFOLIO

Companies that are project-driven organizations must be careful about the type and quantity of projects they work on because of the resources available. Because of critical timing, it is not always possible to hire new employees and have them trained in time or to hire subcontractors who may possess questionable skills.

Figure 14–9 shows a typical project portfolio, which was adapted from the life cycle portfolio model commonly used for strategic planning activities. Each circle represents a project. The location of each circle represents the quality of resources and the life cycle phase that the project is in. The size of the circle represents the magnitude of the benefits relative to other projects, and the pie wedge represents the percentage of the project completed thus far.

In Figure 14–9, project A has relatively low benefits and uses medium-quality resources. Project A is in the definition phase. However, when project A moves into the design phase, the quality of resources may change to low quality or strong quality. Therefore, this type of chart has to be updated frequently.

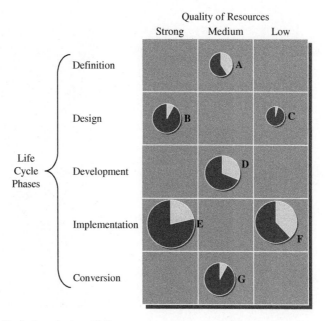

Figure 14–9. Typical project portfolio.

Figures 14–10–14–12 show three types of portfolios. Figure 14–10 represents a high-risk project portfolio where strong resources are required for each project. This may be representative of project-driven organizations that have been awarded large, highly profitable projects. This could also be a company in the computer field that competes in an industry that has short product life cycles and where product obsolescence occurs six months downstream.

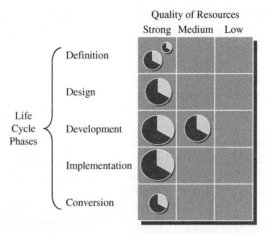

Figure 14–10. High-risk portfolio.

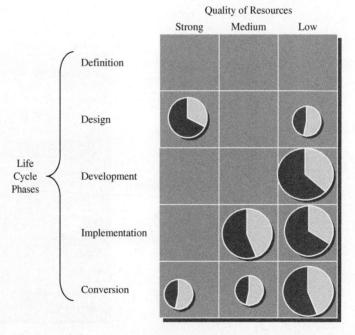

Figure 14–11. Profit portfolio.

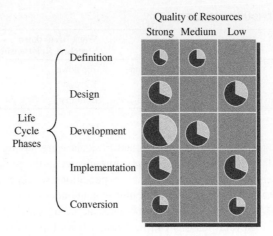

Figure 14–12. Balanced portfolio.

Figure 14–11 represents a conservative, profit portfolio where an organization works on low-risk projects that require low-quality resources. This could be representative of a project portfolio selection process in a service organization or even a manufacturing firm that has projects designed mostly for product enhancement.

Figure 14–12 shows a balanced portfolio with projects in each life cycle phase and where all levels of resources are being utilized, usually quite effectively. A very delicate juggling act is required to maintain this balance.

14.9 PROBLEMS WITH MEETING EXPECTATIONS

Why is it that, more often than not, the final results of either a project or an entire portfolio do not meet senior management's expectations? This problem plagues many corporations, and the blame is ultimately (and often erroneously) rationalized as poor project management practices. As an example, a company approved a portfolio of 20 R&D projects. Each project was selected on its ability to be launched as a successful new product. The approvals were made following the completion of feasibility studies. Budgets and timetables were then established such that the cash flows from the launch of the new products would support the dividends and the cash needed for ongoing operations.

Full-time project managers were assigned to each of the 20 projects and began with the development of detailed schedules and project plans. For eight of the projects, it quickly became apparent that the financial and scheduling constraints imposed by senior management were unrealistic. The project managers on these eight projects decided not to inform senior management of the potential problems but to wait to see if contingency plans could be established. Hearing no bad news, senior management was left with the impression that all launch dates were realistic and would go as planned.

TABLE 14–3. COST/HOUR ESTIMATES

Estimating Method	Generic Type	Work Breakdown Structure Relationship	Accuracy	Time to Prepare
Parametric	Rough order of magnitude	Top down	25% to +75%	Days
Analogy	Budget	Top down	−10% to +25%	Weeks
Engineering (grassroots)	Definitive	Bottom up	−5% to +10%	Months

The eight trouble-plagued projects were having a difficult time. After exhausting all options and failing to see a miracle occur, the project managers reluctantly informed senior management that their expectations would not be met. This occurred so late in the project life cycle that senior management became quite irate, and several employees had their employment terminated, including some of the project sponsors.

Several lessons can be learned from this situation. First, unrealistic expectations occur when financial analysis is performed from "soft" data rather than "hard" data. In Table 14–1 we showed the differences between a feasibility study and a cost-benefit analysis. Generally speaking, feasibility studies are based on soft data.

Therefore, critical financial decisions based on feasibility study results may have significant errors. This can also be seen in Table 14–3, which illustrates the accuracy of typical estimates. Feasibility studies use top-down estimates that can contain significant error margins.

Benefit-to-cost analyses should be conducted from detailed project plans using more definitive estimates. Cost-benefit analysis results should be used to validate that the financial targets established by senior management are realistic.

Even with the best project plans and comprehensive benefit-to-cost analyses, scope changes will occur. Periodic reestimations of expectations must be performed on a timely basis. One way of doing this is by using the rolling wave concept shown in Figure 14–13. The rolling wave concept implies that as you get farther along in the project, more knowledge is gained, which then allows more detailed planning and estimating. The latter then provides additional information from which we can validate the original expectations.

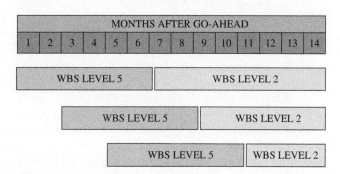

Figure 14–13. Rolling wave concept.
Note: WBS = work breakdown structure

Continuous reevaluation of expectations is critical. At the beginning of a project, it is impossible to ensure that the benefits expected by senior management will be realized at project completion. The length of the project is a critical factor. Based on project length, scope changes may result in project redirection. The culprit is most often changing economic conditions, resulting in invalid original assumptions. Also, senior management must be made aware of events that can alter expectations. This information must be made known quickly. Senior management must be willing to hear bad news and have the courage to possibly cancel a project.

Since changes can alter expectations, project portfolio management must be integrated with the project's change management process. According to Mark Forman, the associate director for IT and e-government in the Office of Management and Budget:

> Many agencies fail to transform their process for IT management using the portfolio management process because they don't have change management in place before starting. IT will not solve management problems—re-engineering processes will. Agencies have to train their people to address the cultural issues. They need to ask if their process is a simple process. A change management plan is needed. This is where senior management vision and direction is sorely needed in agencies.[8]

Although the comments here are from government IT agencies, the problem is still of paramount importance in nongovernmental organizations and across all industries.

14.10 MISALIGNMENT ISSUES

One of the serious problems resulting in not meeting expectations is when decisions made by the project team appear to be in the best interest of that project and not necessarily in the best interest of the entire organization funding the project or even the entire company. Their cause is most often a lack of alignment between project and strategic or organizational objectives [Brito and Medeiros Júnior (2021)].

As project management began transitioning from just traditional projects to the execution of strategic projects including innovation activities, the linkage between the words "strategy" and "projects" became apparent. Strategies are implemented through the execution of the right projects. Project management practices must now deal with strategy formulation and implementation as well as the execution of strategic projects. Project managers are becoming organizational strategists. Unfortunately, as with any new management practices, what also appeared were serious issues that could prevent effective alignment. If corporate-wide adoption of project management techniques is to become relevant for achieving strategic objectives, the alignment issues between project management, business strategy, and innovation activities need to be resolved.

---

8. Stouffer and Rachlin, "A Summary of First Practices and Lessons Learned in Information Technology Portfolio Management," p. 1.

Successful alignment increases the chances that organizations will select the right projects to achieve strategic objectives. Proper alignment will provide organizations more opportunities for making the necessary adjustments to match resources availability to strategic objectives. Effective alignment of project management to business strategy can result in improved organizational communications, especially between senior management and business units, better alignment of project priorities to business strategies, improved synergies between the business units, better monitoring and reporting of project performance, and successful implementation of strategies.

Project Management and Business Strategy

In most project management courses, discussions revolve around project management processes, tools, and techniques. However, there are several other topics that need to be considered when discussing the relationship between projects and business strategy. Project managers today see themselves as managing part of a business rather than just a project. As such, many of the other topics that were not considered as crucial previously for the education of project managers are now becoming significantly important because of their relationship to business strategy. Project and program managers are now becoming significantly more knowledgeable in strategic planning activities. Business and strategic metrics are now being used in addition to the traditional metrics of time, cost, and scope.

Understanding Innovation Project Management

For many companies, strategic planning focuses heavily on the execution of innovation and R&D projects to achieve strategic goals and objectives. Innovation projects often struggle when standard practices used on traditional or operational projects are applied to innovation activities. As such, functional organizations were allowed to manage these types of innovation projects differently using whatever processes they believed would work best even though many companies had methodologies, processes, tools, and techniques that were standardized for use on traditional or operational projects.

Perhaps the most important difference between innovation and traditional projects revolves around project definition. Traditional projects often start with a well-defined business case and SOW, which can minimize alignment issues. Innovation projects can begin with just an idea of a strategic goal. The business case, SOW, work breakdown structure, budgets, and schedules are developed and updated as the innovation project progresses. The direction of an innovation project can change rapidly based upon the results of one test. This can then cause major changes in the innovation project management processes used and changes in the metrics needed for performance measurement and reporting. Alignment must be continuously monitored.

With advances in project management processes, such as using agile or scrum rather than a traditional Waterfall Methodology, emphasis is being placed on applying project management practices to nontraditional projects such as innovation projects. For more than 50 years, there have been articles expounding the benefits of implementing project management practices. There have also been publications discussing the challenges we face with misalignment of project management processes to business objectives.

Today, as we apply project management practices to innovation and strategic business practices, the challenges resulting from misalignments and disconnects are being highly publicized. There still exists a reasonably high failure rate on certain types of projects. On innovation projects, some companies see failure rates or significantly less-than-expected results exceeding 80 percent. More failures than ever before are appearing due to misalignment of critical factors. We have known about the misalignment issues for some time and the devastating results they can bring. Fortunately, application of project management practices to other types of projects, such as business-related or innovation, is expected to generate sufficient research for meaningful solutions to be found.

Understanding Misalignment   Misalignment is the incorrect positioning or use of something in relation to something else. In traditional project management practices, misalignment commonly appeared when project management execution techniques did not appear to fully support the project's objectives as stated in the business case. The cause might be the selection of the wrong processes or selecting metrics that cannot validate that the project's objectives are being met.

Misalignment issues have occurred for some time in project management. But with the application of project management to innovation activities, misalignment issues between business strategy, innovation projects, and project management practices are becoming common. Misalignment does not necessarily result in project failure. The result could be less than optimal achievement of strategic objectives, the wrong outcomes, or the need for additional time to correct the problems.

Senior Management   Senior management is the architect of the corporate strategy.
Misalignment Issues   Unfortunately, they often just establish strategic goals and objectives without fully understanding how projects can be used to achieve these strategic goals and objectives. Strategies are implemented using projects. Senior management must understand how project management can impact business strategies. Misalignment occurs when management has a poor understanding of how to translate strategy into projects [Ansari et al. (2015) and Young and Grant (2015)].

Senior management often delegates the responsibility of project identification to others but retains the authority for project approval. Projects may be approved without careful consideration of their alignment with strategic issues. Managers may be asked to identify projects that are aligned more so to their functional unit strategies rather than critical corporate business strategies.

Priority setting is usually a senior management responsibility. The priorities assigned to projects must be aligned with business strategy [Srivannaboon and Milosevic (2004, 2006)]. Companies usually have more projects in the queue than they can execute with existing organizational resources. Selecting a project and assigning a priority must be based upon capacity planning information that identifies worker skills needed and resource availability. Misalignment occurs when projects are established and prioritized without a clear understanding of resources needed. Project managers can assist with identifying skills needed and determining resource availability. Unfortunately, project managers seem to be brought on board the projects after project approval and prioritization take place.

One of the most significant causes for misalignment between projects and strategy relates to when the project managers are brought on board the projects [Söderlund and Maylor (2012)]. Senior management often creates alignment issues when they believe that "information is power" and refuse to bring project managers on board early to understand the thought processes that were the background to establishing the strategy. Project managers are then assigned to the projects after approval and without understanding the justification for the project, the reasoning behind the assumptions and constraints, and the priority for the project. Project teams may then select an incorrect set of metrics and KPIs for reporting project performance because of different interpretations of the strategy.

Project Management Misalignment Issues

Project managers commonly suffer from a lack of strategic information needed for project decision-making. The problem is a poor line-of-sight between the project team and senior management prior to and during project execution. Project decisions are then made for what might appear to be in the best interest of creating the project's deliverables but may not be the ideal deliverables needed to support the intended business strategy.

Project managers are frequently reassigned to other activities once the project's deliverables are created. It may take months or even years for the real business benefits and value of the deliverables to be achieved. Unfortunately, once the deliverables are created, the project is often turned over to a business benefits harvesting team that may have their own agenda for the deliverables, and the project manager's involvement is no longer seen as being necessary [Kerzner et al. (2022)].

Portfolio PMO Misalignment Issues

Portfolio PMOs can create misalignment issues by selecting the wrong portfolio metrics. As stated by Thorn (2003),

> "Project metrics are not always viewed constructively by the project offices. Their imposition by higher organizational levels can lead to adversarial behavior by a project office when business decisions are viewed as secondary to meeting metric goals."

Metrics are needed to show the value that a portfolio PMO brings to an organization. However, spending an excessive amount of time on these types of metrics can detract from the focus on meeting project expectations asked for in the customer's value objectives. Another issue is that some PMO metrics are largely subjective and can become the driver for the wrong behavior.

There is also the issue of the portfolio PMO acting more so in a monitoring capacity and not having direct control over the projects in the portfolio. This can lead to the selection of metrics that are oriented toward governance rather than project implementation. When this occurs, metrics may show just the outcome of problems rather than sources of the problems.

There is very little disagreement that portfolio PMOs need effective metrics to support their portfolio performance measurement systems. Selecting the right metrics is the issue. Relying on commonly used traditional metrics is a mistake. According to Bhasin (2008), problems with the use of traditional metrics on innovation and R&D activities include:

- Traditional metrics are not suited for strategic decisions.
- Traditional metrics do not measure and report the creation of business value.
- Traditional metrics are not very effective for the evaluation of intangible assets.
- Traditional metrics provide little information on the root cause of problems.
- Portfolio metrics must address all the dimensions of the innovation project management process.

Portfolio PMOs are sometimes seen as an overhead expense. To lower overhead costs, portfolio PMOs generally monitor only high-priority projects, which may be no more than 20 projects. This leads to misalignment issues in that portfolio governance and supervision is being applied to just a few projects rather than a way to maintain a balanced portfolio of all projects.

A Window into the Future   Simply because we understand the existence of misalignments is no guarantee that the problems will be resolved. There are steps however that can be taken:

- Senior management must develop a framework to ensure that selected projects are in line with business strategies.
- Senior management must recognize that some projects, such as innovation projects, may not be able to be managed with the same project management practices used on traditional type projects.
- PMOs can develop their own distinct methodologies and frameworks for the projects under their supervision and governance. They can also develop their own protocols for project selection and prioritization if it correctly supports the organization's strategic direction.
- PMOs must develop business-related and strategic metrics that measure the ongoing alignment of PMO projects to organizational strategic goals and objectives. Senior management must understand the importance of these metrics.
- Metrics must be established that measure business benefits and business value created throughout the life cycle of the project. These metrics must measure and predict both tangible and intangible business value being created.
- Senior management must continuously remind project managers, program managers, and PMOs that they are expected to make decisions that are in the best interest of corporate strategy first rather than other criteria such as functional unit objectives and goals.

Change will not take place quickly. But with the right support and encouragement from executive management change is possible and the potential damage from misalignment issues can be minimized.

REFERENCES

Ansari, R., Shakeri, E., and Raddadi, A. (2015). Framework for Aligning Project Management with Organizational Strategies. *Journal of Management in Engineering*, 31(4), 1–8.

Bhasin, S. (2008). Lean and Performance Measurement. *Journal of Manufacturing Technology Management*, 19(5), 670–684.

Brito, J. V. C. S. and de Medeiros Júnior, J. V. (2021). Alignment Strategic in Project Based Businesses: A Review of the Literature. *Revista Ibero-Americana de Estratégia (RIAE)*, 20(1), 1–25.

Kerzner, H., Zeitoun, A., and Vargas, R. (2022). *Project Management Next Generation: The Pillars of Organizational Success*. Hoboken: John Wiley, Chapter 8.

Söderlund, J. and Maylor, H. (2012). Project Management Scholarship: Relevance, Impact and Five Integrative Challenges for Business Management Schools. *International Journal of Project Management*, 30(6), 686–696.

Srivannaboon, S. and Milosevic, D. Z. (2006). A Two-Way Influence Between Business Strategy and Project Management. *International Journal of Project Management*, 24(6), 493–505.

Srivannaboon, S. and Milosevic, D. Z. (2004). The process of translating business strategy in project actions. *Innovations Project Management Research*. Newtown Square, PA: Project Management Institute.

Thorn, M. E. (2003). Bridge Over Troubled Water: Implementation of a Program Management Office. *SAM Advanced Management Journal*, 68(4), 48–59.

Young, R. and Grant, J. (2015). Is Strategy Implemented By Projects? Disturbing Evidence in the State of NSW. *International Journal of Project Management*, 33(1), 15–28.

14.11 POST-FAILURE SUCCESS ANALYSIS

All projects run the risk of failure. This includes traditional projects that have well-defined requirements based upon historical estimates as well as innovation or strategic projects that may begin with just an idea. The greater the unknowns and uncertainties, such as with innovation projects, the greater the risk of failure.

When innovation projects appear to be failing, there is a tendency for project teams to walk away from the project quickly in hopes of avoiding potential blame and trying to distance themselves from the failure. Project team members usually return to their functional areas for other assignments and project managers move on to manage other projects. Companies most often have an abundance of ideas for new projects and seem to prefer to move forward without determining if failures can be turned into successes.

Part of the problem is with the expression that many project managers follow, namely "Hope for the best, but plan for the worst." Planning for the worst usually means establishing project failure criteria as to when to exit the project and stop squandering resources. When the failure criteria point is reached, all project work tends to cease. Unfortunately, post-failure success may still be possible if the organization understands and implements the processes that can turn failure into success. Post-failure analysis practices are now being taught to PMs and PMO personnel.

There are many sources for failure. The most common cause of perceived failure is not meeting performance expectations. However, failure can also be the result of using the wrong processes, poor project, and/or organizational leadership, making the wrong assumptions, unrealistic expectations, poor risk management practices, and focusing on the wrong strategic objectives. Many of these activities that were seen as the causes of failure can be overcome by reformulation of the project such that post-failure success may be attainable. Successful project management practices must focus on more than just creating deliverables. They must also focus on processes needed for converting failures into successes. Unfortunately, these practices are seldom taught in project management courses until recently.

Sensemaking

Sensemaking is the process of making sense out of something that was novel, heavily based upon uncertainties or ambiguities, and failed to meet expectations. Sensemaking is one of the commonly used techniques that may be able to convert failures into potential successes. Sensemaking cannot guarantee successful recovery but can improve the chances of success.

Turning failure into success requires first, a culture of normalizing failures and second, a process of problem formulation characterized by sensemaking that is both retrospective and prospective [Morais-Storz et al. (2020)]. Retrospective sensemaking addresses the issue of *what happened*, looks at the causes that led up to the potential failure, and try to make sense out of the results. Prospective sensemaking addresses *what to do now* and envisions what the future might look like if we construct and implement a plausible new path (see Figure 14–14).

Figure 14–14. Post-failure success analysis.

The leadership style that the project manager selects determines whether retrospective or prospective sensemaking will be emphasized as well as the approach to reformulating the project, if necessary. The outcome from sensemaking can be a reformulation of the original problem or the development of new plans that focus on somewhat different outcomes. Reformulation may be necessary if the team did not fully understand the way that the problem was initially presented to them, decision-making was being made based upon guesses rather than facts, or if significant changes occurred in the enterprise's environmental factors. The knowledge gained from sensemaking may indicate that the original expectations are still valid, but that project must be reformulated. If the original expectations are no longer valid, then a new trajectory with perhaps modified expectations may be necessary.

The Need for New Metrics   Sensemaking requires more information than may be available in earned value measurement systems. When implementing traditional projects using the Waterfall Methodology and accompanied by well-defined requirements, decision-making is centered around the time, cost, and scope metrics.

On other projects such as those involving innovation, given the high probability of failure, additional metrics may be required to determine the impact of variables that may have changed from the original problem formulation and resulted in a need for post-failure analysis. Some of the new metrics include:

- The number of new assumptions made over the project's life cycle.
- The number of assumptions that changed over the project's life cycle.
- Changes that occurred in the enterprise's environmental factors.
- The number of scope changes approved and denied.
- The number of time, cost, and scope baseline revisions.
- The effectiveness of project governance.
- Changes in the risk level of the critical work packages.

Learning from Failure   Many project management educators advocate that the only true project failures are those from which nothing is learned. The cost of not examining failures can be very expensive if mistakes are repeated. Examination of a failed innovation project usually results in the discovery of at least some intellectual property that can be used elsewhere as well as processes that may or may not have been effective.

Organizations have looked at project failures for the purpose of capturing best practices so that mistakes are not repeated. Some people believe that more best practices can be found from failures than from successes if people are not hesitant to discuss failures. Unfortunately, the best practices discovered from both success and failure analyses are usually related to changes in the forms, guidelines, templates, and checklists. Very little effort is usually expended at post-project reviews to identify behavioral best practices, which may very well have been the root cause of the failure.

Sensemaking allows organizations to address the psychological barriers that may have led up to the failure [Sitkin (1992)]. As stated by Morais-Storz et al. (2020):

"Failure is important for effective organizational learning and adaptation for several reasons. Failure helps organizations discover uncertainties, which are difficult to predict in advance [Sitkin (1992)], creates learning readiness and motivates learning and adaptation [Cyert and March (1992)], increases risk-seeking behavior [Kahneman and Tversky (1979)], and act as a shock trigger to draw organizational attention to problems [Van de Ven et al. (2008)]."

Learning from failures can disrupt an organization if the outcome indicates that a significant change is needed such as in the organization's culture, business model, or processes. Innovation and strategic projects are generally not formulated with the same delicacy as traditional projects. The problem is further compounded by the fact that most innovation project managers are not brought on board the project until after organizational management has approved and possibly prioritized the project, identified the constraints, and stated the assumptions the team should follow. The innovation project manager, therefore, begins implementation with a project formulation that neither he/she nor the team participated in. Sensemaking usually occurs in a collaborative setting, perhaps involving most of the team members. Initial problem formulation, on the other hand, may be accomplished with just a few people, most of whom have never had to manage this type of project.

The literature discusses various ways to formulate projects, but usually traditional or operations projects rather than innovation or strategic projects. There is a lack of literature on the link between failure and how the knowledge gained during retrospective sensemaking can benefit project reformulation on innovation-type projects by challenging complacency, the use of existing processes, and ineffective leadership in dealing with ambiguity.

The Failure of Success

Perhaps the greatest psychological barrier to innovation is when an organization becomes complacent and refuses to challenge its assumptions, business model(s), and the way they conduct business. The organization is usually financially successful and believes that it will remain financially successful for years to come. As such, profitability and market share become more important than innovation, and this can create a significant psychological barrier for innovation teams. Eventually, the marketplace will change, profitability will be eroded, and innovation will become the top priority. But by this time, the failure of success may have taken its toll on the company to a point where the company may never recover. Effective portfolio PMOs can provide the guidance necessary to possibly convert a failure into success.

Characteristics of a potential failure of success environment include:

● Maintaining the status quo is essential.
● Most decisions are made in favor of short-term profitability.

- Maintaining the present market share is more important than investing in opportunities.
- Senior management positions are filled by financial personnel rather than from marketing and sales personnel whose vision created the favorable growth.
- Executives refuse to challenge any of the business assumptions for fear of changing the status quo.
- The guiding principles that led to success are not challenged.
- The VUCA environment is expected to remain stable.
- The company maintains a very low-risk appetite.
- No changes are needed in the organization's business model.
- The company will continue using the same suppliers and distributors.
- No changes are necessary in organizational leadership.
- No plans are made for future generations of managers.
- The organizational culture is based upon command and control from the top floor of the building to the bottom and not depersonalized to support an innovation and free-thinking environment.
- No changes are needed in the organization's reward systems, which most likely are not based upon risk-taking.
- Changes or continuous improvement efforts for project management processes are at a minimum.
- There is not adequate funding for the needed innovation projects, other than perhaps small incremental innovation activities.
- The failure of an innovation project is brushed aside if it has no immediate impact on profitability.
- The use of sensemaking practices is nonexistent.

The critical issue facing innovation and strategic project teams in successful companies is that all of the above characteristics create complex interrelations that team members must deal with. As stated by James O'Toole (1983):

- "Innovation requires the ability to read changes in the environment, and to create policies and strategies that will allow the organization to capitalize on the opportunities those changes create. Ironically, the most successful companies are likely to ignore environmental signals because it seems wildly risky to tamper when things are going well."
- "Corporate reward systems also encourage behavior that is short-term, safe and conservative. The 'punishments' for entrepreneurial failures, even when they are beyond a manager's control, are much greater than the rewards for successful risk taking. Additional sources of discouragement for the innovator are the hurdles and delays associated with too many levels of approval needed for developing a new product or implementing a change in manufacturing processes or administrative practices."

Overcoming the failure of success may require companies to painstakingly change their management systems, culture, business models, and reward systems. Not all companies respond favorably to the risks associated with change management initiatives, even if the necessity exists. The greatest risk is when a change in the organization's business model may be needed, and people may be removed from their comfort zones. Some companies then perform project activities in a stealth mode for fear of upsetting the status quo.

14.12 CONCLUSION

If an organization wishes to excel at project management, it must understand the processes that can turn failures into successes. Discovery is essential. In addition, organizations must foster leadership styles that support failure analyses and accompanying change management efforts needed for retrospective and prospective sensemaking practices.

REFERENCES

Cyert, R. M. and March, J. G. (1992). *A Behavioral Theory of the Firm*. Cambridge, MA: Wiley-Blackwell.

Kahneman, D. and Tversky, A. (1979). Prospect Theory: An Analysis of Decision Under Risk. *Econometrica*, 47(2), 263–91.

Morais-Storz, M., Nguyen, N., and Sætre, A. S. (2020). Post-Failure Success: Sensemaking in Problem Representation Reformulation. *Journal of Product Innovation Management*, 37(6), 483–505.

O'Toole, J. (1983). Declining Innovation: The Failure of Success. *Human Resource Planning*, 6(3), 125–141.

Sitkin, S. B. (1992). Learning Through Failure: The Strategy of Small Losses. *Research in Organizational Behavior*, 14, 231–66.

Van de Ven, A. H., Polley, D., Garud, R., and Venkataraman, S. (2008). *The Innovation Journey*. Oxford: Oxford University Press.

15 Global Project Management Excellence

15.0 INTRODUCTION

In the previous chapters, we discussed excellence in project management (PM) the use of PM methodologies and the hexagon of excellence. Many companies previously described in the book have excelled in all of these areas. In this chapter, we focus on four companies, namely IBM, Deloitte, Comau, and Siemens, all of which have achieved specialized practices and characteristics related to in-depth globalized PM:

- They are multinational.
- They sell business solutions to their customers rather than just products or services.
- They recognize that, in order to be a successful solution provider, they must excel in PM rather than just being good at it.
- They recognize that they must excel in all areas of PM rather than just one area.
- They recognize that a global PM approach must focus more on a framework, templates, checklists, forms, and guidelines, rather than rigid policies and procedures, and that the approach can be used equally well in all countries and for all clients.
- They recognize the importance of knowledge management, lessons learned, capturing best practices, and continuous improvement.
- They understand the necessity of having PM tools to support their PM approach.
- They understand that, without continuous improvement in PM, they could lose clients and market share.
- They maintain a PM office or center of excellence (CoE).
- They perform strategic planning for PM.
- They regard PM as a strategic competency.

These characteristics can and do apply to all of the companies discussed previously, but they are of the highest importance to multinational companies.

15.1 IBM[1]

Overview

The mission of IBMs PMCOE is to provide IBM Project Management Professionals with the skills, education, career progression, community, processes, methods, and tools to be successful in the delivery of IBMs programs and projects. With an outstanding track record of delivering many successful programs during our long history, IBM established the Project Management Center of Excellence (PMCOE) in 1997 to recognize the criticality of strong project execution and the role of the Project Manager in a business' success. It is a testament to the importance IBM places on PM, that the organization still exists today and continues to evolve and innovate.

IBM has recognized that key shifts in the market are causing companies and established professions such as PM to evolve and modernize. In the past, stable and predictable market, organization, and project environments provided the foundation for an effective PM model. This included use of dedicated project teams, support for project office, waterfall and sequential PM approaches, long planning, and a strong focus on process, and risk avoidance.

Today, the world of work has changed. There are shorter cycle times of projects and a need for businesses to act quickly in a more competitive environment. Project success has shifted from achieving predefined metric-driven results to a focus on value and benefits realization. With the pace of technology development and innovation also increasing, these changes are impacting IT projects, project environments, and the PM profession. IBM, no different from other organizations, requires contemporary PM capabilities that embrace ongoing market changes and allow our employees to be recognized for the value they create for ourselves, our clients, and our partners.

Our CoE is at the forefront in preparing our project managers to be successful in meeting this challenge. For example, embedding design thinking and agile ways of working in our accreditation requirements. We also put significant focus on developing soft/behavior skills; strong industry skills and in technologies such as cognitive, AI, and cloud computing.

IBM PM capability is recognized by being a member of the Project Management Institute's (PMI) Executive Council, whose members are elite organizations well-positioned to direct the future of the PM profession and ensure its continued growth and success.

---

1. Material in Section 15.1 provided by Jim Boland, IBM Project Management Center of Excellence (PMCOE) Program Director; Orla Stefanazzi , Global eSharenet and Communications Program Manager, IBM PMCOE; Noel Daly – Worldwide Project Management Profession Leader, IBM PMCOE; Una Duggan – IBM Project Management Method Program Manager, IBM PMCOE; and Michael Coleman – Worldwide Program Manager for Project Management Learning, IBM Learning. © 2022 by IBM Corporation. Reproduced by permission. All rights reserved.

Complexity

When you step back and look at the scale and diversity of the tens of thousands of concurrent projects being managed by our community, it can be truly staggering. We are not only asking our project managers to manage across the traditional boundaries of time, budget, and resourcing but, we need to understand and be able to clearly articulate;

- Enabling technologies (e.g., Cloud, AI, Internet of Things, Cognitive, Quantum, etc.)
- Traditional versus hybrid enterprises (on-premise, off-premise, virtual, etc.)
- Industry-specific solutions (e.g., energy, automotive, public, etc.)
- Platform-specific solutions (e.g., as service offerings)
- Client-specific solutions (e.g., customized solutions, integration across multiple diverse legacy environments, etc.)

Our teams, as well as our client's teams, are invariably global in nature, requiring all of us to ensure diversity and inclusion, cultural differences and sustainability concerns are center stage.

In the following sections, we will outline how the PMCOE enables our PM community, and later on, we will discuss how continuous innovation is shaping how we evolve.

IBM's Project Management Method And Tooling

Successful implementation of projects and programs requires a management system that addresses all aspects of planning, delivering value, and integrating with business and technical processes.

To provide its project teams with consistent methods for implementing PM globally, IBM developed the Worldwide Project Management Method (WWPMM), shown in Figure 15–1, which establishes and provides guidance on the best PM practices for defining, planning, and delivering value on a wide variety of projects and programs.

Client Value Drivers

Flexibility to accelerate delivery
Includes IBM Garage, agile, waterfall and hybrid approaches. Assets enable Project Managers to operate at speed with agillity whilst maintaining a focus on value outcomes for the client

Adaptability to suit client needs
Integrates seamlessly with technical methods and Work Practices, both from IBM's Method Catalog as well as Client-defined approaches

Scalability
Works across the complexity spectrum, from limited duration development projects through ongoing complex programs that span multiple business functions and stakeholders

Continuous Improvement to deliver value
Proactively enhanced based on experience, lessons learned and evolving Industry Standards to improved efficiency, productivity, and measurable value

WWPMM

Provides
100+
Templates

Built on
15
Work Practices

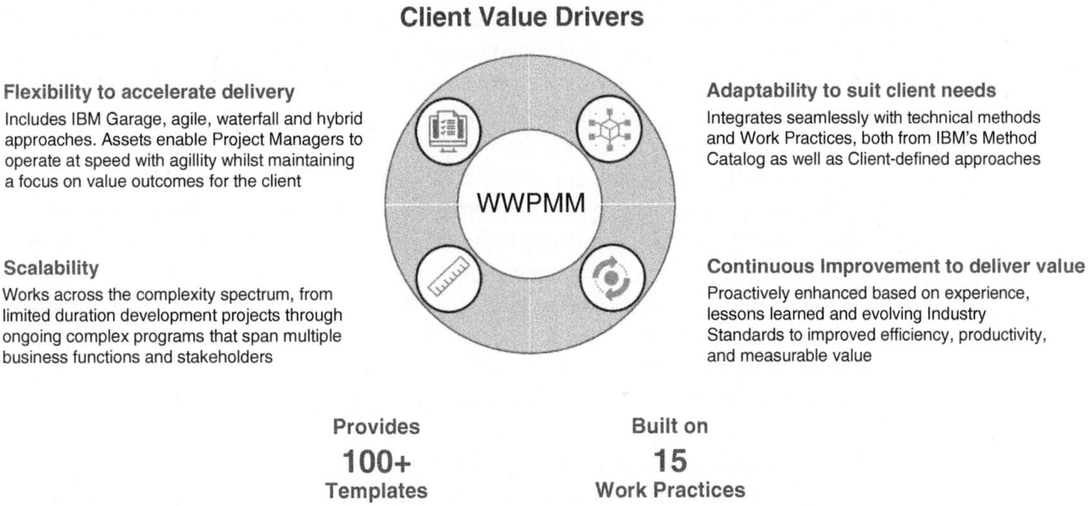

Figure 15–1. Worldwide Project Management Method (WWPMM) client value drivers.

The goal of IBM's PM method is to provide proven, repeatable means of delivering solutions—regardless of geographic location—that ultimately result in successful projects/programs and satisfied clients.

WWPMM expands on the PMI's Project Management Body of Knowledge (PMBOK®) for the IBM environment and specifies PM work product contents.

WWPMM is updated on an ongoing basis, providing support for all types of engagements, ranging from small projects to complex programs using principles and techniques that range from traditional predictive, agile, and hybrid to IBM Garage. The approach taken is to use agile in generic terms and not select a specific agile technique (such as Scrum, Kanban, or XP).

WWPMM describes the way projects and programs are managed in IBM and it includes a number of interrelated components:

- **A library of proven Project and Program Work Practices** that group the tasks, work products, and guidance needed to support a particular area, such as Risk Management.
- **Project and Program Activities** arrange the tasks defined in the PM practices into a series of executable steps designed to meet a particular PM goal or in response to a particular PM situation.
- **Project and Program Work Products** are the verifiable outcomes produced and used to manage a project.

WWPMM includes a set of templates or tool mentors for plans, procedures, and records that may be quickly and easily tailored to meet the needs of each individual project or program.

IBM's structured approach to managing projects and programs includes understanding and adapting to meet our clients' needs and environment. A PM system is the core of this structured approach. The PM system is the way the project will be managed in IBM. It is documented as a collection of plans and procedures that direct all PM activity, and records that provide evidence of their implementation. Project Managers tailor their PM system to adapt to the project context and client needs. Project Managers can choose from agile, traditional, or program work products and activities and can combine approaches to ensure that their project delivers business value.

In order to be generic and applicable across IBM, the PM method does not describe life cycle phases but rather project and program work practices, as shown in Figure 15–2, that can be used repeatedly across any life cycle. This allows the flexibility for the method to be used with any number of technical approaches and life cycles.

Keeping with the need to be flexible, the PM system templates and work products can be tailored to meet geography, business line, or client-specific requirements while still maintaining our commitment for consistent PM.

According to Una Duggan, IBM's WWPMM leader,

The continuous integration of IBM's project management methodology with other IBM initiatives, enhancements from lessons learned and alignment with external standards are necessary to ensure WWPMM will lead to worldwide excellence in the practice of project management.

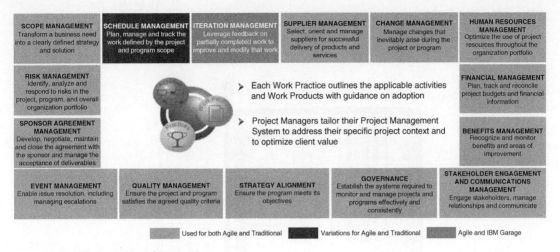

Figure 15–2. Work practices framework.

Agile at IBM is about IBMers thinking and operating differently always focusing on delivering value for our clients. This is where all of IBM, including project managers, can deepen their experience through learning plans, earning badges, and by adopting agile practices to plan, collaborate, and focus on delivering client value.

Since 2006, we have been focusing on agile as a key PM discipline to accelerate our client's transformation outcomes, working with our clients at project, portfolio, and enterprise level.

Examples include:

- For an oil major, we have provided agile coaches (as well as providing training). We also developed a coaching competency framework and helped develop their PM COE
- For a major bank, we helped them kickstart a transformation with a 10-week Agile-DevOps Incubation program, including delivering 7 webinars in 7 months on the topic of agile and growth mindset to over 500+ employees at all levels.
- For a national government agency, we helped by defining and shaping their agile organization, roles and responsibilities, and personnel, before jointly conducting agile stakeholder orientation sessions to build credibility and support for the agile approach

In addition to working with our clients' tools and methods, we have a comprehensive set of internal and external project and program tools (see Figure 15–3). Our Project Managers can choose from a recommended tool set, based on their project profile. Our tools are often shared and aligned to cater to our client's processes and methods. The diagrams below detail just two of our tools that enable our project teams launch and manage their projects seamlessly and efficiently.

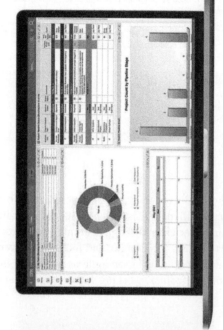

Project Journey Guide

A single reference point to easily navigate through the requirements involved in launching a project. It provides a step-by-step guide on what needs to be done, by when, and how to complete.

Integrated Program Management on IPWC

Manages projects, programs, portfolios and transformations globally. Securely maintains and integrates key program information and automates reporting, improving stakeholder visibility and team member efficiency. Integrate and manage multiple projects within a common workspace

Figure 15–3. Some project management tools.

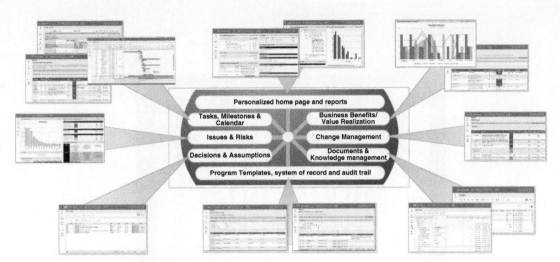

Figure 15–4. ProgramWorkCenter™ functionality.

IBM ProgramWorkCenter™, as shown in Figure 15–4, enables proactive collaboration and secure, integrated management of key program information and documents.

Another example is our Project Health Dashboard shown in Figure 15–5 that enables each project manager to perform a quick and user-friendly HealthCheck on their project.

The project manager is presented with a project health summary view that is easy to understand and take required actions. This view includes:

- Key Project Health Performance metrics in one view (financial, compliance, resource management, etc.).
- Visual and easily consumable data to analyze.
- A predictive Risk Indicator that provides early warning alerts of potential project issues based on the project characteristics and similar project performance.
- Trending views to easily compare progress.

IBMs PMCOE supports project teams in overcoming the numerous challenges they face by focusing on key success factors such as:

1. Ensuring all key stakeholders understand the **value of PM**
2. **Actively engaging sponsors/executives**, addressing their key issues, and generating the support required for the project/program
3. Ensures **strategic alignment between business goals and projects executed** by enabling executive teams to make informed decisions and chose the right projects to achieve business value.
4. **Standardized PM practices** support the enterprise strategy and provide the right level of control to reduce risk and ensure successful delivery.

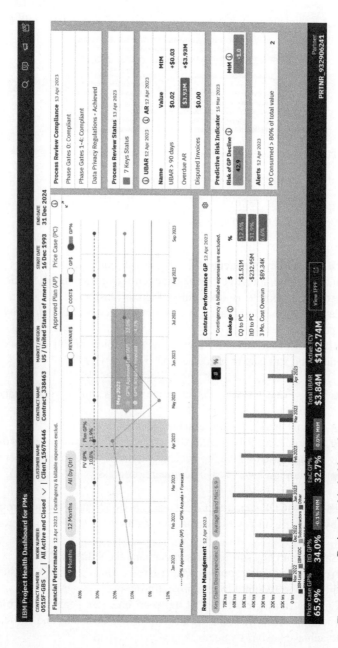

Figure 15-5. Project health dashboard.

5. **PM Best Practices, assets, and intellectual capital** enable organizations to reduce risk and deliver repeatable, high value, and high-quality solutions.
6. Invests in **developing its project manager's talent** to achieve superior project performance and execution of strategic initiatives.

IBM's Business Processes   IBM's basic business processes are seamlessly integrated with our methods. Three of these processes directly benefit many of the projects we manage: Quality, Knowledge Management, and Maturity Assessment.

Quality

IBM's PM method readily conforms to ISO Quality standards. This means that project managers using WWPMM do not have to spend extra time trying to establish a quality standard for their project as the quality standard is already built into a project's management system.

Within IBM Consulting, IBM's business practices require an independent quality assurance review of most projects performed by our worldwide Quality organization. Project reviews play an important role by identifying potential project issues before they cause problems thereby helping to keep projects on time and on budget. The IBM internal reviews and assessments are performed at various designated checkpoints throughout the project life cycle.

The IBM Consulting Quality Delivery Reviews support this holistic quality approach as follows:

- **Project Launch** establishes the necessary support, relationships, agreements, and procedures so projects can be successfully delivered. The output is a collection of plans, processes, procedures, and records, which directs all project team activities and provides the current state and history of the project.
- **Seven Keys to Success**® analyses projects' health in a standard format thereby enabling early warning and course correction. See Figure 15–6.
- The **Phase Gates Review** (PGR) process implements a series of phased reviews at critical milestones during program delivery life cycle. The goal of the PGR process is to ensure that the work products exist, meet the necessary quality standards, and are consistent with the project plan and the service line delivery method.
- **Conditions of Satisfaction** (CoS) are an output from a structured discussion with the client to set and manage their expectations. These conditions describe the client priorities during the sales and delivery phases of the engagement and provide the client with objective criteria to use when assessing their satisfaction with IBM.

A good example of IBMs proven set of methods and work practices is our Seven Keys to Success™ quality process. This simple framework ensures our project teams

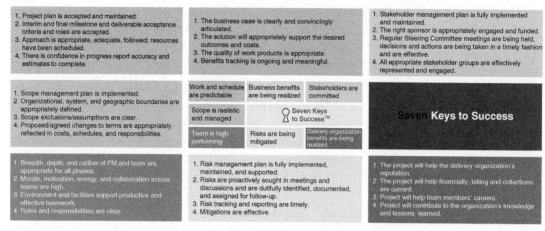

Figure 15–6. Seven keys to Success® quality process.

continue to focus on delivering value to our clients, that the teams continue to be motivated and high performing, stakeholders remain committed and if necessary, recalibrate.

Knowledge Management

IBM PM best practices, assets, and intellectual capital represent the combined expertise of tens of thousands of IBM project managers over decades of work and experience delivering projects and programs.

IBM project managers have access to the best project intellectual capital available including reusable work products such as architectures, designs, and plans. Project managers are encouraged to share their own knowledge and expertise by publishing project work products and experiences. Capturing the best practices and lessons learned on completed projects is fundamental to ensuring future project success. IBM encourages use of IBM Lighthouse—an official Enterprise System of Engagement for crowdsourced delivery excellence documents, coded assets, and thought leadership content. IBM Lighthouse is for finding and sharing knowledge to help IBM employees on their projects and engagements. The tool is built on top of IBM Cloud and Watson AI to fully support IBM Consulting's operating model. IBM Lighthouse is supported by a team of dedicated resources with 20 years of industry-leading best practices, methodologies, and lessons learned. A recent offering from IBM Lighthouse—IBM Account Spaces—allows teams to collaborate throughout the project life cycle, ensuring that team members have the information they need to be structured consistently for improved findability. It anticipates needs and provides personalized recommendations through cognitive tools.

Maturity Assessment

IBM has developed a comprehensive tool and best practice, the Project Management Progress Maturity Guide (PMPMG), to assess its current PM capabilities and the PM services it provides its clients and improve them over time. Studies show that as an organization's PM maturity improves, projects are delivered more effectively and efficiently, customer satisfaction improves, and better business results occur.

PM maturity measures the degree to which elements of a PM process or system are present, are integrated into the organization, and ultimately affect the organization's performance.

Assessments can be undertaken at various levels:

1. Enterprise – across an entire organization
2. Program – across a set of projects
3. Project – individual project
4. Agile – maturity level of using an agile approach

The assessment is performed against a set of best practices through documentation and interviews to look for evidence of deployment, usage, coverage, and compliance. It provides:

- Current capability strengths, weaknesses, and a prioritized list of gaps.
- Improvement action recommendations for high and medium priority gaps.
- An overall maturity level rating for each best practice.

For maximum value, an organization should determine a PM maturity baseline; effectively prioritize, plan, and implement improvement opportunities; and then measure across time to verify consistent improvement in the organization's PM capabilities. By understanding the organization's strengths and weaknesses, actions can be identified for continuously improving PM and achieving business objectives.

As an organization's PM maturity improves, projects are delivered more effectively and efficiently, customer satisfaction improves, and stronger business results are achieved. See Figure 15–7.

IBM's Project Management Skills Development Programs

Enhancing the integration of the methods, business processes, and policies is the ongoing development of IBM's PM professionals through education and certification.

Education

IBM's PM Curriculum is delivered globally and across all lines of business, helping to drive a consistent base of terminology and understanding across the company. Though they are clearly an important audience for the training, attendance is not limited just to Project Managers. Rather, the curriculum is there to meet the PM training needs of all IBMers irrespective of what job role they perform. The curriculum covers traditional and plan-driven approaches as well as agile ones. There is an increasing focus on developing project managers who can take a hybrid approach and draw from the full breadth of PM techniques depending on the specific circumstances they face. A range of delivery modes is utilized depending on the course content and intended audience.

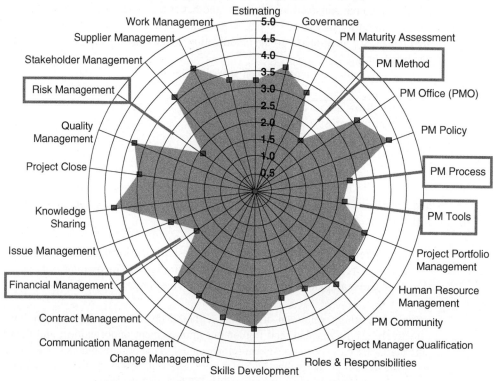

Figure 15–7. Maturity attainment by best practice.

As well as the traditional classroom format, an increasing amount of instructor-led learning is delivered through online virtual classrooms. This was a trend before 2020 but has been accelerated by the global pandemic. Extensive use of self-paced online learning provides easy access to curriculum content at a time and place of the learner's choosing. A Curriculum Steering Committee, composed of representatives from across IBM's lines of business, provides governance of the development of curriculum content. This ensures that the curriculum continues to meet the evolving needs of all parts of the business.

The PM Curriculum is arranged into 4 distinct sections:

(1) The **Core Curriculum** addresses the fundamentals of PM. Employees with limited, or no, prior knowledge can use this section of the curriculum to gain a solid grounding in the disciplines of PM. Introductory courses lay the foundations and more specific courses build on these to develop capabilities in PM Systems, Contracting, Finance, Project Leadership, and IBM's World Wide PM Method. A separate integrative course completes this section of the curriculum by drawing

together the theoretical learning from the earlier courses and blending that with a focus on the practical application of that knowledge.

(2) The **Enabling Education** section provides the opportunity to build on learning from the core curriculum and deepen PM skills in specific areas. This would include more in-depth training on topics such as leadership, training in the use of specific PM tools, and more situational topics such as working across cultural boundaries.

(3) The **PM** section is focused on enhancing general business skills expected of more senior roles and on providing project-based tools and techniques needed to manage large programs with multiple projects and business objectives.

(4) The **Understanding the Basics** section contains courses aimed at employees who support or work on project teams. Basic introductory courses on PM provides them with an understanding of how projects are run and key terms but does not seek to develop them into Project managers.

As we have already noted, the PM Curriculum provides training to people in a wide range of job roles, not just project managers. Conversely it is also the case that the PM Curriculum does not set out to meet all the learning needs of IBM's Project Managers. For example, Project Managers will also require a range of skills specific to their operational context and this will be drawn from IBM's broader learning provision.

Certification The PM profession is one of several IBM global professions established to ensure availability and quality of professional and technical skills within IBM. The PM Professional Development initiative includes worldwide leadership of IBM's PM profession, its qualification processes, IBM's relationship with the PMI, and PM skills development through education and mentoring. These programs are targeted to cultivate project and program management expertise and to maintain standards of excellence within the profession. The bottom line is to develop practitioner competency.

What is the context of a profession within IBM? IBM professions are self-regulating communities of like-minded and skilled IBM professionals and managers who do similar work. Their members perform similar roles wherever they are in the organizations of IBM and irrespective of their current job title. Each profession develops and supports its own community including providing assistance with professional development, career development, and skills development. The IBM professions:

- Help IBM develop and maintain the critical skills needed for its business;
- Ensure IBM clients were receiving consistent best practices and skills in the area of PM; and
- Assist employees in taking control of their career and professional development.

All IBM jobs have been grouped into one of several different functional areas, called job families. A job family is a collection of jobs that share similar functions or skills. If data are not available for a specific job, the responsibilities of the position are compared to the definition of the job family to determine the appropriate job family assignment.

Project managers and, for the most part, Program and Portfolio Managers fall into the PM Job Family. PM positions ensure customer requirements are satisfied through the formulation, development, implementation, and delivery of solutions. PM professionals are responsible for the overall project plan, budget, work breakdown structure, schedule, deliverables, staffing requirements, managing project execution and risk, and applying PM processes and tools. Individuals are required to manage the efforts of IBM and customer employees as well as third-party vendors to ensure that an integrated solution is provided to meet the customer's needs. The job role demands significant knowledge and skills in communication, negotiation, problem-solving, and leadership. Specifically, PM professionals need to demonstrate skill in:

- Relationship management skills with their teams, customers, and suppliers.
- Technology, industry, or business expertise.
- Expertise in methodologies.
- Sound business judgment.

Guidance is provided to management on classifying, developing, and maintaining the vitality of IBM employees. In the context of the PM profession, vitality is defined as professionals meeting PM skill, knowledge, education, and experience requirements (qualification criteria) as defined by the profession, at or above their current level. Minimum qualification criteria are defined for each career milestone and used as individual's business commitments or development objectives, in addition to business unit and individual performance targets.

Skilled project and program management professionals are able to progress along their career paths to positions with more and more responsibility. For those with the right blend of skills and expertise, it is possible to move into program management, project executive, and executive management positions. Growth and progression in the profession are measured by several factors:

- General business and technical knowledge required to be effective in the job role;
- PM education and skills to effectively apply this knowledge;
- Experience that leverages professional and business-related knowledge and skills "on the job;" and
- Contributions to the profession, known as "giveback," through activities that enhance the quality and value of the profession to its stakeholders.

IBM's project and program management profession has established an end-to-end process to "quality assure" progress through the PM career path. This process is called "qualification" and it achieves four goals:

- Provides a worldwide mechanism that establishes a standard for maintaining and enhancing IBM's excellence in project and program management. This standard is based on demonstrated skills, expertise, and success relative to criteria that are unique to the profession.

- Ensures that consistent criteria are applied worldwide when evaluating candidates for each profession milestone.
- Maximizes customer and marketplace confidence in the consistent quality of IBM PM professionals through the use of sound PM disciplines (i.e., a broad range of project and program management processes, methodologies, tools, and techniques applied by PM professionals in IBM).
- Recognizes IBM professionals for their skills and experience.

The IBM project and program management profession career path allows employees to grow from an entry-level to an executive management position. Professionals enter the profession at different levels depending upon their level of maturity in PM. Validation of a professional's skills and expertise is accomplished through the qualification process. The qualification process is composed of accreditation (at the lower, entry levels), certification (at the higher, experienced levels), recertification (to ensure professional currency), and/or level moves (moving to a higher certification milestone). See Figure 15–8.

Accreditation is the entry-level into the qualification process. It occurs when the profession's qualification process evaluates a PM professional for Explorer and Associate levels.

Certification is the top tier of the qualification process and is intended for the more experienced project or program manager. It occurs when the profession's qualification process evaluates a PM professional for Experienced, Expert, and Thought Leader PM levels. These levels require a more formal certification package to be completed by the project manager. The manager authorizes submission of the candidate's package to the Project Management Certification Board. The IBM Project Management Certification Board, comprises professional experts, administers the authentication step in the certification process. The Board verifies that the achievements documented and approved in the candidate's certification package are valid and authentic. Once the Board validates that the milestones were achieved, the candidate becomes certified at the level applied for.

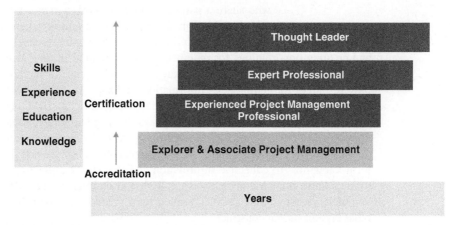

Figure 15–8. IBM's project and program management career growth path.

Recertification evaluates IBM-certified PM professionals for currency at Expert and Thought Leader Project Management levels. Recertification occurs on a three-year cycle and requires preparing a milestone package in which a project manager documents what he/she has done in PM, continuing education, and giveback since the previous validation cycle.

IBM continues to be committed to improving its PM capabilities by growing and supporting a robust, qualified PM profession and by providing quality PM education and training to its practitioners.

Equally important to project manager development and certification is a refinement of the process by which project managers are assigned. Projects are assessed based on size, revenue implications, risk, customer significance, time urgency, market necessity, and other characteristics; certified project managers are assigned to them based on required education and experience factors.

Guidance is provided to management on classifying, developing, and maintaining the vitality of IBM employees. In the context of the PM profession, vitality is defined as professionals meeting PM skill, knowledge, education, and experience requirements (qualification criteria) as defined by the profession, at or above their current level. Minimum qualification criteria are defined for each career milestone and used as individual's business commitments or development objectives, in addition to business unit and individual performance targets.

The PM CoE is chartered to increase Practitioner Competency in project and program management across IBM. This includes worldwide leadership of IBM's PM profession, as shown in Figure 15–9, its Managing Projects and Programs validation processes, and IBM's relationship with PMI, as well as project and program management skills development through education and mentoring. A global team works to cultivate this project and program management expertise and to maintain standards of excellence within the PM profession.

Community

Within IBM, a community is defined as a collection of professionals who share a particular interest, work within a knowledge domain, and participate in activities that are mutually beneficial to building and sustaining performance capabilities. Our community focuses on its members and creating opportunities for members to find meaning in their work, increase their knowledge and mastery of a subject area, and feel a sense of membership—that they have resources for getting help, information, and support in their work lives.

Communities are part of the organizational fabric at IBM but not defined or constrained by organizational boundaries. In fact, communities create a channel for knowledge to cross boundaries created by workflow, geographies, and time and in so doing strengthen the social fabric of the organization. They provide the means to move local know-how to collective information and to disperse collective information back to local know-how. Membership is totally based on interest in a subject matter and is voluntary. A community is NOT limited by a practice, a knowledge network, or any other organizational construct.

The IBM Project Management Community is facilitated by the IBM PMCOE and is one of the largest and active communities across IBM. Membership is open

Badge Earner	IBM	IBM Clients
Broadcast Achievements Signals skills and achievements to peers, potential employers and others	**Generate Qualified Leads** Attracts new candidates to IBM who seek recognition and opportunity	**Verified Skills** Provides a trusted "seal of approval" for employers validating existing talent or potential hires
Motivate Participation Provides recognition for achievement; encourages engagement and retention	**Enhance Our Brand** Significant social media benefit: thousands of brand marks flooding LinkedIn, Twitter, Facebook, blogs and company websites	**Candidate Selection** Provides an easy way to identify candidates to hire or promote
Develop Personal Brand Displays verified achievements across the web. Improves social connections with peers, employers and clients	**Differentiate IBM** IBM leads the industry with a cutting-edge digital credential program	**Improved Company Performance** Motivates employees to drive their own development and improve the performance of the organization
	Track and Nurture Talent Quickly identifies skills gaps and opportunities -- candidate or geo level	

Figure 15–9. IBM project management profession.

to all IBM employees with a professional career path or an interest in PM. The PM community is a self-sustaining community of practice with over seventeen thousand members who come together for the overall enhancement of the profession. Members share experiences, knowledge, and network with fellow project managers. The IBM PMCOE provides various communication and online collaboration platforms to ensure our PM professionals remain connected and engaged. Examples include:

- *Weekly Webinars.* The IBM PMCOE delivers weekly webinars known as "eSharenets." The eSharenet Program provides a wide variety of PM-related topics each year addressing PM best practices and disciplines, PM techniques, methods, tools, and power skill topics such as leadership, team building, and team collaboration. The sessions are delivered virtually to suit the global nature of the audience. The speakers on the program are subject matter experts in their field and there is a mix of internal IBM and external speakers. The majority of these sessions enable IBM's global project managers to claim 1 PDU as part of the PMI recertification requirements.
- Focused PM communications to assist the global PM community in developing their skills and PM career.
- Local community PM events for example locally arranged PMI chapter conferences or PMI award wins.

All IBM'ers are encouraged to be "socially eminent," both inside of IBM as well as externally. According to Orla Stefanazzi, PMP, Communications, and eSharenet Program Manager, *"the IBM Project Management Center of Excellence has driven a strong sense of community for its global project management professionals; this is a best practice among IBM's professions. The IBM PMCOE is ranked one of the top communities across all of IBM, with Slack membership in the top 20."*

Upon entering the PM community, professional hires into IBM are often asked the question: "What is the biggest cultural difference you have found in IBM compared to the other companies in which you have worked?" The most common answer is that their peers are extremely helpful and are willing to share information, resources, and help with job assignments. The culture of IBM lends itself graciously to mentoring. As giveback is a requirement for certification, acting as a mentor to candidates pursuing certification not only meets a professional requirement but also contributes to the community.

IBM's best practices are not just recognized within the company. IBM has a strong track record of being formally recognized for our excellence in PM. Recent awards include:

- *Project Managing Innovation*. In 2021 we won the global PMI Project Excellence Award for a first-of-a-kind project for a commercial bank. In 2019, the government of Taiwan issued its first internet-only banking permits, aiming to bring innovation to the financial industry and create synergistic benefits with tech-empowered services. One of Japan's leading e-commerce giants partnered with IBM on this banking system implementation, which became Taiwan's first internet-only bank. To tackle this first-of-its-kind project and meet our client's goal, IBM leveraged an MVP approach and PMO mechanism. With 75% of the project's timeline falling during COVID, the PMO team responded with agility and took control of the timeline. The project was launched in January 2021 and leads the financial innovation trend in Taiwan.
- *MCA Awards*. In 2021, we won the UK MCA (Management Consultancy Association) award for exceptional support and delivery to a client during COVID-19.

The graphic in Figure 15–10 provides a snapshot of the industry-wide recognition of IBM's PM curriculum, projects, and professionals over the past 25 years.

Innovation   Innovation is essential in the marketplace today. IBM delivers innovation to accelerate industry business transformation with our strategic partners and the broader ecosystem. These projects are often created and scaled by our clients. Some examples are provided below that showcase how IBM innovated to deliver PM solutions that address some of the key challenges faced by project leaders today. These include early project health indication, contract execution, and collaboration between business, IT, and delivery organizations.

- *EPC project health diagnostics*. IBM built an AI solution for a major multinational engineering and construction firm, to predict project health status from structured and unstructured data. The firm uses this to measure the status of megaprojects from inception to completion—predicting (and improving) project outcomes and reducing risk and cost/time overruns.

CONTINUED INDUSTRY RECOGNITION

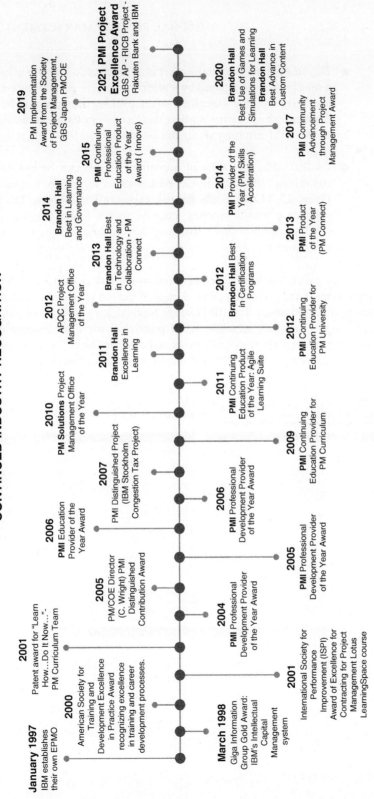

Figure 15–10. Industry recognition for **IBM's** project management curriculum, projects, and professionals.

- *Digital workers.* A digital worker is an executable software solution that applies a set of skills, as required by a given job role, to perform repeatable workflows and drive tangible outcomes, often working alongside its human colleague. There are many examples where this technology has been successfully deployed across IBM resulting in significant savings in PMO effort whilst increasing quality.
- *Contract analyzer.* 40% of a contract's value can be lost due to inefficient contracting. IBM has developed a solution to analyze contracts, as shown in Figure 15–11 and assess the risk exposure within them. The nature of many complex contracts requires intensive review and management—this solution quickly identifies areas that may lead to troubled projects and downstream consequences.

IBM garage. A key aspect of modern-day PM is to handle continuous service improvement and innovation. Our Garage™ method is an end-to-end model for accelerating transformation. The IBM Garage Methodology guides how you work by bringing together an open, seamless set of practices with a human-centric, outcome-first approach. By applying design thinking, agile development, and DevOps tools and techniques, teams learn new skills and master new ways of working. IBM Garage clients

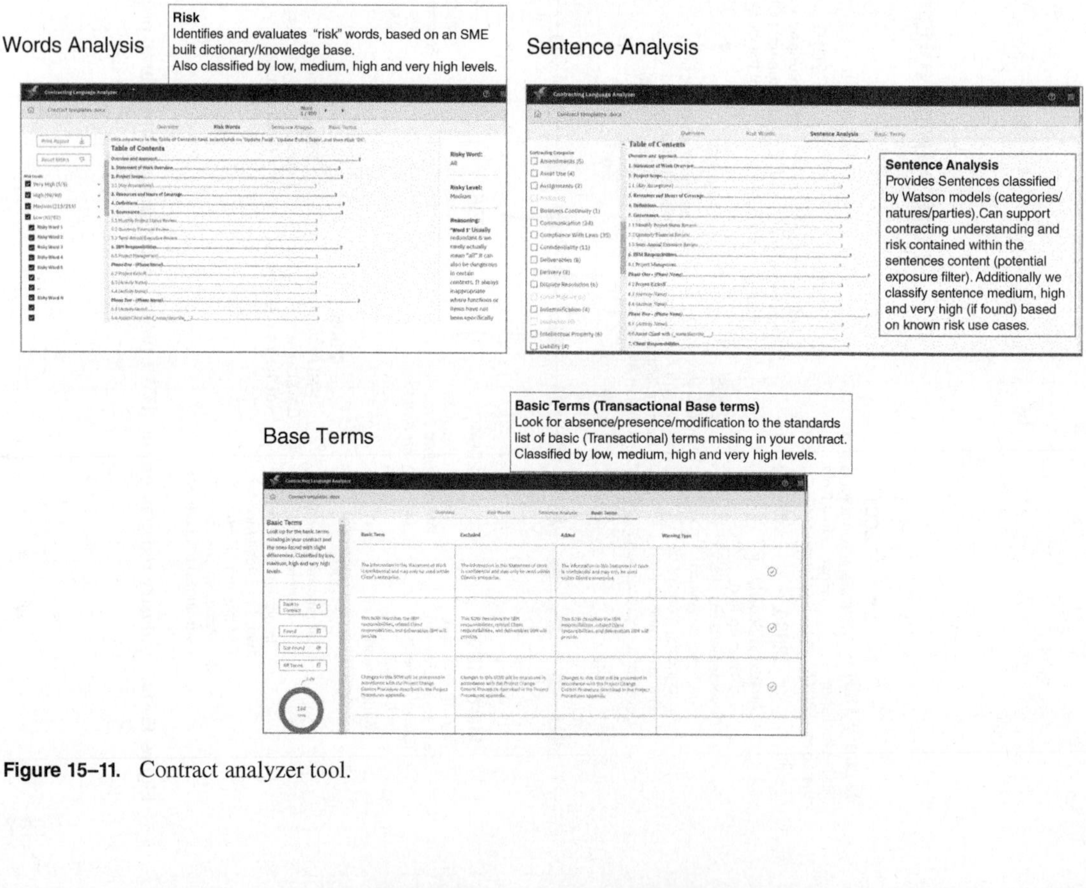

Figure 15–11. Contract analyzer tool.

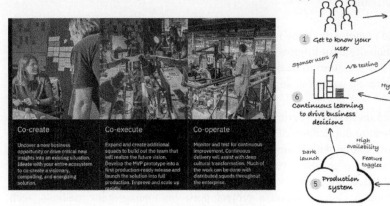

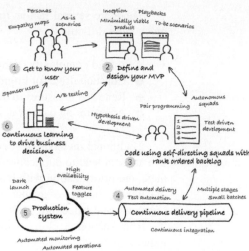

3 phases of IBM Garage Methodology   Example of how it works

Figure 15–12. IBM garage methodology.

experience 10 times more innovative ideas, 67% faster speed to outcomes, 6 times as many projects into production, and 102% ROI. Figure 15–12 demonstrates the three phases of Garage (cocreate, coexecute, and cooperate) and an example of how it really works during an engagement.

15.2 DELOITTE: ENTERPRISE PROGRAM MANAGEMENT[2]

Introduction

Many organizations initiate more projects than they have the capacity to deliver. As a result, they typically have too much to do and not enough time or resources to do it. The intended benefits of many projects are frequently not realized, and the desired results are seldom fully achieved.

---

2. Section 15.2 on Deloitte was provided by Daniel Martyniuk, Christine Lyman, PMP, and Rusty Greer Copyright © 2022 Deloitte Development LLC. All rights reserved. As used in this document, "Deloitte" means Deloitte Consulting LLP, a subsidiary of Deloitte LLP. Please see www.deloitte.com/us/about for a detailed description of the legal structure of Deloitte LLP and its subsidiaries. Certain services may not be available to attest clients under the rules and regulations of public accounting. This publication contains general information only and Deloitte is not, by means of this publication, rendering accounting, business, financial, investment, legal, tax, or other professional advice or services. This publication is not a substitute for such professional advice or services, nor should it be used as a basis for any decision or action that may affect your business. Before making any decision or taking any action that may affect your business, you should consult a qualified professional advisor. Deloitte shall not be responsible for any loss sustained by any person who relies on this publication.

There are several factors that can make delivering predictable project results that much more difficult:

- Added complexity of the transformational nature of many projects.
- Constant drive for improved efficiency and effectiveness.
- Renewed pressures to demonstrate accountability and transparency.
- Accelerating pace of change and constantly shifting priorities.

Traditional methods of coordinating and managing projects are becoming ineffective and can lead to duplication of effort, omission of specific activities, or poor alignment and prioritization with business strategy. Making the right investment decisions, maximizing the use of available resources, and realizing the expected benefits to drive organizational value have never been more important.

This section explores Deloitte's project portfolio management methods, techniques, approaches, and tools ranging from translating organizational strategy into an aligned set of programs and projects, to tracking the attainment of the expected value and results of undertaken transformational initiatives.

Enterprise Program Management

Organizations are facing increased pressures to "do more with less." They need to balance rising expectations for improved quality, ease of access, and speed of delivery with renewed pressures to demonstrate effectiveness and cost efficiency. The traditional balance between managing the business (i.e., day-to-day operations) and transforming the business (i.e., projects and change initiatives) is shifting. For many organizations, the proportion of resources deployed on projects and programs has increased enormously in recent years. However, the development of organizational capabilities, structures, and processes to manage and control these investments continues to be a struggle.

Furthermore, there has been a significant increase in project interdependency and complexity. While many projects and programs likely were confined to a specific function or business area in the past, increasingly we see that there are strong systemic relationships between specific initiatives. Most issues do not exist in isolation and resolutions have links and knock-on impacts beyond the scope of one problem. Not only do projects increasingly span people, process, and technology, but they also have cross-functional, geographical, and often organizational boundaries. Without a structured approach to their deployment, projects, and programs can fail to deliver the expected value. The need for strategic approach to project, program, and portfolio management is great.

Deloitte's approach to project portfolio management is represented by the guiding Enterprise Program Management (EPM) framework that provides a model within which projects, programs, and portfolios fit into a hierarchy where project execution and program delivery are aligned with enterprise strategy and can lead to improved realization of desired benefits. This approach aims to strike a balance between management of results (effectiveness) and management of resources (efficiency) to deliver enterprise value.

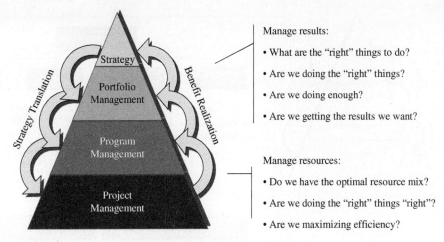

Figure 15–13. Deloitte enterprise program management framework.
Source: © Deloitte & Touche LLP and affiliated entities.

In Figure 15–13, *Strategy* includes the definition of the organization's vision and mission, as well as the development of strategic goals, objectives, and performance measures. The *Portfolio Management* capability translates the organization's enterprise strategy into reality and manages the portfolio to determine effective program alignment, resource allocation, and benefits realization. *Program Management* focuses on structuring and coordinating individual projects into related sets to determine realization of value that may likely not have been attained by delivering each project independently in isolation. Disciplined *Project Management* helps enable that defined scope of work packages are delivered to the desired quality standards.

Strategy and Enterprise Value   Today's business leaders live in a world of perpetual motion, running and improving their enterprises at the same time. Tough decisions need to be made every day—setting directions, allocating budgets, and launching new initiatives—all to improve organizational performance and, ultimately, create and provide value for stakeholders. It is easy to say stakeholder value is important, though it is much more difficult to make it influence the decisions that are made every day: where to spend time and resources, how best to get things done, and, ultimately, how to win in the competitive marketplace or in the public sector and effectively deliver a given mandate.

Supporting the *Strategy* component of the EPM framework, Deloitte's Enterprise Value Map™ (EVM) is designed to accelerate the connection between taking action and generating enterprise value. It facilitates the process of focusing on important areas, identifying practical ways to get things done, and determining if chosen initiatives provide their intended business value. The EVM can make this process easier by accelerating the identification of potential improvement initiatives and depicting how each can contribute to greater stakeholder value.

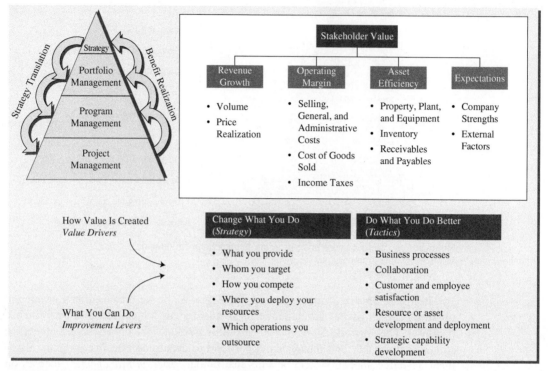

Figure 15–14. Deloitte enterprise value map™ (EVM).
Source: © Deloitte & Touche LLP and affiliated entities.

The EVM, as illustrated at a summary level in Figure 15–14 is powerful and appealing because it strikes a very useful and practical balance between:

● Strategy and tactics.
● What can be done and how it can be done.
● The income statement and the balance sheet.
● Organizational capability and operational execution.
● Current performance and future performance.

Overall, the EVM helps organizations focus on the applicable things and serves as a graphic reminder of what they are doing and why. From an executive perspective, the EVM is a framework depicting the relationship between the metrics by which companies are evaluated and the means by which companies can improve those metrics. From a functional perspective, the EVM is a one-page summary of what companies do, why they do it, and how they can do it better. It serves as a powerful discussion framework because it can help companies focus on the issues that matter most to them.

The EVM is leveraged by Deloitte to help clients:

● Identify things that can be done to improve stakeholder value.
● Add structure to the prioritization of potential improvement initiatives.

- Evaluate and communicate the context and value of specific initiatives.
- Provide insights regarding the organization's current business performance.
- Depict how portfolio of projects and programs aligns with the drivers of value.
- Identify pain points and potential improvement areas.
- Depict past, current, and future initiatives.

Stakeholder value is driven by four basic "value drivers": revenue growth, operating margin, asset efficiency, and expectations:

1. *Revenue growth.* Growth in the company's "top line," or payments received from customers in exchange for the company's products and services.
2. *Operating margin.* The portion of revenues that is left over after the costs of providing goods and services are subtracted. An important measure of operational efficiency.
3. *Asset efficiency.* The value of assets used in running the business relative to its current level of revenues. An important measure of investment efficiency.
4. *Expectations.* The confidence stakeholders and analysts have in the company's ability to perform well in the future. An important measure of investor confidence.

There are literally thousands of actions companies can take to improve their stakeholder value performance, and the EVM, in its full version, depicts several hundred of them. While the actions are quite diverse, the vast majority of them revolve around one of three objectives:

1. Improve the effectiveness or efficiency of a business process.
2. Increase the productivity of a capital asset.
3. Develop or strengthen a company capability.

The individual actions in the Value Map start to identify how a company can make those improvements. Broadly speaking, there are two basic approaches to improvement:

1. Change what you do (*change your strategy*). These actions address strategic changes—altering competitive strategies, changing the products and services you provide and to whom, and changing the assignment of operational processes to internal and external teams.
2. Do the things you do better (*improve your tactics*). These actions address tactical changes—assigning processes to different internal or external groups (or channels), redesigning core business processes, and improving the efficiency and effectiveness of the resources executing those processes.

Portfolio Management

Portfolio Management is a structured and disciplined approach to achieving strategic goals and objectives by choosing the most effective investments for the organization, and determining the realization of their combined benefits and value while requiring the use of available resources.

The Portfolio Management function provides the centralized oversight of one or more portfolios and involves identifying, selecting, prioritizing, assessing, authorizing, managing, and controlling projects, programs, and other related work to achieve specific strategic goals and objectives. Adoption of a strategic approach to Portfolio Management enables organizations to improve the link between strategy and execution. It helps them to set priorities, gauge their capacity to provide, and monitor achievement of project outcomes to drive the creation and delivery of enterprise value.

Deloitte's approach to Portfolio Management can enable an organization to link its strategic vision with its portfolio of initiatives and manage initiatives as they progress. It provides the critical link that translates strategy into operational achievements. As illustrated in Figure 15–15, the Portfolio Management Framework helps to answer these questions: "Are we doing the 'right' things?," "Are we doing enough of the 'right' things?," and "How well are we doing these things?"

Once implemented, the framework helps to transform the business strategy into a coordinated portfolio of initiatives that work together to increase stakeholder value. Additionally, it can provide the tools and techniques to keep projects on track, greatly improving an organization's chances of achieving the desired results. It focuses an

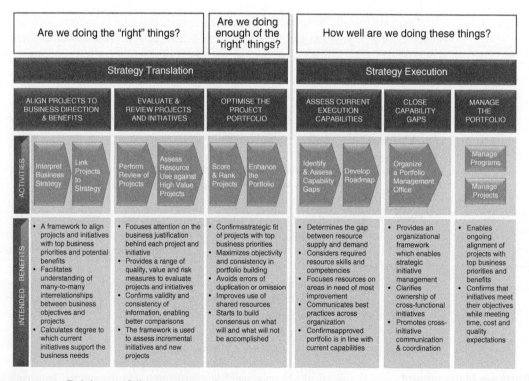

Figure 15–15. Deloitte portfolio management framework.
Source: © Deloitte & Touche LLP and affiliated entities.

organization on initiatives that offer high value-creation opportunities and can also provide a structure and discipline to drive performance improvement initiatives and aid in the identification of continuous improvement opportunities. Last, it confirms that the appropriate resources and budget are made available for critical assignments and provides the tools and techniques to effectively manage an organization's portfolio of initiatives.

The first crucial step in the process of developing an effective project portfolio is the establishment of a method for determining which projects will be within and which will fall outside the scope of the portfolio. A clear definition of what constitutes a "project" in need, as well as identification of the criteria that will be applied to place a particular initiative inside or outside the boundaries of the portfolio. Daniel Martyniuk, a manager in Deloitte Consulting LLP's Strategy & Operations practice that specializes in project portfolio management, highlights:

> While this first step may seem basic insofar as its aim is to provide a basic framework in which to define, sort, and categorize projects, the critical component is carefully capturing all projects that are currently undertaken or proposed for approval. Many hard-to-define projects are often missed, as they may take the form of day-to-day activities or may take place "out of sight." As such, it is essential to define clear boundaries between day-to-day operations and project work; failure to do so may lead to ambiguity and inaccurate representation of the true count of projects in the organization.

A consistent categorization method, such as the Deloitte Investment Framework illustrated in Figure 15–16, helps to answer the question "Why are we allocating resources to this project?" It aims to define the differences between initiatives allowing for immediate recognition and categorization of projects, and it provides the context for comparing projects that are different in nature or scope. It also facilitates allocation of resources by type first, followed by prioritization of projects within a type. Most important, it provides common ground to facilitate dialog and prioritization discussions.

Once the scope of the portfolio has been set, the organization may require a disciplined process to enable continuous alignment of projects to strategic objectives, evaluation, prioritization, and authorization as well as the ongoing management of progress, changes, and realization of benefits.

The Deloitte Portfolio Management Process, as illustrated in Figure 15–17, can serve as a basis for the definition of common portfolio management sequence. It allows for coordination across projects to capitalize on synergies and reduce redundancies. It also helps to outline and identify projects in a comparable format when there are multiple project opportunities and/or organizational pain points to increase the value created by the organization's initiatives while balancing risk and reward.

When the approved list of projects making up the project portfolio is established, project registration and sequencing become the next critical step. Just because a project is now part of the "approved" project registry, it does not mean that it should or will be started right away.

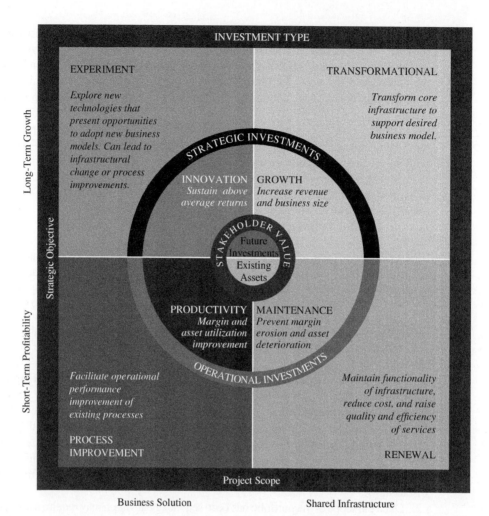

Figure 15–16. Deloitte investment framework.
Source: © Deloitte & Touche LLP and affiliated entities.

A number of factors should be considered when determining the appropriate sequence for project execution. Some of the important decision criteria for project sequencing include:

- *Strategic priority.* The level of importance placed on this project by stakeholders or organizational leadership; fast track the start of those initiatives that directly contribute to the realization of the stated business objectives.
- *Window of opportunity.* Some initiatives need to be completed within a certain period of time in order to gain the desired benefits; give those initiatives the required consideration to help ensure that the opportunity to provide value is not missed.

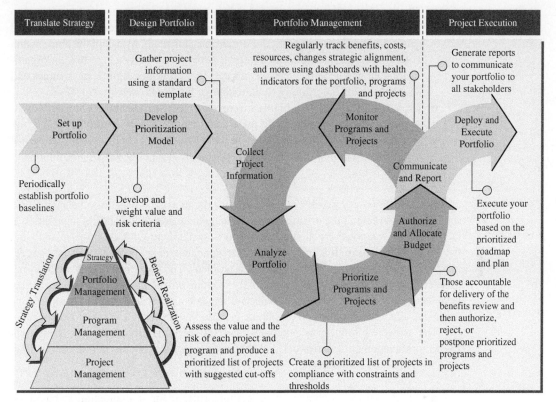

Figure 15–17. Deloitte portfolio management process.
Source: © Deloitte & Touche LLP and affiliated entities.

- *Project interdependencies.* Confirm that all dependencies between related projects have been identified and considered when making project sequencing and initiation decisions; also, consider other dependencies that could impact the start or the completion of projects, such as timing of important decisions, budget cycle, and others.
- *Resource availability.* A project cannot be started until the applicable resources become available to begin working on that particular project; however, remember that "availability" is not a skill, and in addition to getting resources assigned to your projects, make sure that they are the "right" resources in terms of their knowledge, ability, and experience.
- *Risk.* Consider the level of risk being taken on as a result of undertaking a given set of projects; it is a good idea not to initiate high-risk projects all at once, all within the same delivery period; high-risk projects should be closely monitored, and you should strive to find applicable mix of high-risk and low-risk projects; whenever possible, you should consider staggering the execution of high-risk projects and conduct a full risk analysis to determine and agree on appropriate risk mitigation strategies.

- *Change.* Consider the novelty of the undertaking and the amount of change to be introduced into your organization as a result of implementing the proposed set of projects; confirm that your organization is ready to accept the amount and level of changes being created—there is only so much change that an organization can handle; stagger those projects that introduce significant changes and sequence them accordingly to limit change fatigue in your organization.
- *Cost/Benefit.* Different initiatives have varying costs/benefits associated with it; as with risk, it is imperative to understand which projects will provide the biggest benefits for the lowest cost; you do not want to start all of your higher-cost projects at the same time, especially if you are not going to reap all of the benefits up front.

Essential to proper sequencing, and as a result, appropriate portfolio balancing, is a sound understanding of the organization's capacity to deliver as well as the capabilities of its resources. Organizations should know who is available to work on projects and what type of skills they have. It is often easy to determine how many people are there—so creating a resource inventory typically is not a problem. The challenge comes when trying to determine what the resources are currently working on and how much availability they have for project work or additional projects if they are already working on a project. One method available to get that correct picture is to do time tracking of project resources.

The expected long-term outcomes and the benefits of adopting a consistent Portfolio Management Framework and Process can include, but are not limited to, having the ability and capability to:

- Make conscious choices in selecting projects for implementation based on correct and up-to-date information, such as strategic alignment to business priorities, expected benefits, estimated costs, and identified risks.
- Determine capacity (i.e., the number of concurrent projects) for managing small, medium, and large projects to enable prioritization.
- Proactively manage risks associated with small to medium as well as large and complex transformation projects.
- Increase core competencies in PM across the organization and adopt a portfolio management approach to executive decision-making.
- Streamline and standardize processes related to the management of single and related, multiple projects, and project portfolios.
- Maintain a current list of all projects, active and inactive; phase the initiation of projects to match capacity; and improve the delivery in accordance with requirements of approved projects.
- Maximize use of internal resources, and rationalize the use of external resources to supplement internal staff, with greater ability to facilitate value and efficient completion of the approved projects portfolio.
- Measure actual, real-time performance, and track the realization of project and/or program benefits with the ability to identify actual progress made in the achievement of tangible outcomes and real results.

Program Management

In accordance with PMI's practice standards, a program is a group of related projects managed in a coordinated way to obtain benefits and controls not available from managing them individually. Programs may include elements of related work (e.g., ongoing operations) outside the scope of the discrete projects in a program. Some organizations refer to large projects as programs. If a large project is split into multiple related projects with explicit management of the benefits, then the effort becomes a program. Managing multiple projects via a program may enhance schedules across the program, provide incremental benefits, and enable staffing to be optimized in the context of the overall program's circumstances.

As depicted in Figure 15–18, Deloitte's approach to program management highlights four core responsibilities for the program management function: program integration, dependency awareness, standards adherence, and program reporting. The figure further illustrates additional primary and secondary activities that fall within the scope of work for this function.

While time, cost, and scope/quality are important performance measures at the individual project level, coordination, communication, and sequencing are the factors at the program level that help enable the desired results. This is because program

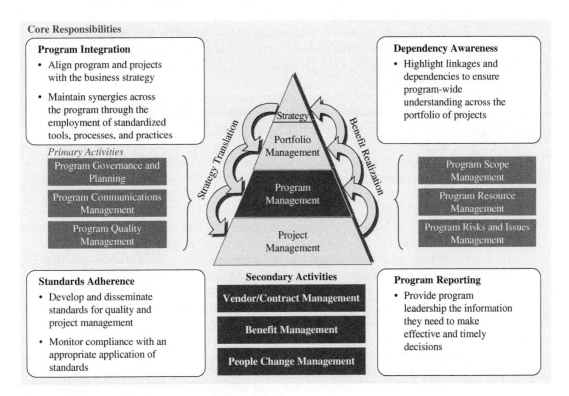

Figure 15–18. Deloitte program management framework.
Source: © Deloitte & Touche LLP and affiliated entities.

management involves grouping and managing a series of projects in an integrated manner, and not just completing individual projects. In the end, good PM can help to deliver the program according to the planned scope. Good program management will also provide a better understanding of the linkages and dependencies between projects and programs across the overall projects portfolio.

Project Management   At the program level, consistency can breed desirable results. This operating rhythm allows for regular reporting and monitoring across multiple projects. At the project level, that consistency does not always make sense. Internal project variance can be a product of a number of factors:

- Type of project (e.g., strategy development, technology implementation, organizational change deployment, etc.).
- Geographical/organizational scope (e.g., single site, single country, global).
- Project implementation model (e.g., agile, waterfall, iterative, etc.).
- Project team size.

These variances lead to different needs and constraints that impact some PM processes. The implication is that enterprise and program guidelines need to be standardized in some areas while retaining flexibility in others. This balance, when appropriately struck, can enable project managers to tailor processes in certain specific areas (e.g., status reporting, work planning) while still supporting minimum standards of performance overall.

Furthermore, other factors external to the project also drive variability and therefore should be considered. Some of these additional factors include:

- Industry.
- Environment (e.g., public sector, regulated, commercial).
- Technology being implemented (e.g., cloud solution, enterprise resource planning, etc.).

Even with so many differentiating elements, a number of processes and guidelines will stay fixed regardless of which PM model is selected. This includes laws or regulations, organizational policies, company standards, project controls, and financial management processes/policies (see Figure 15–19).

It is important that the organization understand where variability is required versus where standardization is required and effective. The objective is to make it easier for project teams to deliver solutions well, not to be dogmatic or overly theoretical, and to help enable the placement of acceptable safeguards to identify and manage out-of-control situations proactively.

As each project is initiated, the particulars of the project are considered. The result is a tailored set of PM processes that align with enterprise and program standards while reflecting the specific nature of the project itself.

Figure 15–19. Comparison of fixed organizational factors versus variable factors.
Source: © Deloitte & Touche LLP and affiliated entities.

The Methodology

A holistic PM solution is concerned with the definition and delivery of specific work streams within an overall EPM framework. It includes standards, processes, templates, training, job aids, and tools. The more that these components can be standardized, the easier it can be to deploy them; teams understand the expectations, know the tools, and have lived the processes.

Enterprise Value Delivery (EVD) for PM is Deloitte's method for delivering consistent PM solutions to our consulting clients. The method systematically addresses specific components of PM and is a common, standards-based approach supported by leading enabling tools, experienced coaching, and training. This method is scalable and flexible; it can be integrated into other Deloitte methods in whole or in part to address relevant PM issues. It incorporates standards while also allowing individual projects to tailor the parts that make sense for their individual situation.

Designed to help Deloitte practitioners manage their projects, EVD for PM is:

- *Scalable*. It uses a modular design to increase its flexibility and can fit the majority of projects accounting for a variety of project sizes or scope.
- *Deliverables-based*. It allows for the iterative nature of PM processes.
- *Prescriptive*. It includes tools, detailed procedures, templates, and sample deliverables specific to the management of the project that help practitioners initiate, plan, execute, control, and close the project.

- *Rich in information*. It houses extensive information about method processes, work distribution, and deliverable creation.
- *Based on experience*. It allows practitioners to leverage reusable material developed through the vast industry experience and knowledge of our practice.
- *Based on leading practices*. It reflects Deloitte leading practices and industry research and experience, allowing Deloitte practitioners to share a common language worldwide.
- *Practical*. It provides realistic and useful information, focusing on what truly works.

Deloitte's PM method content is aligned with the PMI's Project Management Body of Knowledge (*PMBOK® Guide*) and SEI's CMMI. Divided into two disciplines of work, PM and Quality Management, the method includes detailed task descriptions, step-by-step instructions, and considerations for task completion essential in delivering a PM solution. Multiple development aids including guidelines, procedures, and tools supplement each task.

A number of benefits can result from consistently applying the defined PM tasks and deliverables:

- Helps project managers see the big picture and accelerates work.
- Provides a consistent approach and a common language.
- Includes deliverable templates and tools.
- Incorporates quality and risk management, making it easier to improve quality and reduce risk of project deliverables.
- Can be used to manage programs as well as projects.

The Tools

Deloitte has found that leveraging tools focused on enabling PM processes can help drive the adoption of sound PM processes within an organization. There are numerous tools available for organizations to use, and they all have their own sets of advantages and disadvantages (see Figure 15–20). Selecting the appropriate tool can help facilitate management consistency throughout the project, but it is important to differentiate between processes and tools. It is less important which tool is deployed, so long as a rigorous process discipline is retained. Having a ready-to-use and leading-edge tool available is extremely advantageous, but the balance is between flexibility (use the appropriate tool for the job) and standardization (regardless of the tool being used, you should do a prescribed set of tasks).

If an organization does not have a tool available to utilize, Deloitte has a solution that may benefit clients. The customized tool provides sophisticated features while being simplistic enough to deploy to a project team quickly. Security is provided in a multitenant environment and practitioners are trained on the tool prior to engaging in a project.

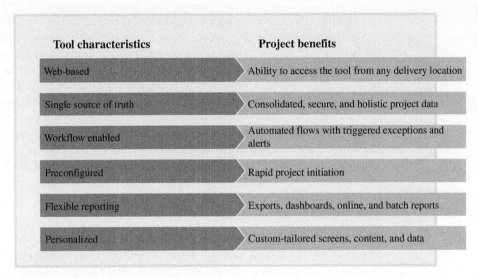

Tool characteristics	Project benefits
Web-based	Ability to access the tool from any delivery location
Single source of truth	Consolidated, secure, and holistic project data
Workflow enabled	Automated flows with triggered exceptions and alerts
Preconfigured	Rapid project initiation
Flexible reporting	Exports, dashboards, online, and batch reports
Personalized	Custom-tailored screens, content, and data

Figure 15–20. Deloitte PM tool characteristics.
Source: © Deloitte & Touche LLP and affiliated entities.

"Agile" PM

Deloitte's EVD solution establishes a foundation for performing common PM tasks while providing the flexibility to tailor processes to address known variability by:

- Defining typical usage models for frequent solution scenarios that incorporate the full spectrum of solution components (documentation, training, samples, etc.).
- Providing guidelines to enable projects to leverage the appropriate processes for the specific project circumstances.
- Allowing flexibility to define an appropriate governance structure to support the specific risk/cost profile for the project.

Emerging market conditions have also driven the need to evaluate standard PM processes, especially in an "agile" environment. Agile approaches are typically used in projects where requirements are unclear or subject to frequent change, and/or where frequent delivery of solution components with the greater value is required. Projects leveraging an agile methodology perform certain PM processes in a manner similar to projects using waterfall or iterative approaches (e.g., risk and issue management, financial management, and some aspects of status reporting), while others are performed in a very different manner. EVD provides guidance for project managers from both perspectives. Furthermore, for those aspects that are different, it prescribes how they can be approached using techniques that are specific to agile.

Specifically, scope management and work planning and tracking in agile are significantly different than in projects using a waterfall or iterative methodology. EVD for agile describes how projects develop and manage the product and sprint backlogs and includes guidelines for documentation of user stories that are sufficiently granular to be addressed within a single sprint. It describes the metrics analysis (using velocity, capacity, and burnup/down) that project teams use to size sprints and user stories and to forecast delivery of components and includes the productivity reports that are used to monitor progress. The important factor is to retain sufficient PM discipline, even if the techniques for managing the work are dramatically different. To summarize, although agile may be vastly different in how a project is executed, basic PM functions still exist, and Deloitte has identified a way to utilize both.

The Project Team

Even with the most intricate method and set of tools, without a dedicated and trained project team, the project can still falter. Resources with significant PM or program management experience typically function at the Program Management level while resources with limited to no project experience can be found at the project level. With that said, having a PM approach that takes the experience level of the resources into account is essential.

To make resources effective throughout the project, there are a few things to consider in tailoring your PM deployment (as shown in Figure 15–21).

- *Understand the internal change management implications.* Change management processes can be more rigid in some places than others. The project team must be aware of what circumstances to go through the change management process and what information is required at what times. This is critical in ensuring that changes get routed correctly and efficiently. Having to review every change request for accuracy slows down the project and can have an unanticipated adverse effect on project timelines.
- *Start basic activities early.* Recognize that while the enterprise and program management layers are typically staffed by dedicated, full-time PM professionals, project teams typically include operational resources who have little to no actual project experience. Make sure the project team knows what is expected from them and knows the basics. Examples include:
  - How often am I reporting status?
  - What is the overall project timeline?
  - Where can I find scope documentation for this project?

Performing a project kickoff and implementing appropriate process rigor early are instrumental in introducing the team to the specific dynamics of that project. This is when the overall timeline, PM protocols, and roles and responsibilities, among other topics, are presented. At this time, the project team can also start other planning activities including template identification or development.

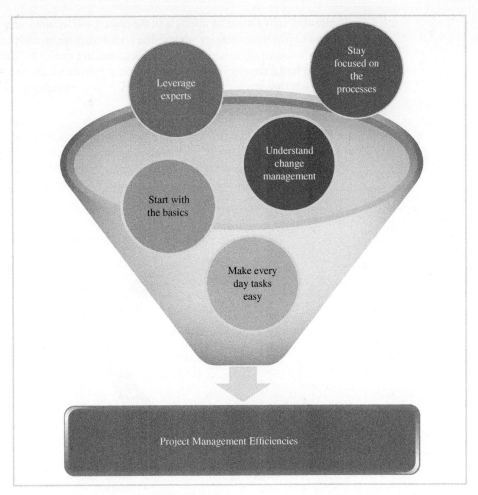

Figure 15–21. Key factors for PM deployment.
Source: © Deloitte & Touche LLP and affiliated entities.

- *Make it easy to do the simple things well.* For everyday tasks, like storing updated deliverables, recording time or status, or updating the work plan, the effort should be minimal regardless of the level of experience. These activities should not cause additional overhead to the project. The more time taken to perform simple things, the less productive time there is for the substantial design or build work on a project.
- *Leverage specialists when more advanced PM activities are required.* Certain activities of PM do require significantly more skill and should not be handled by a project novice. These areas center frequently on work planning efforts. Work

planning activities occur throughout the project and require a deep understanding of managing dependencies, identifying critical path items, and performing meticulous resource allocation. The activities shown in Figure 15–22 are focused only on the initial development of the work plan. Resources also need to understand what adjustments are required to the work plan as changes occur either to scope or resources.

● *Maintain focus on defined PM processes when things get precarious.* As changes occur throughout the project, sometimes it feels like the project can get out of control. One of the ways to avoid this is to uphold the PM processes that were approved at project initiation. A solid PM discipline enables the project

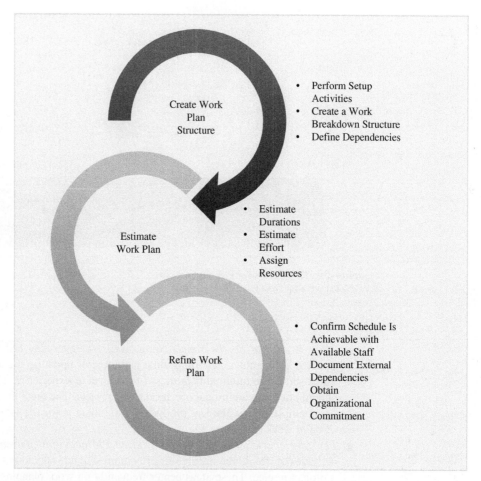

Figure 15–22. Activities required to develop a work plan.
Source: © Deloitte & Touche LLP and affiliated entities.

manager to pull the project out of whatever uncertain situations may arise. If managed diligently, the process can determine the overall effect on a project's scope, timeline, resources, or budget. It is critical to maintain the discipline when things are starting to go awry—just when the average person says "I don't have time for that" is when it becomes more critical.

Excellent PM is a result of a clear understanding of the project context. Deployment of standardized processes and tools can make things easier, but only if balanced to reflect the variations of each particular project. Once this balance is determined, it is primarily a matter of blocking and tackling. Communicate expectations and establish disciplines early, so that it becomes second nature. This allows the project team to focus creative energies on building the best solution, which is, after all, why there is a project in the first place.

Leadership and Governance   Additional factors influence an organization's ability to generate value and deliver transformation results that go beyond having the right project and the right PM and portfolio management processes or templates. According to Daniel Martyniuk:

> The importance of proper governance, leadership, and accountability cannot be underestimated. In my experience implementing project portfolio management, having the right framework to guide project stakeholders through the myriad of decisions that need to be made on a constant basis is a critical differentiating factor between a project's success or failure.

The main purpose of governance is to specify decision rights, clarify accountabilities, and encourage desirable behaviors. Governance is about bringing the appropriate individuals to the table to have the desired conversation under the relevant process to make the preferred decisions given the available information. Governance frameworks depict the structures and processes by which decisions are made, and they define sets of principles and practices for managing:

- *What* decisions need to be made,
- *Who* has the authority and accontability for making decisions, and with whose input, *and*
- *How* decisions get implemented, monitored, measured, and controlled.

As illustrated in Figure 15–23, effective governance requires strong executive sponsorship, clear "business" ownership, and sound technical advisory to facilitate compliance with established regulations, standards, and guidelines. It also requires some type of benefit and value oversight. This can be done through a committee of selected stakeholders who understand the qualitative aspects of a project's value. Most important, each function of the chosen governance framework should be empowered with the required decision-making authority within their area of responsibility.

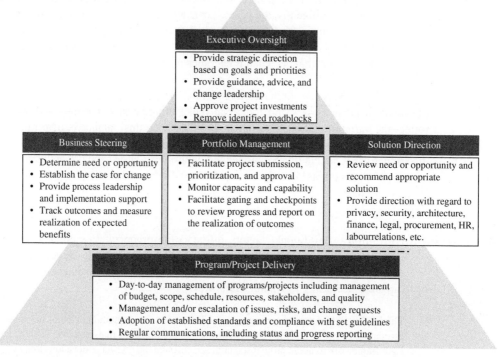

Figure 15–23.   Deloitte project portfolio governance framework.
Source: © Deloitte & Touche LLP and affiliated entities.

People and Organizational Change Management

Last, but most important, it is *people* who are the critical element to achieving project and transformation objectives. They are also a leading cause of transformation results falling short. Integrated people and organizational change management, human resources, and learning services should be delivered across the portfolio at the program and project levels to drive consistency, alignment, and effective delivery across the overall transformation effort.

Illustrated in Figure 15–24, the Deloitte People Dimension of Transformation is a broad framework that aligns with the business strategy and can address everything from risk assessment and leadership alignment to behavioral change, communication, training, organizational design, and more.

One of the major causes of a transformation not achieving its desired objectives is the stakeholders' inability to see and feel the compelling reason for the change. As a result, fear, anger, or complacency can take root and cause resistance. In cases where change is more effective, individuals have a sense of passion. Create compelling, eye-catching, and dramatic situations to help people see and visualize problems, solutions, or progress in addressing complacency, lack of empowerment, or other important issues.

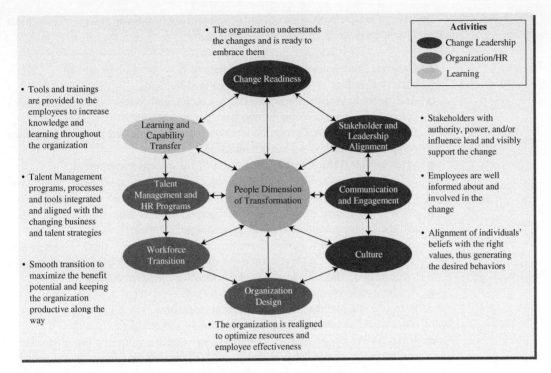

Figure 15–24. Deloitte people dimension of transformation framework.
Source: © Deloitte & Touche LLP and affiliated entities.

Sustained transformation also requires deep, personal commitment at every level of the organization. Some stakeholders will be cocreators who help shape the transformation vision and plans. Some will be interpreters. Other stakeholders will be consumers of the transformation. Effective transformation requires contributions and involvement from all types of players. Alignment and internal commitment start at the top—leaders should be aligned, willing to seek out resistance, and committed to leading the transformation by example.

Transformation projects usually alter structures, work processes, systems, relationships, leadership styles, and behaviors that together create what we know as organizational culture. Creating the culture the organization wants—or preserving the one it already has—may require a deliberate program that aligns with other transformation activities. Without a conscious effort, it is easy to end up with an organization stuck in between new ways of working and old modes of behavior.

To enhance the investment in new business models, technologies, and processes, a formal and deliberate program of education and skills development for the people affected by the transformation is essential. Yet education and training are usually near the bottom of the transformation to-do list.

Select keys to effectively approaching people and organizational change management on transformation projects include those listed next.

- *Get your stakeholders correct.* Understand how the transformation can affect each stakeholder group as well as specific individuals.
- *Anticipate risks.* Identify pockets of resistance before they surface, along with potential business disruptions and risks that might arise.
- *Assess the situation.* Determine whether the magnitude and pace of change are energizing or paralyzing the organization.
- *Set priorities.* Prioritize activities, tackling the critical barriers first.
- *Influence the influencers.* Identify people within each stakeholder group who command respect, and then get them involved as champions for the transformation.
- *Strive for real commitment.* Understand people's circumstances and aspirations—and then make a concerted effort to accommodate them.
- *Equip leaders to drive transformation.* Equip leaders with the knowledge and skills needed to help their people get through this challenging and often traumatic period. Make leaders the role models for the desired behavior.
- *Recognize there may be winners and losers.* The impact of transformation varies from one stakeholder group to the next, and some may not be happy with the outcome. Understand, engage, and inform all stakeholders.
- *Focus on the things that really matter.* An effective culture is one that creates sustainable business value, differentiates the organization from its competitors, supports the specific requirements of the industry, and helps customers get what they really want.
- *Be consistent.* Things that drive behavior and culture should align with one another. Misalignment simply confuses people.
- *Reinforce.* Align people-related initiatives—particularly rewards and incentives—to help foster the new culture. Establish effective leadership models, and introduce new words and vocabulary that highlight the desired behavior.
- *Retain select staff.* Identify top performers and other select staff who are critical to the organization's future results. Let them know they are not at risk.
- *Capture knowledge.* Establish formal processes and systems to transfer and capture organizational knowledge—particularly for sourcing transformations.
- *Be kind but confident.* Decision-makers should be gentle but not show doubts that decisions were required, appropriate, and final.

Conclusion

The adoption and consistent application of standard project, program, and portfolio management frameworks, as well as the implementation of the relevant governance along with effective people and organizational change management techniques, can lead the organization to the realization of a number of benefits, including:

- *Improved executive decision-making.* Enhanced ability to determine which projects to continue/stop, based on correct, up-to-date project status/progress information.

- *Financial transparency and accountability.* Improved ability to manage budget under- and overruns and to shift funds within the portfolio to better manage and respond to unforeseen circumstances and changes in priorities.
- *Enhanced resource capacity management.* Availability of needed information and data to make effective use of available resources and ability to shift resources within the portfolio to enhance resource utilization across projects.
- *Proactive issues and risk management.* Ability to foresee and respond to challenges before they escalate into major problems; a mechanism for bringing select issue resolution or risk mitigation decision/action requirements to the attention of the executives.
- *Standardization and consistency.* Apples-to-apples comparison between projects; improved and more timely internal and external communications with staff, clients, executives, and other stakeholders.
- *Increased collaboration and better results.* Enhanced realization of benefits through the joint management of initiatives as an integrated portfolio; cooperation and improved removal of roadblocks to results.

Although not exhaustive, the topics addressed here outline selected critical factors to achieve the desired results that, based on our practical experience, can guide an organization in the "right" direction as it embarks on the road to implement sustained project portfolio management capability to deliver real, tangible enterprise value.

15.3 COMAU[3]

Innovation in Action: COMAU   Introduction

COMAU is a worldwide leader in the manufacture of flexible, automatic systems and integrating products, processes, and services that increase efficiency while lowering overall costs. With an international network that spans 13 countries, COMAU uses the latest technology and processes to deliver advanced turnkey systems and consistently exceed the expectations of its customers. COMAU specializes in body welding, powertrain machining, and assembly, robotics and automation, electromobility as well as digital initiatives, consultancy, and education for a wide range of industrial sectors. The continuous expansion and improvement of its product range enable COMAU to guarantee customized assistance at all phases of a project—from design, implementation, and installation to production start-up and maintenance services.

Figure 15–25. COMAU contract and project management office.

COMAU innovation is based on manufacturing's digital transformation, added-value manufacturing, and human–robot collaboration; these key pillars allow COMAU to reach new heights in leading the culture of automation.

COMAU Project Management In 2007, COMAU established a contract and PM corporate function, with the aim to create a stronger link between leading roles and PM and to ensure coherence between execution and company strategy. Over time, the contract and PM function grew in responsibility (as shown in Figure 15–25) and modified its structure to reach a configuration based on the coordinated sharing of knowledge and activities between Project Management Office (PME), Risk Management, and the COMAU PM Academy. Another foundation element that sustains COMAU PM framework is represented by the PMI® standard set, recognized as the best practice standard in the Project and Portfolio Management landscape.[4]

Project Management OFFICE (PMO)

In recent years, COMAU started thinking about the industry in a new way, developing new scenarios, designing innovative products, creating ways to streamline production processes, and defining new trends in digital manufacturing, creating a new paradigm to balance collaboration between people and machines.

---

4. Material in this section has been provided by Roberto Guida, COMAU Project Management Vice President, Riccardo Bozzo, COMAU Contract Management, Francesca Gaschino, COMAU Risk Manager, and Paolo Vasciminno, COMAU Project Management Excellence. ©2022 by COMAU. All rights reserved. Reproduced with permission.

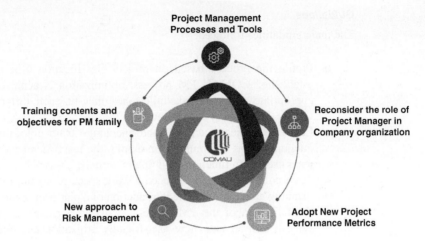

Figure 15–26. COMAU PMO innovation pillars.

We call it HUMANufacturing, the innovative COMAU vision that places mankind at the center of business processes within a manufacturing plant, in full cooperation with the industrial automation solutions and new digital technologies that surround them. The above vision needs changes and innovation in all company organization roles to be sustained.

The COMAU PMO is therefore innovating, expanding its landscape from traditional industrial project portfolio to new forms of projects, such as new product development, digitalization, and industry4.0 projects.

The COMAU PMO through the innovation in each of the five pillars (see Figure 15–26) is now positioned within the organization as a valuable business partner that sharpened its capacity to be adaptive for different forms of projects within the company.

Scenario Analysis

The current context of market evolution highlights new disrupting evidences:

- The degree of innovation is becoming predominant in all COMAU projects.
- The size of traditional EPC projects is continuously decreasing.
- The development of new, innovative, digital products, and solutions is more and more expected by our customer base (traditional automotive, and new customer base: new automotive, general industry, electric vehicle).

Consequently, COMAU began a deep analysis of the current project execution course of action (process, organization, and tools) and of the PM methodologies adopted, to verify the alignment with market needs.

COMAU redesigned its PM process to make it scalable to the degree of innovation typical of each project and adopting where applicable lean and agile solutions.

Guidelines

The main guidelines followed:

- Profile the process based on project classification criteria. From the well-established COMAU PM process, optimization is achieved by selecting the appropriate activities/tasks/milestones that fit each different project, maximizing the business added value.
- Empower project managers and encourage team colocation to improve the integration and communication among the team members, and of the project team itself with functions and finally with the customer.
- Team colocation has introduced a work space re-layout, the integration of the planning activities, and the introduction of a platform organization.
- Digitalization of the PM process, realizing a paperless process in order to achieve a faster and more user-friendly utilization and optimize time/cost of process management.
- Encourage visual management, and exploit the potential of mobile applications (see Figure 15–27). The benefits obtained with these operations have been a reduced complexity, reduced time, and lower effort for milestones and activities/tasks execution and, in general, the reduction of "not value added" operations, maximizing effectiveness and efficiency as requested by the current business scenario.

COMAU Contract Management

Considering the "fixed price" and "turn-key" nature of the most part of the contractual agreements that COMAU establishes with the Customers, the Contract Management

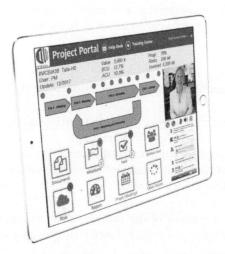
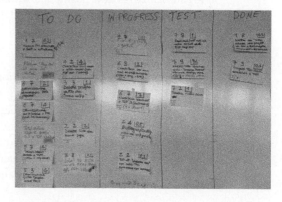

Figure 15–27. COMAU project management digital process and new tools (i.e., agile visual management).

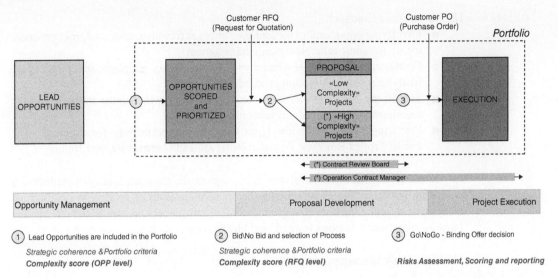

Figure 15–28. Contract management life cycle.

competences applied both in the Contract acquisition and administration phases are deemed crucial by COMAU to sustain its business model.

Project Contracts include Contractual agreements made with Customers but also those executed with Major turn-key Suppliers and, if any, Partners under a consortium or Joint-venture.

The COMAU Contract Management Function is responsible to ensure an effective life cycle management of Project Contracts from creation, administration, changes, and claim management to closeout, during the entire value chain process including Business Development, Order Acquisition, and Operation.

Based on the early evaluation of the complexity rate of a project, the Contract Management is assigned to the project team starting from the proposal phase and continues to be engaged during the execution (See Figure 15–28).

Contract Managers are working closely with Sales and Proposal team as well as PM team to provide support on the following main activities:

During Contract Acquisition
- Analyzing Contracting documents to identify ambiguities, conflicts, and risks (e.g., Force Majeure Clauses and others) and providing ongoing guidance on risks and negotiation strategies.
- Organizing and leading Contract Reviews and Reports preparatory for go/no go decision.
- Ensuring that Contractual documents and information are effectively handed over from the sales phase to the Project Execution Team (for example, through dedicated contract-induction sessions).

During Contract Execution

- Identifying opportunities for customer changes (revenue, cash, and profit enhancement), schedule relief, and claim management.
- Collaborating with legal when disputes with clients, suppliers, and partners are likely to result in litigation proceedings.
- Identifying the customer contractual provisions which must be back-to-back transferred to suppliers and subcontractors.
- Preparing Contract Review reports in an appropriate way for different audiences: Project Reviews, Portfolio Reviews, and Contract Review Board.

In addition to the direct involvement in projects, the Contract Management function is leading initiatives aimed at keeping the appropriate level of awareness on Contract Management, including the delivery of Contract Management courses based on concrete business experience, lessons learned, and business case studies.

COMAU Risk Management

History and Achievement

In 2006, COMAU began to address risk management with a more focused, strategic approach, recognizing it as an essential part of the successful completion of projects. With the increase in organizational complexity and global presence, it was necessary to find more structured and refined tools for managing uncertainties. Consequently, in 2010, a Risk Management Office as part of the PM organization was created.

Specific responsibilities of the risk management office included the definition of a framework (methods, processes, tools) for the management of projects and portfolio risks as well as improving the concrete application throughout the entire project life cycle.

Beginning in 2015, COMAU decided to make a further maturity step, launching an initiative aimed to reinforce and better integrate risk management practices at different company levels (contract sales, project execution, and portfolio management) with a global perspective (see Figure 15–29).

The completion of the initiative provided COMAU with the adequate risk management approach and tools needed to manage the turnkey projects portfolio composed of fixed-price EPC contracts, each characterized by a high degree of complexity and a fixed scope for which a traditional PM predictive approach is to be adopted.

Sales Management   Projects Management   Portfolios Management

Risk Management Process   Risk Management Process

Figure 15–29.   The COMAU Risk Management Initiative.

Risk Management for Innovative Projects

The acceleration impressed by the digitalization era required changes to business models for many companies including COMAU. Risk management as all other functions within companies is required to change as well and to approach new models to cover new forms of projects, those that might be named innovation projects (see Figure 15–30).

Innovation projects such as new product or solution development and digitalization projects have peculiarities that need new risk management approaches. To fit with this requirement, COMAU developed a first "concept" to extend and integrate the well-established traditional risk management.

The concept is based on the following guidelines:

● Risk management is a primary vehicle to reinforce PM "strategic thinking" into project execution.

New risk management approach might bridge the PM perspective from tactic to strategic as it focuses on pursuing a wider business-value objective.

Project managers are now expected to master company business processes in addition to the PM process and to be able to anticipate the implications of project decisions on the different functional objectives, communicate them, and collaborate with functions to set up counteractions.

● New company functions and customers are involved in project risk management.

Accepting that perimeter of project risk management is widening, there are a number of people/structures within the company (i.e., innovation, marketing, sales functions) that become important stakeholders, bearers of interests and objectives that, differently from the past, might be subjected to redefinition while the project is ongoing (agile approach).

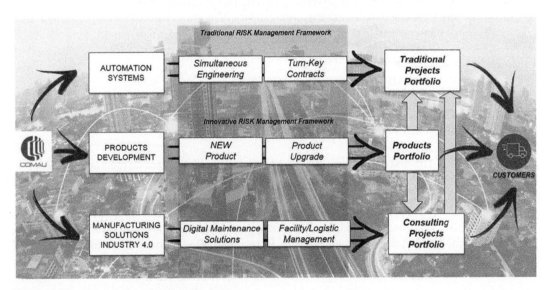

Figure 15–30. New frontiers for Risk Management.

- New metrics to measure project risks.

The value expected and created by innovation projects goes beyond the end of the project itself. The project constraints to be taken into account when assessing the risks are not only limited to scope-time-costs as in a traditional predictive approach. New metrics for the qualitative and quantitative analysis of risks that could include also a wider, more strategic measure of long-term risk impacts are needed. COMAU's new approach to project risk management is therefore based on a hybrid of predictive and adaptive components.

COMAU PM Academy

Developing Innovation Project Managers

The PM Academy in COMAU is responsible to ensure people involved in projects are technically skilled and behaviorally prepared to participate in and manage projects. In front of the need for innovation, PM Academy's challenge is to support project manager transformation in the company—from a person strongly plan and process-oriented to a person more and more aware of its strategic role in the creation of value for the company. And what do we need to reach it? Essentially a project manager must be very flexible, culturally sensitive, political-driven, business-oriented, and a master in communication and leadership.

In this new context PM Academy continues its efforts to provide valuable international PM best practices and support the certification process, but this is integrated with agile and lean concepts and methodologies, and enriched leadership models.

To be an effective PM in the innovation environment, you cannot focus only on certifications. Certifications remain a good method to enlarge people's awareness and knowledge, but now the objective is higher. "New" project managers have to function as part influencers, part psychologists, and part politicians. Not only do they themselves need these soft skills, but they also need to drive teams and stakeholders to a new level of effectiveness in their own soft skills.

PM Academy New Domains

PM Academy started to work on two additional objective areas:

The strategic vision of project managers
The ability of project managers to inspire others

The first domain was developed in parallel with the gradual redefinition of the role of a COMAU project manager, who was becoming central in driving innovation. The second objective is achieved by developing practical tools for soft skill development. But—this is the novelty—these tools are not designed to develop the project manager's soft skills but to provide the PM with tools they can apply to develop soft skills in teams and stakeholders in their projects. In this way, PMs become actors in the process of changing the "culture-of-involvement-in-projects" (in fact, success is strongly associated with the level of engagement in the project and in the team).

Using these tools, team members can:

- Communicate more and better.
- Learn about one another differences.
- Share values and contribute to the definition of "shared values."
- Analyze conflicts and try to solve them.
- Tell their emotions and listen to the other's ones.
- Respect and appreciate differences.
- Feel more satisfied.
- Understand better their role in projects.

In conclusion, the goal is to create PMs that, although they continue to be valuable PM professionals, are at the same time strategic business partners, integration managers, inspirers, and coaches. In order to fulfill this objective, the COMAU PM Academy has enriched its catalog with training in agile and lean PM, PM leadership, and train-the-PM-coach.

15.4 IPLM: ENABLING EXCELLENCE IN A DIGITALLY TRANSFORMED FUTURE OF WORK[5]

Delivering with Speed and Excellence is the new norm

Today's business environment is starting to give us a glimpse of the way we will work in the future. The norm today is operating under a VUCA dynamic environment. When VUCA was first coined, it intended to highlight the volatility, uncertainty, complexity, and ambiguity in the world, yet the future of work expects more disruptions while consumers' demands increase. The risks in innovating and product launches could not be higher and integrated solutions that allow for collaborating and fostering intimate interactions along a product life cycle are a must for tomorrow's organizations.

> *The clearest signs of programs excellence are exhibited in the consistent achievement of speed to market and relatedness to changing consumers' needs*

Having had the opportunity to work in just about every corner of the world, I have seen firsthand the strong interest in programs delivery excellence and how to achieve

---

5. Material in this section has been provided by Dr. Al Zeitoun, *PgMP, PMI Fellow*. He is the Senior Director of Strategy at Siemens Digital Industries Software. Al has had transformation and program management leadership roles spanning a diverse range of industries across global regions, such as with Booz Allen Hamilton, solving customers' pains as industries move to Digital. His experiences included serving on PMI's Global Board of Directors, PM Solutions President, Emirates Nuclear Energy Corporation Executive Director, and International Institute for Learning's Chief Projects Officer. ©2022 Siemens. All rights reserved. Reproduced with permission.

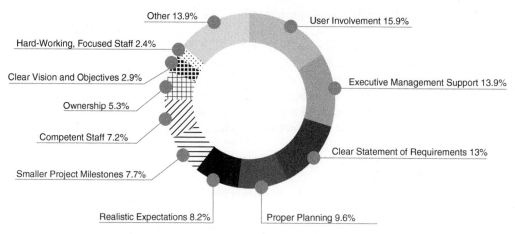

Figure 15–31. Top Causes of Project/Program Failure.

it in the midst of unrelenting uncertainty. The digital movement has only added to the pressures and the promises. As I saw program management mature, one of the biggest gaps continued to be the siloes in the ways of working, missing transparency, starting over anew, extreme wastes of resources and valuable time, and lack of responsiveness to change.

In a time when 10X scaling is a strategic priority for many of our organizations, this is what got me excited about Siemens' digital threads with the End-to-End (E-2-E) strength of Integrated Program & Life cycle Management (IP&LM). Figure 15–31, is adapted from an important Pulse of the Profession In-Depth Report by PMI & BCG in 2014 showed the top 3 enablers of programs success as User Involvement @ 15.9%, Executive Management Support @ 13.9%, and Clear Statement of Requirements @ 13.0%.

I see IP&LM hitting the mark on all these 3 critical differences-making enablers that will continue to dominate in the future of work across global organizations. Whether in consumer goods, manufacturing, energy, financial, pharmaceutical, or enterprise software solutions, the common thread for delivery excellence continues to be identical: bringing the fitting balance between the spheres of process, people, and governance. I will use this proposed delivery excellence model with a focus on these three spheres to draw close analogy to the difference-making that IP&LM brings. Just as in the case of professional tennis players, when the repeatability of healthy patterns is prevalent, excellence is dominant. In order to get us the highest return on our investment in delivery excellence, the digital solutions have to be dynamic and sensitive to our customers' fluid and rising demands.

The 3 spheres in Figure 15–32 also have an integrated effect as they work in tandem to achieve a higher delivery excellence scale. This is what I find to be prevailing in IP&LM. I will explore each of these 3 spheres and draw the analogy to the digital threads that exist in the IP&LM solution.

Figure 15–32. The Three Spheres for Excellence

Process

For the *processing sphere*, the delivery excellence gauge has to take into consideration that working in the future will continue to be in the midst of increasing uncertainty and complexity. The gauge goes from autonomy on one end of the scale to alignment on the other end. The right processes for achieving scale in delivery excellence have to balance the need for strategic alignment across all processes, business units, and products, with the need for local autonomy and adjustments to address local markets.

So, let us look at Figure 15–33 and map that to the future process advantages that IP&LM provides.

Even though the first look at the process journey here looks to be sequential yet it is truly fluid. Speed is the name of the game with the touch points intending to break organizational silos and allowing for the wonderful benefits of cocreating from capturing the idea all across to track. One of the critical delivery gaps for most organizations is that they operate as if every program is the first they have ever done. Becoming a transformed learning organization is such a differential for future of work. As we launch programs for new products in a new region or market, we should capitalize on the great assets that exist and maximize reuse. This frees space and resources to innovate as needed while enough processes for consistency creation remain.

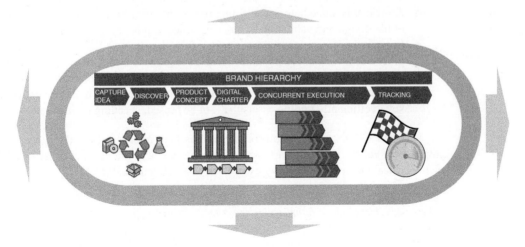

Figure 15–33. IPLM Process View.

I would like to zoom in on a few key elements as examples of where I was observing meaningful processes that enable leveraging assets and workflows:

- A backbone for process consistency is the Digital Twin concept/strategy, which is a virtual entity that mirrors the physical world, turning data into insights to continuously design, build and optimize throughout the product and production life cycle.
- What Siemens did is surround this concept with five digital threads. Siemens considers that the Digital Threads form the neural network of the Digital Twin.
- Comprehending the interactivity across the 5 digital threads helps us understand their impact on the entire transformation mission. These five are: Integrated program and life cycle management (most of my observations focus), smart product and process design, production design and optimization, flexible manufacturing, and traceability and insights
- As I dive into my observations on IP&LM, here is an example that resonated with me while interviewing one of Siemens' subject matter experts: Case in point: the head of marketing for a Hamburg-based company was on a business trip to demonstrate cooling effect to the skin for a new product. He forgot to take a sample along so he stopped by the duty-free and picked one, only to realize upon use that us has zero additional cooling effect. He researched the issue and found that development team canceled the raw material that was in charge of creating this effect. Their answer was: did not know that!
- I cannot emphasize enough how relevant that example is for many of the programs and product releases I have encountered across industries in just around every corner of the globe. This example also speaks to the power of the E-2-E that IP&LM brings to initiatives' repeatable successes. What excited me first about this IP&LM thread is the triple balance of speed, agility, and rigor.
- A key to this immediate power comes from combining the life cycle management of the design and validation of a new product with the program and PM.
- In a good program standard, the overarching level above the specific program processes is a balanced portfolio level. In IP&LM, a brand hierarchy organizes a company's complete product portfolio, including all of the individual products and the innovation assets that define them. Brand Hierarchy management is the key to efficiently finding assets that can be reused, speeding up innovation.

I will come back to specific process features later after I address the people and governance, yet for now to recap some of the reasons why the process consistency is such a critical element for delivery excellence as exhibited by IP&LM:

> *Processes should give a fitting structure yet leave enough room for individual creativity, cocreation, and strengthening ownership.*

- Process fluidity is imperative.
- The key to process excellence here is to sense what are the most suitable types, numbers, and amount of processes for achieving initiatives' outcomes with speed.
- For delivery excellence to happen, and as evident in my review of the IP&LM solution, engagement of stakeholders is critical

*The "**so what**" is: we have to find the right process balance between autonomy and alignment. The organizational culture will show us the degree of fit in choosing the processes that enable both delivery and strategic alignment, yet leave the right size window open for a personal stamp from the different organizational units, thus bringing in their creativity and ownership.*

Scaling process excellence is quite possible using the process richness of IP&LM that demonstrates early adaptors' successes. At that point, it becomes contagious and scaling happens when it clearly becomes attractive for other business units to follow suit.

People

For the *people sphere*, delivery excellence in the future will hinge on transparency and data-driven decisions. The impact of digitalization on running and changing the business represents the competencies required for future excellence. Digital technology contributes strongly as the future of work differentiator, yet at the same time, the reaction to this new revolution has been quite uneven and difficult to sense. As with the process scale, sensing and responding fast to changing needs has to happen at every layer of the organization.

Delivering excellence uses traditional people competencies while being agile in developing competencies for the digital age.

For this people critical sphere to deliver excellence, one has to work on new implementation frameworks that are comprehensively designed around customers. The outward focus on customers that IP&LM creates, circumnavigates internal politics. Typically we use tools like design thinking to bring the best ideas to the mix and to balance machine intelligence with the impact on how our people go about their daily lives.

As I look at the IP&LM and its role in empowering people across the spectrum of the ecosystem, I find it encouraging that there is enough support for building the trust currency that is most vital in achieving successful digital transformations.

- IP&LM enables people by helping each discipline maintain their own set of templates, each being small, clear, and easy to change.
- Product teams can build a choice of approaches that best fit their needs and the circumstances identified by Product Context. The solution works more like a box of building blocks that can be combined very easily and provides an environment that can adapt incredibly fast to any current and future challenges.
- From idea capture to discovery, the solution includes the original intent and purpose of the product, as well as listing all the relevant claims that designs must deliver. This frees people to focus on innovating.
- The Digital Charter helps the right autonomy people need by factoring in the complexity of development and the type of initiative with clarity, such as whether it is a new product variant like a Cold Brew coffee, a product expansion or customization pack, or entirely new product to the world.

The "**so what**" is: *we need new lenses to devise the most appropriate mix of people competencies for us and for our customers' organizations, to excel in delivering complex products and solutions while becoming more innovative and fast.*

I find the clarity of different personas that support different IP&LM critical roles very encouraging. As we know from program delivery challenges, lack of clarity of roles and responsibilities could derail outcomes. Having the right mix such as portfolio manager, marketing manager, governance team, program, project, and product managers properly defined with responsibilities and clear impact on the organization is of utmost value to transformation success.

Scaling the impact of the people sphere requires unleashing the secret recipe for making it work and will result in high collaboration levels across organizational boundaries that we have not yet witnessed. When access to joint trusted data is achieved, such as the marketing brief, which is a critical alignment document that includes the original intent and purpose of the product, as well as listing all the relevant claims that designs must deliver, a space for focusing on collaboration and trust building is created.

> *Excellence in governance is achieved by shifting from the classical steering focus to a heightened delivery focus.*

Governance

For the *governance sphere* in the framework to work, we need a hybrid framework chosen with an eye on the future. The future of work is hybrid and IP&LM exemplifies that. The future of work has governance altered from any classical view of steering committees to a much more fluid approach in order to achieve timely and fast decision-making by executive teams who are hyper-focused on achieving benefits. We have to look holistically at the business and carefully gauge the appetite and readiness for this rethought approach to governance with excellence.

Decision-making success spreads through contagion. As an example of the value of effective decision-making with IP&LM that is core to future governance with speed, let us say a 10K employee organization and 2,000 projects with a setup time of 2 weeks for their new products. So even with a saving of half that time, we save 4,000 person days every year.

The other critical dimension of this sphere is the executive leadership team. Working on complex programs, we have to ensure that the executive group adjusts fast to the new governance with excellence. This requires creating greater transparency in the boardroom and eventually changing the executive role from one of steering to one of experimenting and redirecting. Lightning-speed decision-making requires a massive amount of decision-making muscle building and practice coupled with a growing appetite for taking risks that are enriched by the right data timely.

Here are few specific observations in IP&LM that support successful programs governance in the future of work:

- Enterprise applications, like this, influence a lot of people so it is a competitive advantage to have IP&LM framework enable the governance agility.
- A culture change is required for adaption of IP&LM across industries.

- More interest could exist in consultancy organizations as they support clients to govern and map out business processes.
- Scaling this way of governing to a larger organization is a gap even though some business models are obsolete and could use this data richness to reinvent themselves.
- Cultural understanding would help tremendously in figuring out the governance adjustments impact so resiliency in working across the cultures map and the right degree of risk-taking would be very useful.
- An enabler for future governance is having integrated KPIs: foundation built on data first, build trusted relationships and then work on developing the optimal solution.

The "**so what**" is: *a governance framework for delivery excellence in the future requires the bar to be set quite high in the degree of autonomy modeled by the executive team and cascaded down and across. To achieve a faster organizational cultural tilt, make it safe for teams to have fun experimenting, and count on the ripple effects that successful stories demonstrate to the others. This is core to the power of IP&LM and the possibilities digital twins create for having difficult conversations and decisions early in the life cycle and throughout programs' journeys.*

Scaling the impact of the governance sphere requires a very different way of defining organizational success. Speed is at the center of this new governance and learning is a critical subcomponent of it. Initiatives' delivery success metrics change, behavioral change follows, reflecting how fluid and agile governance needs to look.

> *Excellence is built on increasing process autonomy, digitally capable people, and agile, delivery-focused governance.*

Looking Forward into the Future of Work

Bringing together the 3 delivery excellence spheres creates a multiplier for scaling delivery excellence. Achieving the right balance of process autonomy and alignment, coupled with the right mix of traditional and nontraditional competencies, and integrating that under an umbrella of agile governance anchored in a delivery focus is the secret to expediting and achieving delivery excellence. Integrating the efforts on the 3 critical spheres discussed has the potential of bringing a much higher Return On Investment as compared with a sole focus on any of these 3 spheres individually.

IP&LM hits the mark on a wide spectrum of process, people, and governance interplay. It provides both the balcony view and the very specific details that are needed by the decision-makers of future innovations. The wide perspective that is achieved across the stages of the framework assures that leaders are not lost in the details yet able to zoom in on any required assets. This also frees the creative mindset that drives continual transformative innovations.

Figure 15–34, summarizes some of the key IPLM impact points across the three excellence spheres. Processes are prioritized and focused on adapting with speed while

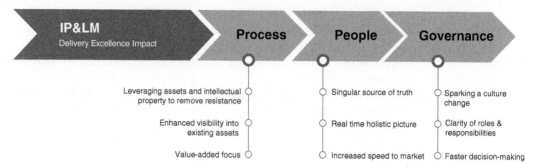

Figure 15–34.   IPLM and driving excellence.

achieving consistent quality. People are empowered to make timely decisions and have a clear organizational role in transforming the ways of working and achieving outcomes. Governance is seamless and objective with the power of data and learning that IP&LM creates.

In this new world, although excellence continues to be a continuing journey, the nature of the road ahead is changing faster than ever predicted, due to the seamless interplay between technology and data. A healthy culture of ongoing excellence focused on the future needs a combination of resilience in pursuing new and different ways of doing things than those we have been comfortable with in the past, together with encouraging experimentation as the vehicle for learning. IP&LM provides a focused approach to delivering new initiatives, innovating and expediting product launches, extracting knowledge from them, and changing fast when it is not working, means that delivery excellence outcomes are within our grasp.

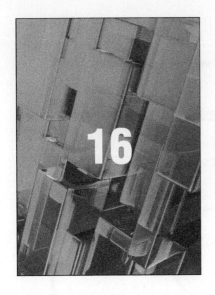

Value-Driven Project Management

16

16.0 INTRODUCTION

Over the years, we have come to accept the traditional definition of project success, namely meeting the triple constraints of time, cost, and scope. More recently, we modified our definition of success by stating that there must be a valid business purpose for working on the project. Success was then recognized as having both a business component and a technical component, as well as meeting strategic business objectives.

Today, we are modifying the definition of success further by adding in a "value" component, as seen in Figure 16–1. In other words, the ultimate purpose of working on the project should be to provide some form of value to both the client and the parent organization. If the project's value cannot be identified, then perhaps we should not be working on it at all.

Value can be defined as what the stakeholders perceive the project's deliverables as being worth. Each stakeholder can have a different definition of value. Furthermore, the actual value may be expressed in qualitative rather than purely quantitative terms. It simply may not be possible to quantify the actual value.

The importance of the value component cannot be overstated. Consider the following statements:

- Completing a project on time and within budget does not guarantee success if you are working on the wrong project.
- Even having the greatest enterprise project management (EPM) methodology in the world cannot guarantee that value will be there at the end of the project.
- Completing a project on time and within budget does not guarantee that project will provide value at completion.

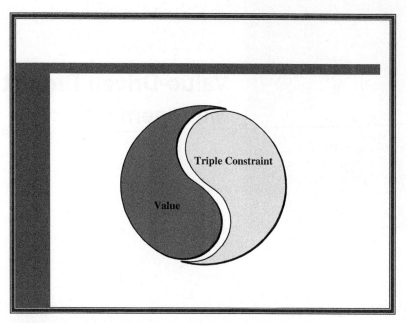

Figure 16–1. Definition of success.

These three statements lead us to believe that perhaps value is now the dominant factor in the selection of a project portfolio. Project requestors must clearly articulate the value component in the project's business case or run the risk that the project will not be considered.

16.1 VALUE OVER THE YEARS

Surprisingly enough, numerous research on value has taken place over the past 20 years. Some of the items covered in the research include:

- Value dynamics
- Value gap analysis
- Intellectual capital valuation
- Human capital valuation
- Economic value-based analysis
- Intangible value streams
- Customer value management/mapping
- Competitive value matrix
- Value chain analysis
- Valuation of information technology (IT) projects
- Balanced scorecard

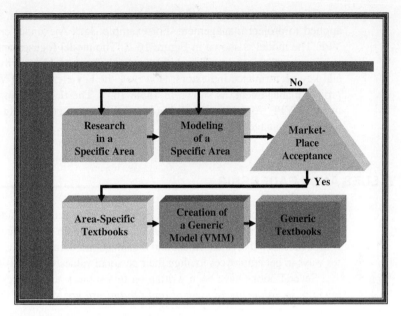

Figure 16–2. Evolution of value-based knowledge.
Note: VMM is a value measurement methodology.

The evolution of value-based knowledge seems to follow the flowchart in Figure 16–2. Research seems to take place in a specific research area, such as calculating the value of software development projects or calculating shareholder value. The output of such research is usually a model that is presented to the marketplace for acceptance, rejection, and/or criticism. Soon others will follow with similar models but in the same research area, such as software development. Once marketplace acceptance concurs on the validity of these models, textbooks appear discussing the pros and cons of one or more of the models.

With the acceptance of modeling in one specific area, modeling then spreads to other areas. The flowchart process continues until several areas have undergone modeling. Once this is completed, textbooks appear in generic value modeling for a variety of applications. The following list contains some of the models that have occurred over the past 15–20 years:

- Intellectual capital valuation
- Intellectual property scoring
- Balanced scorecard
- Future Value Management™
- Intellectual Capital Rating™
- Intangible value stream modeling
- Inclusive Value Measurement™
- Value Performance Framework (VPF)
- Value measurement methodology (VMM)

There is some commonality among many of these models, such that they can be applied to project management. For example, Jack Alexander created a model called VPF. The model is shown in Figure 16–3. The model focuses on building shareholder value rather than creating shareholder value.[1] The model is heavily biased toward financial key performance indicators. However, the key elements of VPF can be applied to project management, as shown in Table 16–1. The first column contains the key elements of VPF from Alexander's book and the second column illustrates the application to project management.

16.2 VALUES AND LEADERSHIP

The importance of value can have a significant impact on the leadership style of project managers. Historically, project management leadership was perceived as the inevitable conflict between individual values and organizational values. Today, companies are looking for ways to get employees to align their personal values with the organization's values.

Several books have been written on this subject, and the best one, in this author's opinion, is *Balancing Individual and Organizational Values* by Ken Hultman and Bill

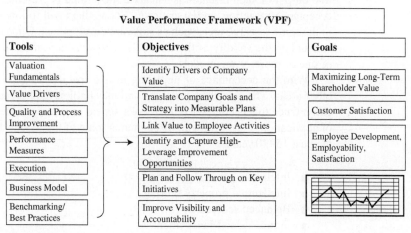

Figure 16–3. The VPF model.

---

1. J. Alexander, *Performance Dashboards and Analysis for Value Creation* (Hoboken, NJ: Wiley, 2007), p. 5. Ibid., pp. 105–106.

TABLE 16–1.   APPLICATION OF VPF TO PROJECT MANAGEMENT

VPF Element	Project Management Application
Understanding key principles of valuation	Working with the project's stakeholders to define value
Identifying key value drivers for the company	Identifying key value drivers for the project
Assessing performance on critical business processes and measures through evaluation and external benchmarking	Assessing performance of the EPM methodology and continuous improvement using the project management office (PMO)
Creating a link between shareholder value and critical business processes and employee activities	Creating a link between project values, stakeholder values, and team member values
Aligning employee and corporate goals	Aligning employee, project, and corporate goals
Identifying key "pressure points" (high-leverage improvement opportunities) and estimating their potential impact on value	Capturing lessons learned and best practices that can be used for continuous improvement activities
Implementing a performance management system to improve visibility and accountability in critical activities	Establishing and implementing a series of or project-based dashboards for customer and stakeholder visibility of key performance indicators
Developing performance dashboards with high-level visual impact	Developing performance dashboards for stakeholder, team, and senior management visibility

Source: J. Alexander, *Performance Dashboards and Analysis for Value Creation*, Hoboken, NJ: Wiley, 2007, p. 6. Reproduced by permission of John Wiley & Sons.

Gellerman.[2] Table 16–2 shows how the concept of value has changed over the years. If you look closely at the items in Table 16–2, you can see that the changing values affect more than just individual versus organizational values. Instead, it is more likely to be a conflict among four groups, as shown in Figure 16–4. The needs of each group might be:

- Project manager
  - Accomplishment of objectives
  - Demonstration of creativity
  - Demonstration of innovation
- Team members
  - Achievement
  - Advancement
  - Ambition
  - Credentials
  - Recognition
- Organization
  - Continuous improvement
  - Learning
  - Quality
  - Strategic focus
  - Morality and ethics
  - Profitability
  - Recognition and image

---

2. K. Hultman and B. Gellerman, *Balancing Individual and Organizational Values* (San Francisco: Jossey-Bass/Pfeiffer, 2002).

TABLE 16–2.   CHANGING VALUES

Moving Away From: Ineffective Values	Moving Toward: Effective Values
Mistrust	Trust
Job descriptions	Competency models
Power and authority	Teamwork
Internal focus	Stakeholder focus
Security	Taking risks
Conformity	Innovation
Predictability	Flexibility
Internal competition	Internal collaboration
Reactive management	Proactive management
Bureaucracy	Boundaryless
Traditional education	Lifelong education
Hierarchical leadership	Multidirectional leadership
Tactical thinking	Strategic thinking
Compliance	Commitment
Meeting standards	Continuous improvements

Source: Adapted from K. Hultman and B. Gellerman, *Balancing Individual Organizational Values*, San Francisco: Jossey-Bass/Pfeiffer, © 2002, pp. 105–106.

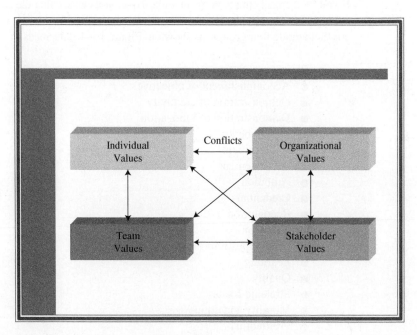

Figure 16–4.   Project management value conflicts.

- Stakeholders
  - Organizational stakeholders: job security
  - Product/market stakeholders: high-quality performance and product usefulness
  - Capital markets: financial growth

There are several reasons why the role of the project manager and the accompanying leadership style have changed. Some reasons include:

- We are now managing our business as though it were a series of projects.
- Project management is now viewed as a full-time profession.
- Project managers are now viewed as both business managers and project managers and are expected to make decisions in both areas.
- The value of a project is measured in business terms rather than solely in technical terms.
- Project management is now being applied to parts of the business that have traditionally not used it.

The last item requires further comment. Project management works well for the "traditional" type of project, which includes:

- Time duration of 6–18 months.
- The assumptions are not expected to change over the duration of the project.
- Technology is known and will not change over the duration of the project.
- People who start the project will remain through its completion.
- The statement of work is reasonably well defined.

Unfortunately, the newer types of projects are more nontraditional and have the following characteristics:

- Time duration over several years.
- Projects can begin with just an idea rather than a business case
- The enterprise environmental factors can change over the duration of the project
- The risks are much greater than with traditional projects
- The assumptions and constraints can and will change over the duration of the project.
- Technology will change over the duration of the project.
- People who approved the project may not be there at its completion.
- The statement of work is ill-defined and subject to numerous changes.

The nontraditional types of projects, such as strategic or innovation projects, have made it clear why traditional project management must change. Three areas necessitate changes:

1. New projects have become:
   - Highly complex and with greater acceptance of risks that may not be fully understood during project approval
   - More uncertain in project outcomes, with no guarantee of value at the end
   - Pressed for speed-to-market regardless of the risks

2. The statement of work (SOW) is:
   - Not always well defined, especially on long-term projects
   - Based on possibly flawed, irrational, or unrealistic assumptions
   - Careless of unknown and rapidly changing economic and environmental conditions
   - Based on a stationary target rather than moving target for final value
3. The management cost and control systems (EPMs) focus on:
   - An ideal situation (as described in the PMBOK® Guide)[3]
   - Theories rather than the understanding of the workflow
   - Inflexible processes
   - Periodically reporting time at completion and cost at completion but not value (or benefits) at completion
   - Project continuation rather than canceling projects with limited or no value

Over the years, we have taken several small steps to plan for the use of project management on nontraditional projects. These include:

- Project managers are provided with more business knowledge and are allowed to provide an input during the project selection process.
- Because of the preceding item, project managers are brought on board the project at the beginning of the initiation phase rather than at the end of the initiation phase.
- Project managers now seem to have more of an understanding of technology rather than a command of technology.

The new types of projects, combined with a heavy focus on business alignment and value, brought with them a classification system, as shown in Figure 16–5.

Operational projects. These projects, for the most part, are repetitive ones, such as payroll and taxes.

They are called "projects," but they are managed by functional managers without the use of an EPM methodology.

- *Enhancements or internal projects.* These are projects designed to update processes, improve efficiency and effectiveness, and possibly improve morale.
- *Financial projects.* Companies require some form of cash flow for survival. These are projects for clients external to the firm and have an assigned profit margin.

---

3. PMBOK is a registered mark of the Project Management Institute, Inc.

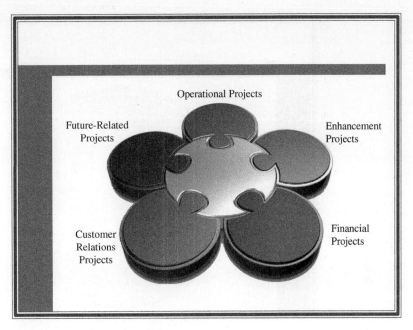

Figure 16–5. Classification of projects.

- *Future-related projects.* These are long-term projects to produce a future stream of products or services capable of generating a future cash flow. These projects may be an enormous cash drain for years, with no guarantee of success.
- *Customer-related projects.* Some projects may be performed, even at a financial loss, to maintain or build a customer relationship. However, performing too many of these projects can lead to financial disaster.

These new types of projects focus more on value than on the triple constraint. Figure 16–6 shows the traditional triple constraints, whereas Figure 16–7 shows the value-driven triple constraints. With the value-driven triple constraints, we emphasize stakeholder satisfaction, and decisions are made around the four types of projects (excluding operational projects) and the value that is expected of the project. In other words, success is when the value is obtained, hopefully within the triple constraints. As a result, we can define the four cornerstones of success using Figure 16–8. Very few projects are completed without some trade-offs. This holds true for both traditional projects and value-driven projects. As shown in Figure 16–9, traditional trade-offs result in an elongation of the schedule and an increase in the budget. The same holds true for the value-driven projects shown in Figure 16–10. The major difference is performance. With traditional trade-offs, we tend to reduce performance to satisfy other requirements. With value-driven projects, we tend to increase performance in hopes of providing added value, and this tends to cause much larger cost overruns and schedule

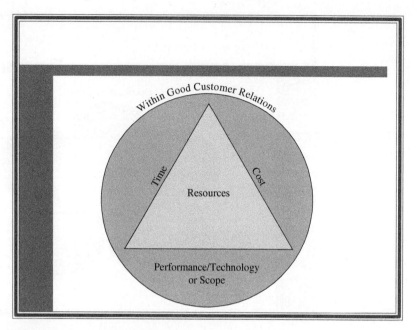

Figure 16–6. Traditional triple constraints.

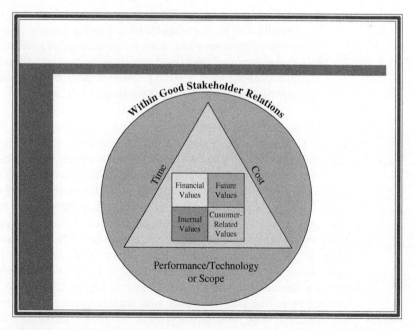

Figure 16–7. Value-driven triple constraints.

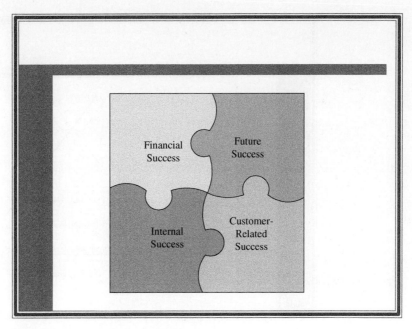

Figure 16–8. Four cornerstones of success.

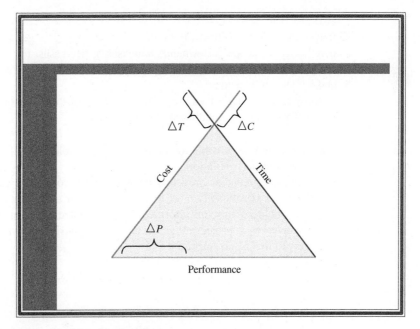

Figure 16–9. Traditional trade-offs.
Note: Δ = deviations from the original plan.

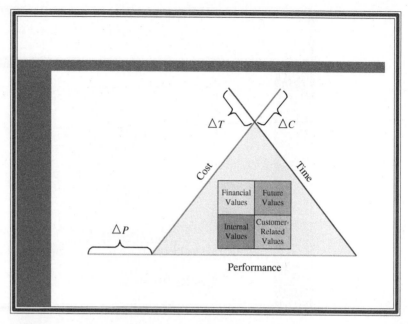

Figure 16–10. Value-driven trade-offs.
Note: Δ = deviations from the original plan.

slippages than with traditional trade-offs. Project managers generally do not have the sole authority for scope/performance increases or decreases. For traditional trade-offs, the project manager and the project sponsor, working together, may have the authority to make trade-off decisions.

However, for value-driven projects, all or most of the stakeholders may need to be involved. This can create additional issues, such as:

- It may not be possible to get all of the stakeholders to agree on a value target during project initiation.
- Getting agreement on scope changes, extra costs, and schedule elongations becomes significantly more difficult the farther along you are in the project.
- Stakeholders must be informed of the targeted value or anticipated value at completion at project initiation and continuously briefed as the project progresses; that is, no surprises!

Conflicts among the stakeholders may occur. For example:

- During project initiation, conflicts among stakeholders are usually resolved in favor of the largest financial contributors.
- During execution, conflicts over future value are more complex, especially if major contributors threaten to pull out of the project.

For projects that have a large number of stakeholders, project sponsorship may not be effective with a single-person sponsor. Therefore, committee sponsorship may be necessary. Membership in the committee may include:

- Perhaps representatives from all stakeholder groups
- Influential executives
- Critical strategic partners and contractors
- Others based on the type of value

Responsibilities for the sponsorship committee may include:

- Taking a lead role in the definition of the targeted value
- Taking a lead role in the acceptance of the actual value
- Ability to provide additional funding
- Ability to assess changes in the enterprise environment factors
- Ability to validate and revalidate the assumptions

Sponsorship committees may have significantly more expertise than the project manager in defining and evaluating the value of a project.

Value-driven projects require that we stop focusing on budgets and schedules and instead focus on how value will be captured, quantified, and reported. Value must be measured in terms of what the project contributes to the company's objectives. To do this, an understanding of four terms is essential.

1. *Benefits.* Advantages.
2. *Value.* What the benefit is worth.
3. *Business drivers.* Target goals or objectives are defined through benefits or value and expressed more in business terms than in technical terms.
4. *Key performance indicators (KPIs).* Value metrics that can be assessed either quantitatively or qualitatively.

Traditionally, business plans identified the benefits expected from the project. Today, portfolio management techniques require identification of the value as well as the benefits. However, conversion from benefits to value is not easy.[4] Table 16–3 illustrates the benefit-to-value conversion. Also, as shown in Figure 16–11, there are shortcomings in the conversion process that can make the conversion difficult.

We must identify the business drivers, and they must have measurable performance indicators using KPIs. Failure to identify drivers and accompanying KPIs may make true assessment of value impossible. Table 16–4 illustrates typical business drivers and the accompanying KPIs.

---

4. For additional information on the complexities of conversion, see J. J. Phillips, T. W. Bothell, and G. L. Snead, *The Project Management Scorecard* (Oxford, UK: Butterworth Heinemann, 2002), Chapter 13.

TABLE 16–3. MEASURE VALUE FROM BENEFITS

Expected Benefits	Value Conversion
Profitability	Easy
Customer satisfaction	Hard
Goodwill	Hard
Penetrate new markets	Easy
Develop new technology	Medium
Technology transfer	Medium
Reputation	Hard
Stabilize workforce	Easy
Utilize unused capacity	Easy

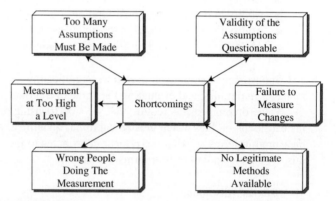

Figure 16–11. Shortcomings.

TABLE 16–4. BUSINESS DRIVERS AND KPI

Business Drivers	Key Performance Indicators
Sales growth	Monthly sales or market share
Customer satisfaction	Monthly surveys
Cost savings	Earned value measurement system
Process improvement	Time cards

KPIs are metrics for assessing value. With traditional project management, metrics are established by the EPM methodology and fixed for the duration of the project's life cycle. But with value-driven project management, metrics can change from project to project, during a life-cycle phase, and over time because of:

- The way the company defines value internally
- The way the customer and contractor jointly define success and value at project initiation

- The way the customer and contractor come to an agreement at project initiation as to what metrics should be used on a given project
- New or updated versions of tracking software
- Improvements to the EPM methodology and accompanying project management information system
- Changes in the enterprise environmental factors

Even with the best possible metrics, measuring value can be difficult. Some values are easy to measure, while others are more difficult. The easy values to measure are often called soft or tangible values, whereas the hard values are often considered intangible values. Table 16–5 illustrates some of the easy and hard values to measure. Table 16–6 shows some of the problems associated with measuring both hard and soft values.

The intangible elements are now considered by some to be more important than tangible elements. This appears to be happening on IT projects where executives are giving significantly more attention to intangible values. The critical issue with intangible values is not necessarily the end result but rather the way that the intangibles were calculated.[5]

Tangible values are usually expressed quantitatively, whereas intangible values are expressed through a qualitative assessment. There are three schools of thought for value measurement:

School 1. The only thing that is important is ROI.

School 2. ROI can never be calculated effectively; only the intangibles are what are important.

School 3. If you cannot measure it, then it does not matter.

TABLE 16–5.   MEASURING VALUES

Easy (Soft/Tangible) Values	Hard (Intangible) Values
Return on investment (ROI) calculators	Stockholder satisfaction
Net present value (NPV)	Stakeholder satisfaction
Internal rate of return (IRR)	Customer satisfaction
Cash flow	Employee retention
Payback period	Brand loyalty
Profitability	Time to market
Market share	Business relationships
	Safety
	Reliability
	Reputation
	Goodwill
	Image

TABLE 16–6.   PROBLEMS WITH MEASURING VALUES

Easy (Soft/Tangible) Values	Hard (Intangible) Values
Assumptions are often not disclosed and can affect decision-making.	Value is almost always based on subjective-type attributes of the person doing the measurement
Measurement is very generic	Measurement is more of an art than a science
Measurement never meaningfully captures the correct data	Limited models are available to perform the measurement

The three schools of thought appear to be an all-or-nothing approach where value is either 100 percent quantitative or 100 percent qualitative. The best approach is most likely a compromise between a quantitative and qualitative assessment of value. It may be necessary to establish an effective range, as shown in Figure 16–12, which is a compromise among the three schools of thought. The effective range can expand or contract.

The timing of value measurements is absolutely critical. During the life cycle of a project, it may be necessary to switch back and forth from qualitative to quantitative assessment, and, as stated previously, the actual metrics or KPIs can change as well. Certain critical questions must be addressed:

- When or how far along the project life cycle can we establish concrete metrics, assuming it can be done at all?
- Can value be perceived simply, and therefore no value metrics are required?
- Even if we have value metrics, are they concrete enough to reasonably predict actual value?
- Will we be forced to use value-driven project management on all projects, or are there some projects where this approach is not necessary?
  - Well-defined versus ill-defined
  - Strategic versus tactical
  - Internal versus external
- Can we develop a criterion for when to use value-driven project management, or should we use it on all projects but at a lower intensity level?

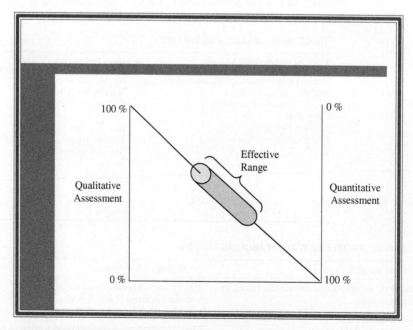

Figure 16–12.   Quantitative versus qualitative assessment.

For some projects, assessing value at closure may be difficult. We must establish a time frame for how long we are willing to wait to measure the value or benefits of a project. This is particularly important if the actual value cannot be identified until sometime after the project has been completed. Therefore, it may not be possible to appraise the success of a project at its closure if the true economic values cannot be realized until sometime in the future.

Some practitioners of value measurement question whether value measurement is better done using boundary boxes instead of life-cycle phases. For value-driven projects, the potential problems with life-cycle phases include:

- Metrics can change between phases and even during a phase.
- Inability to account for changes in the enterprise environmental factors.
- Focus may be on the value at the end of the phase rather than the value at the end of the project.
- Team members may get frustrated at not being able to quantitatively calculate value.

Boundary boxes, as shown in Figure 16–13, have some degree of similarity to statistical process control charts. Upper and lower strategic value targets are established. As long as the KPIs indicate that the project is still within the upper and lower value targets, the project's objectives and deliverables will not undergo any scope changes or trade-offs.

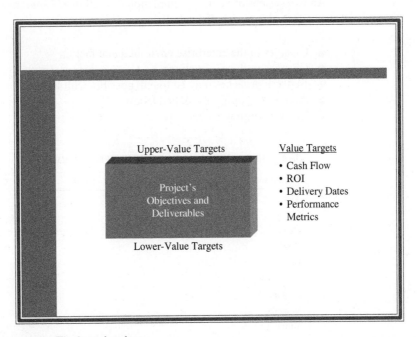

Figure 16–13. The boundary box.

Value-driven projects must undergo value health checks to confirm that they will make a contribution of value to the company. Value metrics, such as KPIs, indicate the current value. What is also needed is an extrapolation of the present into the future. Using traditional project management combined with the traditional EPM methodology, we can calculate the time at completion and the cost at completion. These are common terms that are part of earned value measurement systems. But as stated previously, being on time and within budget is no guarantee that the perceived value will be there at project completion.

Therefore, instead of using an EPM methodology, which focuses on earned value measurement, we may need to create a VMM that stresses the value variables. With VMM, time to complete and cost to complete are still used, but we introduce a new term: "value (or benefits) at completion." Determination of value at completion must be done periodically throughout the project. However, periodic reevaluation of benefits and value at completion may be difficult because:

- There may be no reexamination process.
- Management is not committed and believes that the reexamination process is unreal.
- Management is overoptimistic and complacent with existing performance.
- Management is blinded by unusually high profits on other projects (misinterpretation).
- Management believes that the past is an indication of the future.

An assessment of value at completion can tell us if value trade-offs are necessary. Reasons for value trade-offs include:

- Changes in the enterprise environmental factors
- Changes in the assumptions
- Better approaches may be found, possibly with less risk
- Availability of highly skilled labor
- Breakthrough in technology

TABLE 16–7.   COMPARISON OF EVMS, EPM, AND VMM

Variable	EVMS	EPM	VMM
Time	✓	✓	✓
Cost	✓	✓	✓
Quality		✓	✓
Scope		✓	✓
Risks		✓	✓
Tangibles			✓
Intangibles			✓
Benefits			✓
Value			✓
Trade-offs			✓

As stated previously, most value trade-offs are accompanied by an elongation of the schedule. Two critical factors that must be considered before schedule elongation takes place are:

1. Elongating a project for the desired or added value may incur risks
2. Elongating a project consumes resources that may have already been committed to other projects in the portfolio

Traditional tools and techniques may not work well on value-driven projects. The creation of a VMM may be necessary to achieve the desired results. A VMM can include the features of earned value measurement systems (EVMSs) and EPM systems (EPMs), as shown in Table 16–7. But additional variables must be included for the capture, measurement, and reporting of value.

17

Effects of Mergers and Acquisitions on Project Management

17.0 INTRODUCTION

All companies strive for growth. Strategic plans are prepared, identifying new products and services to be developed and new markets to be penetrated. Many of these plans require mergers and acquisitions to obtain their strategic goals and objectives. Yet even the best-prepared strategic plans often fail. Too many executives view strategic planning as planning only, often with little consideration given to implementation. Implementation success is vital during the merger and acquisition process.

17.1 PLANNING FOR GROWTH

Companies can grow in two ways: internally and externally. With internal growth, companies cultivate their resources from within and may spend years attaining their strategic targets and marketplace positioning. Since time may be an unavailable luxury, meticulous care must be taken to make sure that all new developments fit the corporate project management methodology and culture.

External growth is significantly more complex. External growth can be obtained through mergers, acquisitions, and joint ventures. Companies can purchase the expertise they need very quickly through mergers and acquisitions. Some companies execute occasional acquisitions, whereas other companies have sufficient access to capital such that they can perform continuous acquisitions. However, once again, companies often fail to consider the impact of acquisitions on project management. Best practices in project management may not be transferable from one company to another. The impact on project management systems resulting from mergers and acquisitions is often irreversible, whereas joint ventures can be terminated.

Effect of Mergers and Acquisitions on Project Management

This chapter focuses on the impact on project management resulting from mergers and acquisitions. Mergers and acquisitions allow companies to achieve strategic targets at a speed not easily achievable through internal growth, provided that the sharing of assets and capabilities can be done quickly and effectively. This synergistic effect can produce opportunities that a firm might be hard-pressed to develop itself.

Mergers and acquisitions focus on two components: preacquisition decision-making and postacquisition integration of processes. Wall Street and financial institutions appear to be interested more in the near-term financial impact of acquisitions than the long-term value that can be achieved through better project management and integrated processes. During the mid-1990s, companies rushed into acquisitions in less time than they required for capital expenditure approvals. Virtually no consideration was given to the impact on project management or whether the expected best practices would be transferable. As a result, there have been more failures than successes.

When a firm rushes into an acquisition, very little time and effort appear to be spent on postacquisition integration. Yet this is where the real impact of best practices is felt. Immediately after an acquisition, each firm markets and sells products to the other's customers. This may appease the stockholders, but only in the short term. In the long term, new products and services will need to be developed to satisfy both markets. Without an integrated project management system where both parties can share the same best practices, this may be difficult to achieve.

When sufficient time is spent on preacquisition decision-making, both firms look at combining processes, sharing resources, transferring intellectual property, and the overall management of combined operations. If these issues are not addressed in the preacquisition phase, unrealistic expectations may occur during the postacquisition integration phase.

17.2 PROJECT MANAGEMENT VALUE-ADDED CHAIN

Mergers and acquisitions are expected to add value to the firm and increase its overall competitiveness. Some people define value as the ability to maintain a certain revenue stream. A better definition of value might be the competitive advantages that a firm possesses as a result of customer satisfaction, lower cost, efficiencies, improved quality, effective utilization of personnel, or the implementation of best practices. True value occurs *only* in the postacquisition integration phase, well after the actual acquisition itself.

Value can be analyzed by looking at the value chain: the stream of activities from upstream suppliers to downstream customers. Each component in the value chain can provide a competitive advantage and enhance the final deliverable or service. Every company has a value chain, as illustrated in Figure 17–1. When a firm acquires a supplier, the value chains are combined and expected to create a superior competitive position. Similarly, the same result is expected when a firm acquires a downstream company. But it may not be possible to integrate the best practices.

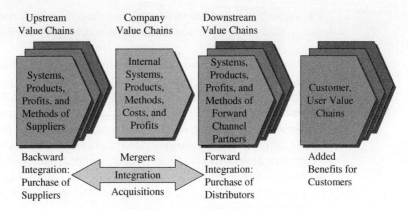

Figure 17–1. Generic value-added chain.

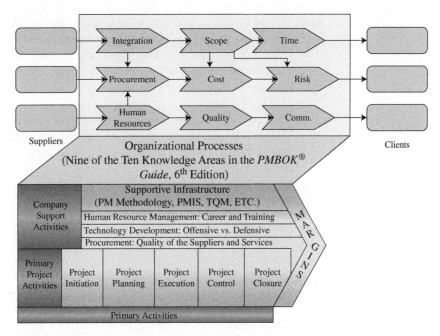

Figure 17–2. Project management value-added chain.

Historically, value chain analysis was used to look at a business as a whole.[1] However, for the remainder of this chapter, the sole focus will be the project management value-added chain and the impact of mergers and acquisitions on the performance of the chain.

Figure 17–2 shows the project management value-added chain. The primary activities are those efforts needed for the physical creation of a product or service. The primary activities can be considered to be the five major process areas of project management: project initiation, planning, execution, control, and closure.

---

1. M. E. Porter, *Competitive Advantage* (New York: Free Press, 1985), Chapter 2.

The support activities are those company-required efforts needed for the primary activities to take place. At an absolute minimum, the support activities must include:

- *Procurement management.* The quality of the suppliers and the products and services they provide to the firm.
- *Effect of mergers and acquisitions on project management.* The ability to combine multiple project management approaches, each at a different level of maturity.
- *Technology development.* The quality of the intellectual property controlled by the firm and the ability to apply it to products and services both offensively (new product development) and defensively (product enhancements).
- *Human resource management.* The ability to recruit, hire, train, develop, and retain project managers. This includes the retention of project management intellectual property.
- *Supportive infrastructure.* The quality of the project management systems necessary to integrate, collate, and respond to queries on project performance. Included within the supportive infrastructure are the project management methodology, project management information systems (PMIS), total quality management system, and any other supportive systems. Since customer interfacing is essential, the supportive infrastructure can also include processes for effective supplier–customer interfacing.

These support activities can be further subdivided into nine of the 10 knowledge areas of the *PMBOK® Guide*. The arrows connecting the nine *PMBOK® Guide* areas indicate their interrelatedness. The exact interrelationships may vary for each project, deliverable, and customer (Figure 17–2).

Each of these primary and support activities, together with the nine process areas, is required to convert material received from your suppliers into deliverables for your customers. In theory, Figure 17–2 represents a work breakdown structure for a project management value-added chain:

Level 1: Value chain
Level 2: Primary activities
Level 3: Support activities (which can include the Stakeholder Management knowledge area)
Level 4: Nine of the 10 *PMBOK® Guide* knowledge areas

The project management value-added chain allows a firm to identify critical weaknesses where improvements must take place. This could include better control of scope changes, the need for improved quality, more timely status reporting, better customer relations, or better project execution. The value-added chain can also be useful for supply chain management. The project management value-added chain is a vital tool for continuous improvement efforts and can easily lead to the identification of best practices.

Executives regard project costing as a critical, if not the most critical, component of project management. The project management value chain is a tool for understanding a project's cost structure and the cost control portion of the project management methodology. In most firms, this is regarded as a best practice. Actions to eliminate or reduce a cost or schedule disadvantage need to be linked to the location in the value chain where the cost or schedule differences originated.

The glue that ties together elements within the project management chain is the project management methodology. A project management methodology is a grouping of forms, guidelines, checklists, policies, and procedures necessary to integrate the elements within the project management value-added chain. A methodology can exist for an individual process, such as project execution, or for a combination of processes. A firm can also design its project management methodology for better interfacing with upstream or downstream organizations that interface with the value-added chain. Ineffective integration at supplier–customer interface points can have a serious impact on supply chain management and future business.

17.3 PREACQUISITION DECISION MAKING

The reason for most acquisitions is to satisfy a strategic and/or financial objective. Table 17–1 shows the six most common reasons for an acquisition and the most likely strategic and financial objectives. The strategic objectives are somewhat longer term than the financial objectives, which are under pressure from stockholders and creditors for quick returns.

The long-term benefits of mergers and acquisitions include:

- Economies of combined operations
- Assured supply or demand for products and services
- Additional intellectual property, which may have been impossible to obtain otherwise
- Direct control over cost, quality, and schedule rather than being at the mercy of a supplier or distributor
- Creation of new products and services
- Pressure on competitors through the creation of synergies
- Cost cutting by eliminating duplicat steps

Each of these can generate a multitude of best practices.

TABLE 17–1.   TYPES OF OBJECTIVES

Reason for Acquisition	Strategic Objective	Financial Objective
Increase customer base	Bigger market share	Bigger cash plow
Increase capabilities	Provide solutions	Wider profit margins
Increase competitiveness	Eliminate costly steps	Stable earnings
Decrease time to market (new products)	Market leadership	Earnings growth
Decrease time to market (enhancements)	Broad product lines	Stable earnings
Closer to customers	Better price–quality–service mix	Sole-source procurement

The essential purpose of any merger or acquisition is to create lasting value that becomes possible when two firms are combined and value exists that would not exist separately. The achievement of these benefits, as well as attainment of strategic and financial objectives, could rest on how well the project management value-added chains of both firms integrate, especially the methodologies within their chains. Unless the methodologies and cultures of both firms can be integrated, and reasonably quickly, the objectives may not be achieved as planned.

Project management integration failures occur after the acquisition happens. Typical failures are shown in Figure 17–3. These common failures result because mergers and acquisitions simply cannot occur without organizational and cultural changes that are often disruptive in nature. Best practices can be lost. It is unfortunate that companies often rush into mergers and acquisitions with lightning speed but with little regard for how the project management value-added chains will be combined. Planning for better project management should be of paramount importance, but unfortunately, it is often lacking.

The first common problem area in Figure 17–3 is the inability to combine project management methodologies within the project management value-added chains. This occurs because of:

- A poor understanding of each other's project management practices prior to the acquisition
- No clear direction during the preacquisition phase on how the integration would take place
- Unproven project management leadership in one or both of the firms
- A persistent attitude of "we–them"

Some methodologies may be so complex that a great amount of time is needed for integration to occur, especially if each organization has a different set of clients and

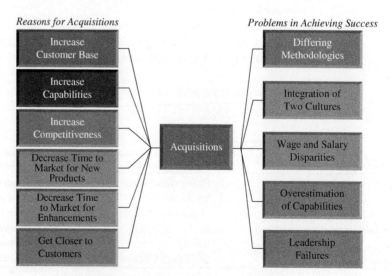

Figure 17–3. Project management problem areas after an acquisition.

different types of projects. As an example, a company developed a project management methodology to provide products and services for large publicly held companies. The company then acquired a small firm that sold exclusively to government agencies. The company realized too late that integration of the methodologies would be almost impossible because of requirements imposed by government agencies for doing business with the government. The methodologies were never integrated, and the firm servicing government clients was allowed to function as a subsidiary, with its own specialized products and services. The expected synergy never occurred.

Some methodologies simply cannot be integrated. It may be more prudent to allow the organizations to function separately than to miss windows of opportunity in the marketplace. In such cases, pockets of project management may exist as separate entities throughout a large corporation.

The second major problem area in Figure 17–3 is the existence of differing cultures. Although project management can be viewed as a series of related processes, it is the working culture of the organization that must eventually execute these processes. Resistance by the corporate culture to support project management effectively can cause the best plans to fail. With opposing cultures, there may be differences in the degree to which each:

- Has (or does not have) management expertise (i.e., missing competencies)
- Resists change
- Resists technology transfer
- Resists transfer of any type of intellectual property
- Allows for a reduction in cycle time
- Allows for the elimination of costly steps
- Insists on reinventing the wheel
- Perceives project criticism as personal criticism

Integrating two cultures can be equally difficult during both favorable and unfavorable economic times. People may resist any changes in their work habits or comfort zones, even when they recognize that the company will benefit from the changes.

Multinational mergers and acquisitions are equally difficult to integrate because of cultural differences. Several years ago, a U.S. automotive supplier acquired a European firm. The American company supported project management vigorously and encouraged its employees to become certified in project management. The European firm provided very little support for project management and discouraged its workers from becoming certified, using the argument that European clients do not regard project management in such high esteem as do General Motors, Ford, and Chrysler. The European subsidiary saw no need for project management. Unable to combine the methodologies, the U.S. parent company slowly replaced the European executives with American executives to drive home the need for a single project management approach across all divisions. It took almost five years for the complete transformation to take place. The parent company believed that the resistance in the European division was more of a fear of change in its comfort zone than a lack of interest by its European customers.

Sometimes there are clear indications that the merging of two cultures will be difficult. When Federal Express acquired Flying Tiger in 1988, the strategy was to merge the two into one smoothly operating organization. At the time of the merger, Federal Express (since renamed FedEx Express) employed a younger workforce, many of whom were part-timers. Flying Tiger had full-time, older, longer-tenured employees. FedEx focused on formalized policies and procedures and a strict dress code. Flying Tiger had no dress code, and management conducted business according to the chain of command, where someone with authority could bend the rules. Federal Express focused on a quality goal of 100 percent on-time delivery, whereas Flying Tiger seemed complacent with a 95–96 percent target. Combining these two cultures had to be a monumental task for Federal Express. In this case, even with these potential integration problems, Federal Express could not allow Flying Tiger to function as a separate subsidiary. Integration was mandatory. Federal Express had to quickly address those tasks that involved organizational or cultural differences.

Planning for cultural integration can also produce favorable results. Most banks grow through mergers and acquisitions. The general belief in the banking industry is to grow or be acquired. During the 1990s, National City Corporation of Cleveland, Ohio, recognized this and developed project management systems that allowed National City to acquire other banks and integrate the acquired banks into National City's culture in less time than other banks allowed for mergers and acquisitions. National City viewed project management as an asset that has a very positive effect on the corporate bottom line. Many banks today have manuals for managing merger and acquisition projects.

The third problem area in Figure 17–3 is the impact on the wage and salary administration program. The common causes of the problems with wage and salary administration include:

- Fear of downsizing
- Disparity in salaries
- Disparity in responsibilities
- Disparity in career path opportunities
- Differing policies and procedures
- Differing evaluation mechanisms

When a company is acquired and integration of methodologies is necessary, the impact on the wage and salary administration program can be profound. When an acquisition takes place, people want to know how they will benefit individually, even though they know that the acquisition is in the best interest of the company.

The company being acquired often has the greatest apprehension about being lured into a false sense of security. Acquired organizations can become resentful to the point of physically trying to subvert the acquirer. This will result in value destruction, where self-preservation becomes of paramount importance to the workers, often at the expense of the project management systems.

Consider the following situation. Company A decided to acquire company B. Company A has a relatively poor project management system in which project management is a part-time activity and not regarded as a profession. Company B, in contrast, promotes project management certification and recognizes the project manager as a full-time, dedicated position. The salary structure for the project managers in company B is significantly higher than for their counterparts in company A. The workers in company B expressed concern that "We don't want to be like them," and self-preservation led to value destruction.

Because of the wage and salary problems, company A tried to treat company B as a separate subsidiary. But when the differences became apparent, project managers at company A tried to migrate to company B for better recognition and higher pay. Eventually, the pay scale for project managers in company B became the norm for the integrated organization.

When people are concerned with self-preservation, the short-term impact on the combined value-added project management chain can be severe. Project management employees must have at least the same, if not better, opportunities after acquisition integration as they did prior to the acquisition.

The fourth problem area in Figure 17–3 is the overestimation of capabilities after acquisition integration. Included in this category are:

- Missing technical competencies
- Inability to innovate
- Speed of innovation
- Lack of synergy
- Existence of excessive capability
- Inability to integrate best practices

Project managers and those individuals actively involved in the project management value-added chain rarely participate in preacquisition decision-making. As a result, decisions are made by managers who may be far removed from the project management value-added chain and whose estimates of postacquisition synergy are overly optimistic.

The president of a relatively large company held a news conference announcing that his company was about to acquire another firm. To appease the financial analysts attending the news conference, he meticulously identified the synergies expected from the combined operations and provided a timeline for new products to appear on the marketplace. This announcement did not sit well with the workforce, who knew that the capabilities were overestimated and that the dates were unrealistic. When the product launch dates were missed, the stock price plunged, and blame was placed, erroneously, on the failure of the integrated project management value-added chain.

The fifth problem area in Figure 17–3 is leadership failure during postacquisition integration. Included in this category are:

- Leadership failure in managing change
- Leadership failure in combining methodologies
- Leadership failure in project sponsorship
- Overall leadership failure
- Invisible leadership
- Micromanagement leadership
- Believing that mergers and acquisitions must be accompanied by major restructuring

Managed change works significantly better than unmanaged change. Managed change requires strong leadership, especially with personnel experienced in managing change during acquisitions.

Company A acquires company B. Company B has a reasonably good project management system, but with significant differences from company A. Company A then decides, "We should manage them like us," and nothing should change. Company A then replaces several company B managers with experienced company A managers. This change occurred with little regard for the project management value-added chain in company B. Employees within the chain in company B were receiving calls from different people, most of whom were unknown to them, and were not provided with guidance on whom to contact when problems arose.

As the leadership problem grew, company A kept transferring managers back and forth. This resulted in smothering the project management value-added chain with bureaucracy. As expected, performance was diminished rather than enhanced.

Transferring managers back and forth to enhance vertical interactions is an acceptable practice after an acquisition. However, it should be restricted to the vertical chain of command. In the project management value-added chain, the main communication flow is lateral, not vertical. Adding layers of bureaucracy and replacing experienced chain managers with personnel inexperienced in lateral communications can create severe roadblocks in the performance of the chain.

Any of the problem areas, either individually or in combination with other problem areas, can cause the chain to have diminished performance, such as:

- Poor deliverables
- Inability to maintain schedules
- Lack of faith in the chain
- Poor morale
- Trial by fire for all new personnel
- High employee turnover
- No transfer of project management intellectual property

17.4 LANDLORDS AND TENANTS

Previously, it was shown how important it is to assess the value chain, specifically the project management methodology, during the preacquisition phase. No two companies have the same value chain for project management or the same best practices. Some chains function well; others perform poorly.

For sake of simplicity, the "landlord" will be the acquirer, and the "tenant" will be the firm being acquired. Table 17–2 identifies potential high-level problems with the landlord–tenant relationship as identified in the preacquisition phase. Table 17–3 shows possible postacquisition integration outcomes.

The best scenario occurs when both parties have good methodologies and, most important, are flexible enough to recognize that the other party's methodology may have desirable features. Good integration here can produce a market leadership position.

TABLE 17–2. POTENTIAL PROBLEMS WITH COMBINING METHODOLOGIES BEFORE ACQUISITIONS

Landlord	Tenant
Good methodology	Good methodology
Good methodology	Poor methodology
Poor methodology	Good methodology
Poor methodology	Poor methodology

TABLE 17–3. POSSIBLE INTEGRATION OUTCOMES

Methodology		Possible Results
Landlord	**Tenant**	
Good	Good	Based on flexibility, good synergy is achievable; market leadership is possible at a low cost
Good	Poor	Tenant must recognize weaknesses and be willing to change; possible culture shock
Poor	Good	Landlords must see present and future benefits; strong leadership is essential for quick response
Poor	Poor	Chances of success are limited; good methodology may take years to get

If the landlord's approach is good and the tenant's approach is poor, the landlord may have to force a solution on the tenant. The tenant must be willing to accept criticism, see the light at the end of the tunnel, and make the necessary changes. The changes, and the reasons for the changes, must be articulated carefully to the tenant to avoid culture shock.

Quite often, a company with a poor project management methodology will acquire an organization with a good approach. In such cases, the transfer of project management intellectual property must occur quickly. Unless the landlord recognizes the achievements of the tenant, the tenant's value-added chain can diminish in performance, and there may be a loss of key employees.

The worst-case scenario occurs when neither the landlord nor the tenant has a good project management system. In this case, all systems must be developed anew. This could be a blessing in disguise because there may be no hidden bias by either party.

17.5 SOME BEST PRACTICES WHEN COMPANIES WORK TOGETHER

The team must be willing to create a project management methodology (and multinational project management value-added chain) that would achieve the following goals:

- Combine best practices from all existing project management methodologies and project management value-added chains.
- Create a methodology that encompasses the entire project management value-added chain from suppliers to customers.
- Meet any industry standards, such as those established by the Project Management Institute (PMI) and the International Organization for Standardization (ISO).

- Share best practices among all company organizations.
- Achieve the corporate launch goals of timing, cost, quality, and efficiency.
- Optimize procedures, deliverables, roles, and responsibilities periodically.
- Provide clear and useful documentation.

At one company, the following benefits were found:

- Common terminology across the entire organization
- Unification of all divisions within the company
- Common forms and reports
- Guidelines for less experienced project managers and team members
- Clearer definition of roles and responsibilities
- Reduction in the number of procedures and forms
- No duplication in reporting

The following recommendations can be made:

- *Use a common written system for managing programs.* If new companies are acquired, bring them into the basic system as quickly as is reasonable.
- *Respect all parties.* You cannot force one company to accept another company's systems. There has to be selling, consensus, and modifications.
- *It takes time to allow different corporate cultures to come together.* Pushing too hard will simply alienate people. Steady emphasis and pushing by management are ultimately the best ways to achieve integration of systems and cultures.
- *Sharing management personnel among the merging companies helps bring the systems and people together quickly.*
- *There must be a common "process owner" for the project management system.* A person on the vice-presidential level would be appropriate.

17.6 INTEGRATION RESULTS

The best-prepared plans do not necessarily guarantee success. Reevaluation is always necessary. Evaluating the integrated project management value added after acquisition and integration are completed can be done using the modified Boston Consulting Group Model, shown in Figure 17–4. The two critical parameters are the perceived value to the company and the perceived value to customers.

If the final chain has a low perceived value to both the company and the customers, it can be regarded as a "dog."

Characteristics of a Dog
- There is a lack of internal cooperation, possibly throughout the entire value-added chain.
- The value chain does not interface well with the customers.

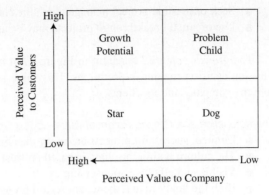

Figure 17–4. Project management system after acquisition.

- The customer has no faith in the company's ability to provide the required deliverables.
- The value-added chain processes are overburdened with excessive conflicts.
- Preacquisition expectations were not achieved, and the business may be shrinking.

Possible Strategies to Use with a Dog

- Downsize, descope, or abandon the project management value-added chain.
- Restructure the company into either a projectized or a departmentalized project management organization.
- Allow the business to shrink and focus on selected projects and clients.
- Accept the position of a market follower rather than a market leader.

The *problem child* quadrant in Figure 17–4 represents a value-added chain that has a high perceived value to the company but is held in low esteem by customers.

Characteristics of a Problem Child

- The customer has some faith in the company's ability to deliver but no faith in the project management value-added chain.
- Incompatible systems may exist within the value-added chain.
- Employees are still skeptical about the capability of the integrated project management value-added chain.
- Projects are completed more on a trial-by-fire basis than through a structured approach.
- Fragmented pockets of project management may still exist in both the landlord and the tenant.

Possible Strategies for a Problem Child Value Chain

- Invest heavily in training and education to obtain a cooperative culture.
- Carefully monitor cross-functional interfacing across the entire chain.

- Seek out visible project management allies in both the landlord and the tenant.
- Use of small breakthrough projects may be appropriate.

The *growth-potential* quadrant in Figure 17–4 has the potential to achieve preacquisition decision-making expectations. This value-added chain is perceived highly by both the company and its clients.

Characteristics of a Growth-Potential Value-Added Chain
- Limited, successful projects are using the chain.
- The culture within the chain is based on trust.
- Visible and effective sponsorship exists.
- Both the landlord and the tenant regard project management as a profession.

Possible Strategies for a Growth-Potential Project Management Value-Added Chain
- Maintain slow growth, leading to larger and more complex projects.
- Invest in methodology enhancements.
- Begin selling complete solutions to customers rather than simply products or services.
- Focus on improving customer relations using the project management value-added chain.

In the final quadrant of Figure 17–4, the value chain is viewed as a star. This has a high perceived value for the company but a low perceived value for the customer. The reason for customers' low perceived value is that you have already convinced them of the ability of your chain to deliver, and your customers now focus on the deliverables rather than the methodology.

Characteristics of a Star Project Management Value-Added Chain
- A highly cooperative culture exists.
- The triple constraints are satisfied.
- Your customers treat you as a partner rather than as a contractor.

Potential Strategies for a Star Value-Added Chain
- Invest heavily in state-of-the-art supportive subsystems for the chain.
- Integrate your PMIS into the customer's information systems.
- Allow for customer input into enhancements for your chain.

17.7 VALUE CHAIN STRATEGIES

At the beginning of this chapter, the focus was on the strategic and financial objectives established during preacquisition decision-making. However, to achieve these objectives, the company must understand its competitive advantage and competitive market after acquisition integration. Four generic strategies for a project management value-added chain are shown in Figure 17–5. The company must address two fundamental questions concerning postacquisition integration:

1. Will the organization now compete on cost or uniqueness of its products and services?
2. Will the postacquisition marketplace be broad or narrow?

The answer to these two questions often dictates the types of projects that are ideal for the value-added chain project management methodology. This is shown in Figure 17–6. Low-risk projects require noncomplex methodologies, whereas high-risk projects require complex methodologies. The complexity of the methodology can have an impact on the time needed for postacquisition integration. The longest integration time occurs when a company wants a project management value-added chain to provide complete solution project management, which includes product and

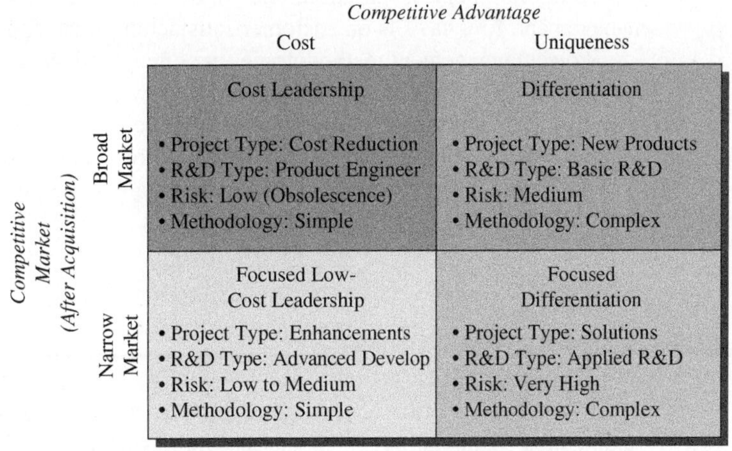

Competitive Advantage

	Cost	Uniqueness
Broad Market	**Cost Leadership** • Project Type: Cost Reduction • R&D Type: Product Engineer • Risk: Low (Obsolescence) • Methodology: Simple	**Differentiation** • Project Type: New Products • R&D Type: Basic R&D • Risk: Medium • Methodology: Complex
Narrow Market	**Focused Low-Cost Leadership** • Project Type: Enhancements • R&D Type: Advanced Develop • Risk: Low to Medium • Methodology: Simple	**Focused Differentiation** • Project Type: Solutions • R&D Type: Applied R&D • Risk: Very High • Methodology: Complex

Competitive Market (After Acquisition)

Figure 17–5. Four generic strategies for project management.

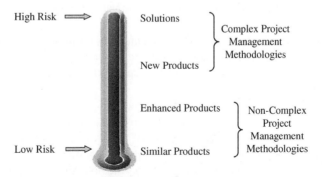

High Risk ⟹   Solutions   } Complex Project Management Methodologies

New Products

Enhanced Products } Non-Complex Project Management Methodologies

Low Risk ⟹   Similar Products

Figure 17–6. Risk spectrum for type of project.

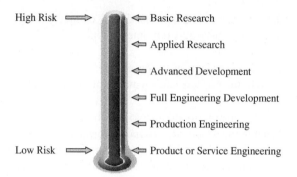

Figure 17–7. Risk spectrum for the types of R&D projects.

service development, installation, and follow-up. It can also include platform project management. Emphasis is on customer satisfaction, trust, and follow-on work.

Project management methodologies are often a reflection of a company's tolerance for risk. As shown in Figure 17–7, companies with a high tolerance for risk develop project management value-added chains capable of handling complex R&D projects and becoming market leaders. At the other end of the spectrum are enhancement projects that focus on maintaining market share and becoming a follower rather than a market leader.

17.8 FAILURE AND RESTRUCTURING

Great expectations often lead to great failures. When integrated project management value-added chains fail, the company has three viable but undesirable alternatives:

1. Downsize the company.
2. Downsize the number of projects and compress the value-added chain.
3. Focus on a selected customer business base.

The short- and long-term outcomes for these alternatives are shown in Figure 17–8. Failure often occurs because the preacquisition decision-making phase was based on illusions rather than facts. Typical illusions include:

- Integrating project management methodologies will automatically reduce or eliminate duplicated steps in the value-added chain.
- Expertise in one part of the project management value-added chain could be directly applicable to upstream or downstream activities in the chain.
- A landlord with a strong methodology as part of its value-added chain could effectively force a change on a tenant with a weaker methodology.

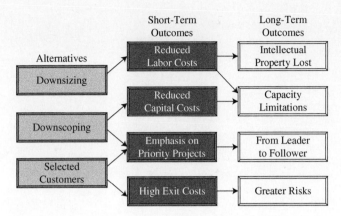

Figure 17–8. Restructuring outcomes.

- The synergy of combined operations can be achieved overnight.
- Postacquisition integration is a guarantee that technology and intellectual property will be transferred.
- Postacquisition integration is a guarantee that all project managers will be equal in authority and decision-making.

Mergers and acquisitions will continue to take place regardless of whether the economy is weak or strong. Hopefully, companies will now pay more attention to post-acquisition integration and recognize the potential benefits.

18 Agile and Scrum

18.0 INTRODUCTION

As project management evolved, new techniques have appeared that are outgrowths of the changes in the project management landscape. Agile and Scrum are two such techniques. Figure 18–1 shows some of the changes that are taking place. Levels 2 and 3, which contain some of the core concepts for agile and Scrum, focus more on the characteristics for growth and maturity of project management whereas Level 1 contains basic principles and focuses on getting the organization to accept and use project management.[1] Because executives initially mistrusted project management and were afraid that project managers might make decisions that were reserved for senior management, we have learned that many of the characteristics of Level 1 were actually detrimental to effective project management implementation and served as significant roadblocks for agile and scrum development. Therefore, the characteristics identified for Level 1 are the reasons why additional levels of maturity were needed, resulting in the development of agile and scrum approaches.

Level 1 (Common Processes)
- Projects are identified, evaluated, and approved without any involvement by project managers.
- Project planning is done by a centralized planning group, which may or may not include the project manager.

---

1. For additional information on Level 1 and Level 2, see Harold Kerzner, *PM 2.0, Leveraging Tools, Distributed Collaboration and Metrics for Project Success* (Hoboken, NJ: Wiley and International Institute for Learning, 2015).

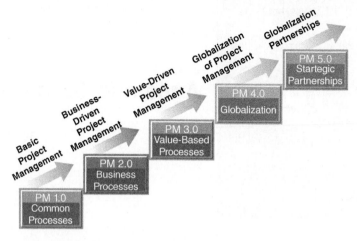

Figure 18–1. Levels of project management maturity.

- Even though the planners may not fully understand the complexities of the project, an assumption is made that the planners can develop the correct baselines and plans, which would remain unchanged for the duration of the project.
- Team members are assigned to the project and expected to perform according to a plan in which they had virtually no input.
- Baselines are established and often approved by senior management without any input from the project team, and again the assumption is made that these baselines will not change over the duration of the project.
- Any deviations from the baselines are seen as variances that need to be corrected to maintain the original plan.
- Project success is defined as meeting the planned baselines; resources and tasks may be continuously realigned to maintain the baselines.
- If scope changes are necessary, there is a tendency to approve only those scope changes where the existing baselines will not change very much.

Level 2 (Business Processes)
- Involvement by the project manager early on in the life cycle.
- Committee sponsorship rather than single-person sponsorship.
- Decentralized project planning.
- A willingness to work with flexible project requirements.
- Using competing constraints rather than the triple constraints.
- A new definition of project success.
- Higher tolerance for scope changes.
- Use of dashboard reporting rather than excessive documentation.
- Heavy usage of virtual project teams.
- High level of customer involvement.
- Capturing lessons learned and best practices.

Level 3 (Value-Based Processes)	● Alignment of projects to strategic business objectives.
	● Traditional life cycle replaced with an investment life cycle.
	● Use of benefits realization planning and value management.

- Going from a methodology to a framework.
- Using metrics that can measure intangible elements.
- Establishing metrics that track assumptions and constraints.
- Having a rapid response to scope changes.
- Use of capacity planning and resource management activities.
- Use of a portfolio project management office (PPMO).

The heaviest users of Levels 1 and 2 appear to be practitioners of agile and scrum. However, because of the importance of these changes, we can expect all forms of project management practices to use these concepts in the future.

18.1 INTRODUCTION TO AGILE DELIVERY[2]

Just like chess, agile is easy to learn but hard to master.

—**Michel Biedermann**

How Do You Answer "Why Agile?" to the CEO?

Here's how you might answer "Why Agile?" to a C-suite executive:

(a) You will reduce the time to first benefits of an awesome idea from months to weeks or even days if necessary.
(b) You will get a higher ROI than funded because you can develop a better product than planned by integrating early and frequent feedback from the business team, end users, and customers in the emerging solution.
(c) You will redefine quality no longer as the absence of defects but instead as to how well your product meets market needs.
(d) Finally, by developing your product using a prioritized backlog, you reduced your risks and ensured that you got the most bang for your buck.

Agile is About Delivering Value Early and Often

Agile delivery is an evolving concept. Unfortunately, over time it has come to mean anything to anyone. We do a daily stand-up meeting? Bingo, we must be Agile! If only. . .

Let's start this introduction to Agile with a very simple definition.

At its essence, Agile is about integrating frequent feedback in the emerging product or service to maximize the value delivered while at the same time reducing time to first benefits. (See Figure 18–2.) Let's dissect this definition.

---

2. Material in Section 18.1 ©2022 by Michel Biedermann, Ph.D. All rights reserved. Reproduced with permission.

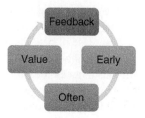

Figure 18–2. The essence of Agile delivery is to integrate feedback in the emerging solution early (a couple of weeks after the start of the project) and often (every couple of weeks) to maximize the value delivered.

- *Feedback.* A key aspect of Agile delivery is to maximize the value delivered by better fulfilling the needs of the customers and end users of the product or service. This is accomplished by iteratively integrating feedback into the emerging solution all along the delivery process. It is best when this feedback is provided by actual customers or end users, but should at least be done by their proxies, such as Product Owners (POs) and the business team.
- *Early.* On a waterfall or plan-driven project methodology the work is completed sequentially, (see Figure 18–3) starting with "requirements" then "analysis" then "design" then "build" then "integrate," then "test" and only then to "deploy." Each phase lasts anywhere from a few weeks to many months. With agile delivery, however, the solution or product emerges within weeks or even days from the start of the project. The team accomplishes this by overlapping most of these phases in parallel (see Figure 18–4). The key is to define just enough of the most important requirements to start the work then deliver the remaining prioritized requirements just in time for the integrated build and test phases to flow smoothly.
- *Often.* The emerging solution is presented to any interested stakeholder, customer, or end user at the end of each iteration or sprint, usually every couple of weeks. The point of this is (a) to validate the health of the project with hands-on testing of the emerging solution and (b) to increase the value of the solution or product delivered by integrating feedback into the requirements and design.

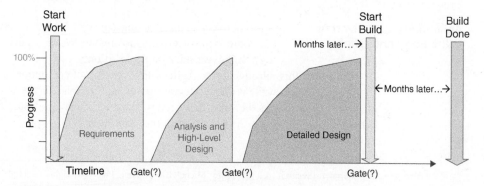

Figure 18–3. Waterfall methodology delays the start of the build phase.

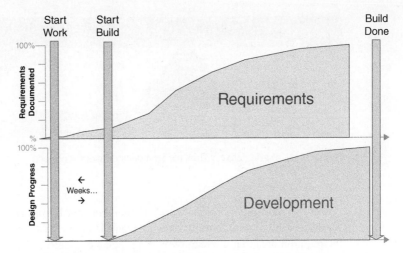

Figure 18–4. Agile delivery overlaps project phases to deliver value sooner.

● *Value.* To maximize early delivery of value, the team must continuously tackle the highest priority requirements first. We define "highest priority" as those requirements which either are (a) the riskiest to the project so that if they cannot be be mitigated, the project will then fail early and thus more cheaply, and (b) the requirements with the highest business value.

It should also be noted that a solution or product only delivers value once it is in production. That is because no end user or consumer would be willing to pay for any of the intermediate work products, such as a fully fleshed-out list of requirements or even the drawings or wireframes of the most beautifully designed product. Only the final product counts.

There you have it! I told you that agile delivery was easy to learn. Now comes the hard part: mastering it. The rest of the content—indeed, the rest of your agile apprenticeship—will focus on the many ways to master the goal of delivering value early and often.

However, before we start this apprenticeship in earnest, we need to understand one more concept, which is the impact of agile on the iron constraints.

Agile Delivery Flips the Iron Constraints

We have all learned about the impact of the three-legged stool or the iron constraints on traditional or waterfall projects: namely, that the scope is fixed by what was funded, while the staffing and schedule needed to deliver that scope are variable.

Agile flips these constraints around (see Figure 18–5). The staffing and schedule are fixed, which makes agile very appealing to management since the budget will not change and, for example, a significant marketing campaign can be reliably scheduled. However, something must clearly give to accomplish this. The variable in the case of agile delivery is the scope of the project.

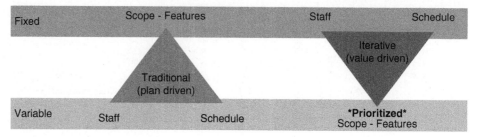

Figure 18–5. Agile delivery flips the iron constraints of a project.

How could this possibly work? Remember that by prioritizing the requirements, we deliver the most important ones up front. If the project runs out of money or time, the requirements that the team would not be able to deliver are the least valuable ones. In other words, life is unsure; eat the dessert first.

Agile Manifesto

Before we dig into scrum, probably the best-known agile delivery framework, let's spend a few minutes reviewing the source of agile, the "Manifesto for Agile Software Development" (see agilemanifesto.org).

Four Values of the Agile Manifesto

The Agile Manifesto starts out with four distinct values:

Individuals and interactions	Over	Processes and tools
Working software	Over	Comprehensive documentation
Customer collaboration	Over	Contract negotiation
Responding to change	Over	Following a plan

Make sure you understand that, as the authors of the manifesto point out, it is not that there is no value in the items on the right; it is just that agile prefers the items on the left. Let's look at them individually.

Individuals and Interactions over Processes and Tools

Agile delivery is not about blindly following frameworks or processes. Though some amount of discipline as represented by a process or framework such as, say, scum is useful, none of these can account for the team failing to own its work and collaborating intensely to solve the problem together. Indeed, the concept that "none of us is as good as all of us" applies very much to agile teams. You will very quickly learn that agile delivery is all about the team, or if you prefer, the "T"eam.

Similarly, agile delivery is not about the tools, since they often get in the way. Choose your preferred project management tool and chances are it requires you to think of the project and manage it in the specific way the tool developers envisioned it.

The takeaway here is that no process can overcome a lack of collaboration between individuals on the team. No process alone can make the team high performing. Once the team members collaborate, then a process helps further accelerate the team's productivity.

Working Software Over Comprehensive Documentation

When thinking about this value, ask yourself the following questions:

1. As a project sponsor, what helps you better understand the health of a project: (a) reading the documentation created by the team or (b) testing the emerging solution yourself on a regular basis, say, after each two-week iteration or sprint?
2. As the end user or consumer of a product, would you pay for documentation beyond a user manual?

We have seen too many teams take this position on the cost of documentation as an excuse for not documenting anything at all. It is not that documentation is not necessary. It is. It is just that it should be limited to what is most valuable. Given this, what documentation is important to agilists? At a minimum, we suggest the following:

1. Product backlog (as a living document, surviving from release to release).
2. RAID (i.e. risks, actions, issues, and decisions) log (as a living document).
3. User manual, especially if the product is not intuitive enough to use without help.
4. Directions on how to promote the solution into production, especially if this process is not fully automated.
5. Directions necessary to support the product once it is in production.

Customer Collaboration over Contract Negotiation

You may be getting the impression that agile delivery requires much more intense collaboration than traditional or waterfall projects. For example, streamlining the documentation must be replaced or at least supplemented by closer collaboration between customer and vendor(s). Instead of documenting, analyzing, and designing all requirements ahead of schedule, even incidentally, those that might end up not being done, the knowledge necessary emerges just in time for its consumption through collaboration.

1. Contract negotiation delays the start of the project and thus the time to first benefit the customer and vendor(s), but especially for the most important entity of all, the consumer or end user of the product.
2. Project delays and surprises have created a culture where both the customer and vendor(s) seek to shift the risks to the other(s). This type of negotiation makes it very difficult and time-consuming to find a win–win situation for all involved. All parties will try to game the system; it is human nature.
3. The team usually cannot be shielded from the stress and the demands of contract negotiations, so its productivity suffers, at least temporarily. Examples include

having to suddenly shift gears to analyze the new requirements to estimate the effort necessary or putting up with the uncertainty of whether team members will be extended on the project.

The net effect of this is that the relationship between the customer and its vendor(s) changes on agile projects from one primarily driven by contracts to one of a tight partnership. What kind of journey are all parties about to embark on if the journey does not start with a win–win for all?

NOTE: Some of you are bound to ask, at this point, how to structure an agile contract. Though this may be much too soon for you to understand all the subtleties of the answer, the takeaway is that we prefer to structure an agile contract with the following phases:

1. [Optional] *Visioning.* This phase is meant to align the strategic business objectives with increasingly refined requirements. It culminates with the definition of the top two or three most important requirements needed to implement the strategic business objectives. More requirements will emerge over time once the work is started.

   Contract type: *Fixed price* because this phase relies on fixed capacity and fixed duration (two to three weeks).

2. *Work kickoff.* The team is assembled and starts working on the single most important requirement until its delivery velocity reaches steady state. This traditionally takes about three to four iterations or sprints. We define velocity as the number of story points accepted by the customer per sprint. Story points are the agile currency for estimating the work.

   Contract type: *Time and Material* because, though this phase relies on fixed capacity (i.e. the team), the duration needed for the team to reach velocity steady state may vary. Depending on setup time, the skills of the team, and the complexity of the business domain, that duration should last about two to four months.

3. *Steady state.* Now that the velocity of the team is known, the duration of the project can be estimated using the size of the product backlog. The currency used to estimate work on agile teams is story points. We will describe story points and how to estimate work in Section 18.2. Suffice it to say that if the sum of the story points for the list of requirements is, say, 200, and the team delivers 20 points on average at a steady state per iteration, then it will take the team 10 iterations to complete the work. This steady-state phase can be done for a fixed price providing the team size, composition, and business domain of the work remain the same. Note that you should not compare velocity between teams lest they use it to game the system. In other words, you cannot say that a team delivering 20 story points per iteration is half as productive as one delivering 40 points. More on this in Section 18.2.

Responding to Change Over Following a Plan

In most projects, if not all of them, the customer's needs evolve as it experiences the emerging solution. This triggers new ideas, new requirements, or even new directions in which to pivot the product. Integrating this feedback so that the final product better meets the demands of the end users is not only acceptable, it is, in fact, very desirable. That is, if an agile delivery project has a requirements documentation phase per se, it is very short, maybe a couple of weeks at most, to let the strategic business objectives guide the identification of the one or two most important requirements that will implement them. Actually, I should have said that the requirements documentation phase extends almost for the entire duration of the project. The intent here is that requirements continuously emerge just in time for consumption by the development team or possibly two to three weeks earlier than that so that the designers and architects can prepare their guidance for the development team.

This value is often the hardest for traditional vendors and customers to embrace. Too many contracts are written in such a way that the project scope must be fully fleshed out and all the effort estimates cemented before they are signed. The irony is that neither the customer nor the vendor(s) knows enough at the beginning of the project to anticipate where the feedback is going to take the product and how much effort truly will be needed. This is reflected in Figure 18–6, the Cone of Uncertainty.

Figure 18–6 shows the amount of information that both the customer and vendor do not know over the duration of a project. The phases documented on the horizontal axis map to typical waterfall or plan-driven delivery. The insidiousness of wanting to fix the project scope and its effort estimates prior to the start of the project is that both sides of the contract divide do not know what they do not know. The customer team only has its best guess of what the requirements should be. In turn, the vendor team

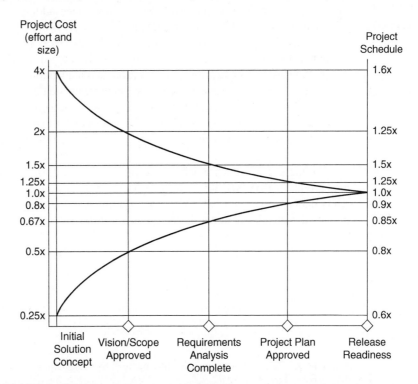

Figure 18–6. Cone of Uncertainty demonstrating what is not known about a project over its various waterfall phases.

tries to estimate what effort will be needed to implement those uncertain requirements. Ironically, both teams will make great effort and spend much time trying to narrow that cone of uncertainty when in fact the best way to do this is to start the work. Thus, agile teams prefer to start working early in the project but respond to change rather than spending time guessing a plan and, more important, dates, only to have them be wrong.

Let's examine why some would say that having a detailed plan is better than knowing just enough to start working. Clearly, there is value in having a plan—a simple, even simplistic one. The key is to not have analysis paralysis, delaying the work while seeking to create the perfect plan. Instead, the business and technical teams should embrace these uncertainties and instead agree on a partnership with the following terms:

1. The business team is welcome to change and reprioritize requirements, provided the technical team continues to have the skills to implement them.
2. If the business domain changes, then the technical team may have to make changes to the skills of the team. This may require a project change request if the seniority or complexity of the skills must change.
3. As the project progresses, but certainly from the midpoint onward, the estimate of number of iterations needed to complete known work can be based on solid data. Thus, the management team should have an increasing comfort about the project duration as it progresses. This will be especially true if the work completed during each iteration meets the definition of done (DoD) to be promoted into production. More on this later.
4. Given the importance of team velocity and since adding individuals to a team changes its velocity, adding resources to a project is done by adding entire teams instead. This can be done relatively painlessly since agile teams are relatively small. Indeed, agile teams should only have between five and nine members. This maximizes communication, collaboration, and innovations. Smaller teams may not have the collective life experiences to generate the quality of ideas to innovate and solve problems that larger ones can. Conversely, since the number of channels of communication increases exponentially, teams larger than about nine members lose much of their effectiveness to communicate. Figure 18–7 illustrates this growth:

For these reasons, agile delivery is the new standard for managing the creation of new products. However, for many people, these values are too broad and difficult to really understand. That is why the creators of the Agile Manifesto then went on to document 12 principles to better define the four values.

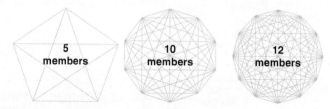

Figure 18–7. Channels of communication in a team increase exponentially based on the formula Team size × (Team size − 1)/2.

12 Principles of Agile Manifesto

The problem with the four values of the Agile Manifesto is that they are too broad, and too abstract for many people to apply effectively. That is why the authors of the manifesto followed up those values with 12 principles to better frame the agile conversation.

1. Early and Continuous Delivery of Valuable Software

The first principle reads: "Our highest priority is to satisfy the customer through early and continuous delivery of valuable software." In other words, not the requirements, the architecture, or the design are as valuable as working software. Think about it this way: As the consumer of product you are about to buy, would you pay just to get a compelling architecture document or awesome wireframes? The answer is no. Do not get me wrong, these work products have some value, but only when they are reflected in working software.

Given the influence of feedback on the quality of the final product, the business team wants to observe, indeed experience, the emerging product as early as possible in the project. By the way, as a project manager, sponsor, or member of the management team, no finely crafted status report gives you as good understanding of the health of the project as experiencing yourself, hands on, the progress of the emerging product.

That is why agile delivery (a) pushes very hard to start the build phase of product as early as possible, usually, a couple of weeks to a month after the start of a project, and (b) demonstrates the emerging product to as broad an audience as possible every few weeks.

2. Welcome Changing Requirements, Even Late

The second principle reads: "Welcome changing requirements, even late in development. Agile processes harness change for the customer's competitive advantage." We already mentioned that integrating continuous feedback into the emerging solution will create a final product that is more valuable than initially planned. So, given the chance to deliver a greater ROI than was originally calculated in the business case and funded, why would not you grab it? This means dealing with uncertainty throughout the project, but this uncertainty is healthy given the better outcome. In other words, embrace uncertainty and be flexible (or dare we say agile) with it.

3. Deliver Working Software Frequently

The third principle reads: "Deliver working software frequently, from a couple of weeks to a couple of months, with a preference to the shorter timescale." Let's see why the manifesto authors prefer a shorter time scale:

1. Given the importance of gathering and integrating feedback into the emerging product, why not often?
2. Shortening the duration between demonstrations instills a sense of urgency to the team. In other words, it reduces procrastination and gives the team permission to say no to extraneous requests. Mind you, it is not as if this no is forever; if the request is legitimate, it will be prioritized for the next iteration, which will start at most in two weeks.

3. Shorter iterations expose the inefficiencies and impediments slowing down the flow of the work. For example, there will be a much bigger sense of urgency to resolve a blocker if the time-boxed iteration lasts two weeks rather than two months. In other words, a two-day delay represents 20 percent loss of productivity in a two-week or 10-day iteration. The same delay will only be 5 percent in a two-month or 40-day iteration. Again, bigger sense of urgency. . .

4. Another typical work inefficiency that is better exposed with shorter iterations is the tendency of new agile teams to "waterfall their iterations." Teams should focus on completing first the iteration's single most important requirements, then focus on the next most important requirement. This ensures that maximum value is generated during the iteration. Teams waterfalling their iterations will instead focus on understanding all the iterations' requirements first, then analyzing them all, then designing them, building them, and finally testing them all at once (see Figure 18–8). If the team is blocked or slowed down during say, the design phase of the iteration, it risks not being able to complete any of the work and thus would deliver no value for this iteration. In other words, completed development work should reach the testing team for validate after early in the iteration, ideally after a couple of days, not late in the iteration (see Figure 18–9).

5. Since continuous improvement is one of the principal tenets of agile delivery, we prefer to reflect on how to improve more often than less.

It is for these reasons that an iteration duration of two weeks has emerged as a best practice.

4. Business and Technical Teams Must Collaborate Daily

The fourth principle reads: "Business people and developers must work together daily throughout the project." It used to be that the business team spent weeks documenting the requirements, then threw them over the wall before having to wait months to see the final product, only to likely be disappointed. One way to prevent this is for both

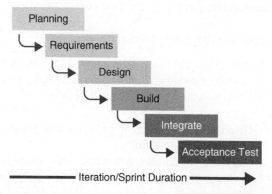

Figure 18–8. Example of a waterfalled iteration where work is done by sequential phases. If time runs out during an iteration, it is likely that little value or working software will be delivered.

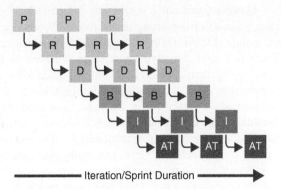

Iteration/Sprint Duration

Figure 18–9. Requirements completed per priority in typical agile delivery fashion. If time runs out during an iteration, the work on the most important requirements will at least have been completed, thus delivering value in the form of working software.

business and technical teams to continuously work together throughout the project. Here are some of the typical tasks that happen during an agile delivery project:

1. The business team documents just enough of the one or two most important requirements for the technical team to go to work. "Just enough" traditionally means two to four sentences describing the desired outcome and the value to the end user. Traditionally, this should fit on an index card (4×5 in.; 10×13 cm). This card represents a starting point from which the technical team can ask questions as they build the solution. Rather than documenting on paper a ton of details, these details emerge just in time, during conversations as the technical team needs the answers. This allows both technical and business teams to clarify each other's assumptions in real time.

2. By collaborating daily, the business team can answer the dozens of questions that undoubtedly emerge from the technical team, regardless of whether requirements were written on an index card or a 200-page document. Without this ongoing dialog, it would be up to the technical team to make these dozens of microdecisions every day. "Should the button go here or there?" "Should the trigger be this or that?" "What should happen if the answer is 'x' and not 'y'?" It is easy to see how having the business team answer these questions as they emerge throughout the day would lead to a solution that better meets the team's vision and needs.

3. The business team refines and reprioritizes the requirements once or twice an iteration. Requirements are prioritized by business value and by risk to the project. The riskiest requirements should be tackled as early as possible in the project so that if they must doom the project, the failure can happen as quickly and therefore as cheaply as possible (i.e. fail early, fail cheap).

Ideally, the most important requirements should always be ready to be consumed by the technical team for the next two to four iterations. This will give more time to the designers (e.g., architects, web designer teams team) and infrastructure engineers to get a jump on completing their work in time for the development and testing team to tackle theirs.

However, without a doubt, this principle has become the hardest one to implement because of the strong desire to reduce product development costs by offshoring the technical team while the business team stays located at the company headquarters. There is a common saying that when it comes to the business and technical teams working together, latitude hurts but longitude kills. Since the most effective form of communication is face-to-face, it is one thing to separate the two teams merely by distance while they work in the same time zone. At least this way, always-on phone lines and video links enable the quick microdecisions from the business team when the technical team needs help or clarifications. The cumulative effect of faulty assumptions, usually by the technical team, can really sidetrack an effective development effort. It is another thing altogether when teams are separated by time zones. The imposition of having your work schedule shifted by as much as 12 hours makes it very difficult to maintain team motivation over time, let alone ensure that only the best resources stay on the project—not just the best available ones. However, given these real-world project constraints, here are some prioritized options:

1. Overlap the two teams, or at least as many members of them as possible, by as many work hours as feasible. A reasonable minimum would be at least two hours. During this overlap, focus primarily on answering the microdecisions and testing the emerging stories.
2. Assign a proxy to the business team who is colocated with the technical team and who is empowered to be the voice of the business and the customer. This proxy is usually a business analyst with a deep understanding of the team's business domain. An additional risk to having a business proxy is that the business team but especially the PO (more on this role in Section 18.2 on scrum) risks abdicating in part or in whole the decisions to the proxy. This is suboptimal for many reasons.
3. Colocate a reduced set of technical resources with the business team. Their responsibility is to be the communication bridge with the offshore resources. The disadvantage of this option is that microdecisions will be delayed by at least one day and are at the mercy of clear and accurate communications between the two technical teams.
4. Document the requirements in as many details as necessary for the technical team to minimize assumptions and request microdecisions. There are at least four disadvantages to this option:
   a. Documenting the requirements to the level necessary will likely become the bottleneck in the development effort, especially if changes or, worse yet, entire pivots are needed since it is not possible to compensate for these through big design up front.
   b. Too many teams have paralysis by analysis when documenting requirements ahead of time and in writing, because they try to anticipate answers to all microdecisions.
   c. For most offshore teams, English is at best a second language, so finding someone who can understand its nuances places another bottleneck on the technical team
   d. Written language is not a very effective form of communication, especially when trying to address complex problems.

5. Team Up Motivated Individuals and Trust and Support Them

The fifth principle reads: "Build projects around motivated individuals. Give them the environment and support they need, and trust them to get the job done." Ideally, the goals of the organization, the project, and the individuals are all aligned. This alignment increases the engagement of the individuals and their ownership of their work. In turn, this helps facilitate the trust and support of the management team. The bottom line is that this principle is ultimately about culture, the hardest part of an agile transformation.

6. Most Effective Way of Communicating with the Development Team Is Face to Face

The sixth principle reads: "The most efficient and effective method of conveying information to and within a development team is face-to-face conversation." It is from this principle that the concept of colocating the business and technical teams evolved. See Principle #4 for pitfalls of not enabling face-to-face communications, especially by offshoring the technical team and potential options to address this problem.

7. Working Software Is the Primary Measure of Progress

The seventh principle reads: "Working software is the primary measure of progress." This principle should go without saying. However, it is amazing the number of managers and leaders who still insist on understanding the health of a project or program via status reports. These reports end up taking the teams many hours, if not days, to carefully craft so as not to unintentionally expose uncomfortable truths. Indeed, it would be much better and more effective for all involved if the managers observed firsthand the sprint or iteration demos. In about a two-hour meeting, every other week or so, they would understand the context behind requests, risks, issues, and blockers—in other words, the true health of the project. It is tough to obfuscate transparency and lack courage when the primary metric used is observing working software.

8. Agile Processes Should Be Able to Maintain a Constant Pace Indefinitely

The eighth principle reads: "Agile processes promote sustainable development. The sponsors, developers, and users should be able to maintain a constant pace indefinitely." Promoting a sustainable development pace and effort makes life easier for all involved. This is accomplished by the following:

1. Dividing the work into sprints or iterations, each of about two weeks in duration, so that all resources, not just the technical team, are under a constant but low-level pressure to get things done, mitigate risks, and address issues before they become blockers, and so on. Losing one day of productivity due to a blocker does not mean much on a plan-driven project but means an instant 10 percent drop in productivity when the team works in 2-week or 10-day iterations.
2. By making sure that each story meets the DoD and is therefore ready to be promoted into production before it is accepted by the business team or the PO at the end of

the iteration, the productivity or velocity (more on this in Section 18.2) of the team becomes predictable. A steady-state velocity becomes an invaluable planning tool. In other words, meeting the DoD eliminates the need for other project phases like Functional Testing, Integration, User Acceptance Testing, and so on. The problem with having these separate and distinct phases is that, as discussed in the section on the "Responding to Change over Following a Plan" value, we cannot accurately anticipate the duration of each of these phases. Without that, it is not possible to effectively manage the expectations of the stakeholders in general and of leadership.

9. Continuous Technical Excellence and Good Design Enhance Agility

The ninth agile principle is: "Continuous attention to technical excellence and good design enhances agility." The corollary to this principle is that technical debt is an agile antipattern. Technical debt is the price we must pay today to correct technical decisions made in the past. A typical example might be that many companies decided a few years ago that they would standardize on a given browser version rather than keeping their web environment up to date. The advent of mobile applications require that these companies first correct this technical debt by upgrading their browsers and their underlying applications to modern versions. Correcting this technical debt often costs millions of dollars and months to implement right at a time when everyone is clamoring for mobile apps.

10. Maximize the Work Not Done

The tenth agile principle is: "Simplicity—the art of maximizing the amount of work not done—is essential." One typical example of this principle is that preparing more of the list of requirements than two to four iterations ahead of the development effort may not minimize the work not done. That is because the feedback received during the sprint or iteration demos will likely shape the rest of the requirements, so much of this additional work may be for nothing.

Another way of looking at this is that ineffective agilists use this principle to paraphrase the first half of Einstein's quote: "Everything should be made as simple as possible, but not simpler." In other words, this principle is worded in such an unusual way to better emphasize that the point is to be deliberate in selecting the work that does not need to be done rather than merely minimizing the work done. As such, this principle is often abused in the following ways:

1. Teams lean on this principle to avoid meeting a minimum amount of standard work. For example, too many teams combine this principle with value of "working software over comprehensive documentation" as an excuse to not document anything. A minimum of documentation is usually warranted, even or maybe, especially on agile projects. It just does not take the form of requirements. Some documents that come to mind include help files, user guides, code promotion direction if this process is not automated, and others.

2. Another way this principle is used is to prevent goldplating, whether of the design or the build and testing. Too many teams design for the what-if scenario. What if we are asked for this scenario or that scenario? Easy enough! Let us design and implement a solution that can accommodate all three, what we are asked for, and the two alternate scenarios. This way, we are covered. Besides designing and coding for options that likely would not see the light of day, especially if the requirements change during the project, the business team may discover a substantial pivot that will make the result so much better. Besides having spent time developing an unnecessary solution, that work will likely have to be undone to meet the new requirements.

It is for these reasons at least that the team should consciously look for the simplest solution that meets the stated requirements. Things may change in the future, but let that need dictate the minimum refactoring needed.

11. The Best Architecture, Requirements, and Designs Emerge from Self-organizing Teams

The eleventh agile principle reads: "The best architectures, requirements, and designs emerge from self-organizing teams." The concept of self-organizing teams is key here. Over time, this concept has been clarified: the teams should be both empowered and cross-functional. The teams are empowered because they are closest and best equipped to quickly make the needed decisions, and they are cross-functional because all skills needed to design, build, and test stories should be included on the team. This minimizes dependencies on other teams and maximizes the chance that a story or requirement will be completed, integrated, and tested end to end before the end of the sprint or iteration. Empowered and cross-functional teams are best equipped to ensure that each story meets the DoD. It should be noted that architecting or designing solutions in a world where requirements often emerge just in time is a difficult problem to solve. This is especially true as the systems become more and more complex. However, taking the story-level view of the design simplifies things, though this is done at the expense of a systemic and systematic view of the entire emerging solution. To address this problem, concepts like tightly aligned but loosely coupled systems or emergent versus intentional design, agile modeling with its automated testing, continuous integration, and continuous delivery, can help.

12. Team Regularly Reflects on Improving and Makes Necessary Adjustments

The twelfth agile principle reads: "At regular intervals, the team reflects on how to become more effective, then tunes and adjusts its behavior accordingly." One of the key pillars of agile is that of continuous improvement. It implies that the teams know best, though possibly with the help of an agile coach, how to work better and faster. As such, following each iteration or sprint, the team and project stakeholders should gather and have the courage to objectively look at their latest slice of work. You will see much more about this in the sprint retrospective in Section 18.2.

18.2 INTRODUCTION TO SCRUM[3]

Overview of Scrum   Scrum is probably the best-known agile framework. Though it started out being used for software development, many of its principles are now applied well beyond IT, to more general purposes, such as developing products or services.

At its core, scrum is a framework to collaboratively develop and deliver products of the highest possible value. The following six aspects best describe it (see Figure 18–10):

1. *Lightweight.* Scrum is defined by three roles, four events, and five artifacts. That is all. We will look at this in detail below.
2. *Easy to understand.* Due to its simplicity, scrum is very easy to explain and learn. As a matter of fact, the "Scrum Bible," which can be found at www.scrumguides.org, has only 17 very easy-to-read pages.
3. *Difficult to master.* As mentioned, scrum is deceptively easy to learn. Mastering it requires a long time. We will see some of the reasons why below.
4. *Transparency.* Scrum, just like agile, aims for transparency. One of the key concepts that everyone must understand and agree to is the Definition of Done, or DoD. These are the criteria by which the work produced by the team will be evaluated for completion before being accepted as done. A side benefit of this is that agilists are pushing for transparency throughout projects, including in the processes the team will follow; the daily progress report; its estimates; its metrics; its velocity (i.e. the speed at which the team completes its work; (more on this below); and the risks, issues, and especially blockers. Being this transparent requires courage—courage, for example, to share the bad news early since not all managers follow the adage that "good news can wait, bad news can't." Courage also to announce what you are committing to

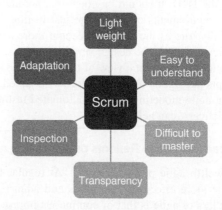

Figure 18–10.   The six components that are the essence of scrum.

delivering in the next sprint; courage, therefore, to be held accountable every couple of weeks for what you delivered. Another form of courage includes seeking frequent feedback on the work produced. These forms of transparency-driven courage are just some examples of why scrum is difficult to master.

5. *Inspection.* At the end of each sprint, the team demonstrates to anyone interested in the work it accomplished. In doing so, it seeks to gather the feedback necessary to make the result better and more valuable than originally planned. Also, the team and stakeholders regularly take a critical look at the people, processes, and tools with an eye toward continuous improvement.

6. *Adaptation.* Based on the inspections just described, the team and stakeholders can freely decide to make the changes needed to increase the quality and value of the product being created and to increase the team's productivity.

Let us now dive deeper into the three roles, four events or ceremonies, and five artifacts or work products that compose scrum.

The Three Roles in Scrum

Scrum projects are characterized by just three roles. They are:

1. Team member
2. Scrum master
3. Product owner

Team Members

Team Members Deliver Value

The team members are only the developers and testers. They are generally the ones delivering value on the project. Though the project may also have additional temporary or part-time help from architects, designers, and infrastructure engineers as needed, agilists consider that it is the former that delivers the bulk of the value.

The team generally constrains the work to be delivered. It is therefore imperative that every effort be made to protect its productivity. For example, meetings involving the team should be minimized both in number and duration.

Cross-Functional and Self-organized

The most effective scrum teams are cross-functional and self-organized. They are cross-functional because the team should have all the skills needed to complete the required work, end to end. For example, a software development team might need to have front-end or user interface (UI) skills, orchestration layer or APIs, and back-end or database skills. Such a team would be able to complete the development, end-to-end integration, and testing of its stories in one sprint. In other words, such a team would make sure each of its stories meets the DoD.

Another advantage to being cross-functional is that this allows the team to minimize the need for part-time specialists but also to cover for when a team member is unavailable, since between them, team members can cover all roles, although maybe not with the same productivity.

It should be noted that many companies prefer a separate team for the back-end or database development, for example. Furthermore, these ancillary teams also often work in a plan-driven or waterfall fashion. This is suboptimal for agile projects since (1) it creates unnecessary dependencies, (2) usually requires project management overhead to work out the rippling side effect of aligning these teams, and (3) likely prevents a story from being completed end-to-end and therefore being deployable in a single sprint.

Besides being cross-functional, scrum teams are also self-organizing. Team members decide how best to tackle the work they committed to delivering in each sprint. For them, the work is less about "I finished my job, now I'm out of here" and more about "I finished my job, now who needs help with theirs so that we can finish in time?" Besides establishing an *esprit de corps* or a team bond, this approach has the additional advantage that everyone gets a chance to learn different roles, technologies, and programming languages and thus grow professionally. A secondary benefit is that this approach further grows the cross-functionality of the team. Generally, this also increases employee satisfaction as team members actively participate in something bigger than themselves.

Team Members Own Estimating the Work

If a team is going to commit to completing in one sprint the work it selected, it must also own estimating that work. This is imperative for two reasons:

1. Estimating the effort needed to complete work engenders a buy-in to that commitment. In other words, how strong would your commitment be to complete a difficult task if someone far more skilled than you decided how long it should take you?
2. The size of estimate varies depending on the experience and skills of those doing the estimating. It may take a junior developer three days to complete a task that would take a senior developer only half a day to complete. How fair or sustainable would it be for the junior developer to continuously struggle to meet estimates decided by a senior developer?

Scrum Master

Servant Leader Helping the Team Deliver High Value

The primary purpose of the scrum master is to be the servant leader of the team. A servant leader is not a traditional manager. They is a servant first and a leader only second. As such, they asks what the team needs to be successful and deliver the most value. It is up to the team to answer that question. The advantage of a servant leader is that they can build trust with the team more rapidly than most managers can.

Addresses Impediments to the Team's Productivity

Effective scrum masters quickly address the impediments slowing the team's productivity. This requires a broad network of people able to help and the experience of getting things done, even in an unconventional way. In addition, managing upward effectively helps address the more difficult organizational blockers.

Ensures that Scrum and Agile Principles Are Followed

The scrum master is the team's primary guide in matters of agile and scrum. During projects, teams will run into problems. Their response to these issues will define how successful their agile transformation will be. Teams with weak scrum masters will fall back on addressing these issues using their old habits of command and control or plan-driven and waterfall methods. It is the responsibility of the scrum master to ensure that the team does not slip backward during these difficult times.

Uses Agile Toolbelt to Increase the Team's Velocity

An effective scrum master draws on much agile and Scrum experience to increase a team's velocity and its effectiveness at delivering better products. Agilists are said to have a large agile toolbelt from which they can draw many tips, tricks, and experiences that go much beyond Scrum and touch many agile frameworks, such as "eXtreme" Programming (XP), disciplined agile (DA), Scrum-of-Scrums (SoS), and even topics like scaling agile using Large Scale Scrum (LeSS), Scaled Agile Framework (SAFe), and Nexus. The best scrum masters even extend their toolbelt to areas such as lean (Green, Black, or Master Black Belt), Six Sigma, Design Thinking, Lean Startup, and DevOps.

Product Owner

Prioritizes Work to Best Achieve Goals and Vision

Whereas the primary purpose of the scrum master is to maximize the team productivity (how it does the work), that of the PO is to ensure that the team delivers the highest possible business value (what it works on). Therefore, the PO needs to understand extremely well the company's strategic goals and vision and translate them into effective tactical requirements or stories.

Responsible for Meeting Business Objectives

To guide the development of the best possible products, applications, or services, the PO should be a very effective voice of the customer or consumer (VoC). They must know intimately the needs, desires, and wishes of the target audience. For this, it is best if the PO comes from the business side of the company.

It is disappointing when companies abdicate the responsibility of the PO to a third-party vendor. This is a sign of weakness from the business leadership of the company. If the company refuses to be accountable for the quality of the products and the experience it will create and sell, why should consumers buy from it?

Ensures Visibility and Transparency of the Backlog

POs often must compete for resources against other teams or programs. One of the best ways to do this is to plan on delivering a compelling business value. This is described by the contents of the product backlog—the list of all the work the team needs to complete. By offering visibility and full transparency of the backlog, the PO shows that they welcomes feedback to deliver an even better product to the consumer than they could have done alone.

Another advantage of a transparent backlog is that the progress to date and planned next steps are easy for all to see. As seen earlier, transparency is one of the key tenets of agile.

Optimizes the Value Delivered by the Team

Ultimately, the success of the PO will be measured by the business value delivered by the team. The idea is that the team should deliver the most valuable features, the "big rocks" first, leaving the least valuable ones for the end of the projects. This way, should funding or time run out, the big rocks were delivered and only grains of sand were missed. This approach should make accepting a variable scope palatable to even the strictest "scope hawks." In other words, a prioritized but variable scope ensures that the most bang for the buck is delivered.

One of the benefits attributed to agile delivery is that of "fail early, fail cheap." What this means is that if the team is doomed to fail because the project risks could not be mitigated, at least let it fail as early as possible so that the lessons learned from this experiment can be gained as cheaply as possible. The way to accomplish this is to prioritize the product backlog not only by business value but also by risk, especially at the beginning of the project. This accomplishes the following:

1. It gives more time to the team, indeed the company, to find a way to mitigate its project risks.
2. If the risks cannot be mitigated, the company might as well learn this as early as possible.

Team Size? Can You Feed It with Two Large Pizzas?

Finally, a word on team size. Scrum teams are purposefully relatively small, between five and nine team members (not counting the PO and the scrum master). As we saw in the agile introduction, the reasons for this are:

1. We prefer at least five team members because in smaller teams the number and quality of ideas and breadth of skills suffer.
2. We prefer not to exceed nine team members to maximize communication effectiveness since the number of channels of communication increases exponentially.

The bottom line is that agilists often say that the right team size is one that can be fed by two large pizzas. Make it anchovies and *bon appétit*!

The Four Events in Scrum

In Scrum, the work is delivered in sprints. Sprints, or iterations, are time-boxed between one to four weeks with two weeks being the emerging best practice. Shorter sprints have the advantage of:

1. Creating a sense of urgency, not just among the team, but also from the supporting cast. For example, in a two-week sprint, a blocker lasting two days because the assignee did not jump on resolving it quickly enough means a 20 percent loss of team productivity, a delay that is hard to overcome.

2. Shorter sprints better expose the weaknesses and inefficiencies of the company's work process. For example, we have seen too many teams using three of the four weeks for development and only the last week for testing. This is called waterfalling your sprints. This process is harder to do with shorter sprints. Instead, the team should complete the development of the most important story first so that it can be tested within the first couple of days of the sprint, not with two days left.

3. Increasing the opportunities for feedback and retrospectives. With two-week sprints, you get twice the opportunities to gain valuable feedback and better your process than with four-week sprints.

4. Making it easier to say no when someone asks the team to insert a forgotten or emergency high-priority item in the sprint since the person will have to wait only at most two weeks for the work to be tackled, assuming it is of high enough priority to jump to the top of the product backlog. Try telling a manager that he will have to wait an entire month when sprints last four weeks.

You may then ask: Why not select sprints lasting only one week? This is a very valid question, but before you try to optimize your work processes this much, try to see how well two-week sprints work for you.

Regardless of sprint duration, scrum has fundamentally four events, also called ceremonies (see Figure 18–11). The entire team, plus the PO and scrum master, participate in the following:

1. Sprint planning
2. Daily Scrum
3. Sprint review or demonstration
4. Sprint retrospective

Some agilists include "Product Backlog Refinement" in this list.

Sprint Planning

The team, PO, and scrum master gather the first morning of the sprint to plan it. This ceremony should last about one to two hour(s) per week of sprint duration. So, the

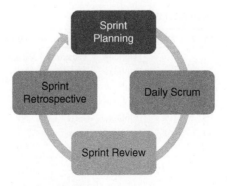

Figure 18–11. The four scrum ceremonies.

planning for a two-week sprint should last about two to four hours. Obviously, this duration is dependent on the agile maturity of the team, the complexity of the stories, and the team's familiarity with them.

The team starts by understanding its bandwidth. Table 18–1 is an example.

This is where things can get complicated very quickly. I have seen some teams calculate the number of available work hours, then calculate ideal time versus work time in the hope of getting very accurate estimates. These calculations can quickly consume the better part of the planning meeting while often giving a false sense of accuracy. Instead, I recommend a much simpler way.

Use Team Velocity as a Planning Tool for the Sprint

Team velocity is defined as the average number of story points the team has delivered per sprint lately. A story point is a measure of the complexity of a story. As described in detail in the introduction to agile, the complexity of a story is assigned a number approximately following the Fibonacci sequence: 1, 2, 3, 5, 8, 13, 20, 40, and 100. A story worth 8 story points is four times as complex as one worth 2 points. Team members can quickly get a gut feeling on the complexity of a story by reading it and asking some clarifying questions of the PO.

Assigning complexity points to stories is traditionally done by playing planning poker. With experience, it takes teams about three to five minutes per story to estimate it. The point here is not to aim for perfect accuracy but instead to quickly get to good enough. When the team cannot decide between neighbors on the scale, settle on the higher number.

Over time, the velocity of a team reaches a steady state. This usually takes between three to five sprints, but it can vary. From there, let us assume that a team can reliably deliver about 35 story points per sprint. Its goal when planning the sprint is therefore to select about 35 points' worth of stories from the top of the prioritized product backlog. Those stories should go into the sprint backlog. Unlike the product backlog that gets continuously reprioritized as needed, the sprint backlog is frozen and becomes the scope of the sprint once the team commits to its delivery.

TABLE 18–1. A SPRINT BANDWIDTH EXAMPLE

Team Member	Day1	Day2	Day3	Day4	Day5	Day6	Day7	Day8	Day9	Day10
Jarry	x	x	x	Holiday	Vacation	Vacation	Vacation	Vacation	x	x
David	x	x	x	Holiday	x	x	x	x	x	x
Mike	x	x	x	Holiday	x	x	x	x	x	x
Johan	x	x	x	Holiday	x	x	x	x	x	x
Fred	x	x	x	Holiday	x	x	x	x	x	x
Andrew	x	x	x	Holiday	x	x	x	x	x	x
Ron	x	x	x	Holiday	x	x	x	x	x	x

Just as with all prioritized work in agile delivery, the team should attack the sprint backlog from the top to make sure that the most valuable, or riskiest, stories are completed first. This way, should there be an unpleasant surprise, at least the highest-value work was completed.

Now, what happens if the team has not yet reached steady-state velocity? It should use the velocity from previous sprints as a loose guideline for the selection of the work.

How about during the very first sprint, when there is no velocity to draw on? Let the team figure out how much to put on its collective plate. It likely will get this estimate wrong, and that is acceptable for now. Accuracy will come with experience. Unlike with waterfall projects lasting months, in this case, if the team is wrong in its estimates, it will only be for one two-week slice of work.

One way to validate the number of story points selected by the team is to ask team members to break down their stories into individual tasks and estimate that duration in hours. The guidance I give for this is that tasks should be between, say, four hours and two days. Tasks shorter than four hours are not worth the overhead needed to manage them. Conversely, with tasks estimated to take longer than two days, the team does not understand early enough if there will be problems completing them. Once the stories have been broken down into their tasks, add their estimated durations. This total should be in line with the team's available bandwidth during the sprint. This also allows the scrum master to validate that no individual team member has too much on her plate. In either case, remember that perfection is the enemy of good enough.

One more thing: Make sure not to compare velocities between teams as an indicator of their productivity. The law of unintended consequences is that teams will start to artificially inflate their estimates to fool the system.

Use Velocity as a Longer-Term Planning Tool for Releases

Another very useful way to use velocity as a longer-term planning tool is to use it to estimate the number of sprints remaining to reach a minimum viable product (MVP) or a release milestone. Start by having the team(s) estimate the work remaining in the product backlog. As an example, let us say that there are 600 story points remaining to deliver. Then total the number of story points a program can deliver per sprint. Note that you are not comparing team velocities here, just adding up their numbers. For example, let us say that the program teams reliably deliver 125 story points per sprint. Therefore, it should take them 600/125 or 4.8 sprints to complete the work. Note that this process assumes that (1) the program has reached steady-state velocity; (2) that team availability over the next five sprints in this case is roughly the same as it has been up to date (in other words, no European-type vacations where everyone leaves for an entire month); and (3) the remaining work and team composition are very like those seen so far.

Note that there is a loss of productivity cost for the team to estimate an entire product backlog. This is especially true if done too early in the project since the backlog is bound to change, possibly even significantly, based on the feedback to be received. So, use this process judiciously.

Daily Scrum

The purpose of the daily scrum is simply to plan the next workday. For this, each team member answers as succinctly as possible the following three prioritized questions:

1. Do you have a blocker or an impediment?
2. What do you plan to accomplish today?
3. What did you accomplish since the last daily scrum?

Note the use of the word "accomplish." This is done on purpose. In the interest of time, the team should be primarily interested in a member's accomplishments. In other words, if the daily scrum was merely a status report meeting, then covering all the work done and meetings attended might be suitable, but it is not. It is about the finished work since this often means a hand-off to a teammate—for testing, for example, or review by the PO for acceptance. It is this focus on the bottom line that drives many teams to have their daily scrum standing up as a reminder to keep this meeting short so that members can return to what they are paid for, which is to add value.

Though anyone is encouraged to attend the daily scrum to keep their finger on the pulse of the team, only team members should be talking. Too often people outside the team, managers in particular, see this team gathering as an opportunity to share their own agenda. The result is that meetings that should last no more than about 15 minutes drag on much longer than that. It is the role of the scrum master to cut short these ancillary discussions. One way to do this with a little humor is to wave an Elmo doll. Elmo is one of the characters from the American TV show called *Sesame Street*. For agilists, Elmo stands for Enough, Let's Move On!

It is perfectly acceptable for team members or even "outsiders" to stay behind after the daily scrum to go in depth on various topics, providing those not interested or involved get to return to work.

Sprint Review

The Sprint review happens on the last day of the sprint. Its purpose is for the team to show its progress "to the world" and to seek feedback on the emerging solution. As such, anyone interested in the project should be invited to attend. This is a perfect opportunity for project sponsors, for example, to experience firsthand the progress and health of the project. This is also the perfect opportunity for future users of the product to guide its development by providing constructive feedback and generating a buzz around it.

The duration of the meeting can vary greatly. We have seen it anywhere from 15 to 20 minutes at the beginning of a project when there is little of the product to show yet, to a couple of hours when the functionality is delivered by an entire program. In the early stages, especially, a team member may need to do the "driving" if the product is just too fragile. If many senior managers are present, the driving is probably best done by the PO as they will be able to best describe the business value being delivered.

The feedback provided by the audience should then be integrated into the product backlog and prioritized for future development.

Sprint Retrospective

Another tenet of agile and scrum is the concept of continuous improvement. Though this improvement can be done at any time, the primary event for this is the sprint retrospective. This scrum ceremony traditionally closes a sprint on its last afternoon and lasts about an hour, possibly more depending on the extent of the root cause analysis of key problems identified. Traditionally, in the spirit of transparency, anyone can attend, but teams starting their agile journey prefer to limit attendance to the core team: team members, scrum master, and PO.

Of the many ways to hold a retrospective, I prefer the following:

1. Start by having attendees answer the following questions using post-it notes. The tone of the conversation should be constructive and in no way accusatory. Everyone is in the same boat, after all.
   a. What went well in this sprint?
   b. What do we need to change or start doing?
   c. What still puzzles you?
2. The scrum master or facilitator then reviews the contributions with the audience, asking for clarifications where appropriate.
3. The audience then votes on the top two or three items the team should tackle in the next sprint. To vote, each audience member or core team member is handed up to three sticky notes that they can place on any of the post-its. They may choose to apply them all on a single item or spread them across three of them.
4. The votes are then tallied, and the team performs a brief root cause analysis on the top two or three vote getters by at least asking the "Five Whys." The team may opt to time-box this exercise to, say, 10 to 15 minutes each.
5. The top vote getters along with potential solutions are added to the product backlog to be tackled in the next sprint.

In addition, it is usually desirable for the team to publish their findings so that these lessons learned can be broadly applied. This is another example of transparency and courage.

Product Backlog Refinement

The purpose of the product backlog refinement meeting is to review new stories with the team so that they can be estimated. Prior to meeting with the team, the PO, possibly with the help of a business analyst, validates that the stories are ready to be reviewed with the team. This means that the acceptance criteria for each story are fully but concisely documented. This is especially important since (1) these have a major impact on the estimation of the complexity of the work, (2) the team needs to know when to stop development to avoid goldplating the work, and (3) the testers will use these criteria to validate the acceptable completion of the work.

Team Estimates the Complexity of the Stories

It is the role of the team members to estimate the complexity of the stories using planning poker, as seen earlier. It is difficult to predict the duration of backlog refinement meetings. Their length depends on the agile maturity of the team, its experience in the business domain in general, and especially understanding and estimating of the requirements. A rule of thumb might be one to two hours per week.

The Five Artifacts of Scrum   Scrum is composed primarily of five artifacts. These are:

1. *Product backlog.* The backlog that holds all the work the team must do.
2. *Sprint backlog. The* subset of the product backlog that the team focuses on during a sprint.
3. *Increment.* The *release* into production of completed work and the associated business value gained.
4. *Stories.* The agile requirements formatted for brevity and clarity.
5. *Definition of Done.* The increment or project-specific criteria that stories must meet before they can be fully accepted and are ready to be deployed

Product Backlog Holds All the Work

The product backlog is an ordered list of everything the team works on. In particular, it contains ideas, features, functions, requirements, enhancements, and fixes that might be needed for the product to be deployed.

The product backlog is a dynamic and living document. As such, it is never complete. As a matter of fact, it may even survive the end of the given project, in the event, not all work could be completed.

It contains two types of stories: business stories and technical ones.

- Business stories are those describing a desired business outcome.
- Technical stories are those describing the technical design, infrastructure, or engineering needed for the deployment of the business stories.

Business stories and their technical dependent ones must be clearly linked so that the PO knows what technical work needs to be completed before development of a given business story can start. If the technical work needed does not precede the business development, technical debt ensues. Technical debt is the price the company will eventually have to pay to correct technical decisions made in the past. An example of technical debt is the decision that many companies made in the past to standardize Internet Explorer version 8. A few years later, the big rage was that everybody wanted to deploy mobile apps. Upon further investigation, it appeared that a mobile platform required a modern browser. It cost companies millions of dollars and months of effort to upgrade their infrastructure to HTML5 and CSS3. As you can well imagine, technical debt is an agile antipattern.

The product backlog is owned by the PO. The technical stories added to the product backlog are typically owned by the enterprise architect or the solution architect.

The team (developers and testers) is responsible for estimating their work. This is usually done using poker planning, as described above.

A healthy product backlog should have enough prioritized stories ready to be consumed by the team over the next two to four sprints or iterations but no more. It serves no purpose to fully refine a product backlog at the beginning of the project since many requirements will evolve or even be eliminated based on the expected continuous feedback. Maximizing the work not done is an agile best practice.

Sprint Backlog Fixes the Scope for a Sprint

The sprint backlog is the subset of the highest-priority or riskiest work remaining in the product backlog. More specifically, it is the forecasted work that the team commits to delivering during the present sprint. Only the team can manage the sprint backlog; for everyone else, it is frozen.

Increments Get Released on Demand

An increment is the increasing sum of all the product backlog items completed over successive sprints and accepted for deployment. As such, at the end of every sprint, the evolving increment must always meet the DoD (see below) and be deployable on demand.

It is now common practice, especially in first-of-a-kind projects to deliver a MVP as early as possible rather than wait to deploy the final product in a big bang approach. As its name implies, an MVP is small subset, some would say a bare-bones subset, of the product's anticipated full functionality. Leveraging MVPs allows the business team to (1) validate key and risky concepts while there still is time to apply changes based on the real-world feedback received, (b) start generating a small revenue stream, and (c) create a buzz of anticipation around the emerging product.

When deciding what features to test in an MVP, the business team should also focus on delivering a minimum delightful experience (MDE). Without delivering a pleasant experience, the potential value of the features shown may be ignored.

The PO decides if, and when, to release the increment in production. After the release of an increment, a new one is created.

Stories Tell the Story

Requirements Are Easier to Manage when Stories Have a Hierarchy

Though scrum technically only mentions one type of story, the user story, many teams wrestle with defining this type of very granular story on large projects. Instead, when a top-down approach to fleshing out requirements is desirable, teams may benefit from the new hierarchy of stories shown here:

1. *Business objectives.* Not technically a story, strategic business outcomes set the "North Star" that all other stories must aim toward. Strategic business outcomes are set by the company leadership. For example, "$500M in new assets under management coming from the accounts of millennials within 12 months of product launch."

2. *Epics*. Epics translate the business objectives into an actionable strategy. They often take months, certainly longer than a quarter, to complete.
3. *Features*. Features translate the strategic epics into a tactical approach. They must be completed in less than a quarter.
4. *User stories*. These stories are the tactical implementation of a feature. User stories must be completed within one sprint. If they cannot be, they should be split into smaller stories.
5. *Tasks*. It is a best practice for the team to break down user stories into the tasks needed to complete that work. As seen above, tasks should last between four hours and about two days. Totaling the duration of the tasks is an easy way to validate how much work the team is committed to delivering in a sprint.

Documenting Stories Is Very Easy

Documenting requirements or stories on an agile project is quite simple. The template usually looks like this:

- As a < user role | actor | system >,
- I want to < desired goal or outcome >,
- So that I may < anticipated value or benefits >.
- I will know I am done when < all the acceptance criteria >

The general guidance is that a story should fit on an index card, maybe front and back if numerous acceptance criteria are needed. Just as with most aspects of agile, the point here is to get to the essence of the story very quickly. The details will be fleshed out later as part of a conversation between the PO and the team.

Examples of Business and Technical Stories

1. As a < student eager to party > I want to < get cash from the ATM machine > so that I can < have a good time tonight >. I will know I am done when < either my bank account is empty or I have retrieved $50 >.
2. As a < blind customer >, I want < the screen reader to allow me to access the shopping cart of a website > so that I may < purchase items >. I will know I am done when < I can add and remove items from my shopping cart >.

Documenting requirements this way can be done very quickly. The stories can also easily be understood, including by team members for whom English is a second language. However, no one should expect that the team can build functionality based on such terse information. Indeed, these stories serve as the start of the conversation between the PO and team. The details emerge just in time through questions and answers during the development and testing effort based on the needs of the team. This is the primary reason why the PO should plan on spending much of her time collaborating with the team.

Make Sure You INVEST in Your Stories

Well-defined stories should exhibit the following characteristics:

1. *Independent*. Stories should be implementable without pulling in other stories. The exception would be if technical stories need to be implemented before a business one. The key is not to get into a chicken-or-egg situation between stories; implementing story A should not depend on story B if story B depends on story A.
2. *Negotiable*. Stories are a starting point, not a contract between the team and the PO. For example, a PO would be very interested to learn that by changing slightly an acceptance criteria, the effort needed to complete the work could decrease significantly.
3. *Valuable*. Stories should define the value delivered to the user.
4. *Estimatable*. Developers and testers must be able to estimate the story's required effort or better yet, its complexity.
5. *Small*. Larger stories are often too complex to estimate and manage. Breaking them down also facilitates their flow through the process and requires less rework should something go wrong.
6. *Testable*. A story must have at least one success criterion; otherwise, how can it be tested and how can the team make sure that it delivers value to the costumer?

Every Story Must Implement a Vertical Slice of Business Functionality

In software development, each business story must implement end-to-end business functionality across all four technical layers.

1. UI layer
2. Validation layer
3. Business logic layer
4. Data access layer

Though focusing on only one or two layers delivers some value, the story would not be fully testable at the end of the sprint. The idea is that by the end of a sprint, the accepted stories should meet the DoD and therefore be deployable.

Definition of Done: It Is in the Details

There are generally two definitions of done. The first one pertains specifically to the stories, the second to the increment.

User-Story-Specific DoD

1. *Unit and automated testing*. Unit testing of stories has become a best practice, especially when facilitating automated testing. Automating testing itself, especially when paired with a test-first methodology, is also becoming standard. One reason for automated testing is that it greatly decreases the risks associated with code refactoring or rework.

2. *All success criteria were met.* It should go without saying that for a story to be accepted, it must meet all success criteria, and it must have been tested end to end.

3. *Deployment scripts tested.* By the time a story reaches the end of a sprint, it likely will have been deployed multiple times. This deployment should be automated to increase reliability of deployment. Also, it is a good idea to be able to roll the deployment back using scripts, just in case.

4. *Infrastructure requirements documented.* The infrastructure on which a business story will land must be documented so that the operations team can validate the readiness of the various environments needed.

5. *Documented workarounds to remaining defects.* Acceptable workarounds to remaining defects must be documented for the operations staff and end users.

6. Etc.

Increment-Specific DoD

1. *Coding and architecture standards.* It is a best practice to audit the code to make sure it meets coding and architecture standards.

2. *Reviews.* Regularly conducting code and architecture reviews is an excellent way for the team to continuously improve its technical proficiency.

3. *End-user documentation.* Even features very intuitive to use still need end-user documentation, for the help functionality, for example.

4. *Notify change management.* The team responsible for managing changes resulting from the deployment of a new product should be involved early and often in the development process. This will allow them to ensure better readiness and a more enjoyable experience for the end users, employees, or consumers.

5. *Cross-stream integration.* Integration between workstreams, projects, or programs is traditionally one of the most difficult phases of a waterfall or plan-driven project. This is primarily due to (a) the often many-months lag between the time when a snippet of code was written and the time it is found to cause an integration defect. (b) Because the integration phase is one of the last gates before deployment in production, teams are often under a lot of pressure to complete this as quickly as possible. Agile and scrum teams address this up front by deploying each build in a "like live" or "pseudoproduction" environment. This environment is often virtual and gathers all the binaries currently in production along with all the builds from teams targeting the same future production environment.

6. Etc.

These two short lists are not exhaustive. You should research the criteria that apply best to your project and your company. Figure 18–12 shows how all of the activities fit together.

Tying All the Above Together

1. The product backlog is prioritized primarily by business value but also by risks to the project, the latter especially at the beginning of the project.

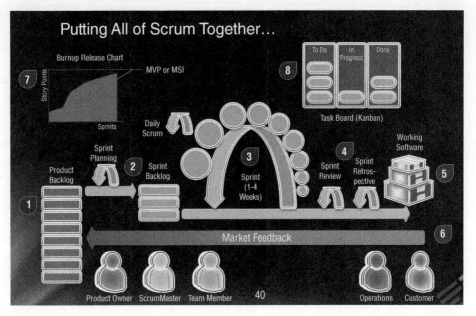

Figure 18–12. How the activities fit together

2. During the sprint planning session, the team selects enough work from the top of the product backlog to fill its plate. It then commits to delivering those requirements during that sprint.
3. Shorter sprints are preferred over longer ones. Try two-week-long iterations. Conduct the daily scrum by having each team member answer the following three questions:
   a. What did you accomplish since the last daily scrum?
   b. What do you plan to accomplish today?
   c. What is blocking or slowing you down?
4. On the last day of the sprint, the team demonstrates its accomplishments during the sprint review and then takes a critical look at ways to increase its productivity in the sprint retrospective.
5. The accepted stories get added to the increment until the business team or the PO decides to deploy it in production.
6. The market feedback is captured and added back into the product backlog to make the next version of the product even better.
7. The cumulative team or program velocity is plotted against a target line representing the number of story points needed for the MVP. This chart is called a burnup chart.

A Kanban board is an excellent way of visualizing the work and its process(es) as we manage upward.

18.3 DELOITTE AND ENTERPRISE VALUE DELIVERY FOR AGILE METHOD[4] _____

Many organizations today seek ever-increasing visibility of progress and speed to market in their software development processes. Their leaders seek more effective means of responding to stakeholder input and demonstrating rapid results. Their customers seek better user experiences and software tailored to their needs. To meet these challenges, projects and organizations as a whole are progressively moving to agile development methodologies.

Agile is a group of software development methods based on iterative and incremental development, where requirements and solutions evolve through collaboration between self-organizing, cross-functional teams. Agile manages complexity, unpredictability, and change through visibility, inspection, and adaptation. Because agile provides greater transparency and visibility to progress, it demands greater discipline than traditional approaches.

The fast-paced and constantly evolving nature of digital technology necessitates a working process that is flexible and adaptable. We have found that an agile development methodology provides an excellent delivery process for early and ongoing evaluation of quality and the creation of cutting-edge software. Small and highly coordinated cross-functional teams release a working build every sprint to ensure that the project is never far off track from a production release as it evolves. Agile allows us to quickly deliver finished products that meet the needs of users and stakeholders.

Deloitte uses an iterative development approach that meets the needs of our clients by leveraging our integrated *Enterprise Value Delivery (EVD) for Agile Method*. EVD for agile is an empirical method that embraces change in the business environment and technology landscape, with a focus on delivering value to our clients quickly and providing discipline and transparency to achieve the desired project results. The method is based on scrum and on the experience and industry-leading practices accumulated from Deloitte's work on agile projects. It includes the tasks needed to develop the product and sprint backlogs and to manage the project through quick iterations (called sprint cycles). The EVD for Agile Method provides processes, templates, samples, and accelerators that promote quality and value and help to manage agile projects through time-boxed sprint cycles. EVD for agile is an empirical method that embraces change in the business environment and technology landscape, with a focus on delivering value to our clients quickly and providing discipline and transparency to achieve the desired project results.

---

4. Material in Section 18.3 on Deloitte was provided by Daniel Martyniuk, Christine Lyman, PMP, and Rusty Greer. © 2022 Deloitte Development LLC. All rights reserved. As used in this document, "Deloitte" means Deloitte Consulting LLP, a subsidiary of Deloitte LLP. Please see www.deloitte.com/us/about for a detailed description of the legal structure of Deloitte LLP and its subsidiaries. Certain services may not be available to attest clients under the rules and regulations of public accounting. This publication contains general information only and Deloitte is not, by means of this publication, rendering accounting, business, financial, investment, legal, tax, or other professional advice or services. This publication is not a substitute for such professional advice or services, nor should it be used as a basis for any decision or action that may affect your business. Before making any decision or taking any action that may affect your business, you should consult a qualified professional advisor. Deloitte shall not be responsible for any loss sustained by any person who relies on this publication.

Key components of Deloitte's approach to agile, reflected in our EVD for Agile Method (see Figure 18–13), include:

- Three phases—*Discovery, Sprint Cycle*, and *Release*, spanning definition of the project *vision* through to product *launch*.
- Focus on *working software* and continuous improvement.

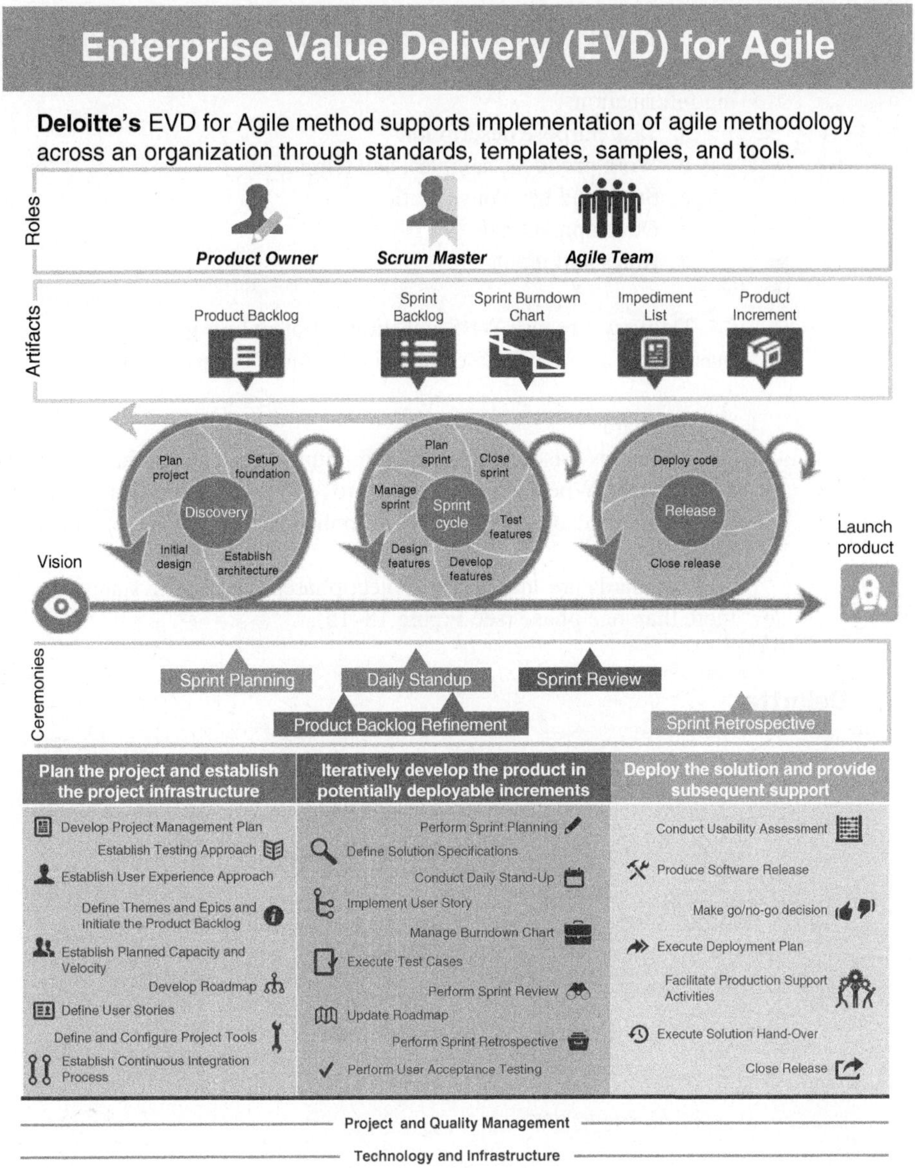

Figure 18–13. Deloitte Enterprise Value Delivery (EVD) for Agile framework.

- Scope defined through the use of *product* and *sprint backlogs* and prioritized by the *PO* to forecast releases with the *roadmap*.
- Work managed through quick iterations called *sprints*, facilitated by the *scrum master*, and progress is visualized on a *task board* (or equivalent tool).
- Team measures include *capacity*, *velocity*, and *burndown*.

Specific EVD resources that support and enable the delivery of agile projects include:

1. Delivery Process Work Breakdown Structure (**WBS**) (see Figure 18–14)

    The WBS includes around 180 tasks, where each task is described with the following information:

    a. Task purpose description
    b. Work products, samples, and templates
    c. Steps and key considerations
    d. Development aids
    e. Roles and inputs

    Delivery Process WBS constitutes not only a great starting point for new scrum masters but also a valuable checklist for more experienced ones.

2. Standards

    Standards are defined as the minimum requirements that new pursuits and projects should meet. When followed, standards provide a more consistent and defined way of delivering in an agreed way. Where applicable, standards are embedded into the method description.

    Standards are mapped to project phases, where one standard can be applicable to more than one phase (see Figure 18–15).

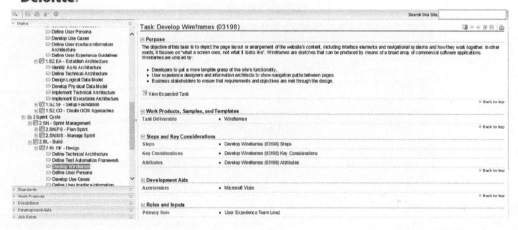

Figure 18–14. Delivery Process Work Breakdown Structure.

Standards	Pre - project	Discovery	Sprint Cycle	Release
	Standards (defined as minimum requirements all new pursuits and projects should meet)			
Estimation	✓ Estimate effort and/or story points	✓ Estimate effort and/or story points	✓ Estimate effort and/or story points	✓ Estimate effort and/or story points
SOW	✓ SOW			
Project Management		✓ **Project Management Plan** ✓ **Product Backlog** • **Team Capacity** • **Team Velocity** ✓ **Roadmap**	✓ **Product Backlog** • Team Capacity • Team Velocity ✓ **Roadmap** ✓ **Sprint Backlog** • Daily Stand-up Updates • Burndown Chart • Sprint Review (Stakeholder Feedback) • Sprint Retrospective (Lessons Learned Report)	✓ **Product Backlog** • Team Capacity • Team Velocity • Daily Stand-Up Updates
Project Health Metrics (PHM)		• Quality Management Plan • Project Status Report	• Project Status Report	• Project Status Report
Requirements Management		✓ **Product Backlog** (Define Themes and Features, Define User Stories)	✓ **Product Backlog** ✓ **Requirements Traceability Matrix**	✓ **Product Backlog**
Code Management		✓ **Continuous Integration Process**		
Test Management		✓ **Test Strategy**	• Test Case • Defect Tracking Log	
Training and Change		• Organizational Change Management Strategy	• Communications Plan • End-User Training Strategy • Training Development Plan	
Deployment			✓ **Deployment Plan**	✓ **Release Go/No-Go Criteria**

Figure 18–15. EVD standards by phase.

641

3. Work products, disciplines, and development aids.

In this section of EVD, samples and templates of all key work products can be found. They are grouped by project tasks:

    a. Project management—e.g., action items log

    b. Quality management—e.g., metrics plan for agile

    c. Requirements—e.g., definitions of ready and done

    d. Analysis and design—e.g., data modeling checklist

    e. Development—e.g., code review checklist

    f. Testing—e.g., defect tracking log

    g. Deployment—e.g., release go/no-go criteria

    h. Technology—e.g., technical architecture

    i. Organizational change management—e.g., change impact assessment report

Repository of sample work products is constantly developed and enriched with new templates. As for development aids, here stakeholders can find: accelerators, guidelines, procedures, and tools.

4. Job roles

The last section of EVD presents description of all important project roles, focusing not only on scrum roles (see Figure 18–16).

Deloitte Experience Addressing Agile Delivery Challenges

Agile projects require the ongoing collaboration and commitment of a wide array of stakeholders, including business owners, developers, and security specialists. Challenges in achieving and maintaining such commitment and collaboration include

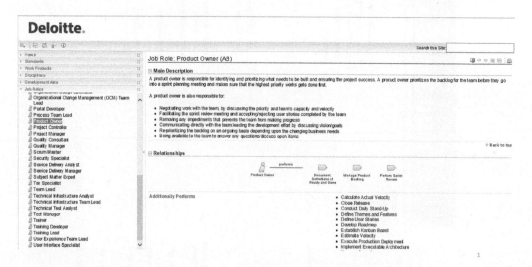

Figure 18–16. Job roles.

teams having difficulty collaborating closely and committing to frequent input or teams having difficulty transitioning to self-directed work.

When an organization following waterfall software development migrates to agile, new tools and technical environments may be required to support that approach, as well as updates to guidance and procurement strategies. Following are the challenges in preparing for agile:

- Timely adoption of new tools is difficult.
- Technical environments are difficult to establish and maintain.
- Procurement practices may not support agile projects.

Agile projects develop software iteratively, incorporating requirements and product development within an iteration. Such requirements may include compliance with legal and policy requirements. Challenges in executing steps related to iterative development and compliance reviews include teams having difficulty managing iterative requirements. Furthermore, compliance and regulatory reviews can be difficult to execute within an iteration timeframe.

Agile advocates evaluation of working software over the documentation and milestone reporting typical in traditional project management. Challenges in evaluating projects related to the lack of alignment between agile and traditional evaluation practices include the fact that some reporting practices do not align with agile, while traditional artifact reviews and status tracking do not align with agile.

Table 18–2 outlines Deloitte's approach to addressing the key agile delivery imperatives.

TABLE 18–2.   KEY AGILE DELIVERY IMPERATIVES

Delivery Imperative	Deloitte Approach
Managing the integrity of the business solution in the highly iterative and parallel agile delivery environment	We work with all parties (client, Deloitte, other vendors) as a single integrated team, actively and continuously involving all stakeholders to guide the project vision, review achieved outcomes, and continuously prioritize business value.
Coordination of multiple delivery teams and dependencies across the program	A scrum of scrums hierarchy should be established with daily cross-team meetings to review team progress and identify any cross-team dependencies or impediments. Experienced scrum masters coordinate issue resolution across teams and escalate to program leadership when necessary.
Robust program governance supporting transparency and visibility into program status and progress across all teams	Deloitte's governance structure supports daily tracking of sprint burndown and user story status, planned and actual team velocity and capacity, burnup of user stories against planned releases or other milestones, and weekly reporting at the program level of issues, risks, and overall program status.
Early establishment and continuous validation of the definition of the minimally viable product (MVP)	The definition of the MVP is established during the discovery phase and will be revisited throughout project execution as epics and user stories are further defined to facilitate the rolling roadmap forecasting and backlog management against prioritized business objectives.

Transitioning from Waterfall to Agile

In today's digitally enabled economy, development organizations must continuously find ways to help their firms meet objectives in an effective and cost-efficient manner. Especially important are responsiveness to key business drivers, the ability to provide greater visibility into the development process, and speed of implementation.

To help meet these objectives, organizations are increasingly moving to agile development. However, as with any change, the move to agile is not always easy. The ease of transition is affected by the culture of the organization adapting agile. Organizations with collaborative, people-focused cultures tend to adapt to agile successfully. The cultural shift from non-agile to agile development places certain requirements on the group making the transition. Those changes include a shift to a mentality of constant production and results, adopting a higher degree of collaboration, validation of estimate accuracy, and ensuring a shared understanding of what is being developed.

To successfully adopt agile, firms must properly frame the agile deployment methods within their organization. The first step involves understanding the business issues being addressed. The second step is to evaluate the organizational culture affected by adoption. The third step is for the organization to create a strategy to deploy agile in the organization. The fourth and final step is to tailor the agile development methodology to the specific project context.

Converting stakeholders to the agile mindset requires understanding the types of stakeholders who play a role in an agile development project, grasping the potential impacts of their various levels of engagement on the project, and persuading them to think of an agile project in terms of capacity rather than in terms of the clear scope commitments that are one of the hallmarks of development under waterfall methodology.

As with stakeholders in an agile project, the planning, design, and build elements of an agile project need to be properly aligned to the agile mindset to succeed. Planning is relatively decentralized on an agile project, with the agile project manager serving as an important leader of a pluralistic decision-making process that the project manager promotes. The design process on successful agile projects tends to be less document heavy than on non-agile projects, but instead emphasizes key concepts and is responsive to ongoing stakeholder feedback. The build process requires discipline to quickly test and deploy code. A mindset of continuous improvement is critical for achieving success in an agile environment.

18.4 BEST PRACTICES IN PROJECT MANAGEMENT BASED ON THE AGILE OPERATING MODEL IMPLEMENTATION BY DELOITTE[5]

Introduction

Since agile principles, a while ago mainly associated with software development, have been steadily adopted by more industries and business sectors (Marketing, Human Resources, Finance), we all need to learn how to navigate in this environment to maintain not only the highest standards of project management but above all, to adapt to such rapidly changing market needs. According to Gartner's future ways of work report, by 2026, 75% or more of organizations will have interdisciplinary teams following agile principles, up from at least 25% in 2020. This is not just a fad for the agile approach, which will pass when a new, more innovative concept of project management emerges on the market. This is by far the most effective approach to cope with a more complex, digitally oriented, post-pandemic world and its constant change.

Following the global trend of agile adoption, we needed to substantially expand both our team (since 2021 it has doubled) and the scope of our services providing support also within non-IT departments of the enterprise. As part of Deloitte, we initiated a cyclical meeting called "Agile Breakfast," during which we exchange experiences from the use of agile approaches. We are also the authors of the Deloitte Agile Organizations Blog and the podcast of the same name.

Agile Operating Model

As a company that delivers value for its clients by providing professional services in areas of audit, consulting, financial advisory, legal, tax and relentlessly broadening our scope, we had faced the problem of delivering a replicable value of our transformation projects, considering the diversity, and complexity of our customers' needs. Therefore, the Deloitte Agile Operating Model was introduced (Figure 18–17) to create a baseline and cover all key aspects of agile transformations. It consists of four areas: Structure, People, Process, and Technology.

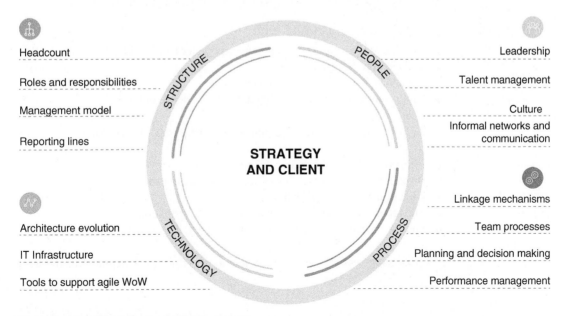

Figure 18–17. Agile Operating Model By Deloitte

This operating model facilitated the implementation of agile practices across all areas of the organization in a very structured and efficient way, mitigating the risk of complications, which occurs during complex transformations. Since the development of the model, we have implemented it in several organizations in various industries.

In the following sections, we will elaborate on our approach, using the case study of a leading insurance company in Poland.

In the past, the client made several unsuccessful attempts to carry out an agile transformation, each of them failed and caused many negative experiences for the employees and executives.

Our project was a part of a larger Digital Transformation Programme (~500 people). Deloitte Digital consultants were responsible for the strategy and the operationalization of strategic aims.

Despite being a market leader, our client lacked the organizational maturity to start the digital transformation on their own. Products were developed using mainly waterfall approaches and only 7% of the employees worked in an agile way (i.e. about 60 people). Most of the processes were *siloed*, i.e. divided into separate organizational areas based on specializations. Hence, the creation of a new product, in this case— creating insurance for diabetics, took about 12 months.

Although the company's management was open to change, they needed guidance in formulating their aims and implementing the new approach in the organization. To facilitate the implementation process, Deloitte established the *Change Team*—a team of employees, empowered to make decisions affecting the design and orchestration of the transformation. Throughout the transformation process, the Change Team worked closely with key SMEs, CXO, and ELT.

At the beginning of the transformation, we formulated the following aims:

- Become more client centric,
- Develop the best digital products and experiences on the Polish insurance market,
- Reduce time-to-market,
- Transform the functional silos, and
- Support a sense of responsibility and independence of teams.

In the first phase, w identified a number of crucial issues to resolve:

- Limited independence below the board level,
- Business and IT representatives were in separate teams,
- Lack of a concise approach, i.e. scrum,
- The repercussions of a previously failed transformations,
- Lack of shared vision and aspirations for an agile transformation.

The following sections will address the above issues in the context of the agile operating model.

Agile Operating Model—Structure

An agile organization should consist of empowered, cross-functional teams focused on delivering products and services with a specific, common goal. Therefore, it is crucial to introduce a flat organizational structure built around customer needs with end-to-end responsibility for the result instead of creating functional silos.

The *structure* component of our operating model is based on four pillars: Headcount, Roles and Responsibilities, Management Model, and Reporting Lines. Each of them covers crucial aspects of the structural transformation within the organization.

Key Elements of Structure Area

- Flattened organizational structure with truly cross-functional, end-to-end teams built around customer needs;
- Self-managing teams of fully committed employees making decisions autonomously as a part of customized Profit and Loss management;
- New roles and responsibilities (e.g., Product Owner, Agile Coach, Chapter Lead);
- Coherent teams in one (virtual) location, enabling effective communication.

Top Challenges—Structure

- Selecting the target archetype of the organization, i.e. archetypes reflecting customer journey, processes, or the client segments;
- Defining the scope of the agile transformation, i.e. deciding which parts of the organization will adopt the agile ways of working and which will continue working in traditional ways.

Top Benefits—Structure

- Effective collaboration and better communication within established units thanks to developed matrix of roles and responsibilities;
- Agile structure was an enabler for a digital transformation;
- Alignment of organizational objectives around end-to-end accountability instead of siloed–department approach.

Case Study

Our client's transformation project started with an in-depth AS-IS analysis focusing on organization's agile maturity and the mapping of its delivery processes. These assessments provided us with the overview of impediments hindering our client's work and informed the next steps of an agile transformation.

In terms of organization's target structure, we first focused on selecting the appropriate archetype. Together with the Leadership and the *Change Team*, we considered a number of options, including customer segment and archetypes based on products or processes. Finally, we chose a hybrid model—a *customer segment—product fit* archetype supported by Centers of Excellence. The main benefits of this structure include customer centricity and shorter time-to-market.

Based on the chosen archetype, we designed the Value Streams, Squads (i.e. teams within Value Streams), and Chapters (see Figure 18–18). What is crucial in the *design phase* is to appoint value stream leaders and technical leaders to facilitate team design and the launch of value streams.

It is important to note that some parts of the organization were not in the scope of the agile transformation and continued working in traditional ways. These include the back office, HR, compliance, and finance.

Agile Operating Model— People
There are four pillars of the *People* component of the agile operating model: leadership, talent management, culture, and informal networks and communication. Each of them is key for a successful and durable change in the human aspects of the organization.

Firstly, the transformation's success strongly depends on the leadership support throughout the transition process. Agile companies need agile leaders, dedicated to empowering and supporting their teams. It is crucial that the leaders are embracing the

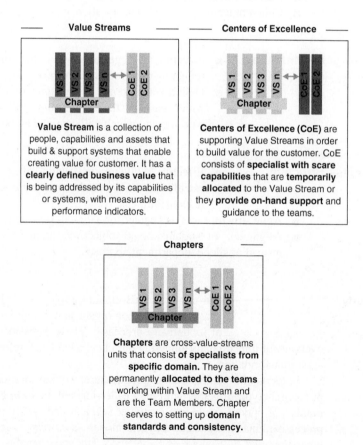

Figure 18–18. The main components of the organization's target structure

agile mindset, becoming role models of desired behaviors, and focus on creating the environment conducive learning and experimentation.

Secondly, a successful agile transformation needs to address the company's approach to talent management, i.e. recruitment, training, and the retention of employees. The aim is to attract and retain employees exemplifying the agile values, such as self-organization, collaboration, willingness to experiment, focus on value, and commitment. The agile-minded employees tend to struggle in strictly hierarchical organizations. Hence, it is crucial to ensure the consistency between the organization's lived and declared values. It is also important to provide opportunities for growth through dedicated coaching and training.

Thirdly, a change in organizational culture is essential for agile transformations. When initiating the cultural change, we mainly focus on promoting agile values, including customer centricity, teams' empowerment, and a sense of responsibility and commitment. In many agile transformations, this step is omitted due to the difficulties in measuring the change in culture as well as a lack of direct impact of culture on value creation.

Lastly, it is crucial to establish frequent and open communication to build a common understanding of the transformation and promote transparency across teams. The employees need to know "the why" behind the change, and clearly see the direction the company is going and how the change process will impact them. A frequent challenge is that employees have difficulty reconciling their personal aspirations with the new organizational structure and ways of working. The leaders have a big role in addressing this challenge by showing career opportunities in the new operating model.

Key Elements of People Area

- Changing the culture (customer centrism, team empowerment, a sense of responsibility, and commitment);
- Tailored approach to talent recruitment, training, and employee retention;
- Leadership focused on empowering teams;
- Communicate frequently and openly about goals and results to support change across the organization.

Top Challenges—People

- Limited engagement as well as insufficient understanding and support on the part of leadership and the management board—this might jeopardize activities at lower levels, i.e. setting goals and requirements that are not aligned with the QBR process and teams work rhythm;
- Failure to measure employees' engagement—this may cause employee rotation during the transformation;
- Team members having low sense of ownership and responsibility for their products;
- Insufficient empowerment of the teams;
- Missed opportunities by disintegrating teams.

Top Benefits—People

- Intentional change of an organizational culture in accordance with agile principles (responsibility, communication, and collaboration);
- Knowledge of actual employee needs, gained by structured surveys;
- Clear division of responsibilities across different roles (PO, VSL, VSTL, etc.).

Case Study

The *People* part of the transformation was conducted in collaboration with a Deloitte Human Capital team, as a part of a larger Digital Transformation of the client. Working together as members of the *Change Team*, we developed the role cards for the newly implemented roles, the competencies matrix, and the clear approach to performance management as well as to objectives management based on the OKR method.

These changes impacted the organization on many levels, which posed a risk of confusion and resistance on the part of the employees. Hence, it was crucial to implement appropriate *change management* actions, e.g., an ongoing multichannel communication of how the transformation impacts individuals and teams. What was helpful in establishing effective communication was the previously developed mindset of "open doors" between the management and their teams. Another key positive factor was the high team involvement in the transformation.

Agile Operating Model—Process

Efficient processes are key, yet difficult to implement for organizations at scale. Organizations with a complex organizational structure frequently struggle with coordinating their processes across different areas. Hence, the main focus of the transformation is to simplify the processes of planning and decision-making, product development, and performance management. It is also key to minimize or avoid interdependencies between organization's areas for the optimal efficiency and shortest possible T2M. For many organizations, this is a far-reaching change that requires an iterative approach, with continuous inspection and adaptation, so as to enable a gradual evolution towards an optimal operating model.

It is key that the chosen model is tailored to the company's particular context. Leaders and consultants frequently approach organization with a "1 fits all" mindset, trying to implement processes by the book. However, only with the deep understanding of organization's ways of working, we can design processes that will bring genuine value to the organization.

Key Elements of Process Area

- Iterative sprint cycles with regular, lightweight agile events (daily standups, portfolio planning, sprint demo, and retrospective)
- Clearly-defined rules of cooperation with other organizational areas (Finance, HR, Purchasing, and Legal);
- Venture capital style program/portfolio management (e.g., Quarterly Business Reviews);

- The implementation of the adequate performance management process based on the team goals, the role of the coaching triad (Product Owner, Agile Coach, and Chapter Lead), and the principles of the Agile Manifesto.

**TOP CHALLENGES—
PROCESS**

- Agility at the level of operational teams;
- Inefficient processes (e.g., overcomplicated, multilayered decision-making processes);
- Long time to market;
- Project-oriented approach as opposed to the focus on products and delivery of value.

TOP BENEFITS—PROCESS

- Creating a coherent governance model (increasing the autonomy and the decision-making power at team level);
- Higher predictability of teams due to the established processes and artifacts, such as the product development roadmap;
- A custom scaling approach facilitating the coordination of multiple teams working on one product.

Case Study

The main aim of designing the processes was to optimize the work in the new operating model. A key step was to choose an appropriate scaling methodology to synchronize and coordinate different organizational units. Based on the analysis of the available scaling frameworks and the client's particular context, we developed a custom recommendation of dedicated roles and events to govern the organization at scale. We also specified the processes of two-way interactions between different organizational units: value stream to value stream, value stream to CoE, and value stream to BAU.

Afterwards, we focused on the optimal process of initiative management. A key challenge was that our client had three types of initiatives, some of them could not be delivered using agile approaches due to outside restrictions from public institutions. Hence, we designed three types of Agile Project & Product Lifecycles for agile, hybrid, and waterfall initiatives. In our deliverable, we provided a set of in-depth guidelines for *discovery* and *delivery phases*.

Lastly, we established a number of processes to monitor the progress toward Objectives and KPIs, manage the budget and resources, mitigate risks, and adjust the scope of the project/product work. Lastly, we performed a tool analysis to develop a recommendation of tools to optimize the designed processes. For the client, the main benefits of the transformation of processes were the improved understanding and transparency of internal processes, increased access to information, and reduced time to market.

Agile Operating Model—
Technology

The agile operating model also requires a technology transformation to support agile ways of working and frequent releases. This involves an appropriate IT infrastructure and architecture as well as automated testing. Lack of synchronized changes in this area may cause:

- Limited operational possibilities of teams to shorten iterations, as well as test and release valuable increments to the customer;
- The duplications and overlapping of features across the IT landscape.

Key Elements of Technology
Area

- Designing the target architecture landscape;
- Automated testing and implementation (Continuous Delivery and DevOps);
- Infrastructure and supporting tools (open source, infra-cloud, etc.).

Top Challenges—Technology

- High systemic complexity within the organization;
- Further development of sales systems—the client was considering two options: developing the custom CRM system or going for Salesforce.

Top Benefits—Technology

- A clear vision of further technology transformation, including the architecture governance model;
- A cost-effective technology model, including the processes for technical debt management;
- The CRM systems analysis, including the comparison of the custom sales system to salesforce;
- A data governance system enabling data-driven decision-making.

Case Study

We started the technology transformation by conducting an in-depth analysis of the *technology operating model* and the *application landscape*. In our research, we looked at the organization from multiple perspectives, including tech competencies, service orchestration, tech talents, data governance, architecture, and centers of excellence. We also listed all applications the company's using, including their costs and number of users within the company.

Based on our analysis, we divided the target architecture landscape in a way that enables efficient development of different products by the value streams. Afterward, we ensured that the teams are equipped with all necessary tools to monitor their applications. Lastly, we designed the processes of interaction between development, security, and infrastructure teams. A key here was to balance efficiency and stability so as not to disrupt business continuity.

Technology transformation also involved implementing DevOps, along with the optimal processes and tools to enable product development and maintenance within the same team. We also established a DevOps guild, which took care of process automation.

One of the main challenges of technology transformation was the choice between developing the custom CRM system and going for salesforce. Finally, the client opted for the custom solution, which guided the next technology transformation steps.

Summary

The agile operating model by Deloitte facilitated a number of successful agile transformations of companies in different business sectors. The main benefit of the model is that it captures all key elements of an agile transformation, thereby enabling a structured and efficient change. At the same time, it provides room for attending the company's particular context and tailoring the recommendations to the client's needs and business aims, which was crucial in this case—a leading insurance company.

Thanks to the client's commitment and openness to change, our team of specialists could smoothly guide and support the client through the transformation project, achieving the goals formulated at the beginning.

18.5 THE RISK OF METRIC MANIA

During the past decade, there has been a rapid growth in agile project management practices, not just in IT but in other types of projects as well. Most of the principles of agile project management practices have provided beneficial results when applied to non-IT projects. While all of this sounds good, there are also some challenges that accompany the growth.

There is an old adage used in project management, namely "You cannot manage what you cannot measure." Therefore, to manage projects using agile techniques, you must establish metrics to confirm that the benefits are being realized and agile practices are being executed correctly. Fortunately, accompanying the growth in agile practices has been a companion growth in metric measurement techniques whereby today we believe we can measure just about anything. There are good metrics for reporting performance.

Another aspect to be considered as part of the Agile Manifesto: Individuals and interactions over processes and tools, and working software (or product) over comprehensive documentation. Considering the scrum framework, the team should provide artifact transparency; in other words, scrum requires transparency to support decision-making based on perceived state of the artifacts. This points to the direction that agile principles would use just the absolutely necessary metrics, as long as they add value to deliver the items on the product backlog.

Metric Mania

"Metric mania" is the insatiable desire to create metrics for metrics' sake alone rather than for what is really needed. There are disadvantages to having too many metrics and confusion about what metrics to choose.

The result of having too many metrics is that:

● We steal time from important work to measure and report these metrics.
● We provide too much data, and the stakeholders and decision-makers find it difficult to determine what information is in fact important.
● We provide information that has little or no value.
● We end up wasting precious time doing the unimportant.
● Too many metrics can open the door for unnecessary questions from stakeholders and business owners and eventually lead to a micromanagement environment.

In traditional project management, which uses waterfall charts, reporting has always been done around the metrics of time, cost, and scope. With the use of the Earned Value Measurement System, the number of metrics can increase to 12 to 15. As companies become mature in using a new approach, the number of metrics reported is usually reduced.

Metric Management

Having a good metric management program can minimize the damage of metric mania but cannot always eliminate it. There are four steps included in typical metric management programs:

1. Metric identification
2. Metric selection
3. Metric measurement
4. Metric reporting

Metric identification is the recognition of those metrics needed for fact-based or evidence-based decision-making.

Metric selection is when you decide how many and which of the identified metrics are actually needed. Metric selection is the first step in resolving metric mania issues. Ground rules for metric selection might include the following:

● There is a cost involved to track, measure, and report metrics even if we use a dashboard reporting system rather than written reports.
● If the intent of a good project management approach such as agile or scrum is to reduce or eliminate waste, then the number of metrics selected should be minimized.
● Viewers of metrics should select the metrics they need, not the metrics they want. There is a difference!
● Asking for metrics that seem nice to have but provide no informational value, especially for decision-making, is an invitation to create waste.

Paperwork is the greatest detriment to most project managers. The future of project management practices is to create a paperless project management environment. This does not mean that we are 100 percent paperless, since some reporting is mandatory, but it does mean that we recognize that unnecessary paperwork is waste that needs to be eliminated. In doing so, we have gone to dashboard project performance reporting.

Dashboard reporting systems can force viewers to be selective in the metrics they wish to see on the dashboard. A typical dashboard screen has a limited amount of real estate, namely the space for usually only six to 10 metrics that are aesthetically pleasing to the eyes and can be easily read. Therefore, telling stakeholders or business owners that we wish to provide them with one and only one dashboard screen may force them to be selective in determining the metrics they actually need.

Graphic Displays of Metrics

Dashboard performance reporting systems have made it easy to report information. Typical metrics for agile and scrum include stories committed × completed; team velocity and acceleration; sprint, epics, and release burndown rate; and net promoter score; just to name a few.

Several metrics that are common to traditional project management practices which might also be useful in agile and scrum are listed next.

1. Resource Management

    Shows the amount of time people are committed to working on projects. agile and Scrum recommend that the team is fully dedicated to the project. Resource utilization is critical. Without effective resource management people may spend only 50 percent of their time doing productive work on projects. The remaining time could be devoted to rework or succumbing to time robbers such as unnecessary meetings, phone calls, multitasking, and other such activities.

2. Impediments, Defects, and Scope Changes

    An impediment is anything that can block or slow down progress. Impediments require action items to resolve them and should be taken by the leadership of the project and/or team. If the impediments are not resolved in a timely manner, then the fault is usually with the leadership. The metrics can show the impediments that occurred in each month of the project as well as how many impediments were discovered, how many were resolved and removed, and how many had to be escalated to higher levels of management.

    Changes in scope are considered normal on agile and scrum projects and are dealt with at the end of each sprint or interaction. Some people believe that scope changes occur because of poorly defined requirements and faulty planning. While this argument may have merit, most scope changes occur because market conditions have changed or the business model requires reconfiguration. Care must be taken that the team can absorb all of the changes.

3. Value Management

    For more than 50 years, we have defined project success as completing the project within the time, cost, and scope constraints of a project. On the surface, this appears to be a good definition. But what this definition omits is the importance of "value." Any company can complete a project within the constraints of time, cost, and scope and end up providing no business value. A better definition of project success is to create sustainable business value while meeting the competing constraints.

Value can come in many forms, such as economic or business value, social value, political value, religious value, cultural value, health and safety value, and aesthetic value. All of these are important, but generally, we focus on economic or business value.

Another form of value can appear in improvements in customer satisfaction. If customer satisfaction continuously improves, the chances for additional work from customers and stakeholders can be expected to increase as well. Improvements in customer satisfaction can also be used as a team motivational factor. The argument holds true for providing business owners more value than they had expected.

Improving customer satisfaction and/or giving the business owner added value may not happen all of the time. Project failure will happen. A good metric to use in this regard is to track the percentage increase in project success rate or the decrease in project failure rate. This metric is important to the senior levels of management because it could provide an indication as to how successful are the projects.

Love/Hate Relationship

Too many companies end up with a love/hate relationship with metrics, especially metrics related to agility. Metrics can be used to shine a light on the accomplishments of the team by tracking performance, reporting the creation of business value, and identifying ways to reduce waste. Metrics can also be used to identify "pain points," which are situations that bring displeasure to business owners, stakeholders, and clients. The team then looks for ways to reduce or eliminate the pain points.

The hate relationship occurs when metrics become a weapon used to enforce a certain behavior. While good metrics can drive the team to perform well, the same metrics can create a hate relationship if management uses the metrics to pit one team against another. Another hate relationship occurs when the metrics are used as part of an employee's performance review. Reasons for this type of hate relationship are the result of the following:

- Metrics are seen as the beginning of a pay-for-performance environment.
- Metrics are the results of more than one person's contribution, and it may be impossible to isolate individual contributions.
- Unfavorable metrics may be the result of circumstances beyond the employee's control.
- The employee may fudge or manipulate the numbers in the metrics to look good during performance reviews.

18.6 CONCLUSIONS AND RECOMMENDATIONS

We have just scratched the surface in the identification of metrics. Metrics are a necessity with any and all project management approaches, including agile and scrum. However, given the number of possible metrics that can be identified, companies must establish some guidelines to avoid metric mania conditions and love/hate relationships. Possible recommendations include those listed next. This list certainly is not exhaustive but rather is a starting point.

- Select metrics that are needed rather than what people think they might want without any justification.
- Select metrics that may be useful to a multitude of stakeholders, clients, and business owners.
- Make sure the metrics provide evidence and facts that can be used for decision-making.
- Make sure the metrics are used rather than just nice to have displayed.
- Do not select metrics where data collection will be time-consuming and costly.
- Do not select metrics that create waste.
- Do not use metrics where the sole purpose is for performance reviews and comparing one team against another.
- Make sure that the metrics selected will not demoralize the project teams.

Benefits Realization and Value Management

19.0 INTRODUCTION

Organizations in both the public and private sectors have been struggling with the creation of a portfolio of projects that would provide sustainable business value. All too often, companies add all project requests to the queue for delivery without proper evaluation and with little regard if the projects were aligned with business objectives or provided benefits and value upon successful completion. Projects were often submitted without any accompanying business case or alignment to business strategy. Many projects had accompanying business cases that were based on highly exaggerated expectations and unrealistic benefits. Other projects were created because of the whims of management, and the order in which the projects were completed was based on the rank or title of the requestor. Simply because an executive says, "Get it done" does not mean it will happen. The result was often project failure, a waste of precious resources, and, in some cases, business value was eroded or destroyed rather than created.

19.1 UNDERSTANDING THE TERMINOLOGY

It is important to understand the definitions of benefits and value.

A *benefit* is an outcome from actions, behaviors, products, or services that is important or advantageous to specific individuals, such as business owners, or to specific groups of individuals, such as stakeholders. Generic benefits might include:

- Improvements in quality, productivity, or efficiency
- Cost avoidance or cost reduction
- Increase in revenue generation
- Improvements in customer service

Benefits are derived from the goals of strategic planning activities. In the past, traditional business goals were customer satisfaction, cost reduction, and profits, and they focused on near-term targeted savings and deliverables rather than long-term benefits. As such, too much emphasis was placed on the outcome of projects, which on their own may not necessarily deliver long-term benefits. Today, strategic goals and objectives seem to focus on:

- Productivity
- Efficiency
- Performance improvements
- Quality
- Customer service
- Rework
- Cost avoidance
- Revenue generation

Benefits, whether they are strategic or nonstrategic, are normally aligned with the organizational business objectives of the sponsoring organization that will eventually receive the benefits. The benefits appear by harvesting the *deliverables or outputs* that are created by the project. It is the responsibility of the project manager to create the deliverables.

Benefits are identified in the project's business case. Some benefits are tangible and can be quantified. Other benefits, such as an improvement in employee morale, may be difficult to measure and therefore may be treated as intangible benefits. Intangibles may be tough to measure, but they are not immeasurable. Some tough benefits to measure include:

- Collaboration
- Commitment
- Creativity
- Culture
- Customer satisfaction
- Emotional maturity
- Employee morale
- Image/reputation
- Leadership effectiveness
- Motivation
- Quality of life
- Stress level
- Sustainability
- Teamwork

There can also be dependencies between the benefits, where one benefit is dependent on the outcome of another. As an example, a desired improvement in revenue generation may be dependent on an improvement in quality, or better marketing may be needed to attract more tourists.

When scoping out a project, we must agree on the organizational outcomes or benefits we want, and they must be able to be expressed in measurable terms. This is necessary because improvements are usually expressed in financial terms to justify the investment in the business. Typical generic benefits metrics might include:

- Increase in market share
- Reduction in operating costs
- Reduction in waste
- Increase in profitability
- Improvements in productivity and efficiency
- Increase in quality
- Increase in customer satisfaction
- Improvement in employee morale
- Increase in employee retention and reduction in employee turnover

The metrics are needed for feedback to revalidate performance, measure success, investigate anomalies, and decide if health checks are needed.

Benefits realization management (BRM) is a collection of processes, principles, and deliverables to effectively manage the organization's investments and to make the benefits happen.[1] Project management is the vehicle for producing the outcomes that create benefits delivery. Project management focuses on maintaining the established baselines, whereas BRM analyzes the relationship that the project has to the business objectives by monitoring for potential waste, acceptable levels of resources, risk, cost, quality, and time as they relate to the desired benefits. The ultimate goal of BRM is not merely to achieve the benefits but to sustain them over the long term.

Organizations that are reasonably mature at BRM:

- Enjoy better business/strategic outcomes
- Have a much closer alignment of strategic planning, portfolio management, BRM, and business value management
- Use project management successfully as the driver or framework for BRM
- Capture best practices in the BRM activities

Decision makers must understand that, over the life cycle of a project, circumstances can change, requiring modification of the requirements, shifting of priorities, and redefinition of the desired outcomes. It is entirely possible that the benefits can change to a point where the outcome of the project provides detrimental results and the project should be canceled or backlogged for consideration at a later time. Some of the factors that can induce changes in the benefits and resulting value include:

- *Changes in business owner or executive leadership*: Over the life of a project, there can be a change in leadership. Executives who originally crafted the project may have passed it along to others who have a tough time understanding

---

1. For additional information on benefits realization management, see Craig Letavec, *Strategic Benefits Realization* (Plantation, FL: J. Ross, 2004), and Trish Melton, Peter Iles-Smith, and Jim Yates, *Project Benefits Management; Linking Projects to the Business* (Burlington, MA: Butterworth-Heinemann, 2008).

the benefits, are unwilling to provide the same level of commitment, or see other projects as providing more important benefits.

- *Changes in assumptions*: Based on the length of the project, the assumptions can and most likely will change, especially those related to enterprise environmental factors. Tracking metrics must be established to make sure that the original or changing assumptions are still aligned with the expected benefits.
- *Changes in enterprise environmental factors*: Changes in market conditions (i.e., markets served and consumer behavior) or risks can induce changes in the constraints. Legislation and elections can also impact the enterprise environmental factors. Companies may approve scope changes to take advantage of additional opportunities or reduce funding based on cash flow restrictions. Metrics must also track for changes in the constraints and the enterprise environmental factors.
- *Changes in resource availability*: The availability or loss of resources with the necessary critical skills is always an issue and can impact benefits if a breakthrough in technology is needed to achieve the benefits or to find a better technical approach with less risk.

Project *value* is what the benefits are worth to someone. Project or business value can be quantified, whereas benefits are usually explained qualitatively. When we say that the return on investment (ROI) should improve, we are discussing benefits. But when we say that the ROI should improve by 20 percent, we are discussing value. Progress toward value generation is easier to measure than progress toward benefits realization, especially during project execution. Benefits and value are generally inseparable; it is difficult to discuss one without the other.

19.2 REDEFINING PROJECT SUCCESS

For more than five decades, we have erroneously tried to define project success in terms of only the triple constraints of time, cost, and scope. We knew decades ago that other metrics should be included in the definition, such as value, safety, risk, and customer satisfaction, and that these were attributes of success. Unfortunately, our knowledge of metrics measurement techniques was just in its infancy stage at that time, and we selected only those metrics that were the easiest to measure and report on: time, cost, and scope.

For decades, we defined value as:

$$Value = Quality/Cost$$

If we wanted to increase the perceived value, we had to either increase the quality or lower the cost. This equation unfortunately implied that quality and cost were the only components of value.

Today, metric measurement techniques are maturing to the point where we believe that we can measure just about anything.[2] Perhaps the greatest level of research has been in measuring and reporting business value. During the past two decades, research has been conducted in the following areas:

- Value dynamics
- Value gap analysis
- Intellectual capital valuation
- Human capital valuation
- Economic value-based analysis
- Intangible value streams
- Customer value management/mapping
- Competitive value matrix
- Value chain analysis
- Valuation of information technology projects

The output of the research has created value measurement models and metrics:

- Intellectual capital valuation
- Intellectual property scoring
- Balanced scorecard
- Future Value Management™
- Intellectual Capital Rating™
- Intangible value stream modeling
- Inclusive Value Measurement™
- Value measurement methodology (VMM)

Value could very well become the most important word in the project manager's vocabulary, especially in the way that we define project success. In the glossary to the fifth edition of the *PMBOK® Guide*,[3] a project is defined as a temporary endeavor undertaken to create a unique product, service, or result. The problem with this definition is that the unique product, service, or result might not create any business value after the project is completed. Perhaps a better definition of a project might be:

- A collection of sustainable business value scheduled for realization

The definition of project success has almost always been the completion of a project within the triple constraints of time, cost, and scope. This definition likewise must change because it lacks the word "value," and it does not account for the fact that

---

2. For additional information, see Douglas W. Hubbard, *How to Measure Anything; Finding the Value of Intangibles in Business,* 3rd ed. (Hoboken, NJ: John Wiley & Sons, 2014).

3. PMBOK is a registered mark of the Project Management Institute, Inc.

today we have significantly more than three constraints, which we refer to as competing constraints. Therefore, the future definition of success might be:

● Achieving the desired business value within the competing constraints

A definition of project success that includes reference to value becomes extremely important when reporting on the success of benefits realization and value management activities. With traditional project management, we create forecast reports that include the time at completion and cost at completion. Using the new definition for success, we can now include in the forecast report benefits at completion and value at completion. This reporting of benefits and value now elevates project performance reporting to the corporate boardroom.

There is another inherent advantage to using value as part of the project's success criteria. We can now establish termination or pull-the-plug criteria defined in terms of value or benefits that tell us when we should consider canceling a project before additional funds and resources are squandered. All too often, projects are allowed to linger on and continue wasting valuable resources because no one has the heart to cancel the failing project. Establishing cancellation criteria in the business case or benefits realization plan may resolve this issue.

19.3 VALUE-DRIVE PROJECT MANAGEMENT

With the recognition of the importance of value, we are now focusing on value-driven project management activities. Value-driven project management focuses on the delivery of business value outcomes rather than simply deliverables that come from traditional project management practices. Value-driven project management requires an easily understood business case that includes the specific benefits desired.

Project management is now the vehicle for delivering benefits and value. Companies that are mature in BRM also appear to be reasonably mature in project management. In these companies, both the project management approach and the corporate culture are value-driven.

However, some risks need to be considered in value-driven project management:

● Possibility of endless changes in the requirements if not controlled
● Creation of a culture that fosters (possibly unnecessary) scope creep on all projects
● The value determination is made by different people over the project's life cycle
● Refusal to forecast the true value for fear of project cancellation
● Refusal to believe the forecasted value

Benefits desired must be defined at project initiation. But how do we define value in the early life-cycle phases of a project when value may be just a perception?

TABLE 19–1. HARD AND SOFT VALUE METRICS

Easy (Soft/Tangible) Values	Hard (Intangible) Values
Return on investment (ROI) calculators	Stockholder satisfaction
Net present value (NPV)	Stakeholder satisfaction
Internal rate of return (IRR)	Customer satisfaction
Opportunity cost	Employee retention
Cash flow	Brand loyalty
Payback period	Time-to-market
Profitability	Business relationships
Market share	Safety
	Reliability
	Reputation
	Goodwill
	Image

TABLE 19–2. PROBLEMS WITH VALUE METRICS MEASUREMENTS

Easy (Soft/Tangible) Values	Hard (Intangible) Values
Assumptions are often not disclosed and can affect decision making	Value is almost always based on subjective-type attributes of the person doing the measurement
Measurement is very generic	Measurement is more of an art than a science
Measurement never meaningfully captures the correct data	Limited models are available to perform the measurement

We would like to define value as well, but value is what the benefits are worth. The hardest part of value determination is defining the metrics so that measurements can be made. Table 19–1 shows some of the easy and hard metrics that are often used to measure value (and possibly benefits as well), and Table 19–2 shows several of the problems that can be encountered with the measurements. The metrics are needed to validate or revalidate not only benefits and value creation but also the business case, assumptions, and constraints. Decision makers must understand that, over the life cycle of the project, circumstances can change, requiring modification of the requirements, shifting of priorities, and redefinition of the desired outcomes.

Without proper metrics, we tend to wait until the project is way off track before taking action. By that time, it may be too late to rescue it, and the only solution is to pull the plug and cancel a project that possibly could have been saved.

19.4 BENEFITS HARVESTING

Benefits harvesting is the most difficult part of BRM. The problem is not with identifying the benefits or managing the projects to create them. The real issue is harvesting the benefits and managing the transition once the projects are over. The project team

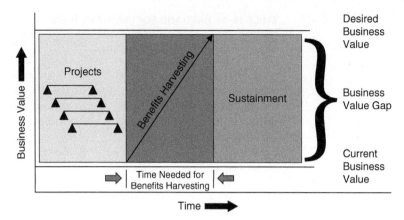

Figure 19–1. Creation of business value over time.

produces the deliverables but may have no control over how the business uses those deliverables to create benefits and value.

The benefits of a project are typically realized over time—sometimes years after the project has been completed and the project team has been disbanded. Some benefits will be near-term, midterm, and long-term. Someone must take ownership of the harvesting process.

Figure 19–1 shows how benefits and value are created over time. The unknown in Figure 19-1 is the amount of time needed to harvest the benefits and the amount of time necessary to sustain the benefits. There must be long-term adoption consideration to maintain benefits sustainment. Organizational change may be needed, and people may have to be moved out of their comfort zones. This is accomplished by people experienced in organizational change management.

19.5 THE BUSINESS CASE

Benefits realization and value management begin with the preparation of the business case. There are six major players in benefits realization and value management projects:

1. A portfolio governance committee composed of members that possess at least a cursory level of knowledge of project management
2. The benefits of business owner
3. The change management owner, if organizational change management is necessary to harvest the benefits at project completion
4. A sustainability owner to assure that the benefits harvested are sustainable
5. A portfolio project management office (PPMO) to assist with metric identification, measuring, and reporting
6. Project and/or program managers

The business owner is responsible for the preparation of the business case and for contributing to the benefits realization plan. Typical steps that are included as part of business case development are:

- Identification of opportunities such as improved efficiencies, effectiveness, waste reduction, cost savings, new business, and others.
- Benefits defined in both business and financial terms
- A benefits realization plan
- Estimated project costs
- Recommended metrics for tracking benefits and value
- Risk management
- Resource requirements
- High-level schedules and milestones
- Degree of project complexity
- Assumptions and constraints
- Technology requirements—new or existing
- Exit strategies if the project must be terminated

Templates can be established for most of the items in the business case. A template for a benefits realization plan might include the following:

- A description of the benefits
- Identification of each benefit as tangible or intangible
- Identification of the recipient of each benefit
- How the benefits will be realized
- How the benefits will be measured
- The realization date for each benefit
- The handover activities to another group that may be responsible for converting the project's deliverables into benefits realization

Well-written benefit realization plans, usually prepared by the business owner, tell what is included and excluded from the scope. Poorly written benefit realization plans imply that everything must be done and can lead to numerous and often unnecessary scope changes. Benefit realization plans are not statements of work. Therefore, there will always be some ambiguity in how the expected benefits of a strategic initiative are defined. Types of ambiguity appear in Table 19–3.

TABLE 19–3. TYPES OF AMBIGUITY

Ambiguity	Description
Expectations	Based on the number of stakeholders and the business owner's previous experience, the benefits realization plan may have vague wording open to an interpretation of the expected outcome.
Priority	Each stakeholder and business owner can have a different interpretation of the priority of the project. The project team may not know the real priority.
Processes	There are numerous processes to select from as part of execution. Process flexibility will be necessary. There are also several forms, guidelines, checklists, and templates that can be used.
Metrics/Key Performance Indicators	There are numerous things that can be measured based on expectations.

19.6 TIMING FOR MEASURING BENEFITS AND VALUE

The growth in metric measurement techniques has made it possible to measure just about anything, including benefits and value. But currently, since many of the measurement techniques for newer metrics are in their infancy, there is still difficulty in obtaining accurate results. Performance results will be reported both quantitatively and qualitatively. There is also difficulty in deciding when to perform the measurements: incrementally as the project progresses or at completion. Measurements of benefits and value are more difficult to determine incrementally as the project progresses than at the end.

Value is generally quantifiable and easier to measure than benefits. On some projects, the value of the project's benefits cannot be quantified until several months after the project has been completed. As an example, a government agency enlarges a road with the aim of reducing traffic congestion. The value of the project may not be known for several months after the construction project has been completed and traffic flow measurements have been made. Value measurements at the end of the project, or shortly thereafter, are generally more accurate than ongoing value measurements as the project progresses.

Benefits realization and business value do not come from simply having talented resources or superior capabilities. Rather, they come from how the organization uses the resources. Sometimes even projects with well-thought-out plans and superior talent do not end up creating business value; they may even destroy existing value. An example might be a technical prima donna who views this project as his or her chance for glory and tries to exceed the requirements to a point where the schedule slips and business opportunities are missed. This occurs when team members believe that personal objectives are more important than business objectives.

19.7 INVESTMENT LIFE-CYCLE PHASES

For years, academia taught that traditional project life-cycle phases begin once the project is approved and a project manager is assigned and end after the deliverables have been created. However, when benefits realization and value management become important, additional life-cycle phases must be included, as shown in Figure 19–2. Project managers are now being brought on board earlier than before and remaining after the deliverables have been produced to measure the business value created. Figure 19–2 is more representative of an investment life cycle than a traditional project life cycle. If value is to be created, then the benefits must be managed over the complete investment life cycle. The traditional project life cycle falls within the investment life cycle. More than six life-cycle phases could have been identified in the investment life cycle, but only these six will be considered here for simplicity.

The first phase, the *Idea Generation (IG) Phase*, which often includes a feasibility study and a cost–benefit analysis, is where the idea for the project originates. The idea can originate in the client's or business owner's organization, within the senior levels or lower levels of management in the parent company or the client's firm, or within the

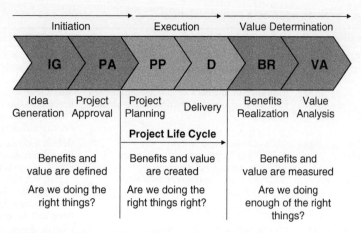

Figure 19–2. Investment life cycle.

organization funding the project. The output of the IG Phase is usually the creation of a business case.

Although the idea originator may have a clear picture of the ultimate value of the project, the business case is defined in terms of expected benefits rather than value. Value is determined near the end of the project based on the benefits that are actually achieved and can be quantified. The benefits actually achieved may be significantly different from the expected benefits defined at project initiation because of many reasons discussed earlier that can lead to changes.

Not all projects require the creation of a business case. Examples might include projects that are mandatory for regulatory agency compliance and are well understood, or simply to allow the business or part of the business to continue more efficiently.

Once the business case is prepared, a request is sent to the PPMO for project approval. Companies today are establishing a PPMOs to control the second phase, the *Project Approval (PA) Phase*, and to monitor the performance of the portfolio of projects during delivery.

The PPMO must make decisions that are in the best interest of the entire company. A project that is considered extremely important to one business unit may have a low priority when compared to all of the other corporate projects in the queue. The PPMO must maximize the business value of the portfolio through proper balancing of critical resources and proper prioritization of projects. The PPMO must address three critical questions, as shown in Table 19–4.

The activities identified with the third question in Table 19–4 are usually part of the PPMO's responsibility for determining if all of the benefits were captured or if additional projects need to be added to the queue.

Most companies tend to believe that project managers should be brought on board after the project has been approved and added to the queue. The argument is that project managers are not businesspeople, have limited information that could help in the approval process, and are paid to make project-based decisions only. This is certainly not true today. In today's world, project managers view themselves as managing part of

TABLE 19–4.   TYPICAL ROLE OF A PORTFOLIO PMO

Critical Questions	Areas of Consideration	Portfolio Tools and Processes
1. Are we doing the right things?	Alignment to strategic goals and objectives, such as shareholder value, customer satisfaction, or profitability Evaluation of internal strengths and weaknesses Evaluation of available and qualified resources	Templates to evaluate rigor of business case Strategic fit analysis and linkage to strategic objectives Matrix showing relationships between projects Resources skills matrices Capacity planning templates Prioritization templates
2. Are we doing the right things right?	Ability to meet expectations Ability to make progress toward benefits Ability to manage technology Ability to maximize resource utilization	Benefit realization plans Formalized, detailed project plans Establishing tracking metrics and key performance indicators Risk analysis Issues management Resource tracking Benefits/value tracking
3. Are we doing enough of the right things?	Comparison to strategic goals and objectives Ability to meet all customers' expectations Ability to capture all business opportunities that are within capacity and capability of company's resources	Overall benefits tracking Accurate reporting using project management information system

a business rather than just managing a project. Thus, project managers are paid to make both project-based and business-related decisions on their projects.

When project managers are brought on board after project approval, they are at the mercy of the information in the business case and benefits realization plan. Unfortunately, these two documents do not always contain all of the assumptions and constraints, nor do they discuss the thought process that went into creating the project.

Perhaps the most important reason for bringing the project manager on board early is for resource management. Projects are often approved, added to the queue, and prioritized with little regard for the availability of qualified resources. Then, when the benefits are not delivered as planned, the project manager is blamed for not staffing the project correctly.

Some of the critical staffing issues that need to be overcome include:

- Management does not know how much additional work can be added to the queue without overburdening the labor force
- Projects are added to the queue with little regard for (1) resource availability, (2) skill level of the resources needed, and (3) level of technology needed
- No central repository exists solely for staffing strategic projects
- Project prioritization is based on guesses rather than evidence or facts
- There are no techniques for understanding how a scope change on one project may affect workloads on other ongoing projects
- Resource decisions are made before project approval and before the project manager is brought onboard
- Lack of understanding of how project managers can assist early on in capacity planning and resource management
- Critical resources are assigned to failing or non-value-added projects

Project managers may very well be the best people qualified to critically identify the number of resources needed and the skill levels of the assigned staff. The ability to bring a project manager on board early makes it easier for the portfolio governance personnel to perform effective resource management practices, according to Figure 19–3.

Even when assigning project managers early in the investment life cycle, resource management shortcomings can occur. These shortcomings include:

- Not capturing all resource demands
- Lacking knowledge of the resource skill levels needed
- Changing resource needs on a project due to scope changes
- Not accounting for the resources that may be needed if transformational activities are required
- Shifting priorities due to firefighting on other critical projects
- Having unrealistic benefit and value estimates

If the shortcomings are not identified and properly managed, the results can be:

- A failure of benefits realization planning
- No maximization of portfolio business value
- Continuous changes to the portfolio
- Continuous reprioritization
- Continuous conflicts over manpower

The benefits of effective resource management are well known:

- Balancing workloads among the most critical projects
- Improvements in resource usage efficiencies by assigning resources with the right skills

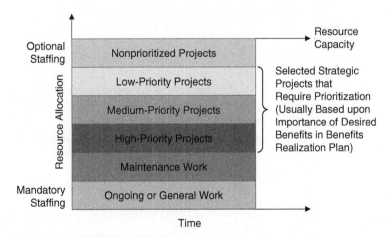

Figure 19–3. Resource management activities.

- Better planning and control of projects
- Better selection of a portfolio of projects that can maximize business value

The third life-cycle phase is the *Project Planning (PP) Phase*. This phase includes preliminary planning, detailed planning, and updates to benefits realization planning. Although the business case may include assumptions and constraints, the PPMO may provide additional assumptions and constraints related to overall business objectives and the impact that enterprise environment factors may have on the project. The benefits realization plan that may have been created as part of the business case may undergo significant changes in this phase.

The benefits realization plan is not the same as the project plan but must be integrated with the project plan. The benefits realization plan and the accompanying project plan may undergo continuous changes as the project progresses based on changing business conditions.

The fourth life-cycle phase is the *Delivery (D) Phase*. This phase, as well as the PP Phase, are most commonly based on the domain areas of the *PMBOK® Guide*. Traditional project management methodologies are used. In this phase, the project manager works closely with the PPMO, the business owner, and the steering/governance committee to maximize the realization of the project's benefits.

Performance reporting must be made available to the PPMO as well as to the appropriate stakeholders. If the project is no longer aligned with business objectives, which may have changed during delivery, the PPMO may recommend that the project be redirected or even canceled so that the resources can be assigned to other projects that can maximize portfolio benefits.

The fifth and sixth life-cycle phases in Figure 19–3 are the *Benefits Realization (BR) Phase* and the *Value Analysis (VA) Phase*. The benefits realization plan, regardless of the life-cycle phase in which it is prepared, must identify the metrics that will be used to track the benefits and accompanying value. Benefits and value metrics identification are the weak links in benefits realization planning. Much has been written on the components of the plan, but very little appears on the metrics to be used. However, companies are now creating value metrics that can be measured throughout the project rather than just at the end.[4]

These last two life-cycle phases are often called benefits harvesting phases, which refer to the actual realization of the benefits and accompanying value. Harvesting may necessitate the implementation of an organizational change management plan that may remove people from their comfort zones. People must be encouraged to make the changes permanent and not revert to their old ways when the projects end.

The people responsible for benefits harvesting need to consider:

- Organizational restructuring
- New reward systems
- Changing skills requirements

---

4. For information on creating and reporting value metrics, see Harold Kerzner, *Project Management Metrics, KPIs and Dashboards*, 4th ed. (Hoboken, NJ: John Wiley & Sons and International Institute for Learning, 2022), Chapter 5.

- Records management
- System upgrades
- Industrial relations agreements

Full benefit realization may face resistance from managers, workers, customers, suppliers, and partners. There may be an inherent fear that change will be accompanied by loss of promotion prospects, less authority and responsibility, and possibly loss of respect from peers.

Benefits harvesting may also increase benefits realization costs because of:

- Hiring and training new recruits
- Changing the roles of existing personnel and providing training
- Relocating existing personnel
- Providing additional or new management support
- Updating computer systems
- Purchasing new software
- Creating new policies and procedures
- Renegotiating union contracts
- Developing new relationships with suppliers, distributors, partners, and joint ventures

19.8 CATEGORIES OF BENEFITS AND VALUE

Part of strategic planning is to create a balanced portfolio of projects. For simplicity's sake, we use the four categories of projects shown in Figure 19–4. These same four categories can then be used to identify the categories of benefits and value. Numerous benefits, values, and accompanying metrics can be used for each category. Only a few appear here as examples.

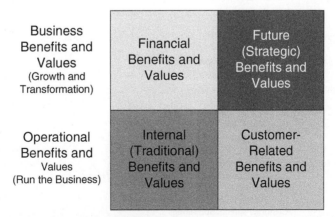

Figure 19–4. Categories of benefits and value.

Metrics must be established in each quadrant to serve as early warning signs of possible problems. Some examples of metrics that can identify benefit erosion problems are:

- Metrics on the number of scope changes, which identify the possibility of a schedule slippage and cost overrun
- Metrics on the number of people moved to put out fires elsewhere, which also indicate the possibility of a schedule slippage and cost overrun
- Metrics on excessive overtime, which could indicate serious unresolved issues
- Metrics on missed deadlines, which could indicate that the time to market may slip and opportunities may be lost

Table 19–5 shows typical benefits for each of the four categories. The metrics in the last column can be used to track the benefits.

The portfolio governance committee exists for the entire investment life cycle. Its role includes:

- Establishing the right priorities
- Eliminating surprises
- Building contingencies into the portfolio
- Maintaining response flexibility
- Controlling scope creep

TABLE 19–5. BENEFITS IN EACH CATEGORY

Category	Benefits	Project Tracking Metrics
Internal benefits	Processes for adherence to constraints Templates for identifying objectives, sign-offs, and capturing best practices Maintaining a best practices and metrics library Control of scope changes Control of action items Reduction in waste	Time Cost Scope Quality Number of scope changes Duration of open action items Number of resources Amount of waste Efficiency
Financial benefits	Improvements in ROI, NPV, IRR, and payback period Cash flow Improvements in operating margins Maintaining or increasing market share	Financial metrics ROI calculators Operating margin
Future (strategic) benefits	Reducing time to market Image/reputation Technical superiority Creation of new technology or products Maintaining a knowledge repository Alignment of projects to strategic objectives	Time Surveys on image and reputation Number of new products Number of patents Number of retained customers Number of new customers
Customer-related benefits	Customer loyalty Number of customers allowing you to use their name as a reference Improvements in customer delivery Customer satisfaction ratings	Loyalty/customer satisfaction surveys Time to market Quality

- Trying to do more with less
- Ensuring informed decisions using metrics
- Capturing best practices
- Understanding future resource needs

The portfolio governance committee must make strategic decisions and metrics assist in the process. Types of strategic decisions include the need to:

- Verify that value is being created
- Know the risks and how the risks are being mitigated
- Know when to intervene
- Predict future corporate performance
- Confirm that projects are still aligned with strategic objectives
- Perform resource reoptimization if necessary

The role of the PPMO is to work with the governance committee and determine the optimal resource mix for project delivery and benefits realization while honoring the imposed constraints. The PPMO also supports metric identification, measurement, and reporting. The PPMO supports the governance committee by addressing the following questions:

- Do we have any weak investments that need to be canceled or replaced?
- Must any of the programs and/or projects be consolidated?
- Must any of the projects be accelerated or decelerated?
- How well are we aligned to strategic objectives?
- Does the portfolio have to be rebalanced?

Sometimes the benefits result in best practices that can be applied to other projects. Table 19–6 illustrates benefits from several companies and in which quadrant the benefits appeared. Some benefits can be attributed to more than one quadrant.

As mentioned previously, it is important to know whether the measurements of benefits and value should be done incrementally or at the end of the project. Examples of incremental versus end point measurements are shown in Table 19–7. As mentioned, end-of-project measurements are generally more accurate, but some measurements may also be made incrementally.

TABLE 19–6. COMPANY-SPECIFIC BENEFITS

Company	Benefit Category	Benefit
General Electric	Future	Improving productivity
Motorola	Financial	Control of scope creep
Computer Associates	Internal	Better handling of customer expectations
ABB	Future	Project audits to seek out continuous improvement opportunities
Westfield Group	Internal	Development of an online intranet enterprise project management system
Antares Solutions (Medical Mutual)	Customer related	Customer-focused change control process

TABLE 19–7. EXAMPLES OF BENEFITS

Benefit Category	Benefit	Measured Incrementally	Measured at End
Internal	Speed up sign-offs	Yes	
Financial	Improving ROI, NPV, IRR, and shortening payback period	Yes	Yes
Future (Strategic)	Speed up product commercialization process		Yes
Customer related	Improving customer satisfaction	Yes	

19.9 CONVERTING BENEFITS TO VALUE

Value is what the benefits are worth either at the end of the D Phase or sometime in the future. Even though the benefits may be on track for achievement, the final value may be different from the planned value based on the deliverables produced and the financial assumptions made. Here are two examples of converting benefits into value:

1. A company approved the development of a customized software package with the expected benefit of reducing order entry processing time, which would be a savings of approximately \$1.5 million annually. The cost of developing the package was estimated at \$750,000. The value calculation was as follows:

$$\text{Value} = (60 \text{ workers}) \times (5 \text{ hours/week}) \times (\$100/\text{hour}) \times (50 \text{ weeks})$$
$$= \$15 \text{ million in yearly savings}$$

2. A company decided to create a dashboard project performance reporting system to reduce paperwork and eliminate many nonproductive meetings. The value calculation was made as follows:

- Eliminate 100 pages or reports and handouts each month at a fully burdened cost of \$1,000/page, or a savings of \$1.2 million.
- Eliminate 10 hours of meetings per week for 50 weeks, with 5 people per meeting, at \$100 per hour, or a savings of \$250,000.

$$\text{Value} = \$1,200,000 + \$250,000 = \$1.45 \text{ million in yearly savings}$$

In both cases, the companies received multiyear benefits and value from the projects.

19.10 GO-LIVE PROJECT MANAGEMENT

One of the challenges facing executives is determining who is best qualified to function as the leader for benefits harvesting. Some people argue that the project manager should remain on board even after project is ready to "go live." In this case, because benefits harvesting could require a great deal of time, the project manager may very well be functioning

TABLE 19–8. CHANGE IN SKILLS FOR A GO-LIVE PROJECT MANAGER

Traits	Differences
Authority	From leadership without authority to significant authority
Power	From legitimate power to judicious use of power
Decision making	From some decision-making to having authority for significant decision making
Types of decisions	From project-only decisions to project and business decisions
Willingness to delegate	Length and size of project will force project managers to delegate more authority and decision-making than they normally would
Loyalty	From project loyalty to corporate vision and business loyalty
Social skills	Strong social skills are needed since we could be working with the same people for years
Motivation	Learning how to motivate workers without using financial rewards and power
Communication skills	Communication across the entire organization rather than with a select few
Status reporting	Status of strategic projects cannot be made from time and cost alone
Perspective/outlook	Having a much wider outlook, especially from a business perspective
Vision	Must have same long-term vision as the executives and promote that vision throughout the company
Compassion	Must have much stronger compassion for workers than in traditional or short-term projects since the team members may be assigned for years
Self-control	Must not overreact to bad news or disturbances
Brainstorming and problem-solving	Must have very strong brainstorming and problem-solving skills
Change management	Going from project to corporate-wide change management
Change management impact	Going from project to organizational change management effects

as a functional manager, in which case the skills needed could be different from those required for traditional project management. This is shown in Table 19–8. A project manager may not be qualified to assume the role of a go-live project manager on all projects.

19.11 PORTFOLIO BENEFITS AND VALUE

The project tracking metrics identified in Table 19–5 are designed to track individual projects in each of the categories. However, specific metrics can be used to measure the effectiveness of a portfolio of projects. Table 19–9 shows the metrics that can be used to measure the overall value created by project management on individual projects, a traditional PMO and a PPMO. The metrics listed under project management and many of the metrics under the traditional PMO are considered micro-metrics focusing on tactical objectives. The metrics listed under the PPMO are macro-level metrics that represent the benefits and value of the entire portfolio. These metrics can be created by grouping together metrics from several projects. Benefits and value metrics are also used to help create the portfolio metrics.

Both the traditional and PPMOs are generally considered as overhead and subject to possible downsizing unless the PMOs can show through metrics how the organization benefits from their existence. Therefore, metrics must also be established to measure the value that the PMO brings to the parent organization.

TABLE 19–9. METRICS FOR SPECIFIC TYPES OF PMOS

Project Management	Traditional PMO	PPMO
Adherence to schedule baselines	Growth in customer satisfaction	Business portfolio profitability or ROI
Adherence to cost baselines	Number of projects at risk	Portfolio health
Adherence to scope baselines	Conformance to the methodology	Percentage of successful portfolio projects
Adherence to quality requirements	Ways to reduce number of scope	Portfolio benefits realization
Effective utilization of resources	changes	Portfolio value achieved
Customer satisfaction levels	Growth in yearly throughput of work	Portfolio selection and mix of projects
Project performance	Validation of timing and funding	Resource availability
Total number of deliverables produced	Ability to reduce project closure rates	Capacity and capability available for portfolio
		Utilization of people for portfolio projects
		Hours per portfolio project
		Staff shortage
		Strategic alignment
		Business performance enhancements
		Portfolio budget versus actual
		Portfolio deadline versus actual

TABLE 19–10. INTERPRETATION OF THE METRICS

Benefit Metric	Project Manager's Interpretation	Customer's Interpretation	Consumer's Interpretation
Time	Project duration	Time to market	Delivery date
Cost	Project cost	Selling price	Purchasing price
Quality	Performance	Functionality	Usability
Technology and scope	Meeting specifications	Strategic alignment	Safe buy and reliable
Satisfaction	Customer satisfaction	Consumer satisfaction	Esteem in ownership
Risks	No future business from this client	Loss of profits and market share	Need for support and risk of obsolescence

It is important to understand that some of the micro-metrics used for tracking benefits may have different meanings for customers or ultimate consumers. As an example, let us assume that you are managing a project for an external client. The deliverable is a component that your customer will use in a product he or she is selling to customers (i.e., your customer's customers or consumers). Table 19–10 shows how each of the metrics may be interpreted differently. It is important to realize that benefits and value are like beauty; they are in the eyes of the beholder. Customers and contractors can have different perceptions of the meaning of benefits and value, as well as of the associated metrics.

19.12 ALIGNMENT TO STRATEGIC OBJECTIVES

Because of advances in metric measurement techniques, models have been developed by which we can show the alignment of projects with strategic business objectives. One such model appears in Figure 19–5. Years ago, the only metrics used were time, cost, and scope. Today we can include metrics related to both strategic value and business value. This allows us to evaluate the health of the entire portfolio of projects as well as individual projects.

Since all metrics have established targets, we can award points for each metric based on how close we come to the targets. Figure 19–6 shows that the project identified in Figure 19–5 has thus far received 80 points out of a possible 100 points.

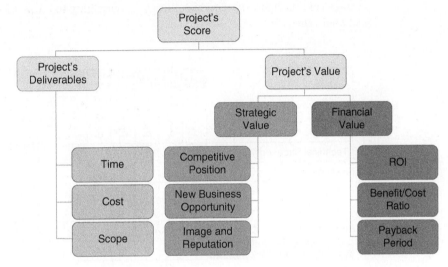

Figure 19–5. Project scoring model.

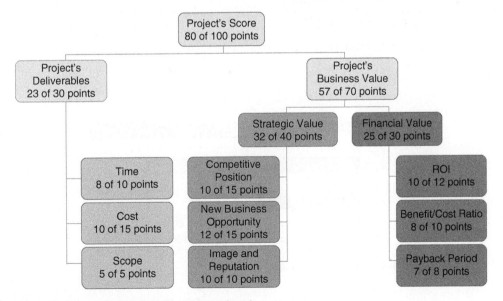

Figure 19–6. Project scoring model with points assigned.

Figure 19–7 shows the alignment of projects with strategic objectives. If the total score in Figure 19–6 is between zero and 50 points, we would assume that the project is not contributing to strategic objectives at this time, and this would be shown as a zero or blank cell in Figure 19–7. Scores between 51 and 75 points would indicate a "partial" contribution to the objectives and are shown as a 1 in Figure 19–7. Scores between 76 and 100 points would indicate fulfilling the objective and are shown as a 2 in Figure 19–7. Periodically we can summarize the results in Figure 19–7 to show management in Figure 19–8, which illustrates our ability to create the desired benefits and final value.

Strategic Objectives:	Projects								Scores
	Project 1	Project 2	Project 3	Project 4	Project 5	Project 6	Project 7	Project 8	
Technical Superiority	2		1			2		1	6
Reduced Operating Costs				2	2				4
Reduced Time to Market	1		1	2	1	1		2	8
Increase Business Profits			2	1	1	1		2	7
Add Manufacturing Capacity	1		2	2		1		1	7
Column Scores	4	0	6	7	4	5	0	6	

	No Contribution
1	**Supports Objective**
2	**Fulfills Objective**

Figure 19–7.   Match projects to strategic business objectives.

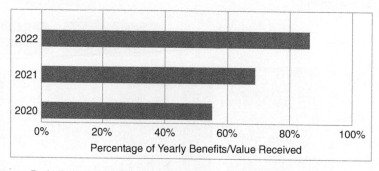

Figure 19–8.   Periodic benefits and value achieved.

19.13 CAUSES OF COMPLETE OR PARTIAL BRM FAILURE

No matter how hard we try to become good at benefits realization and value management, there are always things that can go wrong and lead us to disaster. Fourteen such causes of failure that can occur along the entire investment life cycle include:

1. No active involvement by the business owner or stakeholders.
2. Decision makers are unsure about their roles and responsibilities, especially in the early life-cycle phases.
3. The project is approved without a business case or benefits realization plan.
4. A high level of uncertainty and ambiguity exists in defining the benefits and values such that they cannot be described adequately in a document such as a benefits realization plan.
5. Highly optimistic or often unrealistic estimates of benefits are made to get project approval and a high priority.
6. Failing to recognize the importance of effective resource management practices and the link to BRM.
7. Maintaining a heavy focus on the project's deliverables rather than on benefit realization and the creation of business value.
8. Using the wrong definition of project success.
9. Managing the project with traditional rather than investment life-cycle phases.
10. Using the wrong metrics, unreliable metrics, or simply lacking metrics to track benefits and value.
11. Failing to track benefits and value over the complete life cycle.
12. Not having any criteria established for when to cancel a failing project.
13. Having no transformational process if necessary where the benefits and value can be achieved only from necessary organizational changes in the way the firm must now conduct business.
14. Failing to capture lessons learned and best practices, thus allowing mistakes to be repeated.

Item 14 is often the solution to correct the first 13 problems from recurring.

19.14 CONCLUSION

Because of the importance of benefits and value, today's project managers are more business managers than the pure project managers of the past. Today's project managers are expected to make business decisions as well as project-based decisions. Project managers seem to know more about the business than their predecessors.

With the growth in measurement techniques, companies will begin creating metrics to measure benefits and value. While many of these measurement techniques are still in their infancy, the growth rate is expected to be rapid.

Index